Functional Thin Films Technology

Advances in Materials Science and Engineering
Series Editor
Sam Zhang

Biological and Biomedical Coatings Handbook: Processing and Characterization
Sam Zhang

Aerospace Materials Handbook
Sam Zhang and Dongliang Zhao

Thin Films and Coatings: Toughening and Toughness Characterization
Sam Zhang

Semiconductor Nanocrystals and Metal Nanoparticles: Physical Properties and Device Applications
Tupei Chen and Yang Liu

Advances in Magnetic Materials: Processing, Properties, and Performance
Sam Zhang and Dongliang Zhao

Micro- and Macromechanical Properties of Materials
Yichun Zhou, Li Yang, and Yongli Huang

Nanobiomaterials: Development and Applications
Dong Kee Yi and Georgia C. Papaefthymiou

Biological and Biomedical Coatings Handbook: Applications
Sam Zhang

Hierarchical Micro/Nanostructured Materials: Fabrication, Properties, and Applications
Weiping Cai, Guotao Duan, and Yue Li

Biological and Biomedical Coatings Handbook, Two-Volume Set
Sam Zhang

Nanostructured and Advanced Materials for Fuel Cells
San Ping Jiang and Pei Kang Shen

Hydroxyapatite Coatings for Biomedical Applications
Sam Zhang

Carbon Nanomaterials: Modeling, Design, and Applications
Kun Zhou

Materials for Energy
Kun Zhou

Protective Thin Coatings and Functional Thin Films Technology, Two-Volume Set
Sam Zhang, Jyh-Ming Ting, Wan-Yu Wu

For more information about this series, please visit: www.crcpress.com

Series: Advances in Materials Science and Engineering
Series Editor: Sam Zhang

Advances in Materials Science and Engineering
Series Statement
Materials form the foundation of technologies that govern our everyday life, from housing and household appliances to handheld phones, drug delivery systems, airplanes, and satellites. Development of new and increasingly tailored materials is key to further advancing applications with the potential to dramatically enhance and enrich our experiences.

The *Advances in Materials Science and Engineering* series by CRC Press/Taylor & Francis is designed to help meet new and exciting challenges in Materials Science and Engineering disciplines. The books and monographs in the series are based on cutting-edge research and development, and thus are up-to-date with new discoveries, new understanding, and new insights in all aspects of materials development, including processing and characterization and applications in metallurgy, bulk or surface engineering, interfaces, thin films, coatings, and composites, just to name a few.
The series aims at delivering an authoritative information source to readers in academia, research institutes, and industry. The Publisher and its Series Editor are fully aware of the importance of Materials Science and Engineering as the foundation for many other disciplines of knowledge. As such, the team is committed to making this series the most comprehensive and accurate literary source to serve the whole materials world and the associated fields.

As Series Editor, I'd like to thank all authors and editors of the books in this series for their noble contributions to the advancement of Materials Science and Engineering and to the advancement of humankind.

Sam Zhang

Functional Thin Films Technology

Edited by Sam Zhang, Jyh-Ming Ting and Wan-Yu Wu

CRC Press
Taylor & Francis Group
Boca Raton London New York

CRC Press is an imprint of the
Taylor & Francis Group, an **informa** business

First edition published 2022
by CRC Press
6000 Broken Sound Parkway NW, Suite 300, Boca Raton, FL 33487-2742

and by CRC Press
2 Park Square, Milton Park, Abingdon, Oxon, OX14 4RN

© 2022 Taylor & Francis Group, LLC

CRC Press is an imprint of Taylor & Francis Group, LLC

Reasonable efforts have been made to publish reliable data and information, but the author and publisher cannot assume responsibility for the validity of all materials or the consequences of their use. The authors and publishers have attempted to trace the copyright holders of all material reproduced in this publication and apologize to copyright holders if permission to publish in this form has not been obtained. If any copyright material has not been acknowledged please write and let us know so we may rectify in any future reprint.

Except as permitted under U.S. Copyright Law, no part of this book may be reprinted, reproduced, transmitted, or utilized in any form by any electronic, mechanical, or other means, now known or hereafter invented, including photocopying, microfilming, and recording, or in any information storage or retrieval system, without written permission from the publishers.

For permission to photocopy or use material electronically from this work, access www.copyright.com or contact the Copyright Clearance Center, Inc. (CCC), 222 Rosewood Drive, Danvers, MA 01923, 978-750-8400. For works that are not available on CCC please contact mpkbookspermissions@tandf.co.uk

Trademark notice: Product or corporate names may be trademarks or registered trademarks and are used only for identification and explanation without intent to infringe.

ISBN: 978-0-367-54177-4 (hbk)
ISBN: 978-0-367-54178-1 (pbk)
ISBN: 978-1-003-08808-0 (ebk)

Typeset in Times
by SPi Global, India

Contents

Preface to Protective Thin Coatings and Functional Thin Films Technology – 2-Volume Set ix

Editors xi

Contributors xiii

Chapter 1 Combinatorial Synthesis Applied to the Development of Thin Film Materials for Nanoelectronics 1

Takahiro Nagata and Toyohiro Chikyow

Chapter 2 Sputter Deposited Nanostructured Coatings as Solar Selective Absorbers 21

Jyh-Ming Ting and Yi-Hui Zhuo

Chapter 3 Novel Mode-Locked Fiber Lasers with Broadband Saturable Absorbers 47

Lu Li and Wei Ren

Chapter 4 Thin Coating Technologies and Applications in High-Temperature Solid Oxide Fuel Cells 83

San Ping Jiang and Shuai He

Chapter 5 Mutually Influenced Stacking and Evolution of Inorganic/Organic Crystals for Piezo-Related Applications 127

Jr-Jeng Ruan and Kao-Shuo Chang

Chapter 6 Single-Atom Catalysts on Nanostructure from Science to Applications 151

Yi-Sheng Lai, Anggrahini Arum Nurpratiwi, and Yen-Hsun Su

Chapter 7 Growth, Characteristics and Application of Nanoporous Anodic Aluminum Oxide Synthesized at Relatively High Temperature 165

Chen-Kuei Chung, Ming-Wei Liao, and Chin-An Ku

Chapter 8 Multifunctional Superhydrophobic Nanocomposite Surface: Theory, Design, and Applications 199

Yongquan Qing and Changsheng Liu

Chapter 9 Solution-Processed Oxide-Semiconductor Films and Devices 225

Bui Nguyen Quoc Trinh, Endah Kinarya Palupi, and Akihiko Fujiwara

vii

viii Contents

Chapter 10 Gold Nanocrystal-built Films for SERS-based Detection of Trace
Organochlorine Pesticides 253

Xia Zhou, Hongwen Zhang, and Weiping Cai

Chapter 11 The Effect of Deposition Parameters on the Mechanical and Transport
Properties in Nanostructured Cu/W Multilayer Coatings 287

Alexander M. Korsunsky and León Romano Brandt

Index 319

Preface to Protective Thin Coatings and Functional Thin Films Technology – 2-Volume Set

Over the decades, films and coatings have been developed and applied in industries that spread all over people's life of the current society and the defense. Films and coatings also evolved from single compound to multicompound to multilayer and to nanostructures and nanocomposites. All the portable electronic devices, such as cell phone and iPad, and removable storage media, such as memory cards and USB flash drives, involve, heavily, the use of micro-/nanoscaled films and coatings. In all, films or coatings either provide protection of the surface they are attached to or provide certain functionality through the film itself. To capture the most recent advances in both aspects, we put together a 2-volume set, one on protection and other on functional applications:

Protective Thin Coatings Technology

Functional Thin Films Technology

Films and Coatings, in essence, are two different things. Coatings have to realize their usefulness through attaching on to the surface of a substrate, as in the case of hard coatings on a mode or drill to provide lubrication for antisticking or hardening of the drill surface to lengthen drill life. Films on the other hand provide their functionality by standing alone with or without substrate. Even in the case of having a substrate, the substrate is there to provide a backing support, not necessarily having too much to do with the functionality the films are there to provide. As such, for protective applications, we use "coatings" (thus "thin" coatings are used only to differentiate from "thick" coatings) and for functional applications, we use "films".

Protective Thin Coatings Technology focuses on deposition/processing technologies and property characterizations in protective applications. *Functional Thin Films Technology* deals with deposition/processing technologies and property characterizations in functional applications.

In some fields, deposition and processing carry somewhat different meanings. For instance, in device-making technologies, processing includes assembly. However, in other fields, it simply means making or fabrication of films or coatings. As here, we do not focus on assembly of devices, but give ourselves the freedom of using both deposition and processing to refer to fabrication.

To be specific, *Protective Thin Coatings Technology* covers technologies for Sputtering of Flexible Hard Nanocoatings, Deposition of Solid Lubricating Films, Multilayer Transition Metal Nitrides, Integrated Nanomechanical Characterization of Hard Coatings, Corrosion and Tribo-Corrosion of Hard Coating, High-Entropy Alloy films and Coatings, Thin Films And Coatings for High-Temperature Applications, Nanocomposite Coating on Magnesium Alloys and the Correlation Between Coating Properties and Industrial Applications.

Functional Thin Films Technology deals with technologies aiming at functionality when used in Nanoelectronics, Solar Selective Absorbers, Solid Oxide Fuel Cells, Piezo Applications, Sensors, Absorbers, Catalysts, Anodic Aluminum Oxide, Superhydrophobics, Semiconductor Devices, etc. Also included is a chapter that deals with the Transport Phenomena in Nanostructured Coating.

In summary, these two books highlight the development and advance in the preparation, characterization, and applications of protective and functional micro-/nanoscaled films and coatings. People working in areas related to semiconductor, optoelectronics, plasma technology, solid-state energy storages, 5G, etc., and students studying electrical, mechanical, chemical, materials, etc., engineering will find these books useful. To be specific, these include Senior Undergraduate Students, Graduate Students, Industry Professionals, Researchers and Academics.

The editors would like to thank all chapter authors for their active contribution and timely effort to ensure the smooth publication of these high-quality books to catch the new trend and development in the topical matters. The editors thank Allison and Gabrielle of the publisher for their support along the way. Sam would also like to acknowledge the Fundamental Research Funds for the Central Universities SWU118105.

Editors

Prof. Sam Zhang Shanyong (張善勇), academically better known as Sam Zhang, was born and brought up in the famous "City of Mountains" Chongqing, China. He received his Bachelor of Engineering in Materials in 1982 from Northeastern University (Shenyang, China), Master of Engineering in Materials in 1984 from Iron & Steel Research Institute (Beijing, China) and PhD degree in Ceramics in 1991 from The University of Wisconsin-Madison, USA. He was a tenured full professor (since 2006) at the School of Mechanical and Aerospace Engineering, Nanyang Technological University. Since January 2018, he joined School of Materials and Energy, Southwest University, Chongqing, China and assumed duty as Director of the Centre for Advanced Thin Film Materials and Devices of the university.

Prof. Zhang was the founding Editor-in-Chief for *Nanoscience and Nanotechnology Letters* (2008 to December 2015) and Principal Editor for the *Journal of Materials Research* (USA) responsible for thin films and coating field (since 2003). Prof. Zhang has been serving the world's first "Thin Films Society" (www.thinfilms.sg) as its founding and current president since 2009. Prof. Zhang has authored/edited 13 books, of which 12 are published with CRC Press/Taylor & Francis. Of these books, *Materials Characterization Techniques* has been adopted as core textbook by more than 30 American and European universities since October 2015. That book had also been translated into Chinese and published by China Science Publishing Co in October 2010 and distributed nationwide in China (available online at Amazon.cn). Very recently, Prof. Zhang has just published his book Materials for Energy (Sam Zhang (ed.), October 2020: 6-1/8 x 9-1/4: 528pp Hb: 978-0-367-35021-5 eBook: 978-0-429-35140-2. URL:https://www.routledge.com/Materials-for-Energy/Zhang/p/book/9780367350215. Prof. Zhang's new book on "Materials for Devices", to be published by CRC Press, is also in the pipeline. Meanwhile, Prof. Zhang is the Series Editor for Advances in Materials Science and Engineering Book Series published by CRC Press/Taylor & Francis.

Prof. Zhang has been elected as Fellow of Royal Society of Chemistry (FRSC) and Fellow of Thin Films Society (FTFS) in 2018, and Fellow of Institute of Materials, Minerals and Mining (FIoMMM) in 2007. Prof. Zhang's current research centers on areas of Energy Films and Coatings for solar cells, Hard yet tough nanocomposite coatings for tribological applications by physical vapor deposition; Measurement of fracture toughness of ceramic films and coatings, and Electronic/Optical Thin Films. Over the years, he has authored/co-authored over 360 peer reviewed international journal papers. As of December 8, 2020, as per Web of Science (https://publons.com/researcher/2817766/sam-zhang/), the sum of the times cited: 10342, average citations per article: 29.1, h-index: 54.

Prof. Zhang holds Guest Professorship at Shenyang University (2002), Institute of Solid State Physics, Chinese Academy of Sciences (2004), Zhejiang University (2006), Harbin Institute of Technology (2007), Xian Jiaotong University (2018), Kashi University (2018), and Xiamen University of Science and Technology (2020). Prof. Zhang also serves as International Advisor to Shenzhen Association for Vacuum Technology Industries.

Dr. Jyh-Ming Ting is an Advanced Semiconductor Engineering Chair Professor of the Department of Materials Science and Engineering at National Cheng Kung University (NCKU) in Taiwan. He received a BS degree in Nuclear Engineering from National Tsing Hua University in Taiwan 1982, and completed MS and PhD works at the Department of Materials Science and Engineering in University of Cincinnati in 1987 and 1991, respectively. From 1990 to 1997, Dr. Ting worked at Applied Sciences, Inc. (ASI), as a Scientist and then the R&D Director. In August 1997, Dr. Ting became a faculty in NCKU working on various carbon materials and low-dimensional materials, and their (nano)composites. At the early stage, nanoscaled (composite) thin films and CNT were of interest. In particular, Prof. Ting developed and patented several novel techniques that allow the growth of aligned CNTs on metallic substrates, also holding the record of the growth rate until now. In recent years, in response to the serious issues related to environment, his research focuses on applying low-dimensional materials, mainly nanoparticles and 2D layered structures, and their composites, to energy generation/storage and photodegradation. More recently, Prof. Ting is working in high-entropy oxides. Prof. Ting has authored over 170 journal articles, given more than 30 invited talks and a few keynote speeches, and holds more than 30 patents.

Dr. Wan-Yu Wu was born in Taipei, Taiwan. She obtained her Bachelor's degree in 2000 and PhD degree in 2008 from the Department of Materials Science and Engineering of National Cheng Kung University, Tainan, Taiwan. In 2010, Dr. Wu became an Assistant Professor in MingDao University in Taiwan where she was also a key member of the Surface & Engineering Research Center. In 2014, she joined the Department of Materials Science and Engineering in Da-Yeh University in Taiwan as an Associate Professor. Due to her well-recognized expertise in thin film and coating technology, Prof. Wu was commissioned to establish a Coatings Technology Research Center in Da-Yeh university and became a full professor in 2019. Over the years, she has focused her research on thin film and nanostructure materials synthesized using High-Power Impulse Magnetron Sputtering (HiPIMS) deposition and wet chemistry processes, respectively. Prof. Wu also actively participates in professional society. She is in the board of directors in Taiwan Association for Coating and Thin Film Technology (TACT). She also serves as a symposium chair in several international conferences including International Conference on Metallurgical Coatings and Thin Films (ICMCTF), ThinFilms Conference, and TACT International Thin Films Conference.

Contributors

León Romano Brandt
University of Oxford
Oxford, UK

Weiping Cai
Key Lab of Materials Physics, Anhui Key Lab
of Nanomaterials and Nanotechnology
Institute of Solid State Physics, HFIPS, Chinese
Academy of Sciences
Hefei, P.R. China

Kao-Shuo Chang
National Cheng Kung University
Tainan, Taiwan

Toyohiro Chikyow
Research and Services Division of Materials
Data and Integrated System
National Institute for Materials Science
Ibaraki, Japan

Chen-Kuei Chung
Department of Mechanical Engineering
National Cheng Kung University
Tainan, Taiwan

Akihiko Fujiwara
Department of Nanotechnology for Sustainable
Energy, School of Science and Technology
Kwansei Gakuin University
Hyogo, Japan

Shuai He
School of Chemistry
University of St Andrews
St Andrews, Fife, UK

San Ping Jiang
Fuels and Energy Technology Institute &
Western Australian School of Mines:
Minerals, Energy and Chemical Engineering
Curtin University
Perth, Australia

Alexander M. Korsunsky
University of Oxford
Oxford, UK

Chin-An Ku
Department of Mechanical Engineering
National Cheng Kung University
Tainan, Taiwan

Yi-Sheng Lai
Department of Materials Science &
Engineering
National Cheng Kung University
Tainan, Taiwan

Lu Li
School of Science
Xi'an University of Posts and
Telecommunications
Xi'an, China

Ming-Wei Liao
Department of Materials Science and
Engineering
National Tsing Hua University
Hsinchu, Taiwan

Changsheng Liu
School of Materials Science and Engineering
Northeastern University
Shenyang, China

Takahiro Nagata
Research Center for Functional Materials
National Institute for Materials Science
Ibaraki, Japan

Anggrahini Arum Nurpratiwi
Department of Materials Science &
Engineering
National Cheng Kung University
Tainan, Taiwan

Endah Kinarya Palupi
Department of Nanotechnology for Sustainable
Energy, School of Science and Technology
Kwansei Gakuin University
Hyogo, Japan

Yongquan Qing
School of Materials Science and Engineering
Northeastern University
Shenyang, China

Wei Ren
School of Science
Xi'an University of Posts and
Telecommunications
Xi'an, China

Jr-Jeng Ruan
National Cheng Kung University
Tainan, Taiwan

Yen-Hsun Su
Department of Materials Science & Engineering
National Cheng Kung University
Tainan, Taiwan

Jyh-Ming Ting
Department of Materials Science and Engineering
National Cheng Kung University
Tainan, Taiwan

Bui Nguyen Quoc Trinh
Vietnam National University, Hanoi, VNU
Vietnam-Japan University, Nanotechnology
Program, Luu Huu Phuoc, Nam Tu Liem
Hanoi, Vietnam

Vietnam National University, Hanoi, VNU
University of Engineering and Technology
Key Laboratory for Micro-Nano Technology
Hanoi, Vietnam

Hongwen Zhang
Key Lab of Materials Physics, Anhui Key Lab
of Nanomaterials and Nanotechnology
Institute of Solid State Physics, HFIPS, Chinese
Academy of Sciences
Hefei, P.R. China

Xia Zhou
Key Lab of Materials Physics, Anhui Key Lab
of Nanomaterials and Nanotechnology
Institute of Solid State Physics, HFIPS, Chinese
Academy of Sciences
Hefei, P.R. China

Yi-Hui Zhuo
Department of Materials Science and
Engineering
National Cheng Kung University
Tainan, Taiwan

1 Combinatorial Synthesis Applied to the Development of Thin Film Materials for Nanoelectronics

Takahiro Nagata and Toyohiro Chikyow
National Institute for Materials Science, Ibaraki, Japan

CONTENTS

1.1 Introduction: Background and Driving Forces 1
1.2 Development of Materials Using Combinatorial Synthesis.. 2
1.3 High-Throughput Material Development Based on Combinatorial Technology..................... 4
1.4 Thin Film Combinatorial Synthesis.. 5
1.5 Combinatorial Characterization.. 8
1.6 Control of Electrical Properties in Binary Alloy Systems 10
1.7 Control of Electrical Properties in Binary Oxide Systems 13
1.8 Process and Composition Optimization for Complex Oxides... 13
1.9 Requirements and Future Issues 15
References... 17

1.1 INTRODUCTION: BACKGROUND AND DRIVING FORCES

This chapter briefly introduces combinatorial thin film synthesis techniques with applications to the development of new nanoelectronic materials. These methods are of interest because they allow the high-throughput synthesis of materials as a component of systematic investigations. Combinatorial synthesis is a general term for techniques capable of rapidly generating information regarding the properties of a material in conjunction with systematic compositional changes, together with variations in fabrication parameters such as sintering temperature and pressure. Combinatorial chemistry is concerned with techniques and methodologies for the fabrication of compound libraries containing information on chemical and pharmacological substances (Figure 1.1) [1, 2]. Researchers in this field focus on the development and application of efficient systematic syntheses of chemical compounds using combinatorial theory and statistical calculations, together with the use of specific evaluation methods. This can include the development of nanoscale materials such as inorganic and organic thin films in conjunction with the creation of databases for these substances. Combinatorial thin film synthesis in particular is a relatively new research field that has been growing for the past 20 years, with significant progress [3, 4]. As a result, libraries of thin film compositions based on inorganic and organic materials have been constructed. Combinatorial technologies together with

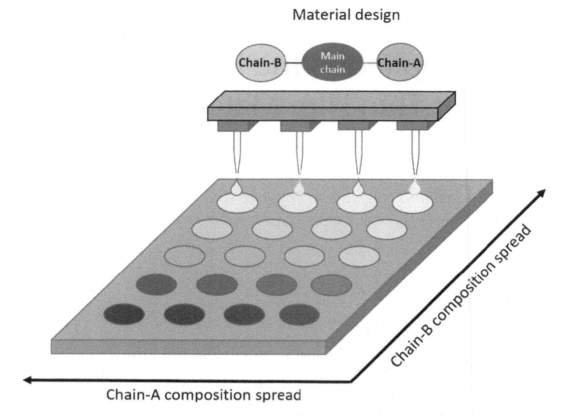

FIGURE 1.1 A diagram showing the construction of a combinatorial chemical library based on varying the side chains of an organic compound to obtain a variety of materials. Side chains A and B are synthesized in specific combinations via the automatic feeding of varying concentrations of raw materials such that different possible compositions are realized.

new high-throughput evaluation and computing methods are vital to the preparation of these libraries, which are the basis of materials informatics.

1.2 DEVELOPMENT OF MATERIALS USING COMBINATORIAL SYNTHESIS

In recent years, materials informatics has attracted increasing attention with regard to the development of solid-state thin film materials [5–7]. Figure 1.2 shows the development cycle of a thin film material, including combinatorial synthesis. This process comprises the following basic stages.

(1) Material design

To obtain materials having targeted properties, both the composition and structure of the substances have to be considered. In the case of commercially available materials, appropriate candidates can be selected using databases such as the Materials Project [8], NIMS-MatNavi [9], AtomWork-adv [10], and Springer Materials [11] databases, as well as values obtainable from academic journals. Theoretical predictions can also be helpful when choosing combinations of material candidates.

(2) Synthetic design

Depending on the possible variations of the target composition, both the deposition method and the starting materials may be considered. If the initial raw material is an oxide, the oxide itself is typically used as the starting compound rather than generating the starting material via

Combinatorial Synthesis

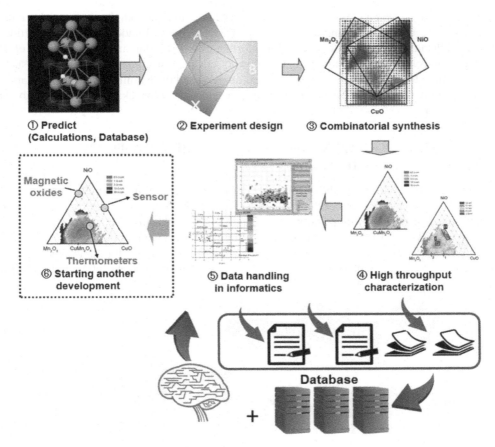

FIGURE 1.2 A diagram showing the combinatorial technology cycle as applied to the development of a new material.

the reactive sputtering of a metal target, so as to maximize efficiency. In this stage, the composition distribution and sample geometry are also predetermined according to the equipment specifications.

(3) Combinatorial synthesis

Preliminary studies of fabrication parameters such as deposition speed, mask speed, and substrate temperature are performed based on the experimental design, followed by combinatorial thin film synthesis.

(4) Combinatorial measurements

Samples having various compositions are evaluated using different measurement techniques that either provide rapid results or allow for multipoint analyses.

(5) Data aggregation, analysis, and re-measurement

The results of the evaluations are aggregated into a compositional chart and the data are analyzed. The points, at which the target characteristics are most prominent or where unique characteristics appear, are identified. If necessary, additional measurements are performed at the extracted points.

(6) Discovery of new materials for practical applications

The optimal composition determined by this process is incorporated into a test device and verification experiments are conducted. This process is used to verify the potential of the new material and establish potential practical applications in cooperation with industrial partners.

Informatics can be expected to be particularly effective in optimizing both the design and synthesis of new materials, and in improving the efficiency of data analysis. In addition, the acquisition of experimental data that will serve as a basis for informatics is also vital, and the generation of data will be based on combinatorial methods. In recent years, advances in computational science, such as machine learning, have accelerated the pace of data analysis and the learning efficiency of materials analysis, although the systematic accumulation of data can be a bottleneck. Fortunately, combinatorial synthesis is a powerful tool for this purpose.

1.3 HIGH-THROUGHPUT MATERIAL DEVELOPMENT BASED ON COMBINATORIAL TECHNOLOGY

Figure 1.3 presents an outline of the development of films having compositional gradients as part of a materials research process. In 1965, the first technique was developed to spontaneously generate a distribution of thin film compositions on a single substrate via simultaneous deposition using multiple sources [12]. In 1995, a second-generation combinatorial chemistry technique was reported in which the composition was varied over a patterned substrate [13]. At the same time, researchers at Symyx published a paper on the combinatorial synthesis of solid phosphors [14]. The third

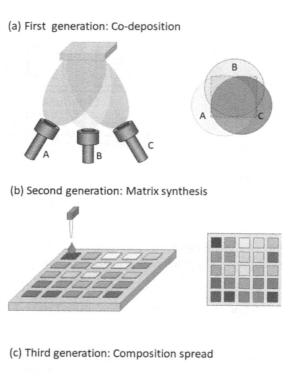

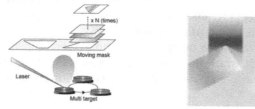

FIGURE 1.3 Progress in combinatorial thin film synthesis. The left-hand images are schematics of typical synthesis methods while the right-hand images show the resulting samples.

Combinatorial Synthesis

generation of combinatorial technology, which was first developed in 1996, further advanced this process, and various groups developed new techniques for film synthesis as well as novel measurement equipment. In this time frame, the National Institute of Standards and Technology (NIST) Center for Combinatorial Methods was also established, and played an important role in the development of combinatorial technologies and in the transfer of technology to various companies. As a result, companies such as GE and DuPont began to use combinatorial technologies to develop new materials. In addition, educational institutions such as the University of Maryland and the University of Illinois performed research in this field, along with venture enterprises such as Intermatex and Symyx [15]. Southampton University established a combinatorial research center in 1999, primarily to develop catalytic materials, and started the venture company Ilika Technologies [16], which has been involved in the development of solid-state batteries. Since then, the use of combinatorial technologies in chemistry and other fields has become increasingly popular, and this has led to the development of partnerships between Japanese and European organizations. Since the early days of combinatorial technologies, discussions concerning data mining, common databases, and materials design using computational science have been held continuously, and international conferences have been held regularly since 1999. These techniques have not yet become well established in materials design, primarily due to issues related to common data formats and measurement devices, which are gradually being addressed. In addition, recent developments in computing technology and increases in research and development based on large datasets have led to new combinatorial technologies. Consequently, the materials development process has continually evolved into the field referred to as materials informatics (or material genomics in the United States). Historically, Japan has lagged behind Europe and the United States in combinatorial chemistry as applied to the development of chemicals and pharmaceuticals. However, the Japan Science and Technology Agency's Strategic Basic Research Project (JST-CREST) led by Koinuma et al. (then at the Tokyo Institute of Technology (TITech)) and initiated in 1996 has performed work in the combinatorial synthesis of inorganic materials and in metrology. In addition, the Comet Project based on a collaboration between the National Institute for Materials Science (NIMS) and TITech and initiated in 1999 has developed combinatorial materials science and technology as well as synthetic techniques such as bulk synthesis and physical vapor deposition. The research and development of comprehensive combinatorial technologies integrating a wide range of fields, including process technologies such as ion implantation, evaluation methods, database creation, and computational science, has led to the creation of many new processes [17–21]. As a result of this work, Comet Inc. was established in 2007 as the first company in Japan to provide combinatorial technologies [22]. The following section introduces the combinatorial thin film synthesis technologies primarily developed by NIMS and TITech, and discusses potential applications of these techniques.

1.4 THIN FILM COMBINATORIAL SYNTHESIS

The use of combinatorial synthesis techniques has become increasingly common as a means of optimizing numerous multicomponent functional materials [13, 23, 24]. In the case of the development of thin film materials, combinatorial pulsed laser deposition (combinatorial PLD) is one of the most widely used combinatorial synthesis methods developed in the above-mentioned collaborative project. As shown in Figure 1.4(a), the PLD method uses an excimer laser (typically a KrF or ArF laser with λ=248 or 193 nm, respectively) or a Q-switched Nd:YAG laser (typically employing the third or fourth harmonic at 355 or 266 nm, respectively) to irradiate a target material (generally a sintered ceramic pellet). The irradiation of this target with high energy, short wavelength lasers generates a plasma that is subsequently deposited on a substrate material. A combinatorial thin film synthesis system consists of mechanisms for exchanging multiple targets and for rotating the substrate to achieve a process for the fabrication of the entire film, controlled by a computer [25]. Using this mechanism, films having multiple compositions on the same substrate can be fabricated in a single pass. Figure 1.4(b) summarizes a binary combinatorial thin film fabrication procedure comprising

(a) System setup (pulsed laser deposition system)

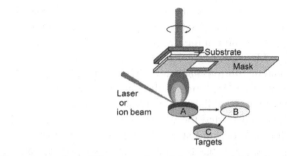

(b) Base procedure: binary system fabrication (side view)

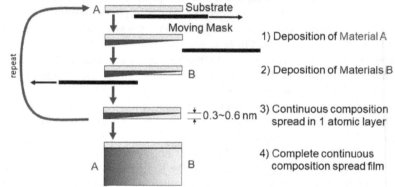

(c) Ternary system fabrication (bottom view)

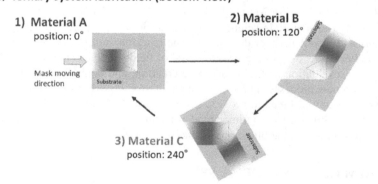

FIGURE 1.4 (a) A schematic diagram of a combinatorial thin film deposition system. (b) A schematic diagram of a compositional gradient thin film sample fabrication procedure based on a binary material deposition cycle. The process involves the deposition of materials A and B. During each deposition, a mask moves at a constant speed from one side of the substrate to the other, after which the target is changed. Alternating between steps (1) and (2) creates binary compositional gradient samples. (c) A schematic diagram of the fabrication of a ternary composition gradient thin film sample. In this process, materials A, B, and C are deposited. In each deposition, after rotating the substrate by 120°, a mask moves at a constant speed from one side of the substrate to the other, after which the target is changed.

Combinatorial Synthesis

two steps. In each step, a mask with a square-shaped hole moves at a constant speed from one side of the substrate to the other, after which the target is changed. The composition and thickness of the film can be precisely controlled using the moving mask so as to obtain a uniform variation in the composition of the film. The mask speed is calibrated in advance to adjust the amount of target material and give a uniform film thickness, and the mask moves faster or slower depending on the growth rate of the material. The maximum film thickness obtainable from a single cycle is typically one- or one-half unit cells of the film material (0.3–0.6 nm), and alternating between steps (1) and (2) creates variations in the film composition. In the case of a ternary system, the substrate can also be rotated, based on a process developed by the NIMS group [26, 27], as shown in Figure 1.4(c). During this process, the substrate is rotated by 120° after the deposition of each material. The advantage of this method is the accuracy of the composition of each component, as demonstrated in Figure 1.5.

The PLD thin film growth technique is useful for growing oxides and nitrides. However, this technique is unsuitable for metal thin films because of problems such as the deposition of evaporated metal on the substrate in the form of droplets due to the thermal effect of the laser irradiation. The NIMS group developed a combinatorial metal film synthesis method as a means of mitigating this issue, based on using a focused Ar ion beam sputtering technique [28, 29]. This approach employs separate ion generation and film forming chambers so that a high vacuum can be maintained in the latter to prevent oxidation of the metal film. The ion gun generates an ion current of 20–50 µA in this process, which is a modification of the low-power ion gun used in conventional combinatorial PLD systems and has a small irradiation area of 1 mm². The gun can be used to produce specimens having ternary composition gradients, as shown in Figure 1.5(c).

Based on the materials and the stage of research and development, different synthesis methods can be applied. Molecular beam epitaxy (MBE), PLD, and ion beam sputtering can all be used as film formation methods during basic research where compositional accuracy is required. In contrast, sputtering is applicable in conjunction with limited compositional ranges, as might be associated with the initial stages of practical device fabrication.

Some current combinatorial synthesis methods use masks in conjunction with vapor deposition, and these techniques have advantages with regard to the study of numerous thin film materials, including oxides, nitrides, metals, and organic materials. These techniques can be used to produce high-quality, high-precision samples by selecting the most appropriate method for a specific film material and application. The advantages and disadvantages of these processes are summarized in Table 1.1. As an example, multitarget sputtering is a physical vapor deposition method that is

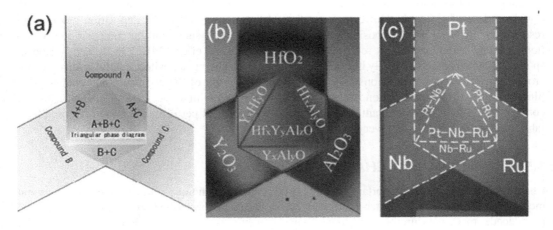

FIGURE 1.5 (a) A schematic diagram of a ternary compositional gradient thin film structure. Photographic images of (b) a ternary oxide and (c) a ternary alloy compositional gradient thin film sample.

TABLE 1.1
Advantages and disadvantages of physical vapor deposition methods used in combinatorial methods

	Metal	Oxide/ Nitride	Large-area deposition	Growth rate controllability	Handleability	Low energy (low damage)	Co-deposition
MBE	◎	○	*	◎	Δ	◎	◎
PLD (Laser MBE, UV laser)	*	◎	*	◎	○	Δ	*
Ar focused ion beam sputtering	◎	Δ	*	◎	○	○	*
Multitarget sputtering	◎	◎	◎	○	◎	*	○
Thermal evaporation (EB)	◎	*	◎	*	○	○	Δ

* poor, < Δ fair, <○ good, < ◎ excellent

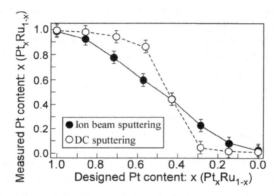

FIGURE 1.6 Results from the XRF analyses of Pt-Ru alloy films with compositional gradients deposited by focused ion beam sputtering (solid symbols) and DC-sputtering (open symbols).

commonly used to fabricate industrial production equipment because it can be used together with a wide variety of materials. However, spreading of the plasma during the deposition process can result in the material being deposited moving around the mask such that the compositional variations are less accurate. This is referred to as the wraparound effect. MBE, PLD, and ion beam sputtering, which generate highly linear flows of the material being deposited, exhibit reduced wraparound effects. Figure 1.6 demonstrates the improved linearity of the compositional changes in a film deposited by focused Ar ion beam sputtering compared with a specimen produced using the conventional DC magnetron sputtering method. This improved performance makes it possible to fabricate ternary samples with very high precision.

1.5 COMBINATORIAL CHARACTERIZATION

Characterization is also an important combinatorial synthesis technique, and advanced measurement methods combined with computer-controlled stage systems can be applied for this purpose. Over the past decade, many measurement techniques have been employed in conjunction with combinatorial thin film syntheses. These have included two dimensional X-ray diffraction (2D-XRD) and atomic force microscopy (AFM) in addition to several optical characterization techniques. Specimens with

Combinatorial Synthesis

variable compositions are initially evaluated to assess their basic composition and structure, after which the electrical and optical properties of the target materials are investigated. In the early stages of combinatorial technology development, the time required for sample evaluation was relatively long due to insufficient positional resolution and the large number of measurement points required for such analyses. However, advances in measurement instrumentation that enable the examination of small areas and improvements in equipment control, data processing, and computer processing speed now allow variable composition specimens to be evaluated as fast as they can be prepared.

One example of such technological advancement is the structural analysis of thin films by 2D-XRD. Using the Bruker 2D-XRD detector, Debye–Scherrer rings can be analyzed based on data having 2θ and χ-axes over a range of $30°$, which can be as high as $60°$ depending on the conditions, in an image frame obtained from a single scan, as shown in Figure 1.7(a). Using a high-power X-ray source with a rotating Cu anode system, it is possible to obtain results within several minutes from an extremely thin film (with a thickness of approximately 10 nm), providing the film is crystalline. Figure 1.7(b) presents an example of an image acquired using the 2D-XRD method, which allows an intuitive assessment of the crystallization state of the film as well as the crystal quality and orientation. Specimens with compositional variations are examined using an X-ray beam focused to a diameter of 50 μm together with the automatic mapping function of the software associated with the instrument. A distribution map corresponding to the structure of the sample can then be generated by computer, as shown in Figure 1.7(c), such that the crystallization point and singularities associated with compositional changes can be identified. Reductions of the measurement area and computer-controlled automation have been used to improve the investigation of composition as well as electrical, optical, and magnetic properties, enabling the distributions of these characteristics to be visualized along with the X-ray structural analysis.

As noted, AFM has also been employed for the analysis of variable composition samples. AFM is commonly used for the surface imaging of materials such as ceramics and polymers, as well as biological specimens. An AFM apparatus incorporates a cantilever with a probe tip that is used

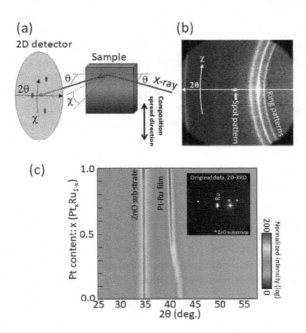

FIGURE 1.7 (a) The experimental setup for 2D-XRD analyses and (b) a typical 2D-XRD image obtained from a Pt-Ru alloy film with a compositional gradient on a ZnO substrate. The inset shows a 2D-XRD image for the Ru side (x = 0).

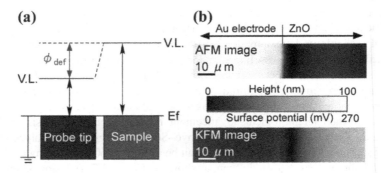

FIGURE 1.8 (a) A schematic diagram summarizing the KFM technique. In this process, the sample surface is grounded, which is equivalent to the Fermi level, and the surface potential is estimated from the difference between the potentials of the probe tip and the sample (φ_{def}). (b) A KFM image of an Au/ZnO stack structure.

to scan the sample surface. When the tip is close to the surface, the force between the tip and the sample deflects the cantilever according to Hooke's law [30]. Typically, the extent of this deflection is monitored using a laser spot reflected from the top surface of the cantilever onto a photodiode array. By using the probe tip and the sample as a top electrode and an electrical ground, respectively, an AFM system can be used to map electrical properties. Some techniques related to this concept are scanning non-linear dielectric microscopy (for the analysis of dielectric properties) [31], spreading resistance, and conductive AFM (for the analysis of resistivity). One of the advantages of AFM is that it is a non-destructive technique. Typical electrical property assessments such as current–voltage (I-V, to assess resistivity) and capacitance–voltage (C-V, to assess dielectric properties) require electrodes that cover the film surface. After the device pattern fabrication, it is difficult to investigate the properties of the entire area.

Kelvin probe force microscopy (KFM) is also a useful means of investigating a metal/oxide interface. This technique measures the local contact potential difference between a conducting AFM tip and the sample, as shown in Figure 1.8(a), thereby mapping the work function or surface potential of the material. KFM is also used to study the band offset and/or electrical properties of a heterointerface, such as that between a metal and semiconductor (see Figure 1.8(b)). In such cases, the Schottky barrier height (Φ_B) can be obtained from the difference between the potential of the metal and that of the oxide. AFM-based electrical property measurements combined with a source measurement unit, we can generate an electrical properties map, and other physical characteristics such as magnetic properties, hardness, coefficient of friction, and viscoelasticity can also be obtained.

Currently, automated mapping techniques are also employed in conjunction with analytical methods, including X-ray, Raman, and photoluminescence spectroscopy. The large amounts of data generated by these methods are modified and analyzed based on informatics so as to accelerate the material development process.

1.6 CONTROL OF ELECTRICAL PROPERTIES IN BINARY ALLOY SYSTEMS

The primary goal of combinatorial technology is the creation of libraries of thin film materials. Existing bulk databases provide the physical properties of materials synthesized in an energetically stable state (that is, their equilibrium state). However, nanoscale thin films may be in a non-equilibrium state and may also undergo mechanical distortions caused by the formation of junctions with other materials. These phenomena can lead to differences between the properties of thin film and bulk databases, although comparison with bulk databases is also important to understand the unique properties of the thin films.

The advantage of thin film synthesis is that it allows the use of single-crystal substrates and provides the ability to synthesize thin films in a non-equilibrium state. This makes it possible to

Combinatorial Synthesis

synthesize materials that are difficult to obtain with bulk materials. The preparation of alloys with systematic compositional variations as thin films can also yield a phase diagram different from that of bulk materials. This approach can provide new insights into the properties of the bulk substance and assist in the development of new materials. There are several examples in the literature of the synthesis of thin films on single-crystal substrates in conjunction with alloying or the formation of solid solutions.

The synthesis of Pt-Ru alloy thin films on single-crystal oxide substrates has been reported, in which the substrates exhibited similar changes in crystal orientation to that of the bulk material while having controlled crystal orientations [32, 33]. Metal electrodes are an important fundamental technology associated with the fabrication of electronic devices. As such, the author's group has studied the Schottky behavior of metal alloy thin film electrodes formed on oxides by means of combinatorial techniques. The aim of such work was to understand the phenomena occurring at the metal/oxide interface in oxide junctions and to control the electrical properties of these materials. Some of this work involved the deposition of Pt-Ru alloys on ZnO. Pt has a high work function and forms Schottky junctions with ZnO. In contrast, Ru is a low work function material but also forms a Schottky bond with ZnO based on work function values reported to date. In prior research, a Pt-Ru alloy was deposited on a ZnO substrate using an ion beam sputtering combinatorial method, and the structure and electrical and optical properties of this alloy were evaluated. Prior to the experimental trials, the bulk phase diagrams were examined. Our required diagram should indicate that no or a few points representing chemically stable intermediate phases were most likely to provide continuous phase and physical properties changes, and so Pt-Ru and Pt-W systems were selected. The Pt-W system has no phase boundary, while the Pt-Ru system has two phase boundaries [34]. However, the Pt-W alloy was found to exhibit oxidation of the W when deposited on ZnO. A structural analysis indicated that a crystal structure was maintained throughout the transition from pure Pt to Ru and that the material also obeyed Vegard's law (see Figure 1.7(c)). An XRD pole figure assessment demonstrated that a Pt-Ru thin film was epitaxially grown on the ZnO substrate and that the variations in the crystal structure as the composition changed from pure Pt to pure Ru were consistent with those observed in the bulk phase diagram.

The results of electrical characterization are presented in Figure 1.9, and indicate that the Pt and Ru sides showed Schottky and ohmic characteristics, respectively. Although the height of the Schottky barrier changed continuously, the difference in the barrier on going from pure Pt to pure Ru was less than that estimated from work function data reported in the literature. These data, together with the results of ellipsometry analyses, suggested that the changes in the electrical properties of

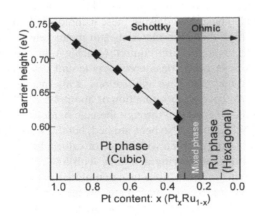

FIGURE 1.9 The effect of Pt content on the SBH values (Φ_B: squares) of the Pt-Ru alloy films on ZnO substrates as calculated from I-V characteristic. The lower figure is the Pt-Ru alloy phase diagram. XRD analysis showed that the Pt-Ru alloy films exhibited the same trends as the bulk phase diagram.

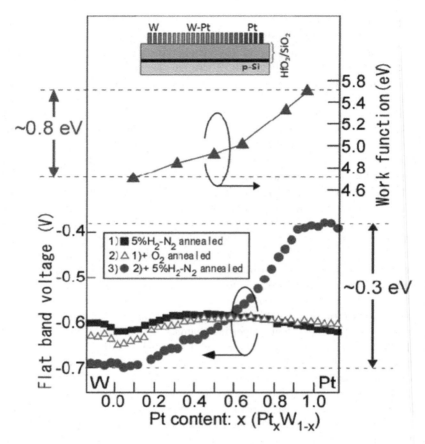

FIGURE 1.10 The effect of composition on the work functions of Pt-W binary composition tilt films and on the flat band shift of Pt-W applied as a gate material to a high dielectric material. The inset shows the sample structure.

the Pt-Ru alloy were correlated with the crystal structure of the alloy, which may have been responsible for the lower than expected rate of change of the Schottky barrier.

Metal gate materials can also be assessed using these techniques. In the field of nanoelectronics, the practical applications of high-dielectric materials and metal gates have been realized over the past decade, although new technologies are required to achieve higher functionality and integration. In this respect, nitrides have been considered as potential gate materials for high-dielectric compounds such as HfO_2, and may also have applications as alternatives to conventional polycrystalline Si gates. These materials have a work function of about 4.5 eV and a Fermi level located at approximately the center of the Si band gap through the gate oxide film. Similar materials, such as metallic silicides (including $NiSi_2$), have also been studied, but the work functions of these new gate materials must be precisely controlled over a wide range of values. In prior work, the author's group attempted to tune work functions by alloying materials with low and high work functions and by stacking materials with different work functions using combinatorial methods.

Figure 1.10 summarizes the results of a systematic evaluation of the effects of compositional changes in a Pt-W alloy on HfO_2, and of annealing, on the flat band shift of a metal organic semiconductor capacitor. The work functions plotted here were estimated on the basis of KFM measurements, by combining reported work function values for Pt [35, 36]. These values are seen to have continuously increased with increases in the Pt content, indicating that the work function could be

Combinatorial Synthesis 13

adjusted by changing the particular combination of metals. In addition, reversible changes in the flat band voltage and with composition after annealing under 5% hydrogen affected the potential state at the interface and thus the metal gate characteristics.

These studies of alloy films with compositional gradients based on bulk phase diagrams for crystal structures have provided new information concerning the electrical properties of such films with applications in nanoelectronics. The data demonstrate that compound thin films are more complex than bulk metal alloys.

1.7 CONTROL OF ELECTRICAL PROPERTIES IN BINARY OXIDE SYSTEMS

It is possible to synthesize certain materials as thin films, such as oxides, that are difficult to synthesize in bulk. In this regard, combinatorial thin film synthesis is a useful means of obtaining systematic variations in composition so as to assess the effects on physical properties. The binary oxide system In_2O_3-Ga_2O_3 is a good example of one such application of this technique [37].

Recently, Ga_2O_3 and In_2O_3 have attracted attention as wide band gap semiconductor channel materials. However, it is challenging to control the concentrations of carriers in these materials by doping, and there are also issues related to controlling their electronic states. These are crystalline polymorphic compounds and so the formation of metastable phases is expected to improve the control of surface electronic states and charge carriers in semiconductors. Furthermore, the production of solid solutions is expected to permit modulation of the band gaps of these materials, and doping efficiency may be improved by controlling the semiconductor carrier concentrations and the crystal structure. Bulk crystals of these oxides cannot be obtained from highly solubilized solid solutions owing to the large differences in formation energies and vapor pressures in such systems.

Thin films of $(Ga_xIn_{1-x})_2O_3$ having compositional gradients have been prepared on single crystal oxide substrates by combinatorial laser deposition, and XRD analyses have revealed the high solubility of In_2O_3 in Ga_2O_3. In the case of thin films with high In_2O_3 concentrations (x < 0.4), the epitaxial growth of an In_2O_3 (111) structure was observed. Upon decreasing the In_2O_3 content, a polycrystalline structure was observed, including In_2O_3 (100) and β-Ga_2O_3 phases. X-ray photoelectron spectroscopy (XPS) demonstrated that, at high In_2O_3 concentrations (x < 0.4) and using an epitaxial In_2O_3 structure, a surface electron accumulation layer (SEAL) appeared, as shown in Figure 1.11(a). In contrast, at low In_2O_3 concentrations (x > 0.4) the material had a polycrystalline structure, the dominant valence band structure was attributed to Ga_2O_3, the SEAL disappeared, and the sheet resistance was increased by at least five orders of magnitude, as shown in Figure 1.11(b). The in-gap state and valence band structure of the $(Ga_xIn_{1-x})_2O_3$ solid solution system were greatly affected by the presence of Ga_2O_3, and the position of the valence band maximum was shifted to a higher binding energy. These results suggest the possibility that the electrical properties of this material could be adjusted by varying the composition of the $(Ga_xIn_{1-x})_2O_3$.

1.8 PROCESS AND COMPOSITION OPTIMIZATION FOR COMPLEX OXIDES

Combinatorial thin film synthesis is often employed during the optimization of film composition and growth conditions. When developing thin film dielectric materials for use at high temperatures, two combinatorial optimizations have been applied: compositional and dopant concentration optimizations. Generally, a base material with a variable composition is used as the target for PLD or sputtering techniques so as to ensure low concentrations of specific elements. Here, we discuss the development of a high temperature capacitor film made of a complex oxide as an example.

Silicon carbide (SiC) has emerged as an important semiconductor in power devices, especially high-temperature electronics. SiC-based active devices such as transistors and Schottky diodes are capable of operating at temperatures above 500°C [38, 39] and are currently at the advanced manufacturing stage. In addition to SiC, several other nitride [40, 41] and oxide [42] semiconductors with wide band gaps have been used in various devices. In sharp contrast to this progress in the

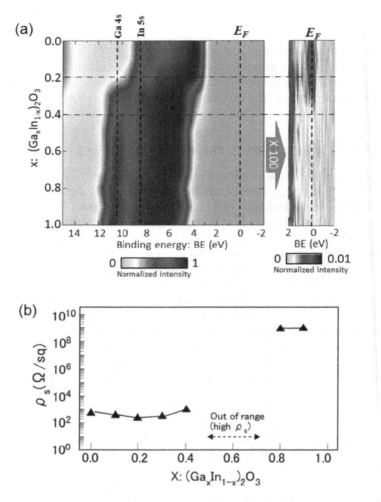

FIGURE 1.11 (a) The valence band spectra and (b) sheet resistance of $(Ga_xIn_{1-x})_2O_3$ films.

design of active devices, capacitors (which are critical passive components in electronic devices) presently function efficiently only below 175°C. Thus, advances in such devices will require both active and passive devices to be improved. As an example, the automobile industry requires power systems that function up to 250°C while the aviation and space sectors may require temperatures as high as 500°C. In addition to the present 175°C limitation, capacitors are currently bulky and occupy a sizeable proportion of the space in electronic modules. Therefore, thin film capacitors that can be operated at elevated temperatures and can also be readily integrated with active devices are required to permit the fabrication of compact units with low inductive losses and faster switching. In this regard, the primary challenge is to find a high-temperature dielectric for use in capacitors. The author's group developed a series of $BaTiO_3$-based thin film materials having high dielectric constants, $(1-x)[BaTiO_3]-x[Bi(Mg_{2/3}Nb_{1/3})O_3]$–(BT-BMN), using a combinatorial technique [43–45]. Prior to this experimentation, the bulk data were surveyed. The requirements for this new material included a high dielectric constant (above 200), a highly stable dielectric constant with temperature changes (typically a change of less than 15% up to 400°C), low dielectric loss and leakage current, an absence of hazardous components, and, most importantly, simple processing technology. Relaxor ferroelectric materials fulfilled most of these criteria and were investigated extensively, although many of the lead-free dielectric systems were previously examined in the form of bulk

Combinatorial Synthesis

ceramics. Among the possible candidates, the $BaTiO_3$-based bulk relaxor BT-BMN was identified as promising due to its high dielectric constant (over 700) in its bulk form and its temperature stability [46]. Curie temperatures were found to vary with the ratio of $BaTiO_3$ to $Bi(Mg_{2/3}Nb_{1/3})O_3$, and the variations in the dielectric constant were below 10% above 100°C. In the bulk form, the dielectric constant variation was higher and the film structure exhibited significant currier temperature dispersion with changes in the stacking of the modified composition layers and in the crystal orientation.

The first challenge was to control of the Bi concentration in the material, as a result of the high vapor pressure of Bi. A high growth temperature was required to obtain a crystalline oxide film, and even the PLD method (which is known to be a highly efficient means of compositional variation) produced shifts from the stoichiometric amount of Bi. Combinatorial BT-BMN thin films were grown on $Pt/SiO_2/Si$ substrates using PLD, and deviated from stoichiometry, using targets enriched in Bi to optimize the Bi content, as illustrated in Figure 1.12. XPS analyses confirmed a linear Bi concentration gradient in the product. The as-deposited films were post-annealed at high temperatures under oxygen to obtain crystalline materials. The crystallinity of each film (as determined from the full width at half maximum values in XRD patterns) was found to improve as the Bi content was increased. Figure 1.12 summarizes the electrical properties of these films, and indicates that the dielectric constant increased along with the Bi concentration up to 7 wt% Bi. Scanning non-linear dielectric microscopy, which is an extension of AFM and can detect ferroelectric behaviors, revealed that the ferroelectric phase distribution was improved in the vicinity of 7 wt% Bi, at which point the leakage current was also minimized. These results suggest that a non-stoichiometric Bi concentration in the PLD target affected the thin film electrical properties. The dielectric constant stability for this material ($\Delta\varepsilon$), defined as $\Delta\varepsilon = 100 \times (\varepsilon_{max} - \varepsilon_{min})/\varepsilon_{RT}$ (%), is plotted in Figure 1.12. Here ε_{max}, ε_{min}, and ε_{RT} are the maximum, minimum, and room temperature dielectric constants. The dielectric constant is seen to have been close to 275 while the stability was below 16% from room temperature to 400°C. The dielectric loss was also below 0.1 over the entire temperature range. The dielectric constant plateaued from 150°C to 350°C, indicating high stability of the maximum dielectric constant. This was observed for all Bi concentrations in the films, although the leakage current remained high at over 10^{-7} A/cm^2.

In subsequent work, the dopant concentration was optimized. To reduce the leakage current and improve the temperature stability of the dielectric constant, Ta was added as a dopant, based on an existing materials database [47]. Ta was able to reduce oxygen vacancies owing to its low oxidization energy, and was also expected to inhibit Bi migration in the BT-BMN layer at high temperatures on the basis of thermal diffusion theory [48–50]. To investigate the Ta doping effects, a specimen having a Ta concentration gradient was fabricated using a combinatorial method, in which the Ta concentration ranged from 0 wt% to 13 wt%. Figure 1.13 shows the effects of the Ta concentration on the dielectric properties and leakage current. A small amount of Ta doping greatly reduced the leakage current and improved the thermal stability of the dielectric constant, although the dielectric constants decreased gradually with increases in the Ta concentration. These results suggest that the dielectric constant can be adjusted in conjunction with low leakage and high stability by Ta doping. The electrical properties of these films were relatively good and so this material shows promise with regard to high-temperature capacitor applications [51, 52].

1.9 REQUIREMENTS AND FUTURE ISSUES

The combinatorial method can provide information regarding the properties of nanoscale materials in a systematic manner, including the areas of the nanoscale material that show a tendency to correspond to the bulk properties and information specific to newly developed nanoscale thin films. This technique can also provide guidelines for the development of advanced materials. Combinatorial synthesis methods consist of two components: a software component related to the development and use of the synthesis method and a hardware component related to the device itself.

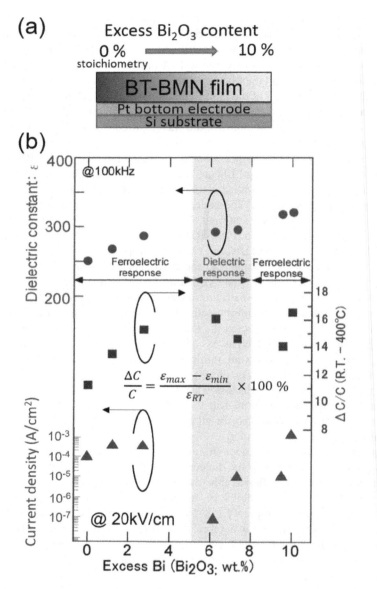

FIGURE 1.12 (a) A schematic diagram of a Bi_2O_3 film having a compositional gradient. (b) The effect of Bi concentration on the dielectric constant, temperature stability, and leakage current for combinatorial thin films annealed at 900°C.

In materials informatics, it is important to employ computational sciences and to design materials in a highly efficient manner so as to build a database. In this regard, combinatorial technology can increase efficiency through a systematic experimental process that provides information regarding the material. However, in future, the commonality of data and the seamless movement and integration of information between different platforms will be important. In the case of synthesis technology, it would be desirable to devise a data format that is versatile and allows easy handling of synthesis blueprint information and ready transfer to the next stage of evaluation and analysis. As an example, user interfaces of combinatorial PLD and sputtering methods have already been developed in a manner that provides intuitive information. The integration of the compositional

Combinatorial Synthesis

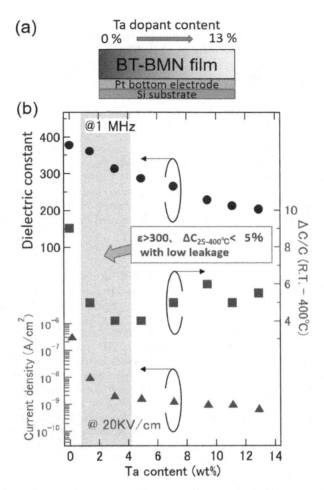

FIGURE 1.13 (a) A schematic diagram of a film with a Ta dopant gradient. (b) The effect of the Ta doping concentration on the dielectric constant, temperature stability, and leakage current for combinatorial thin films annealed at 900°C.

gradient blueprint and the mapping information provided by measurement results during the synthesis of specimens is currently performed manually using graphing software. Standardizing these data would allow integration into a common platform for informatics, which would be expected to speed up the synthesis and analysis processes.

REFERENCES

[1] Furka, Á. 2002. Combinatorial chemistry 20 years on… *Drug Discov. Today* 7: 1–4.
[2] Rasheed, A., and R. Farhat. 2013. *Combinatorial chemistry: a review. Int. J. Pharm. Sci. Res.* 4: 2502–2516.
[3] Takeuchi, I., R. B. van Dover, and H. Koinuma. 2002. Combinatorial Synthesis and Evaluation of Functional Inorganic Materials Using Thin-Film Techniques. *Mater. Res. Soc. Bull.* 27: 301–308.
[4] Koinuma, H., and I. Takeuchi. 2004. Combinatorial solid-state chemistry of inorganic materials. *Nat. Mater.* 3: 429–438.
[5] Cui, J., Y. S. Chu, O. Famodu, Y. Furuya, J. Hattrick-Simpers, R. D. James, A. Ludwig, S. Thienhaus, M. Wuttig, Z. Zhang, and I. Takeuchi. 2006. Combinatorial search of thermoelastic shape-memory alloys with extremely small hysteresis width. *Nat. Mater.* 5: 286–290.

[6] Ward, L., A. Agrawal, A. Choudhary, and C. Wolverton. 2016. A general-purpose machine learning framework for predicting properties of inorganic materials. *Npj Comput. Mater.* 2: 16028–16034.

[7] Ren, F., L. Ward, T. Williams, K. J. Laws, C. Wolverton, J. Hattrick-Simpers and A. Mehta. 2018. Accelerated discovery of metallic glasses through iteration of machine learning and high-throughput experiments. *Sci. Adv.* 4: 1566.

[8] https://materialsproject.org/

[9] https://mits.nims.go.jp/en/

[10] https://atomwork-adv.nims.go.jp/en/service.html

[11] https://materials.springer.com/

[12] Kennedy, K., T. Stefansky, G. Davy, V.F. Zacky, and E. Parker. 1965. Rapid method for determining ternary-alloy phase diagrams. *J. Appl. Phys.* 36: 3808–3810.

[13] Xiang, X. D., X. Sun, G. Briceño, Y. Lou, K. A. Wang, H. Chang, W. G. Wallace Freedman, S. W. Chen, and P. G. Schultz. 1995. A combinatorial approach to materials discovery. *Science* 268: 1738–1740.

[14] Danielson, E., J. H. Golden, E. W. McFarland, C. M. Reaves, W. H. Weinberg, and X. D. Wu. 1997. A combinatorial approach to the discovery and optimization of luminescent materials. *Nature* 389: 944–948.

[15] https://www.3ds.com/

[16] https://www.ilika.com/

[17] Ohtani, M., T. Fukumura, M. Kawasaki, K. Omote, T. Kikuchi, J. Harada, A. Ohtomo, M. Lippmaa, T. Ohnishi, D. Komiyama, R. Takahashi, Y. Matsumoto, and H. Koinuma. 2001. Concurrent x-ray diffractometer for high throughput structural diagnosis of epitaxial thin films. *Appl. Phys. Lett.* 79: 3594–3596.

[18] Fujimoto, K., K. Takada, T. Sasaki, and M. Watanabe. 2003. Combinatorial approach for powder preparation of pseudo-ternary system $LiO_{0.5}$–X–TiO_2 (X: $FeO_{1.5}$, $CrO_{1.5}$ and NiO). *Appl. Surf. Sci.* 223: 49–53.

[19] Inoue, S., S. Todoroki, T. Matsumoto, T. Honda, T. Araki, and T. Tsuchiya. 2002. Development of the combinatorial glass formation tester. *Mater. Res. Soc. Symp. Proc.* 700: 201–208.

[20] Yanase, I., T. Ohtaki, and M. Watanabe. 2002. Application of combinatorial process to $LiCo_{1-X}Mn_XO_2$ ($0 \leq X \leq 0.2$) powder synthesis. *Solid State Ionics* 151: 189–196.

[21] Okazaki, N., H. Odagawa, Y. Cho, T. Naganuma, D. Koiyama, T. Koida, H. Minami, P. Ahamet, T. Fukumura, Y. Matsumoto, M. Kawasaki, T. Chikyow, H. Koinuma, and T. Hasegawa. 2002. Development of scanning microwave microscope with a lumped-constant resonator probe for high-throughput characterization of combinatorial dielectric materials. *Appl. Surf. Sci.* 189: 222–226.

[22] https://www.comet-nht.com/index-e.html

[23] Schneemeyaer, L.F., R.B. van Dover, and R.M. Fleming. 1999 High dielectric constant Hf-Sn-Ti-O thin films. *Appl. Phys. Lett.* 75: 1967–1969.

[24] Chikyow, T., T. Nagata, P. Ahmet, T. Hasegawa, D. Kukuznyak, and H. Koinuma. 2010 Combinatorial oxide film synthesis and its application to new materials discovery. Oxide Thin Film Technology-Growth and Applications. pp. 37–57 (ISBN: 978-81-7895-468-4, Editor(s): Tomoyasu Inoue) (Transworld Research Network, 2010, India)

[25] Hasegawa, K., P. Ahmet, N. Okazaki, T. Hasegawa, K. Fujimoto, M. Watanabe, T. Chikyow and H. Koinuma. 2004 Amorphous stability of HfO_2 based ternary and binary composition spread oxide films as alternative gate dielectrics. *Appl. Surf. Sci.* 223: 229–232.

[26] Ahmet, P., Y.Z. Yoo, K. Hasegawa, H. Koinuma and T. Chikyow. 2004. Fabrication of three-component composition spread thin film with controlled composition and thickness. *Appl. Phys. A Mater. Sci. Process.* 79: 837–839.

[27] Chikyow, T., P. Ahamet, K. Hasegawa, and H. Koinuma. 2003. Multi-element compound manufacturing apparatus. Japan patent, 2003-277914,A

[28] Ahmet, P., T. Nagata, D. Kukuruznyak, S. Yagyu, Y. Wakayama, M. Yoshitake, and T. Chikyow. 2006. Composition spread metal thin film fabrication technique based on ion beam sputter deposition. *Appl. Surf. Sci.* 252: 2472–2476.

[29] Nagata, T., M. Haemori, and T. Chikyow. 2009. Capability of focused Ar ion beam sputtering method for combinatorial synthesis of metal films. *J. Vac. Sci. Technol. A*, 27: 492–495.

[30] Binnig, G., C.F. Quate, and Ch. Gerber. 1986. Atomic force microscope. *Phys. Rev. Lett.* 56: 930–933.

[31] Cho, Y., S. Kazuta, and K. Matsuura. 1999. Scanning nonlinear dielectric microscopy with nanometer resolution. *Appl. Phys. Lett.* 75: 2833–2835.

[32] Nagata, T., J. Volk, M. Haemori, Y. Yamashita, H. Yoshikawa, R. Hayakawa, M. Yoshitake, S. Ueda, K. Kobayashi, and T. Chikyow. 2010. Schottky barrier height behavior of Pt-Ru alloy contacts on single-crystal n-ZnO. *J. Appl. Phys.* 107: 103714.

[33] Nagata, T., J. Volk, M. Haemori, Y. Yamashita, H. Yoshikawa, S. Ueda, K. Kobayashi, R. Hayakawa, M. Yoshitake, and T. Chikyow. 2009. Interface structure and the chemical states of Pt film on polar-ZnO single crystal. *Appl. Phys. Lett.* 94: 221904.

[34] Hutchinson, J. M. 1972. Solubility relationships in the ruthenium-platinum system. *Platin. Met. Rev.* 16: 88–90.

[35] Derry, G. N. and Z. Ji-Zhong. 1989. Work function of Pt(111). *Phys. Rev. B* 39: 1940.

[36] Gu, D. and S. K. Dey. 2006. Effective work function of Pt, Pd, and Re on atomic layer deposited HfO_2. *Appl. Phys. Lett.* 89: 082907.

[37] Nagata, T., T. Hoga, A. Yamashita, T. Asahi, S. Yagyu, and T. Chikyow. 2020. Valence band modification of a $(Ga_xIn_{1-x})_2O_3$ solid solution system fabricated by combinatorial synthesis. *ACS Comb. Sci.* 22: 433–439.

[38] Neudeck, P. G., D. J. Spry, L.Y. Chen, G. M. Beheim, R. S. Okojie, C. W. Chang, R. D. Meredith, T. L. Ferrier, L. J. Evans, M. J. Krasowski, and N. F. Prokop. 2008. Stable electrical operation of 6H–SiC JFETs and ICs for thousands of hours at 500 °C. *IEEE Electron Dev. Lett.* 29: 456–459.

[39] Neudeck, P. G., D. J. Spry, L.-Y. Chen, C. W. Chang, G. M. Beheim, R. S. Okojie, L. J. Evans, R. D. Meredith, T. L. Ferrier, M. J. Krasowski, and N. F. Prokop. 2009. Prolonged 500 °C Operation of 6H-SiC JFET Integrated Circuitry. *Mater. Sci. Forum* 615–617: 929–932.

[40] Baliga, B. J. 2013. Gallium nitride devices for power electronic applications. *Semicond. Sci. Technol.* 28: 074011.

[41] Xu, Z., J. Wang, Y. Cai, J. Liu, Z. Yang, X. Li, M. Wang, M. Yu, B. Xie, W. Wu, X. Ma, J. Zhang, and Y. Hao. 2014. High temperature characteristics of GaN-based inverter integrated with enhancement-mode (E-Mode) MOSFET and depletion-mode (D-Mode) HEMT. *IEEE Electron Dev. Lett.* 35: 33–35.

[42] Higashiwaki, M., K. Sasaki, A. Kuramata, T. Masui, and S. Yamakoshi. 2014. Development of gallium oxide power devices. *Phys. Status. Solidi. A.* 211: 21–26.

[43] Kumaragurubaran, S., T. Nagata, K. Takahashi, S-G. Ri, Y. Tsunekawa, S. Suzuki, and T. Chikyow. 2015. Combinatorial synthesis of $BaTiO_3$-$Bi(Mg_{2/3}Nb_{1/3})O_3$ Thin-films for High-temperature Capacitors. *Jpn. J. Appl. Phys.* 54: 06FJ02.

[44] Kumaragurubaran, S., T. Nagata, K. Takahashi, S-G. Ri, Y. Tsunekawa, S. Suzuki and T. Chikyow. 2015. $BaTiO_3$ based relaxor ferroelectric epitaxial thin-films for high-temperature operational capacitors. *Jpn. J. Appl. Phys.* 54: 04DH02.

[45] Kumaragurubaran, S., T. Nagata, Y. Tsunekawa, K. Takahashi, S-G. Ri, S. Suzuki, and T. Chikyow. 2015. Epitaxial growth of high dielectric constant lead-free relaxor ferroelectric for high-temperature operational film capacitor. *Thin Solid Films*, 592: Part A, 29–33.

[46] Malhan, R. K., N. Sugiyama, Y. Noguchi, and M. Miyayama, 2012. US Patent 8, 194, 392, B2.

[47] Nagata, T. 2020. Development of new high-dielectric constant thin films by combinatorial synthesis, *ECS Trans.* 97: 61–66.

[48] Yoshitake, M., Y.-R. Aparna, and K. Yoshihara. 2001. General rule for predicting surface segregation of substrate metal on film surface. *J. Vac. Sci. Technol. A* 19: 1432–1437.

[49] Yoshitake, M., 2012 Prediction of influence of oxygen in annealing atmosphere on surface segregation behavior in layered materials *Jpn. J. Appl. Phys.* 51: 085601.

[50] Nagata, T., S. Kumaragurubaran, Y. Tsunekawa, Y. Yamashita, S. Ueda, K. Takahashi, S.-G. Ri, S. Suzuki, and T. Chikyow. 2016. Interface stability of electrode/Bi-containing relaxor ferroelectric oxide for high-temperature operational capacitor. *Jpn. J. Appl. Phys.* 55: 06GJ12.

[51] Yuan, Y., C. J. Zhao, X. H. Zhou, B. Tang, and S. R. Zhang. 2010. High-temperature stable dielectrics in Mn-modified $(1-x)Bi_{0.5}Na_{0.5}TiO_{3-x}CaTiO_3$ ceramics. *J. Electroceram.* 25: 212–217.

[52] Raengthon, N., and D. P. Cann. 2011. High-K $(Ba_{0.8}Bi_{0.2})(Zn_{0.1}Ti_{0.9})O_3$ ceramics for high-temperature capacitor applications. *IEEE Trans. Ultrason. Ferroelectr. Freq. Control*, 58: 1954–1958.

2 Sputter Deposited Nanostructured Coatings as Solar Selective Absorbers

Jyh-Ming Ting and Yi-Hui Zhuo
National Cheng Kung University, Tainan, Taiwan

CONTENTS

2.1 Introduction to Solar Thermal System 21
2.2 Classification of Solar Selective Absorber Coatings 22
    2.2.1 Intrinsic Absorber 24
    2.2.2 Semiconductor–Metal Tandem Absorber 25
    2.2.3 Multilayered Absorber 25
    2.2.4 Cermet Absorber 27
    2.2.5 Textured Absorber 34
2.3 High-Temperature Absorber Coatings 36
2.4 Conclusion 44
References 44

2.1 INTRODUCTION TO SOLAR THERMAL SYSTEM

In order to decrease the discharge of carbon dioxide and combat the energy crisis, renewable energies, especially solar energy, continue to be at the central stage. The total solar power incident on the earth is 166 PW per year, which is 10,000 times more than mankind's energy consumption. Among all of the renewable energy sources, solar power generates more than 99% of the electricity [1]. Two of the most important ways to harvest the solar power are through concentrating solar power (CSP) and photovoltaics (PV) systems. PV converts the sunlight directly into electricity. A CSP system concentrates the sun light on an absorber that converts the solar radiation into heat. The heat then drives a heat engine connected to an electrical power generator to generate electricity. Figure 2.1 shows the typical configuration of a CSP power plant. The solar beam first radiates at the optical heliostat's arrays. In order to increase the energy density, these mirror arrays are made to reflect the solar beam right to the solar absorber to heat up the heat transfer fluid in the absorber. The fluid is transferred to the thermal storage tank where the thermal energy is stored. Thereafter, the fluid is transferred to the steam turbine and generate the electric power. Afterword, the waste steam is recycled and condensed in the condenser. In a thermal storage tank, solar selective coating (SSC) is deposited on the surface of a solar receiver/absorber in order to increase the solar absorptance and decrease the thermal emittance, and hence for enhanced solar-thermal transformation efficiency. The best-performing CSP system provides 30% energy conversion efficiency, which is higher than that of a PV system (~18%). Moreover, the heat storage system makes CSP functional even without sun light. Due to the SSC, the thermal emission of the tank is minimized so that the heat transfer fluid temperature is well maintained and therefore provides enough energy to steam up the water during the night. In order to maximize the harvest of the solar power, the location of a CSP power plant is also critical. Most of the CSP plants lie in the low-latitude areas, especially in the desert areas.

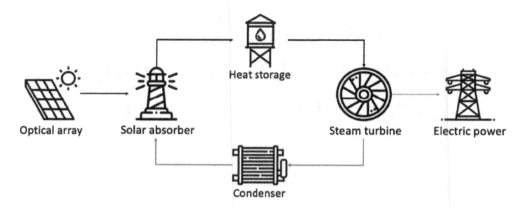

FIGURE 2.1 A schematic of the CSP system.

Based on the design of the optical path, there are four major types of CSP systems, which include solar tower, parabolic trough, linear Fresnel, dish Stirling types [2]. In the parabolic trough CSP, the parabolic surface reflectors reflect the solar beam to the tube. The tube containing heat transfer fluid lies right on the focal point, as shown in Figure 2.2(a). The fluid is then transferred to a turbine that subsequently generates electricity. Normally, the total energy transfer efficiency in the CSP power plant is defined as the output electricity power over the solar radiation power. The total energy transfer efficiency, that is, the percentage of absorbed energy utilized, is up to 15%, which is similar to that of the PV technique. The parabolic trough is the most mature type of CSP around the world and has been applied in many power plants in, for example, the US and Spain. The working temperature is around 400°C which is the lower than that of other optical concentrating designs. The optical design of the parabolic trough CSP gives the lowest solar concentration, leading to its lowest operating temperature and low energy conversion efficiency. The linear Fresnel type is similar to the parabolic trough type. The major difference is the reflector mirrors. In the Linear Fresnel system, the mirrors are flat (some are slightly curved) and lay down on the ground at different angles so that they concentrate the solar beam to the absorber tube, as shown in Figure 2.2(b). Thus, the cost and the efficiency are also similar. The solar tower provides a better solar-thermal efficiency because of the higher working temperature. The working temperature can be up to 800°C. A flat heliostat array, consisting of thousands of heliostats, surrounds one tower. The flat mirrors cost less than that used in the parabolic trough for curvature is no longer considered. The mirrors concentrate the solar beam on the top of the tower, where the solar receiver is located, as shown in Figure 2.2(c). The SSC is around the absorber on the top of the tower. The heat transfer fluid in the receiver then flows through the pipe and reach the turbine. A challenge to this type of optical path design is its high operating temperatures, exceeding 800°C, which is too high for the traditional SSCs. Degradation of the SSC occurs, leading to reduced optical performance. Unlike the other CSP systems, the dish Stirling system is equipped with an individual engine for each set of parabolic-dish reflectors, as shown in Figure 2.2(d). The travel distance of the heat transfer fluid is hence shortened and the energy loss during the transfer process therefore decreases. The energy conversion efficiency (~30 %) is the highest among the all four types of CSP. Despite its immaturity and high cost, the dish Stirling design represents an important method to harvest the solar energy. Its working temperature is even higher than that of the solar tower system, which posts even stringent requirements for the high-temperature SSCs.

2.2 CLASSIFICATION OF SOLAR SELECTIVE ABSORBER COATINGS

SSC is required to have high solar absorptance (α) and low thermal emittance (ε_T). The solar absorptance (α) is defined as the fraction of the solar energy absorbed by a surface in the solar radiation spectrum from 0.3 to 2.5 μm, as shown in Equation. (2.1).

Sputter Deposited Nanostructured Coatings

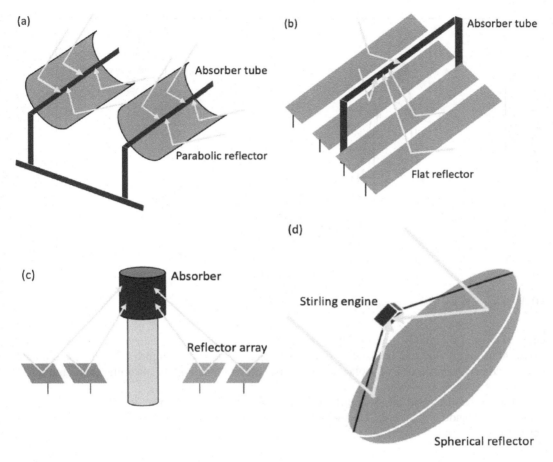

FIGURE 2.2 Four types of optical concentrating designs. (a) Parabolic trough, (b) linear Fresnel reflector, (c) solar tower, and (d) dish Stirling [3].

$$\alpha = \frac{\int_{0.3\,\mu m}^{2.5\,\mu m} I_{sol}(\lambda)(1-R(\lambda))d\lambda}{\int_{0.3\,\mu m}^{2.5\,\mu m} I_{sol}(\lambda)} \quad (2.1)$$

where the $I_{sol}(\lambda)$ is the solar radiation power at air mass (AM) 1.5 [4] and R(λ) is the spectral reflectance. The thermal emittance (ε_T) is defined as a ratio of the radiant emittance of the surface heat to that of the blackbody between 2.5 and 25 μm as shown in Equation (2.2).

$$\varepsilon_T = \frac{\int_{2.5\,\mu m}^{25\,\mu m} I_b(\lambda)(1-R(\lambda))d\lambda}{\int_{2.5\,\mu m}^{25\,\mu m} I_b(\lambda)} \quad (2.2)$$

where the $I_b(\lambda)$ is the spectral radiance of a black body. In the traditional CSP system, the solar absorber works around 100°C. Therefore, the thermal emittance is normally evaluated at 100°C and the emittance is designated ε_{100}. To maximize the harvest of the solar energy, an ideal SSC is required to exhibit a solar absorptance (α) of 1 in the solar region (0.3–2.5 μm) and a thermal

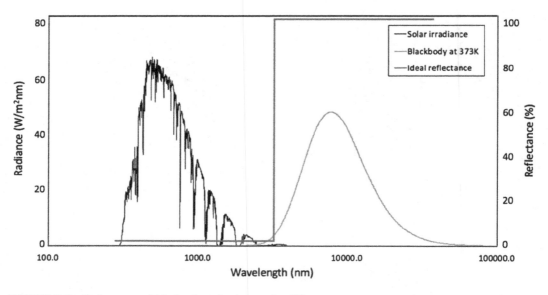

FIGURE 2.3 Reflectance of ideal solar selective coating [5].

emittance (ε) of 0 in the IR region (2.5–20 μm). In other words, this is realized by zero reflectance in the solar region and one reflectance in the IR region, as shown in Figure 2.3. Also, to evaluate the optical performance of SSC, solar selectivity is defined as the ratio of absorptance to emittance, that is, α/ε_T.

SSCs can be classified into five different types, which are (A) intrinsic absorber, (B) semiconductor metal tandem absorber, (C) multilayered absorber, (D) cermet absorber, and (E) textured absorber, as shown in Figure 2.4.

2.2.1 Intrinsic Absorber

An intrinsic absorber is a material whose selective absorption is an intrinsic characteristic. It is more stable and can be fabricated with simple process. However, it often suffers optically less effective

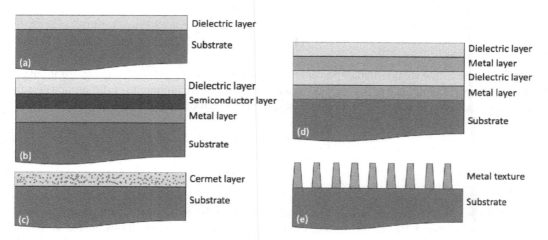

FIGURE 2.4 Five different types of solar selective coatings. (a) Intrinsic absorber, (b) semiconductor metal tandem absorber, (c) cermet absorber, (d) multilayered absorber, and (e) textured absorber.

Sputter Deposited Nanostructured Coatings

problem for there is no ideal selectivity in nature substance. Most of the intrinsic absorbers are made out of transition metal compounds. The d orbital of a transition metal ion is not fully filled. While bonding with oxygen, nitrogen, and boron, hybridized orbitals of the resulting oxides, nitrides, boride, and oxynitride, respectively, exhibits enhanced light absorption. These materials also have high reflectance in the IR band, indicating its low emittance. Various compounds, such as TaC, TiC, CuO, ZrB_2, SnO_2, In_2O_3, V_2O_5, and LaB_6, all possess good intrinsic solar selectivity. The advantage of the intrinsic absorbers is its structural simplicity, which often contributes to the thermal stability. The single-layered design gives no inter-layer diffusion and unexpected phase transformation during the high temperature and long-term duty. For the intrinsic absorber, the coating thickness is one of the major concerns [6]. When the thickness increases from 20 to 200 nm, the absorptance of the thin film increases linearly from 10% to 91%. Further thickness increase from 200 to 2000 nm gives no obvious change in the absorptance (i.e., 91% → 95%). In the meantime, the emittance also increases with the thickness. Therefore, a solar selective absorber needs to be thick enough in order to achieve high solar absorptance, but not too thick to avoid high thermal emittance. However, the intrinsic absorbers still exhibit less than satisfactory optical efficiency. For example, without any other modification, the famous black chrome possessed 86.8% of absorptance and 8.8% of emittance. The black nickel exhibits 86.7% of absorptance and 10.9% of emittance. Because there is no ideal intrinsic material, different designs are required to improve the spectral selectivity. The concept of using an anti-reflective layer, tandem absorber, composite materials, etc., have therefore been realized.

2.2.2 Semiconductor–Metal Tandem Absorber

Generally speaking, the bandgap width of a semiconductor is less than 3 eV. The correlation between the bandgap, E_g, and absorption wavelength minimum, λ_m, is

$$E_g = h\nu_M = \frac{hc}{\lambda_m} \tag{2.3}$$

where h is Plank constant and c is the speed of light. As mentioned above, the solar irradiations are mostly in between 300 and 2500 nm. To maximize the light harvest, semiconductor having a small enough E_g that allows the absorption to extend into the IR light region is desirable. Popular semiconductors silicon (Si) and germanium (Ge) possess bandgaps of 1.12 eV and 0.67 eV, respectively, and are great solar absorber candidates. However, most of these low E_g semiconductors exhibit a high refractive index, which leads to a large optical loss during the absorption. A popular method to modify the coating structure, called "tandem", is to add different layers for different purposes. A semiconductor–metal tandem absorber is made of one mirror-like metal layer, as the IR reflector, and a semiconductor layer. The reflection of the light in IR band is enhanced so that the thermal emittance is decreased. Moreover, an anti-reflection (AR) layer is desired in order to decrease the reflectance, as shown in Figure 2.4 (b). Normally, the AR layer is made of dielectric material to avoid reflection. The layer is so thin that it is almost transparent. The AR layer allows more light to penetrate the surface of the coating and entering the absorber. The solar selectivity will be enhanced not just because of the semiconductor absorber property, but also due to the addition of the AR layer and the IR reflector.

2.2.3 Multilayered Absorber

The multilayered absorber is consisted of several layers. The dielectric material and metal are deposited on the substrate layer by layer. A semitransparent metal, which reflects and transmits the light, is deposited in between two dielectric layers. A second or bottom metal layer is used when the substrate is not metallic, as mentioned above. The thickness of the metal layer is critical. If it is too

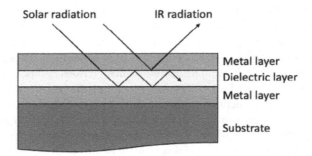

FIGURE 2.5 Schematic of multilayered absorber.

thick, the light will not be able to travel through it. Normally, 10–30 nm thick coatings are used. The dielectric material possesses high selectivity and largely absorbs the radiations, giving an identical function as the intrinsic absorber. The solar light is trapped in the dielectric layer between the two mirror-like metal layers as shown in Figure 2.5. With the multiple reflections, the light absorption efficacy is enhanced, thus giving improved solar selectivity. The enhanced solar selectivity due to the light trapping/reflections can be understood by considering the effect of interface. When the light irradiates across an interface, the reflectance and the transmittance can be approximated using the following equations.

$$R = \frac{|n_1 - n_2|^2}{|n_1 + n_2|^2} \quad (2.4)$$

$$T = \frac{4n_1 n_2}{|n_1 + n_2|^2} \quad (2.5)$$

where n_1 is the refractive index of the top layer and n_2 is the second layer. It is therefore necessary to have an n_2 that is slightly larger than n_1 to maximize the light transmittance and minimize the reflectance. In the multilayer model, the light will be trapped in the dielectric layer because the TWO adjacent metal layers possess lower refractive index. Therefore, multiple light reflections occur between the two metal layers, thus enhancing the absorptance.

For multilayered absorbers, several popular oxide dielectric materials are Al_2O_3, Cr_2O_3, and SiO_2-based absorbers. Following these oxides is the use of aluminum nitride and the combination of oxide and nitride. In 1980, J. A. Thornton sputter deposited a 3-layered absorber, that is, $Mo/Al_2O_3/Mo$ on stainless steel and an Al_2O_3 was used as the AR layer [7]. The solar absorptance was 94% and the thermal emittance was 7%. The molybdenum (Mo) layer is not used as an IR reflector but serves as a buffer layer to block the impurity diffusion from the stainless steel. As shown in Figure 2.6, the compositions at the interface between the Mo layer and the substrate before and after the annealing at 500°C are very identical. It appears that there is no Fe and Cr diffusion across the interface. Actually, the addition of a buffer layer is a very common method to improve the thermal stability. It will be elaborated in the next section.

Aluminum nitride has been used in multilayered absorbers. Multilayered absorber coatings consisting of alternating layers of aluminum nitride (AlN) and Ti were deposited on quartz substrates using a co-sputter deposition technique [8]. A copper layer was first deposited on the substrate as the IR reflector since quartz exhibits poor IR reflectance. It is noted that in the study, the solar absorptance of the multilayered absorber coating is correlated to the color of the coating. The broadened peak in the visible light band of the reflectance spectrum gives rise to the color change. The darker-colored coatings, that is, black and purple, exhibit better light absorptance than the

Sputter Deposited Nanostructured Coatings

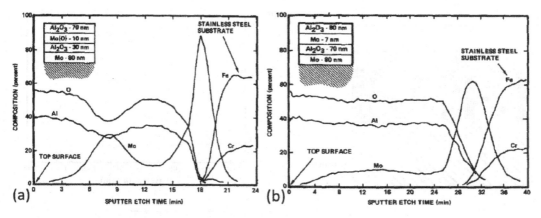

FIGURE 2.6 Depth profiles of multilayered SS/Mo/Al$_2$O$_3$/Mo/Al$_2$O$_3$ (a) before and (b) after annealing [7].

TABLE 2.1
Correlation of the color and optical performance of Ti/AlN-based multilayered solar absorber coatings [8]

Sample color	Black	Purple	Yellowish green	Red	Yellowish orange
α	0.92	0.94	0.88	0.82	0.92
ε	0.05	0.05	0.26	0.27	0.06

---

lighter-colored coatings, that is, red, orange, and green. The results are summarized in Table 2.1. The comparison can only serve as a reference from which the solar absorptance is estimated from the appearance or color.

Another interesting design of multilayered absorber is shown in Figure 2.7(a) [9]. The reflective index of each layer was first determined as shown in Figure 2.7(b). It is seen that the refractive index of AlTiN is larger than that of AlTiON, which is larger than that of AlTiO. According to their reflective index, each layer was deposited in sequence, as show in Figure 2.7(a). The light then can travel through the whole multilayered structure under minimized reflection. The multilayered Ti/AlTiN/AlTiON/AlTiO coating was deposited on stainless steel substrate. In spite of the complex structure design, only two targets, namely, Al and Ti, were used. The sputter gases used were oxygen and nitrogen. The optimized absorptance was 93% and the emittance was 16 %. Thermal stability was tested at 850°C in vacuum for different times. After the annealing, the reflectance spectra are almost identical, as shown in Figure 2.7(c).

2.2.4 Cermet Absorber

Cermet is a metal–dielectric composite. The "cer" stands for ceramics and the "met" stands for metals. A cermet consists of metal nanoparticles (NPs) homogeneously distributed in a dielectric ceramic matrix. The optical performance of the dielectric ceramic is enhanced due to the light scattering from the metal NPs. These films are transparent in the IR band region but have high absorption in the UV and visible band. Moreover, the enhancement of the absorptance is also contributed by surface plasmon resonance from the NPs. The cermet concept offers a high degree of flexibility, and the solar selectivity can be optimized by proper choice of constituents, thin film thickness, nanoparticle concentration, nanoparticle size, and orientation [10]. For the cermet absorbers,

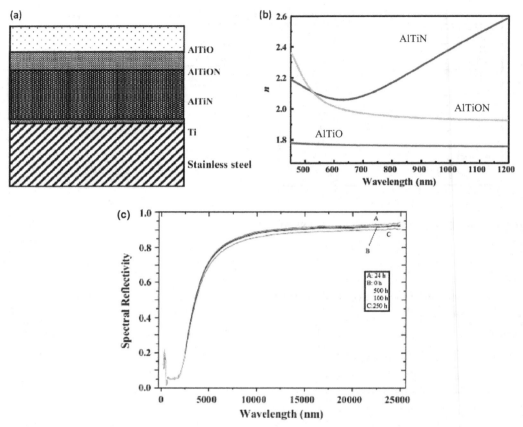

FIGURE 2.7 (a) Multilayered design of Ti/AlTiN/AlTiON/AlTiO, (b) refractive index of each layer, and (c) optimized reflectance spectrum under different annealing duration [9].

conventional black materials, such as Au-MgO and Cr-Cr_2O_3 (as known as the black chrome) were used in the beginning. Popular dielectric matrix materials also include carbon and aluminum oxide (Al_2O_3). This led to the introduction of Ni-Al_2O_3 and Mo-Al_2O_3 cermet. In addition to oxide, AlN was also commonly used for its stability.

In the beginning, some black materials were considered as good candidates for SSCs. Chromium oxide, also known as the black chromium, was often used as the cermet matrix in the early days [11]. John C. C. Fan investigated Cr-Cr_2O_3 cermet on stainless steel substrate in 1977. A co-sputtering technique, employing a Cr_2O_3 and Cr targets, was used to prepare the coating. The effect of the Cr volume fraction was addressed. The Cr volume fraction was controlled to vary from 0% to 35% by adjusting the Cr target power. As shown in Figure 2.8(a), the refractive index shifts up with increasing Cr volume fraction. The refractive index of a material is a dimensionless number that describes how fast light travels through a material. A higher refractive index indicates that light is more likely to be trapped in the material. In the Cr-Cr_2O_3 cermet, the light is trapped because of the scattering of the NPs. This means that the optical properties can be easily manipulated by adjusting the volume fraction of the metal NPs. For the surface AR layer, the matrix Cr_2O_3 is conveniently used. The adhesion is better and the thermal expansion coefficient is comparable. The resulting structure gives 92% absorptance and 8% thermal emittance. The whole reflectance curve is shown in Figure 2.8 (b). Further material characterizations were investigated by S. Khamlic [12]. The Cr-Cr_2O_3 cermet was deposited on molybdenum substrate. Figure 2.8(c) shows the diffraction pattern of the cermet. The diffraction peaks at (0 1 2), (1 0 4), (1 1 0), (1 1 3), and (0 2 4) indicates the formation of rhombohedral structured α-Cr_2O_3. No Cr metal diffraction peak is seen because the metal concentration was

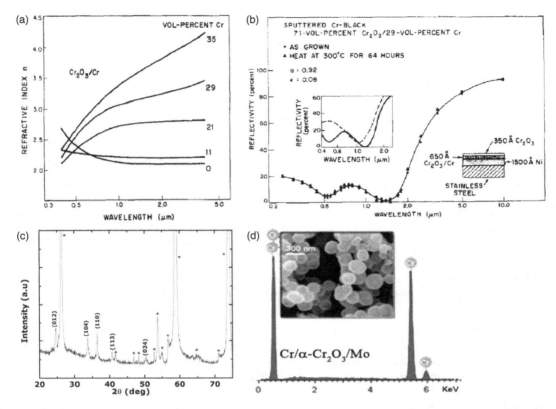

FIGURE 2.8 (a) Correlation of the Cr metal volume fraction in and refractive index for Cr-Cr$_2$O$_3$ cermet. (b) Reflectance spectrum of optimized SS/Ni/Cr-Cr$_2$O$_3$/Cr$_2$O$_3$. (c) XRD pattern and (d) SEM image and EDS analysis of Cr-Cr$_2$O$_3$ on Mo substrate [11, 12].

way too low. Figure 2.8(d) shows the SEM image and the EDS analysis. The sphere-shaped NPs are dispersed in the matrix and their average diameter is about 200 nm.

Black cermet Au-MgO has also been investigated in 1977 [13]. The Au-MgO cermet was deposited on stainless steel substrate using an RF sputtering process. A single composite target of MgO (75%) and Au (25%) was used. Mo was used as the IR reflector. As shown in Figure 2.9(a), 93% absorptance and 9% emittance were obtained in the as-deposited state and the optical performance was nearly identical after 400°C air-annealing for 64 h. The clear image of Au NPs in the MgO matrix was taken by L. K. Thomas in 1989 as shown in Figure 2.9(b) [14]. A DC generator was used and the substrate was copper. A comparable optical performance was achieved: 91% absorptance and 4.8% emittance.

Amorphous carbon-based materials are widely used in various fields. Among them, amorphous hydrogenated carbon (a-C:H) is a black material. The structural and optical properties of a-C:H thin films have been studied. a-C:H thin film has been used in SSC. H. Y. Cheng deposited a single-layered Cr-containing a-C:H cermet on the silicon substrate using DC and RF co-sputtering [15]. Methane was used and varied to control the absorptance and the thermal emittance of the resulting coatings. The thickness of each sample was controlled to be about 200 nm. The optical absorptance spectra of Cr-containing a-C:H cermets are shown in Figure 2.10(A). While the CH$_4$ concentration varies from 2% to 8%, the solar absorptance varies from 68% to 33%. The absorptance curves of pure Cr and a-C:H are also shown for comparison. The thermal emittance spectra of Cr-containing a-C:H cermets are shown in Figure 2.10(B). The spectra are almost unchanged with the CH$_4$ concentration. The thermal emittance is about 7%. The solar selectivity was thus optimized to be 10.6.

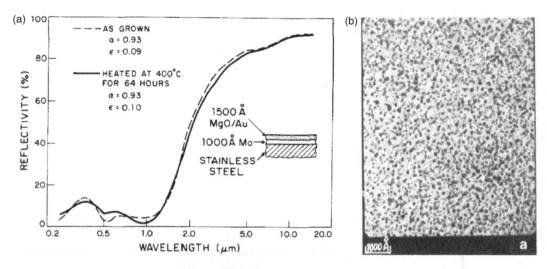

FIGURE 2.9 (a) Reflectance spectra of optimized Mo/Au-MgO (dashed line: as-deposited; solid line: after 400°C annealing). (b) TEM image of Au nanoparticles in Au-MgO cermet [13, 14].

According to the XPS analysis shown in Figure 2.10(C), the five peaks centered at 283.49, 284.67, 285.46, 286.24, and 287.53 eV are indicative of C/Cr, C=C (sp2), C-C (sp3) or C-H, C-O, and C=O, respectively. In Figure 2.10(D), an HR-TEM image shows crystalline Cr NPs, with an average size of 1.58 nm, distributing in the amorphous carbon matrix.

Cermet based on Al_2O_3 has also been widely investigated due to its excellent optical property and thermal stability. $Ni-Al_2O_3$ cermet was deposited on stainless steel substrate using an RF planar magnetron sputtering technique [16]. A hot-pressed $Ni-Al_2O_3$ target was used. A solar absorptance of 84% was obtained using only a single $Ni-Al_2O_3$ absorber layer. The absorptance was improved to 91% with a SiO_2 surface AR layer, as shown in Figure 2.11(a). Moreover, the absorber coating survived 550°C air annealing for 6 h. After the annealing, the solar absorptance is 91% and thermal emittance is 7%, as shown in Figure 2.11(b). In Figure 2.11 (c), the depth profile shows that there is no change in the composition.

Although the use of metal NPs increases the absorptance of the coating, the refractive index of the coating also increases. The idea of "graded cermet" is then used [17, 18]. Graded cermet absorber coating consists of a high metal volume fraction (HMVF) cermet layer and a low metal volume fraction (LMVF) cermet layer as shown in Figure 2.12. Compared to a homogeneous cermet single-layered coating, the interference at the interface between the HMVF cermet layer and LMVF cermet layer increases the solar absorptance. The HMVF cermet possesses a higher refractive index than the LMVF cermet due to its higher metal content. Thus, the LMVF cermet layer is deposited on the top of the HMVF cermet layer to allow effective light transmission from the LMVF cermet layer to the HMVF cermet layer.

Pt-containing a-C:H has also been investigated [17]. XRD patterns of single-layered Pt-containing a-C:H coatings deposited at different platinum target powers and a fixed carbon target power of 60 W are shown in Figure 2.13(a). Only the crystalline Pt is seen but not the amorphous carbon. The intensity of the Pt diffraction peak increases with the Pt target power, indicating increasing Pt concentration with the Pt target power. The optical absorptance of single-layered coating as a function of the platinum target power is shown in Figure 2.13(b). The absorptance improves with the platinum target power due to increasing Pt concentration. It was found that at a higher Pt concentration, the dispersion of PT NPs in the a-C:H matrix is more desirable. An optimized single-layered Pt-containing a-C:H gives 72% absorptance. To improve the optical performance, double graded cermet absorber coatings were prepared. The top LMFV layer is a Pt-containing a-C:H deposited

Sputter Deposited Nanostructured Coatings

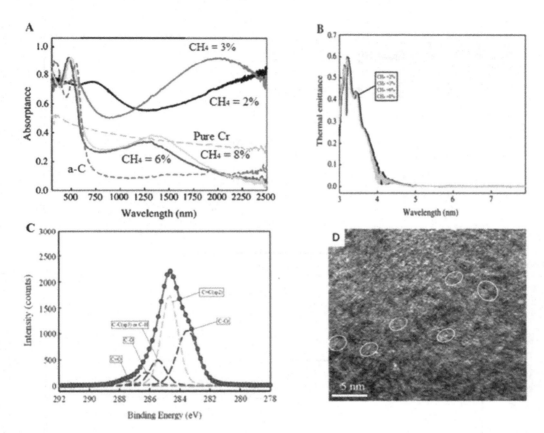

FIGURE 2.10 (A) Optical absorptance and (B) thermal emittance spectra of Cr-(a-C:H) obtained at different methane concentrations. (C) XPS spectrum (D) HRTEM image of a-C:H/Cr coating obtained at 4% methane concentration [15].

at a Pt target power of 60 W. Two different Pt-containing a-C:H layers were used as the HMLV layers, obtained using Pt target powers of 100 and 80 W. In all Pt-containing a-C:H layers, the carbon target power was fixed 60W. The absorptance reaches 94% and 90% for the 100 W and 80 W HMVF samples, respectively, as shown in Figure 2.13(c). A low thermal emittance has been obtained as shown in Figure 2.13(d).

Based on the encouraging result of the Al_2O_3-based cermet SSCa, 4-layered design was used [19], as shown in Figure 2.14. The AR layer, having an even lower refractive index, allows more light to enter the coating and increases the absorptance. In order to simplify the design, the ceramic matrix in the cermet is often used as the AR layer material. Moreover, an IR metal layer beneath the absorber coating is used. The IR layer decreases the thermal emittance from the substrate and hence improves the solar selectivity.

X. K. Du deposited a 4-layered graded cermet SSC of Mo/LMVF Mo-Al_2O_3/HMVF Mo-Al_2O_3/Al_2O_3 on stainless steel using a magnetron sputter deposition technique [20]. With increasing layer number, the reflectance in the UV band becomes lower, indicating a better solar absorptance (from 0.36 to 0.92). Afterward, a further improvement of the thermal stability was made by adding an Fe_2O_3 layer. It is noted that the Fe_2O_3 layer is deposited on the substrate and a buffer layer is used, which will be further discussed later. The optical performance of the progressive layer-by-layer deposited coatings is shown in Table 2.2. From Table 2.2, it was clear that the absorptance is primarily contributed by the cermet layers. Moreover, it is the graded cermet design that enhances the absorptance; however, the thermal emittance remains high. The reflectance is shown in Figure 2.15.

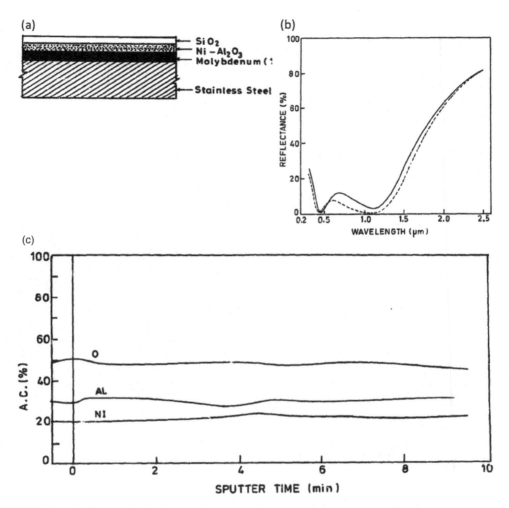

FIGURE 2.11 (a) Schematic of Ni-Al$_2$O$_3$ SSC, (b) reflectance spectra of optimized Mo/Ni-Al$_2$O$_3$/SiO$_2$ (solid line: as-deposited; dashed line: after 550°C anneal), and (c) Auger depth profile of Ni-Al$_2$O$_3$ [16].

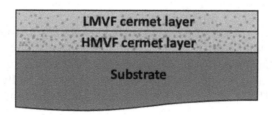

FIGURE 2.12 Schematic of double graded cermet absorber coating.

The solar absorptance and the thermal emittance of the final optimized 4-layered cermet coating, either with or without an Fe$_2$O$_3$ layer, are the same. The entire SSC exhibits 92% absorptance and 19% emittance in both structures.

In addition to oxides, different ceramic such as nitride was studied as well. Aluminum nitride serves as desired matrix in cermet because of the high thermal and chemical stability. Q. C. Zhang deposited the classic 4-layered structure with the AlN as the matrix [21]. A reactive sputtering

Sputter Deposited Nanostructured Coatings

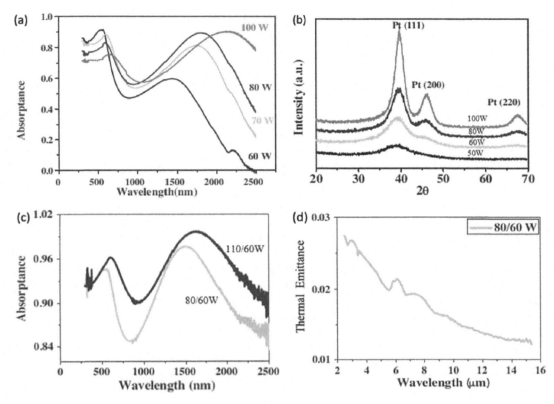

FIGURE 2.13 (a) GIXRD spectra and (b) Optical absorptance of Pt-containing a-C:H SSC deposited at different Pt target powers varying from 50 W to 100 W. (c) Optical absorptance and (d) emittance of double-layered Pt-containing a-C:H coatings [17].

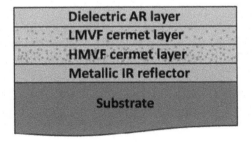

FIGURE 2.14 Schematic of 4-layered cermet-based absorber coating.

TABLE 2.2
Optical performance of various SSCs, including the substrate [20]

Samples	Absorbance	Emissivity
SS	0.36	0.08
SS–(Fe$_2$O$_3$)	0.72	0.08
SS–(Fe$_2$O$_3$)/Mo	0.58	0.12
SS–(Fe$_2$O$_3$)/Mo/HMVF	0.84	0.17
SS–(Fc$_2$O$_3$)/Mo/HMVF/LMVF	0.88	0.18
SS–(Fe$_2$O$_3$)/Mo/HMVF/LMVF/Al$_2$O$_3$	0.92	0.19

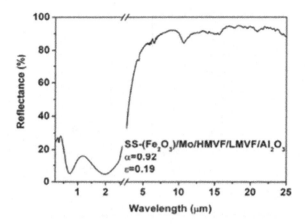

FIGURE 2.15 The reflectance curves of SS-Fe$_2$O$_3$/Mo/ HMVF/LMVF/Al$_2$O$_3$ [20].

technique was used with Al as the target and nitrogen as the reactive gas. Metals including stainless steel, Ni$_{80}$Cr$_{20}$, TZM (consisted of 99% of Mo, 0.5% of Ti, and 0.1% of Zr), and tungsten (W) were used. The four metals were chosen because they were all unreactive to the nitrogen, so that no other nitride except for AlN was formed. In the sputter deposition chamber, the N$_2$ inlet (reactive gas) and Ar inlet (working gas) are separated in order to create the different atmospheres. A screen is placed in between the two gas inlets to prevent contamination. After the optimization of the thickness, the absorptance of those four different AlN-based cermets were 93–96%, and the emittance of which were 3–4%. The near-zero reflectance valley located at 0.6 μm and 1.3 μm also appeared in the previous research [22]. It was the interference of the double graded cermet that made the reflectance so low, which is beneficial to the absorptance.

2.2.5 Textured Absorber

It was known that surface roughness of a material is able to enhance the spectral selectivity when the surface roughness is comparable to the wavelength of the light [23]. Figure 2.16 shows how roughness affects the spectral absorptance of bulk materials. The higher the surface roughness, the more the light absorptance [24]. This concept has also been applied to SSC.

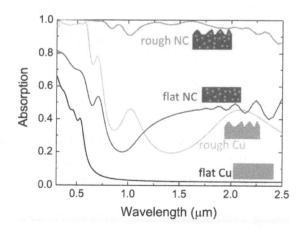

FIGURE 2.16 Comparison of the absorption of rough Cu, flat Cu, rough NC, and flat NC (NC: Cu-SiC nanocomposite) [24].

Sputter Deposited Nanostructured Coatings

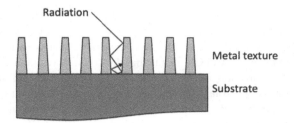

FIGURE 2.17 Schematic of textured absorber coating.

In 1975, J. J. Cuomo came up with a new concept to improve the absorptance. A dense array of whiskers was grown on the surface of tungsten hexafluoride (WF_6) and achieved 98% solar absorptance and 26% thermal emittance at 550°C. The so-called textured absorber is shown in Figure 2.17. The trenches trap the radiation and enhance the reflection and absorptance. The distance between the trenches are hundreds of nanometers. The pattern on the surface is in the nanoscale but is, however, very fragile. The surface of the absorber therefore must be protected.

K. K. Wang investigate the optical constants of $C-TiO_2$ composite and found that the extinction coefficient is pretty high, which could lead to a high absorptance while exposed by the solar radiation [25]. In order to further understand the characteristics of the $C-TiO_2$, textured and cermet types of $C-TiO_2$ were investigated. Four types of SSCs having dielectric carbon matrix were compared [26]. In Figure 2.18(a), Wang took the optical constants of $C-TiO_2$ in his previous work and

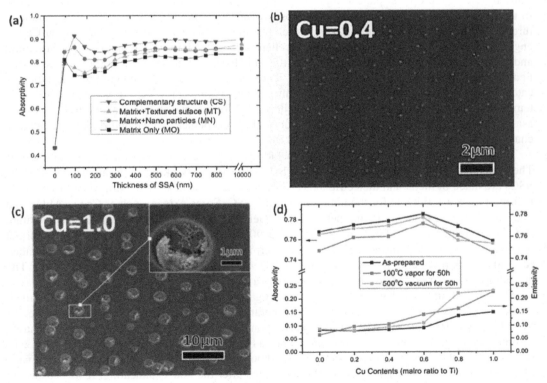

FIGURE 2.18 (a) Calculated solar absorptance of four different types of $C-TiO_2$-based solar selective coatings on copper substrates. Surface morphology of $Cu-(C-TiO_2)$ having Cu/Ti ratio = (b) 0.2 and (c) t1.0. (d) Solar absorptance and emittance of the as-prepared coating, and 100°C vapor-aging and 500°C vacuum-aging coatings [26].

calculated the following four absorptance curves. MO means the pure and homogeneous C-TiO$_2$ without any other modification. MN means the cermet film. The metal nanoparticles concentration was considered as 20% here. MT means that the voids on the surface were considered. The voids concentration was set to be 20%. And CS means that the coating was made of 60% of C-TiO$_2$, 20% of nanoparticles, and 20% of voids on the surface. It is obvious that both cermet and texture are effective method to improve the solar absorptance. Using two methods at once is even better. The Cu-(C-TiO$_2$) cermet coating was then deposited on the copper substrate to prove this. The sol-gel method was used, and the main variable was the Cu molar ratio to Ti of the precursor solution. While the Cu molar ratio increased from 0 to 1.0, except for the increase of the Cu concentration in the cermet coating, the voids on the surface also increased. The size, depth, and the amount of the voids all increased. In Figure 2.18(b), while the Cu ratio was 0.4, there were only few tiny holes that could be barely recognized. However, while the Cu ratio was 1, there were plenty of holes separated around the surface as shown in Figure 2.18(c). These holes were able to trap the light and increase the absorptance. In Figure 2.18(d), it shows the absorptance and the emittance of the coating. It seems that the coating was optimized while the Cu content was 0.6. The absorptance was 80% and the emittance was 8%. The coating was also sent to test the thermal stability. It survived the 500°C vacuum for 50 h. Although the result was not good, the 100°C water vapor test was proceeded, and the result was shown as well. The positive effect of the surface textures was proven.

2.3 HIGH-TEMPERATURE ABSORBER COATINGS

The efficiency of photothermal conversion depends on the solar receiver, typically steel, and the SSC. Based on the optical path design of the CSP and heat transfer fluid's nature, the roadmap to next-generation CSP plants anticipates a progression to central towers with working temperatures in excess of 650°C. Normally, a higher temperature in the heat reservoir gives a better Carnot engine efficiency. Thus, simply exhibiting excellent solar selectivity is way not enough. Structural and chemical stability at the work temperature is equally important. Solar absorber can be classified by considering the working temperature. Three different working temperatures are typically categorized: low temperature (T < 100°C), middle temperature (100°C < T < 400°C), and high temperature (T > 400°C). The performance criteria (PC) of thermal stability is therefore expressed as: $PC = -\Delta\alpha + 0.25\Delta\varepsilon < 0.05$ [27], where $\Delta\alpha$ is the absorptance change and $\Delta\varepsilon$ is the emittance change between that measured at room temperature and an elevated temperature.

The general requirements for the high-temperature absorber coatings is the high thermal stability. The structure and phase do not change in the high-temperature working environment. Most of the SSCs are made of metal and dielectric material. Some metals, such as W or Mo, with extreme high melting points are suitable for the high-temperature SSC. Some oxides, such as Cr$_2$O$_3$, Al$_2$O$_3$, and SiO$_2$ already possess great thermal performance. Generally speaking, a qualified high-temperature SSC fits the requirements of high absorptance (α > 90%), low emittance (ε < 10%), and low thermal degradation (PC < 5%). In the early days, the thermal test was carried out in a vacuum chamber. Recently, some researches test the thermal stability with lower level of vacuum, or even in air. The annealing time also extends from a couple hours to a couple weeks.

Al$_2$O$_3$-based materials have been widely investigated for temperature SSCs. One of the critical factors that make a SSC stable in elevated temperatures is the use of a barrier layer. The Fe$_2$O$_3$/Mo/Mo-Al$_2$O$_3$ (HMVF)/Mo-Al$_2$O$_3$(LMVF)/Al$_2$O$_3$ on the stainless steel mentioned above substrate exhibits good high-temperature performance when a Fe$_2$O$_3$ diffusion barrier is used [20]. Figure 2.19 shows XPS analysis of Fe$_2$O$_3$/Mo-Al$_2$O$_3$ deposited on stainless steel after vacuum annealing at various temperatures for 1 h. The peak of Mo disappears after the 800°C annealing. It was suspected that the Mo diffused through the Fe$_2$O$_3$ layer, and the Mo concentration decreased so that the signal was lost. It is a typical interface diffusion in which atoms migrate through layers [28–30]. The use of a diffusion barrier is a common method to improve the thermal stability of SSCs. In this study, the stainless steel substrate was annealed before the sputtering process in order to grow a Fe$_2$O$_3$

Sputter Deposited Nanostructured Coatings

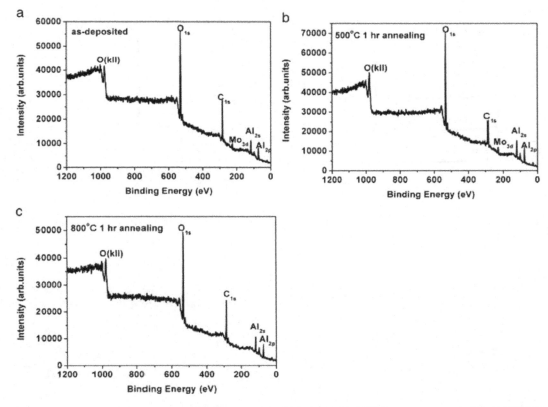

FIGURE 2.19 XPS spectra (a) as-deposited, and (b) 500°C and (c) 800°C vacuum annealed Fe_2O_3/HMVF on stainless steel [20].

layer as a diffusion barrier on the surface. Figure 2.20 compares the XRD patterns of Mo and Fe_2O_3/Mo on stainless steel substrates before and after heat treatment at 800°C for 2 h. In Figure 2.20(a), various phases are seen after the 800°C annealing, indicating the occurrence of interface diffusion between the substrate and Mo layer. The Mo reacts with Fe and C in the stainless steel substrate. On the other hand, in Figure 2.20(b), these phases do not appear, indicating the interface diffusion has been prevent by the Fe_2O_3 barrier layer. As a result, the diffusion issue must be considered for a SSC so that the undesirable interface diffusion is prevented during high working temperatures. However, despite that the desirable interface diffusion is prevented, surface cracks have been found after the annealing, as shown in Figure 2.21 [20].

Another diffusion issue is related to the cermet materials. Cermet is one of the best materials for high-performance SSC. However, diffusion between the pure metal NPs and the matrix in a cermet can occur at elevated. The use of bimetallic alloy NPs therefore has been studied. J. H. Gao deposited a AgAl-Al_2O_3 cermet on the stainless steel substrate and obtained improved thermal stability, as compared to Ag-Al_2O_3 cermet [31]. In this study, three targets of Al_2O_3, Al, and Ag, were used. After the co-sputter deposition, the sample was pre-annealed subsequently at 200°C and 300°C for 2 h each in air, followed by 500°C annealing in nitrogen at for 18 h. These series of heat treatments are the key to modify the NPs and hence improve the thermal stability. To test the advantage of the NPs modifications, the samples were sent to the thermal stability test. Figure 2.22(a) and (b) shows the cross-sectional TEM images of Ag-Al_2O_3 cermet and AgAl-Al_2O_3 cermet, respectively. There is no significant trace of diffusion in Figure 2.22(a) for there is not NPs modification after the deposition. In Figure 2.22(b), the interfaces in AgAl-Al_2O_3 cermets are more blurred because of the diffusion

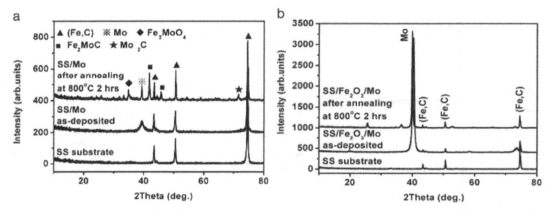

FIGURE 2.20 XRD patterns of (a) Mo and (b) Fe$_2$O$_3$/Mo on stainless steel substrates before and after heat treatment at 800°C for 2 h [20].

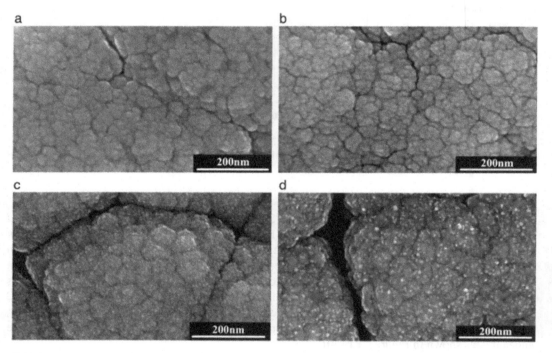

FIGURE 2.21 Surface morphologies of Al$_2$O$_3$ AR layer in (a) as-deposited, and (b) 500-, (c) 650-, and (d) 800-°C heat-treated samples (3 h) [20].

during the NPs modification. The Ag and Al atoms in cermet layers diffused to the adjacent Al$_2$O$_3$ layer. In Figure 2.22(c), the reflectance curve before and after thermal stability test of Ag-Al$_2$O$_3$ cermets is exhibited. The absorptance is 91.7% and the emittance is 11.7%. After the thermal stability test, the absorptance degraded to 84%. In the meanwhile, the Nps-modified AgAl-Al$_2$O$_3$ performed much better. In Figure 2.22(d), instead of degradation, the solar selectivity increased after the thermal stability test, that is, $\alpha = 89\%$, $\varepsilon = 33\% \rightarrow \alpha = 94.2\%$, $\varepsilon = 15.4\%$. The success of the NPs modification was proven. The effect of heat treatment of the bimetallic alloy NPs is depicted in Figure 2.23. The as-deposited cermet coating consists of composite NPs in which Al atomic species

Sputter Deposited Nanostructured Coatings

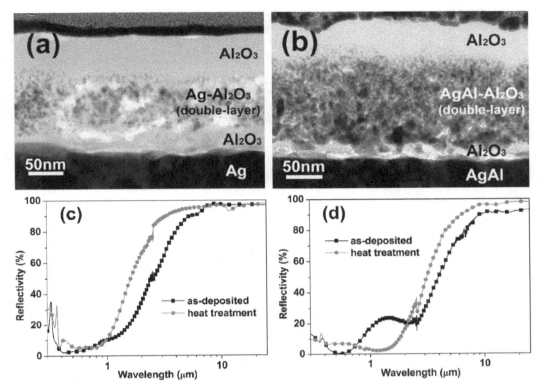

FIGURE 2.22 TEM cross section of (a) Ag-Al$_2$O$_3$ cermet and (b) AgAl-Al$_2$O$_3$ cermet. Spectral reflectance of (c) Ag-Al$_2$O$_3$ cermet and (d) AgAl-Al$_2$O$_3$ cermet [31].

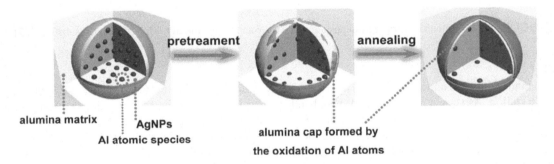

FIGURE 2.23 Schematic of the mechanism of the AlAg NPs [31].

are homogeneously distributed in Ag. During the heat treatment, the Al atoms diffuse to the surface of the NP and are oxidized to form alumina. After the heat treatment, the alumina covers the entire NP surface to form a cap that protects the NP from interacting with the matrix in the cermet. Several bimetallic NPs of WTi, AgAl, and WTa have been investigated in the last few years, and have been proven to improve the thermal stability due to the prevention of the interface reaction [30, 32].

For use as a dielectric in cermet, HfO$_2$ is one of the most focused materials. HfO$_2$, having a melting point of 2758°C, has been used as a refractory ceramic in the early days. N. Selvakumar deposited a Mo/HfO$_x$/Mo/HfO$_2$ SSC on copper substrate and obtained good optical performance, with $\alpha = 91.8\%$ and $\varepsilon = 8\%$, in the as-deposited state [33]. After the post annealing at 800°C for 2 h in vacuum, the optical performance slightly dropped, with $\alpha = 89.3\%$ and $\varepsilon = 8\%$). However, the SSC

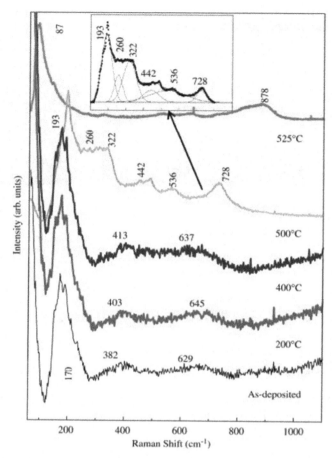

FIGURE 2.24 Raman spectra of as-deposited and 2 h air-annealed Mo/HfO$_x$/Mo/HfO$_2$ on Cu substrate [33].

cannot withstand annealing in air. The solar selectivity drops to 2.2 (α = 76.1% and ε = 35%) due to oxidation. As shown in Figure 2.24, new bands appear in the 500°C and 525°C annealed samples. The peaks, centered at the 442 cm^{-1} and 728 cm^{-1}, represent the appearance of MoO$_2$. Oxidation of molybdenum happens at 500°C, leading the formation of MoO$_2$. Afterward, the reaction between MoO$_2$ and HfO$_2$ takes place at 525°C and HfMo$_2$O$_8$ forms. On the other hand, there is no significant difference among the patterns of samples annealed at different temperatures, even at 800°C, as shown in Figure 2.25. It is therefore the above compounds that cause the serious degradation of the optical performance. Oxidation represents a basic reason for the degradation of optical performance at high temperatures.

In addition to oxides, some nitrides are also of great interest. For example, aluminum nitride, which has high thermal and chemical stability, has been used as a dielectric in cermet, useful as high-temperature SSC. Shen Yue deposited AlN/Al$_x$O$_y$ on aluminum substrate and obtained a very high α of 97% in the as-deposited state [34]. After 500°C vacuum annealing, the absorptance decreases to 94%. The SSCs were examined using AES depth profiling analysis. In Figure 2.26(a), the concentration profiles of oxygen and nitrogen indicate clear 3-layered structure in the as-deposited state. However, for the annealed sample, diffusion occurs. It is obvious that the nitrogen and oxygen atoms diffuse through layers, as shown in Figure 2.26(b). The concentrations of each element in the absorber layer and AR layer are almost identical, indicating the occurrence of diffusion.

Sputter Deposited Nanostructured Coatings

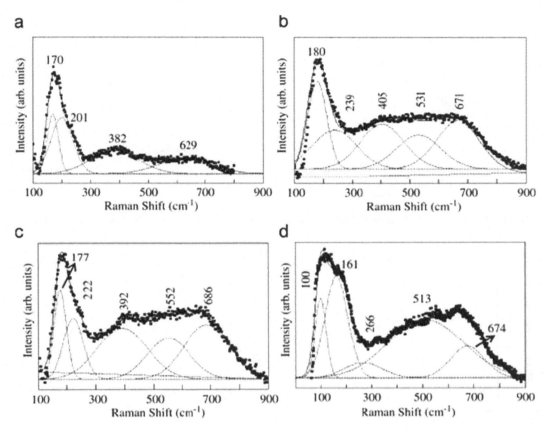

FIGURE 2.25 Raman spectra of (a) as-deposited, and (b) 600-, (c) 700-, and (d) 800-°C vacuum-annealed Mo/HfO$_x$/Mo/HfO$_2$ [33].

Another dielectric for cermet is Si$_3$N$_4$. It is the most thermodynamically stable form of the silicon nitride family. It is used as bearings in spacecraft because of the high thermal stability. A. Rodriguez-Palomo deposited an Ag/MoSi$_2$-Si$_3$N$_4$/Si$_3$N$_4$ multilayered structure on stainless steel substrate [35, 36]. The silver layer serves as the IR reflector layer and successfully decreases the emittance to only 2%. The solar selectivity is 40.5 (i.e., $\alpha = 81\%$, $\varepsilon = 2\%$), which is pretty outstanding. Figure 2.27(a) shows the reflectance curves obtained after vacuum-annealing at different temperatures. The optical performance remains similar (i.e., $\alpha = 80\%$, $\varepsilon = 2\%$) up to 650°C vacuum annealing. However, after air annealing, the coating could not survive at 450°C, as shown in Figure 2.27(b). The solar selectivity drops to 10.33 ($\alpha = 62\%$, $\varepsilon = 6\%$). This is attributed to the Ag degradation, which makes its reflectance to decay in the IR band. On the other hand, with the removal of the Ag layer (MoSi$_2$-Si$_3$N$_4$/Si$_3$N$_4$), the absorptance increases to 91% but the emittance also increases to 13%, as shown in Figure 2.27(c). The optical performance appears to be similar up to 700°C Figure 2.27(c). The result indicates that the removal of the metallic Ag prevents the formation of nitride or even sulfide at high temperatures, however, with the price of having a higher emittance.

Titanium nitride has been widely used as a hardening coating for cutting tools. This implies that it has high temperature stability, which is desired for a dielectric in cermet. CY Li deposited TiN multilayered structure (i.e., TiN$_x$/TiN$_x$O$_y$/N-doped TiO$_2$/SiO$_2$) on stainless steel substrate and achieved an excellent solar selectivity of 60.7 ($\alpha = 97.1\%$ and $\varepsilon = 1.6\%$) [37]. SiO$_2$, as an AR layer, effectively increases the absorptance. The absorptance was enhanced from 88% to 97% as a result (Figure 2.28(a)). In Figure 2.28(b), the reflectance curves of the TiN$_x$/TiN$_x$O$_y$/N-doped TiO$_2$/SiO$_2$

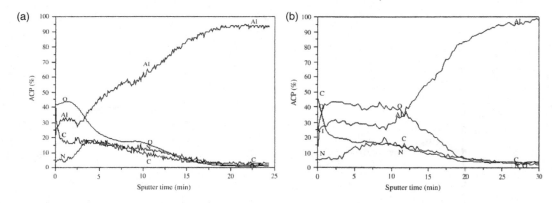

FIGURE 2.26 AES depth profiles of AlN/Al$_x$O$_y$ (a) before and (b) after 550°C 40 h vacuum annealing [34].

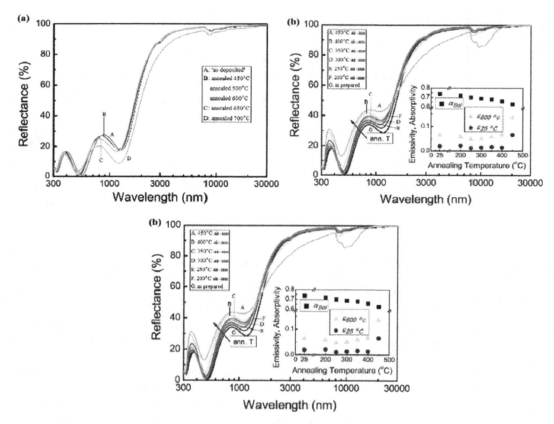

FIGURE 2.27 Reflectance curves obtained at different annealing temperatures. (a) Ag/MoSi$_2$-Si$_3$N$_4$/Si$_3$N$_4$, vacuum annealing, (b) Ag/MoSi$_2$-Si$_3$N$_4$/Si$_3$N$_4$, air annealing, and (c) MoSi$_2$-Si$_3$N$_4$/Si$_3$N$_4$, air annealing [35, 36].

structure after different annealing tests are shown. The coating works just fine even at 600°C. The absorptance only decreases slightly from 94.6% to 93.1% and the emittance increases insignificantly from 5% to 5.8%. However, after vacuum annealing at 800°C, the absorptance decreased to 72.6%. According to the XRD patterns, as shown in Figure 2.28(c), no extra peak is seen after the high-temperature annealing. This indicates no chemical reaction occurs during the annealing. However, the SEM analysis, as shown in Figure 2.29, indicates that there are surface cracks on the

Sputter Deposited Nanostructured Coatings 43

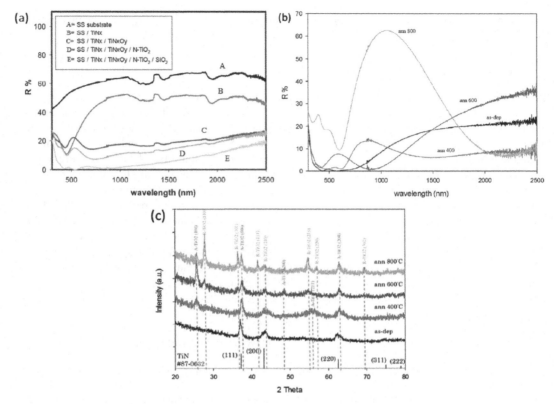

FIGURE 2.28 (a) Reflectance curves of TiN_x/TiN_xO_y/N-doped TiO_2/SiO_2 stacking up layer by layer; (b) reflectance curves of TiN_x/TiN_xO_y/N-doped TiO_2/SiO_2 after different temperature annealing; (c) XRD patterns of as-deposited and annealed multilayer coatings on Si substrate [37].

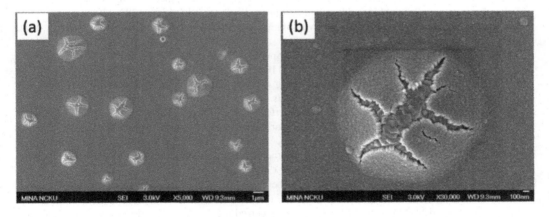

FIGURE 2.29 Surface morphologies of TiN_x/TiN_xO_y/N-doped TiO_2/SiO_2 at different magnifications after 2 h 800°C annealing [37].

800°C annealed sample. This is blamed for the degraded optical performance. Therefore, the SiO_2 was replaced by HfO_2. It is noted that they both have similar optical constants (n, k values). HfO_2 is a well-known refractory material for its high melting point (2758°C) and thermal stability, which is effective as an AR layer for SSC used in high-temperature SCP application [38]. After 800°C annealing, no surface damage was observed. The as-deposited TiN_x/TiN_xO_y/N-doped TiO_2/HfO_2

gives $\alpha = 89.9\%$; the annealed sample performs almost identically with $\alpha = 87.2\%$. The degradation was limited, and HfO_2 was proven to be a better choice for thermal stability.

2.4 CONCLUSION

Recent research efforts have mostly focused on the improvement of the spectral selectivity. Various structures have been designed and fabricated, and different material combinations have been investigated. Dielectric materials, including Al_2O_3, Cr_2O_3, MgO, SiO_2, and AlN, are desirable materials for intrinsic spectral selectivity. As a metal inclusion, Cu, Al, Ag, Cr, Ti, W, Ni, and Mo are commonly used due to the IR reflectance. Additionally, the need for thermal stability is becoming more important. The targeted working temperatures are higher than 650°C. Current "high-temperature" solar selective absorbers cannot provide satisfied thermal stability as a result. The high-temperature thermal environment causes oxidation and diffusion, thus degrading the optical performance. In order to improve the thermal stability, novel SSCs are made of highly stable dielectric material, such as TiC, HfO_2, and Si_3N_4. The metal inclusions also turn out to be more complicated. Examples are WTi, AgAl, and WNi. The use of a bimetallic alloy inclusions represents a high-temperature solution. In other words, solar selectivity and high-temperature ($> 650°C$) thermal stability must be considered at the same time for the future development of SSC materials.

REFERENCES

[1] Abbott, D., Keeping the energy debate clean: How do we supply the world's energy needs? *Proceedings of the IEEE*, 2009. 98(1): p. 42–66.

[2] Fuqiang, W., et al., Progress in concentrated solar power technology with parabolic trough collector system: A comprehensive review. *Renewable and Sustainable Energy Reviews*, 2017. 79: p. 1314–1328.

[3] Drosou, V., P. Kosmopoulos, and A. Papadopoulos, Solar cooling system using concentrating collectors for office buildings: A case study for Greece. *Renewable Energy*, 2016. 97: p. 697–708.

[4] Kasten, F. and A.T. Young, Revised optical air mass tables and approximation formula. *Applied Optics*, 1989. 28(22): p. 4735–4738.

[5] Zhang, K., et al., A review on thermal stability and high temperature induced ageing mechanisms of solar absorber coatings. *Renewable and Sustainable Energy Reviews*, 2017. 67: p. 1282–1299.

[6] Trotter, D. and A. Sievers, Spectral selectivity of high-temperature solar absorbers. *Applied Optics*, 1980. 19(5): p. 711–728.

[7] Thornton, J.A., A.S. Penfold, and J.L. Lamb, Sputter-deposited Al2O3/Mo/Al2O3 selective absorber coatings. *Thin Solid Films*, 1980. 72(1): p. 101–110.

[8] Wu, Y., et al., *Colored solar selective absorbing coatings with metal Ti and dielectric AlN multilayer structure. Solar Energy Materials and Solar Cells*, 2013. 115: p. 145–150.

[9] Barshilia, H.C., Growth, characterization and performance evaluation of Ti/AlTiN/AlTiON/AlTiO high temperature spectrally selective coatings for solar thermal power applications. *Solar Energy Materials and Solar Cells*, 2014. 130: p. 322–330.

[10] Cheng, H.-Y., Y. Tzeng, and J.-M. Ting, Characteristics of sputter deposited Pt/Al2O3 thin films. *Ceramics International*, 2016. 42(16): p. 19393–19396.

[11] Fan, J.C. and S.A. Spura, Selective black absorbers using rf-sputtered Cr2O3/Cr cermet films. *Applied Physics Letters*, 1977. 30(10): p. 511–513.

[12] Khamlich, S. and M. Maaza, Cr/α-Cr2O3 monodispersed meso-spherical particles for mid-temperature solar absorber application. *Energy Procedia*, 2015. 68: p. 31–36.

[13] Fan, J.C. and P.M. Zavracky, Selective black absorbers using MgO/Au cermet films. *Applied Physics Letters*, 1976. 29(8): p. 478–480.

[14] Thomas, L. and T. Chunhe, Microstructure and reflectance of sputtered Au/MgO films on Cu. *Solar Energy Materials*, 1989. 18(3–4): p. 117–126.

[15] Cheng, H.-Y., et al., Reactively co-sputter deposited aC: H/Cr thin films: Material characteristics and optical properties. *Thin Solid Films*, 2013. 529: p. 164–168.

[16] Sathiaraj, T.S., R. Thangaraj, and O. Agnihotri, Ni-Al2O3 cermet solar absorbers by RF planar magnetron sputtering for high temperature applications. *Solar Energy Materials*, 1989. 18(6): p. 343–356.

[17] Lan, Y.-H., et al., Platinum containing amorphous hydrogenated carbon (aC: H/Pt) thin films as selective solar absorbers. *Applied Surface Science*, 2014. 316: p. 398–404.

[18] Thornton, J.A. and J.L. Lamb, Sputter-deposited Pt Al2O3 graded cermet selective absorber coatings. *Solar Energy Materials*, 1984. 9(4): p. 415–431.

[19] Zhang, Q.C. and D.R. Mills, New cermet film structures with much improved selectivity for solar thermal applications. *Applied Physics Letters*, 1992. 60(5): p. 545–547.

[20] Xinkang, D., et al., Microstructure and spectral selectivity of Mo–Al2O3 solar selective absorbing coatings after annealing. *Thin Solid Films*, 2008. 516(12): p. 3971–3977.

[21] Zhang, Q.-C., Metal-AlN cermet solar selective coatings deposited by direct current magnetron sputtering technology. *Journal of Physics D: Applied Physics*, 1998. 31(4): p. 355.

[22] Zhang, Q.C. and D.R. Mills, Very low-emittance solar selective surfaces using new film structures. *Journal of Applied Physics*, 1992. 72(7): p. 3013–3021.

[23] Kennedy, C.E., *Review of mid-to high-temperature solar selective absorber materials*. 2002, National Renewable Energy Lab., Golden, CO.(US).

[24] Bellas, D.V. and E. Lidorikis, Design of high-temperature solar-selective coatings for application in solar collectors. *Solar Energy Materials and Solar Cells*, 2017. 170: p. 102–113.

[25] Wang, K., et al., A one-step method to prepare carbon and dielectric composited films featuring a skeleton–skin structure. *Carbon*, 2015. 94: p. 424–431.

[26] Wang, K., et al., A facile one-step method to fabricate multi-scaled solar selective absorber with nanocomposite and controllable micro-porous texture. *Solar Energy Materials and Solar Cells*, 2017. 163: p. 105–112.

[27] Selvakumar, N. and H.C. Barshilia, Review of physical vapor deposited (PVD) spectrally selective coatings for mid-and high-temperature solar thermal applications. *Solar Energy Materials and Solar Cells*, 2012. 98: p. 1–23.

[28] Cheng, J., et al., Improvement of thermal stability in the solar selective absorbing Mo–Al2O3 coating. *Solar energy materials and solar cells*, 2013. 109: p. 204–208.

[29] Meng, J.-P., et al., Microstructure and thermal stability of Cu/Zr0. 3Al0. 7N/Zr0. 2Al0. 8N/Al34O60N6 cermet-based solar selective absorbing coatings. *Applied Surface Science*, 2018. 440: p. 932–938.

[30] Wang, X., et al., High-temperature tolerance in WTi-Al2O3 cermet-based solar selective absorbing coatings with low thermal emissivity. *Nano Energy*, 2017. 37: p. 232–241.

[31] Gao, J., et al., Silver nanoparticles with an armor layer embedded in the alumina matrix to form nanocermet thin films with sound thermal stability. *ACS Applied Materials & Interfaces*, 2014. 6(14): p. 11550–11557.

[32] Wu, Z., et al., Toward versatile applications via tuning transition wavelength of the WTa-SiO2 based spectrally selective absorber. *Solar Energy*, 2020. 202: p. 115–122.

[33] Selvakumar, N., et al., Structure, optical properties and thermal stability of pulsed sputter deposited high temperature HfOx/Mo/HfO2 solar selective absorbers. *Solar Energy Materials and Solar Cells*, 2010. 94(8): p. 1412–1420.

[34] Yue, S., S. Yueyan, and W. Fengchun, High-temperature optical properties and stability of AlxOy–AlNx–Al solar selective absorbing surface prepared by DC magnetron reactive sputtering. *Solar Energy Materials And Solar Cells*, 2003. 77(4): p. 393–403.

[35] Rodríguez-Palomo, A., et al., High-temperature air-stable solar selective coating based on MoSi2–Si3N4 composite. *Solar Energy Materials and Solar Cells*, 2018. 174: p. 50–55.

[36] Hernández-Pinilla, D., et al., MoSi2–Si3N4 absorber for high temperature solar selective coating. *Solar Energy Materials and Solar Cells*, 2016. 152: p. 141–146.

[37] Li, C.-Y., F.N.I. Sari, and J.-M. Ting, Reactive magnetron sputter-deposited TiNxOy multilayered solar selective coatings. *Solar Energy*, 2019. 181: p. 178–186.

[38] Bronson, A., et al., Compatibility of refractory metal boride/oxide composites at ultrahigh temperatures. *Journal of the Electrochemical Society*, 1992. 139(11): p. 3183.

3 Novel Mode-Locked Fiber Lasers with Broadband Saturable Absorbers

Lu Li and Wei Ren
Xi'an University of Posts and Telecommunications

CONTENTS

3.1 Introduction...47
3.2 Mode-Locked Fiber Laser 49
    3.2.1 Conventional Soliton Fiber Laser 49
    3.2.2 Stretched Pulse Fiber Lasers...49
    3.2.3 Self-Similar Pulse Fiber Lasers 49
    3.2.4 Mode-Locking Technology..50
3.3 Broadband Saturable Absorbers 51
    3.3.1 CNTs...51
    3.3.2 Graphene..52
    3.3.3 Topological Insulators..52
    3.3.4 Transition Metal Dichalcogenides 52
    3.3.5 BP 53
    3.3.6 SA Preparation, Packaging, and Application in Fiber Lasers.......................53
        3.3.6.1 Preparation Methods 53
        3.3.6.2 Package Method 54
3.4 Application of Novel SAs in Mode-Locked Fiber Lasers 54
    3.4.1 Yb-Doped Mode-Locked Fiber Lasers with DF-SWCNT-D_2O SA 54
    3.4.2 Yb-Doped Mode-Locked Fiber Lasers with Bi_2Te_3-based SAs.................58
        3.4.2.1 Yb-Doped Mode-Locked Fiber Laser with Bi_2Te_3/PVA SA.........58
        3.4.2.2 Yb-Doped Passive Mode-Locked Fiber Laser with Bi_2Te_3 SA....61
    3.4.3 Yb-Doped Mode-Locked Fiber Laser Based on WS_2/FM............................68
        3.4.3.1 Yb-Doped Mode-Locked Fiber Laser Based on WS_2/FM...........68
        3.4.3.2 Erbium-Doped Mode-Locked Fiber Laser Based on DF-WS_2 SA 71
    3.4.4 Erbium-Doped Mode-Locked Fiber Laser Based on DF-BP SSA 75
3.5 Conclusions..77
References...78

3.1 INTRODUCTION

The fiber laser is arguably one of the greatest inventions of the 20th century. The fiber laser has greatly improved people's lives and promoted social progress. T. H. Maiman produced the world's first solid-state laser in 1960, and then, fiber lasers were also proposed [1–4]. In the 1980s, the emergence of diode pump lasers brought a new generation of fiber lasers with the advantages of intense and fast heat dissipation, low loss, and high conversion efficiency. Also, the characteristic high output power reaches the order of 10,000 W, which makes the application of fiber lasers suitable in various fields. Depending on the different rare-earth ions that are doped in fiber lasers, the working wavelengths of the lasers cover the wavelength range from near infrared to mid-far infrared [5–8]. Among these lasers, the mid-infrared laser can affect the vibrational energy spectrum of most gases and organic molecules, and it is often used in precision laser spectroscopy, identification of gas, and organic molecular structure dynamics. Lasers also play an irreplaceable role in the fields of lidar, industrial manufacturing, laser-marking welding, molecular optical spectroscopy, laser medical treatment, environmental monitoring, military countermeasures, space communications, semiconductor micromachining, and terahertz generation.

The technology for fiber lasers involves applied optics. By amplifying and filtering light of a specific wavelength, the output light has particular characteristics, such as high brightness, high intensity, high coherence, and directivity [9–12]. At the same time, because of the broad fluorescence spectrum of rare earth ions in fiber lasers, most fiber lasers are tunable. After decades of development, various lasers have been proposed. At present, lasers are mainly divided into Q-switched lasers and mode-locked lasers according to the pulse width [13–17]. According to the gain medium, they are divided into gas lasers, liquid lasers, solid-state lasers, fiber lasers, and semiconductor lasers. Figure 3.1 shows a classification of several major fiber lasers.

Mode-locked fiber lasers are a type of pulsed laser and have been widely studied because of their important applications in nonlinear optics, fiber communication, material processing, and frequency combing. In the field of fiber lasers, a variety of nonlinear materials that combine the two advantages of broadband nonlinear modulation and ultrafast optical response have led to the creation of a new era of saturation absorbers (SAs). In the field of fiber lasers, passive mode-locking (ML) technology using an SA is also an effective method for generating light pulses that range from picoseconds (ps) to femtoseconds (fs) [18–20]. High-performance SAs have the characteristics of wavelength independence, high heat dissipation, and high laser damage threshold, and thus, the exploration of new broadband SA devices with stable performance, simple production, low cost, and a high damage threshold for use in ultrafast lasers has been a research hot spot in technology. Semiconductor saturable absorption mirrors (SESAMs) are one of the most effective pulse operation triggers and are widely used in environmentally robust and stable ML laser. However, commercialized SESAMs have been restricted because of complex manufacturing, narrow bands, and high prices. The new broadband SA can be used to modulate the pulse in a wide wavelength range of 1–3 μm because it combines two advantages: (1) the broadband nonlinear modulation effect and (2) easy integration into the fiber laser system. SAs have aroused widespread interests. Novel reported SAs mainly includes carbon nanotubes (CNTs)[21–24], graphene (GF)[25, 26], topological insulators [27, 28], transition metal sulfides [29, 30], and black phosphorus (BP).

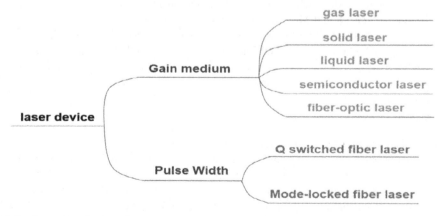

FIGURE 3.1 Several main categories of fiber lasers.

3.2 MODE-LOCKED FIBER LASER

It is well-known that, when the accumulated nonlinear phase shift is large enough, the frequency spectrum and time domain distribution of the output pulse is significantly affected because the core of the fiber is very small. If the accumulated nonlinear phase shift is too large, it causes a multi-pulse phenomenon. To avoid this phenomenon, achieving reasonable and effective control of the nonlinear effects is reasonable compensation for dispersion in the laser cavity.

Mode-locked fiber lasers can be divided into traditional soliton fiber lasers, stretched pulse fiber lasers, and self-similar pulse fiber lasers according to the difference in laser cavity dispersion [31–36].

3.2.1 CONVENTIONAL SOLITON FIBER LASER

The traditional soliton pulses produced in the 1980s were in a 1550 nm fiber laser, which used a stepped single-mode fiber (SMF) to provide negative dispersion. However, because of the nonlinear effect and the balance of negative dispersion, the total dispersion is negative, and it is very close to the transmission variation limit. The $Sech_2$ function is usually used to fit the pulse shape [37]. The pulse energy in the cavity remains unchanged, and there is no abnormal sound [38]. There are two reasons that the shape of the soliton pulse can be maintained: (1) the low-frequency component of the pulse propagates slower than the high-frequency component and (2) the nonlinear effect can cancel this effect.

However, there are loss and gain fluctuations in the actual fiber laser cavity, and hence, the actual soliton does not exist [39]. The soliton loses energy in the form of a dispersive wave. At certain frequencies, the resonance of the dispersive wave causes the spectral sidebands of the Kelly sideband [40]. This observation is interpreted to mean that some of the energy of the soliton is outputted at the coupler, and when the pulse passes through the cavity again, the soliton must recover its energy. Because the peak power changes throughout the cavity, a nonlinear refractive index grating forms, and this changes the nature of the soliton. The soliton stores energy in the form of scattered waves, thereby generating soliton sidebands [41].

In the past few decades, traditional soliton fiber lasers have been extensively and deeply researched, and related fiber laser products have also continuously been put on the market. However, the pulse energy of this type of laser is usually limited to the order of 0.1 nJ. With a continued increase in the pulse energy, a multi-pulse operation is easily obtained.

3.2.2 STRETCHED PULSE FIBER LASERS

Widened pulsed fiber lasers are a method that researchers use to control dispersion. Two fibers with opposite dispersions are used to form the laser, and thus, there is about zero net dispersion in the cavity [42, 43]. The decentralized management method avoids the accumulation of nonlinear pulses and causes the pulses to work in a breathing state [44]. It effectively reduces the peak power of the laser pulse and reduces the nonlinear effects. The maximum and minimum pulse width scan is output at both ends of the cavity. Compared with traditional soliton lasers, widened pulsed fiber lasers produce shorter pulses, generate Gaussian pulses, and have no sidebands in the pulse spectrum.

At present, there has been great progress in research on broadened pulse fiber lasers. Tamura and other researchers have obtained a 77 fs widened pulse output in erbium-doped fiber lasers [45], and Zhipei Sun obtained a pulse output of 113 fs in a CNT broadened pulse fiber laser [21].

3.2.3 SELF-SIMILAR PULSE FIBER LASERS

The net dispersion of a self-similar pulse is slightly greater than zero. During propagation, self-similar pulses accumulate positive frequencies in the cavity, partially compensated by negative dispersion fibers, and the pulse spectrum and time domain remain parabolic without light wave splitting [46, 47]. Parabolic self-similar pulses have unique advantages in pulse amplification systems. When the dispersion of the seed source pulse is widened, the parabolic intensity distribution produces a parabolic nonlinear phase shift. This nonlinear phase shift is consistent with the phase shift of the cumulative dispersion, and thus, it can be easily compensated for by the dispersion delay line [48]. Self-similar pulses can withstand higher nonlinearities, and hence, light wave splitting can be avoided. The F. Röser research team built a stable self-starting, doped-self-similar fiber laser with a pulse output of 100 fs and a single pulse energy up to 14 nJ. After compression, the peak power is as high as 100 kW [49]. The single-pulse energy and peak power of self-similar pulses are 5–6 times higher than the best results obtained before the pulsed fiber lasers are widened. Also, the results have been close to the output parameters of the titanium sapphire laser.

The dissipative soliton pulsed fiber laser works under the circumstance of large positive dispersion or all positive dispersion [18, 31, 50]. Different from the formation of traditional soliton pulses, the formation of this dissipative soliton has the comprehensive effects of gain dispersion, gain saturation, positive dispersion, and nonlinear loss. Among these effects, gain and loss play a leading role, and hence, the soliton is classified as a dissipative type.

At present, research on pulsed fiber lasers based on dissipative solitons has mainly focused on ytterbium-doped 1 μm totally positive dispersion fiber lasers and erbium-doped large positive dispersion fiber lasers at 1.55 μm. W. H. Wise et al. achieved an ultrashort pulse output of 80 fs in a total positive dispersion laser with a peak power as high as 200 kW, which is the highest index of this type of laser and has reaches the level of solid-state lasers [32]. At the same time, dissipative soliton fiber lasers using SAs (such as CNTs and graphene) as mode-locked devices have also been extensively studied [33–36, 51].

3.2.4 Mode-Locking Technology

Depending on the ML mode, the mode-locked fiber lasers can be divided into active mode-locked fiber lasers and passive mode-locked fiber lasers. Depending on the different modulation objects, the active ML can be divided into amplitude modulation ML and phase modulation ML. Active mode-locked fiber lasers usually need a modulated electro-optic modulator or acousto-optic modulator to be combined with the resonator. Also, an externally injected control signal is equal to an integer multiple of the mode interval Δv frequency fm. In this way, it is possible to synchronously obtain a pattern that oscillates within the width of the gain line in the same phase. Strong coherent superposition occurs between these modes to form a mode-locked sequence output. Passive ML technology is an all-optical nonlinear method. A nonlinear device can modulate different intensity parts of an optical pulse. Commonly used passive ML methods include the following technologies: nonlinear polarization evolution (NPE), nonlinear loop mirror (NOLM), cross-phase modulation (XPM), and saturated absorption (SA). The following section focuses on SA technology.

SA is a phenomenon whereby the absorption of light by a material decreases with an increase in the incident light intensity. When the substance absorbs incident light, the electrons in the substance transition from the ground state to a higher energy level, and the electrons of the higher energy level relax to the lower energy level over a period of time. When the incident light is strong enough, the ground state electrons are completely excited to a higher energy level, and the excited state electrons cannot return to the ground state during the relaxation time. Because the ground state electrons are depleted, they can no longer absorb photons to produce a transition, and thus, they show saturable absorption characteristics. With an increase in the incident light intensity, the edge of the valence band is filled with holes, and edge of the conduction band is filled with electrons. Because of the Pauli exclusion principle, two electrons cannot have the same electronic state, and this hinders the continued absorption of light.

In the 1990s, U. Keller et al. invented the semiconductor saturation absorption mirror (SESAM), opening a new era of ultrafast lasers [52]. SESAM is a typical saturable absorbent, mainly because

Novel Mode-Locked Fiber Lasers

the lower carrier is quickly evacuated, and the upper layer is partially occupied. A typical SESAM is grown on a GaAs substrate using molecular beam epitaxy (MBE). A SESAM board needs to be pasted to the heat sink to increase the damage threshold. When used in an all-fiber laser, SESAM is attached to the end face of the fiber.

SESAM is characterized by common parameters including modulation depth (MD), saturation flux, nonsaturable loss, recovery time, and threshold damage [53]. MD characterizes the maximum change in reflectivity when the pulse is radiated on SESAM, and this reflects the bleaching ability of SESAM by the light pulse. When SESAM is irradiated with laser light and its reflectivity reaches 1/e of the MD, the photon energy per unit area on the absorption cross section is called the saturation flux. In general, when the MD is greater, the pulse obtained in the time domain is narrower; when the saturation flux is lower, it is easier to reach saturation. Under normal circumstances, to effectively achieve mode-locked operation, the saturation flux must be 5 times higher than that of SESAM. Nonsaturable loss refers to the loss that still exists when SESAM is saturated with bleach. Nonsaturable loss is an important parameter to measure the performance of SA. Damage threshold refers to the incident light intensity flux density when SESAM is damaged. That is, when the damage threshold is higher, the better the flux density.

At present, SESAM is a mature SA and is widely used in mode-locked fiber lasers to achieve ultrashort pulse output at the femtosecond level [54–56]. However, with advances in science and technology, the requirements for SAs are increasing. In recent years, researchers have actively explored new SAs and applied them in mode-locked lasers. In 1997, quantum dot (QD) materials were used in lasers for the first time, and they successfully achieved pulse output[57]. QDs are considered to be quasi-zero-dimensional nanomaterials. Generally, semiconductor nanocrystals (such as CdSe) are incorporated into glass. Glass has a filtering effect, and the filtering effect of colored glass is used to produce saturated absorption characteristics. Since then, people have been studying nanomaterials as SAs.

In 2004, S. Y. Set et al. used CNTs as SAs for the first time and successfully applied them in ytterbium-doped fiber mode-locked lasers [58]. Since then, a new chapter in the study of new saturable absorbents has begun. Graphene has attracted the attention of researchers because of its unique zero-band gap structure. In 2009, researchers first used the saturation absorption characteristics of graphene in research to achieve laser mode-locked pulse output [59]. Since then, graphene has been used for pulse output of 1–3 μm wavelength [60–66], indicating its broadband saturable absorption characteristics. However, graphene materials are 2D materials. There are van der Waals forces between layers. Since 2012, 2D materials (such as topological insulators (TIs), transition metal dichalcogenides (TMDs), and BP have been considered as SAs because of their unique and complementary properties. Absorbents have thus become the subject of research [67–71]. 1D CNTs and 2D materials have a wide saturation absorption spectrum, and they can modulate a laser of 1–3 μm wavelength to achieve ML.

3.3 BROADBAND SATURABLE ABSORBERS

3.3.1 CNTs

CNTs are 1D nanomaterials that have many excellent characteristics. The structure of CNTs is a hollow tube, in which the carbon atoms are sp^2 and sp^3 hybrid structures. CNTs can be divided into single-walled carbon nanotubes (SWCNTs) and multiwalled carbon nanotubes (MWCNTs). The typical diameter of SWCNTs is 0.6–2 nm. The innermost layer of MWCNTs can reach 0.4 nm, and the thickness can be hundreds of nanometers, but the typical diameter of the tube is 2–100 nm.

Absorption in the range of 500–2500 nm was reported [72]. The saturation threshold was about 10 MW/cm^2, and the MD was about 10%. In addition, MWCNTs have also been found to have saturable absorption properties and are used as SAs.

3.3.2 GRAPHENE

Graphene has extremely excellent physicochemical properties. Graphene is exfoliated from graphite materials, and it is a 2D crystal composed of a single layer (only one atom thick) of sp^2 hybrid carbon atoms. The distance between graphene layers is only 0.34 nm.

As a zero-band-gap material, the conduction band and the valence band of graphene intersect at one point, and there is no forbidden band. When graphene absorbs photons, the electrons in the valence band are excited to the conduction band. Then, these excited hot electrons quickly cool and form a Fermi-Dirac distribution in a very short time. These newly formed electron–hole pairs form a stop band in the value of $-E/2$ from the Fermi level and reduce the absorption rate of photon energy. Next, because of the scattering of phonons, the excited carriers continue to cool, and electron–hole pair recombination dominates. When light continues to shine on graphene, the distribution of carriers reaches a balance. When the incident light intensity is large enough, the electrons are continuously excited to the conduction band, and finally, sub-bands in the conduction band are completely filled with electrons. Because of the Pauli exclusion principle, the transition between the bands is blocked, and graphene reaches saturation. At this point, it does not absorb light and exhibits the characteristics of an SA. The reported maximum incident peak power was 337 MW/cm^2 and the saturation peak power threshold was about 100 MW/cm^2, and the MD is 2% [72].

3.3.3 TOPOLOGICAL INSULATORS

Topological insulators are protected by time reversal symmetry. At a certain energy, there is always a pair of energy states that have opposite spin and momentum. Therefore, the support between these energy states can be suppressed to a large extent. These energy states are characterized by topological indices.

In theory, the energy band structure of this material is a typical insulator type, and there is an energy gap at the Fermi level. However, on the surface of this material, there are always Dirac-type electronic states in the energy gap, and this results in the surface always being metal [72]. Therefore, the information on the surface of a topological insulator can be transferred via the spin of electrons rather than via electric charges as in traditional materials. Among many topological insulator materials, Bi_2Se_3-type topological insulators have a large band gap and the simplest surface Dirac energy spectrum; thus, this type of topological insulator has become a hot spot as a research object in the field of condensed matter physics and materials.

An example is that in a laser system, the self-mode-locked onset threshold was reported to be 70 mW[72]. The center wavelength of the mode-locked pulse spectrum was 1558.4 nm, the 3-dB bandwidth was 2.69 nm, and the symmetrical Kelly sidebands were distributed on both sides of the spectrum. The measured full width at half maximum (FWHM) was 1.86 ps, and when a hyperbolic positive cut was used to fit the data, the pulse width was 1.21 ps. The experimental results provide strong evidence that topological insulators can be used as SA materials, laying the foundation for the use of topological insulator materials for use in ultrafast optics.

3.3.4 TRANSITION METAL DICHALCOGENIDES

TMDs are important 2D semiconductor nanomaterials similar to graphene. The chemical formula of transition metal sulfides is MX_2, where M is a group IV, V, or VI element and X is a chalcogen element. MoS_2 and WS_2 are typical examples of metal sulfides.

TMDs have a layered structure [72]. Each layer has an X-M-X structure; the chalcogen elements are separated by the metal element in two hexagonal planes, and the layers are combined via weak Van der Waals forces. For a layered transition metal sulfide, there is a transition from an indirect band gap to a direct band gap depending on the number of material layers. The bulk MoS_2 exhibits an indirect bandgap structure with a forbidden bandwidth of 1.3 eV, whereas the single-layer MoS_2

Novel Mode-Locked Fiber Lasers

transforms into a direct bandgap with a forbidden bandwidth of 1.9 eV. Interestingly, WS_2 also exhibits this feature.

3.3.5 BP

BP is a layered semiconductor material too. The energy band spacing depends largely on the number of layers. BP is a black crystal with metallic luster and has good electrical transport characteristics. It results from the transformation of white phosphorus under high temperature and high pressure. BP has a layered structure that is similar to the structure of graphite. The layers are combined via van der Waals bonds, and each phosphorus atom is covalently bonded to three adjacent phosphorus atoms. BP has a direct band gap structure in both multilayer and single layer forms. The forbidden bandwidth of BP varies with the number of BP nanolayers. Bulk BP has a forbidden bandwidth of 0.3 eV, and the single layer of BP has a forbidden bandwidth of 2 eV[73].

The experimental results of BP materials as SA erbium-doped fiber lasers [72] was reported with a center wavelength of 1560.5 nm and a 3-dB bandwidth of 10.2 nm. The FWHM was 420 fs. When hyperbolic secant fitting was used, the pulse width was 272 fs, which is the shortest pulse width achieved to date.

3.3.6 SA Preparation, Packaging, and Application in Fiber Lasers

3.3.6.1 Preparation Methods

In general, new broadband SAs in fiber laser system usually include substrate-based SAs, solution-based SAs, evanescent-field based SAs, or SAs film pasted on fiber ferrules. From the processing point of view, each of these SAs has disadvantages that limit widespread industrial applications. Substrate-based SAs are usually fabricated by chemical vapor deposition (CVD) method. When used in fiber laser cavity, the layered materials are required to transfer from substrate to fiber surface and contact with the fiber surface by Van der Waals' force, which leads to the loose contact between the absorber and fiber surface. Solution-based SAs require a host solution with low optical loss and appropriate refractive index, which is hard to find. Evanescent field based SAs employed with tapered fiber or side-polished fiber (SPF) entail extra optical loss and cost. SAs film method is a simple and cost-effective alternative, which possesses the potential for practical applications. Embedding SAs in polymer host materials has been used to form a thin-film composite and widely been used. Many kinds of polymer composites have been used as hosts, such as polymethylmethacrylate (PMMA), polycarbonate (PC), polyimide, and polyvinylalcohol(PVA). However, optical damage threshold of SA/polymer composite film is mostly determined by the host polymer. Attributing the reason that polymer is a kind of organic material, it generally has a low laser damage threshold.

According to the report [74], spraying method involves spraying a new broadband SA solution onto a substrate or the end face of an optical fiber. This absorber solution is an SA solution that is prepared via liquid-phase exfoliation, that is, the broadband SA material powder is put into a specific solvent, which is usually an organic solvent. Then, a surfactant is added to the solution to help the SA material powder become dispersed in the solvent. To improve the full dispersion of the SA material powder in the solvent, an ultrasonic cleaner is used for a period of time. After ultrasonication, the SA material dispersion is subjected to a centrifugal process, mainly to separate out larger pieces of material, and thus, an evenly distributed absorber dispersion can be obtained. This method can produce high-quality SAs, but it requires a lot of time, and the output is low.

The laser absorption method is to immerse the end surface of the fiber through the laser in these new broadband SA dispersions. Because of the absorption of the laser, the SA material is deposited in the core area of the fiber end [75]. The advantage of this method is that less material solution is needed to avoid wasting material, and the SA film only forms in the required specific areas.

In the polymer composite method, the absorber material solution is prepared via the liquid phase stripping method and then mixed with the polymer composite material solution; finally, the polymer composite type SA is obtained by processes such as evaporation and drying [76, 77].

In the mechanical peeling method, which is simple and easy to implement, 2D materials (such as graphene, transition metal sulfides, BP) can be prepared via the mechanical peeling [78]. Transparent tape is used to peel the relatively thin sheet off the piece of 2D material, and this is repeated to make the peeled sheet of 2D material thin enough. Then, the end of a standard fiber jumper is pressed down. Because of the strong adhesion between the 2D material and the ceramic, the thinner 2D material can be adsorbed onto the end face of the fiber jumper head. This is then connected with another standard optical fiber jumper via a flange adapter to produce a 2D material SA device with a full optical fiber structure. Using this method, the defect-free 2D absorber material can be easily transferred to a substrate or to a fiber end face [77].

The last method is CVD, which can be used for direct synthesis of single-layer or multilayer 2D SA materials. Also, vertically aligned high-quality CNT materials can be grown directly on a substrate or fiber end face [79]. Vertically aligned CNT materials have strong polarization-dependent absorber characteristics.

3.3.6.2 Package Method

Six common fiber-type SAs packaged into the fiber laser cavity has been summarized[80]. The simplest and most commonly used type is the sandwich structure, which encapsulates the new broadband SA film between two fiber jumpers[81, 82]. The absorber film may be a pure absorber material film (such as a graphene film or a 2D material film), or an absorber polymer composite film. This type of structure is simple and convenient for manufacturing and is low cost, but it often easily causes damage. In 2013, to prevent the oxidation of this sandwich-structured CNT absorber, Japanese researchers sealed it with nitrogen; this technique avoids oxidation of the structure and increases its damage threshold and working time[83].

The microstructure of the fiber was used for injecting an SA solution into the fiber[80]. This type of method increases the distance between the light and the absorber and improves the nonlinear effect, and it increases the damage threshold of the absorber. The evanescent wave method is also a common and effective method. The typical approach is to use a D-type fiber and a tapered fiber.

3.4 APPLICATION OF NOVEL SAs IN MODE-LOCKED FIBER LASERS

3.4.1 Yb-Doped Mode-Locked Fiber Lasers with DF-SWCNT-D_2O SA

CNTs are the first materials that have been used as novel SAs in mode-locked fiber lasers. For example, CNT/PVA film-type SAs are usually manufactured by embedding CNT materials in a polymer host material. This type of absorber is convenient and flexible to use. However, easy oxygenation affects the damage threshold and long-term stability of the absorber. Therefore, researchers sealed the absorber in nitrogen to improve its performance in a gaseous environment, but this method is complicated to use and difficult for practical application [83]. At the same time, researchers also increased the damage threshold of the absorber via the evanescent wave method. Among all fiber lasers, it is most effective to use tapered fibers and D-shaped fibers (DFs) [84, 85].

We presented a method for retarding the degradation of SAs by immersing DF in SWCNT heavy water (D_2O) solution. This is ideal technology for the evanescent field interaction method and also avoids oxidation from the air. Organic solutions are not chosen because they have two shortcomings: (1) they evaporate too quickly and (2) nonsaturable absorption losses too much between 1 and 2 μm wavelength. The SWCNT-D_2O solution is a better choice because it has lower absorption losses between 1 and 1.3 μm. The SWCNT-D_2O-DF SA does not cause extra nonsaturable losses to a fiber laser system. Incorporating this kind of SA into an Yb-doped fiber (YDF) laser cavity can be used to achieve a long-term stability mode-locked fiber laser. The repetition rate and pulse duration were measured to be 23 MHz and 194 ps, respectively. The solution method with DF has potential for use in mode-locked fiber laser applications in practice.

The band gap energy between the conduction and valence bands of SWCNTs varies with respect to diameter. For SWCNTs, the absorption peak wavelength depends on the averaged diameter of the SWCNTs, and the absorption bandwidth depends on the distribution of the nanotube diameters. The typical absorption bandwidth is in the range of 200–500 nm. Such a wide absorption bandwidth also means that for a specific operating wavelength, only a small part of the sample works, while the other part causes scattering loss. Also, if the arrangement direction of the CNT molecules is ordered, the absorption response is related to polarization. Although the absorbent of the single-walled CNT polymer film type has the advantages of simplicity and flexibility, it has the disadvantage of oxidative aging over long-term use. Therefore, to manufacture a high-performance SA device, these shortcomings must be overcome. Therefore, we have proposed a SWCNT-D_2O solution SA (SSA) device.

D_2O is called "heavy water" because it is composed of deuterium and oxygen. The main reason for choosing heavy water as the solvent for SWCNTs is because the absorption of light by heavy water is less (in the range of 1–1.55 μm) than that of water, and this can further reduce the absorption loss. Figure 3.2 shows the light absorption of an aqueous solution and of a heavy water solution in different wave bands. From the comparison in the figure, it is seen that the absorption of heavy water solution at 1 and 1.55 μm is very small compared to that of the aqueous solution.

SWCNT-D_2O-DF SA was proposed and fabricated in three steps. First, the SWCNT-D_2O solution that was used in experiments was prepared as a reference, and the concentration of the SWCNT-D_2O solution was 1 mg/ml [24]. Figure 3.2 shows the Raman spectrum of SWCNTs at 633 nm. The radial breathing mode (168 cm^{-1}) and G mode (1586 cm^{-1}) show that the material is an SWCNT. To achieve an efficient evanescent field interaction as well as to protect the DF, the distance from the fiber core boundary to the D-shaped surface was optimized to 5μm. According to the literature, less than 10% of the optical power interacts with SWCNT-D_2O-DF SA [87]. The interaction length of the D-shaped area was estimated to be 5 mm.

Finally, DF was sealed in a box that was full of SWCNT-D_2O solution, as shown in Figure 3.3. The SWCNT solution method was used to alleviate thermal damage caused by physical contact with ferrules and also to avoid oxidation. When DF was immersed in SWCNT-D_2O solution, SWCNTs were easily attracted to the surface of the DF as the laser is injecting through the DF [75]. The SWCNTs in D_2O solution can thus interact with the evanescent field that surrounds the D-shaped surface. The SWCNT solution continuously supplies SWCNTs to the surface of DF so that the

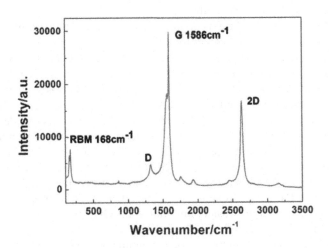

FIGURE 3.2 Raman spectrum of SWCNTs on quartz [86].

FIGURE 3.3 Schematic diagram of a SWCNT-D$_2$O-DF SA [86].

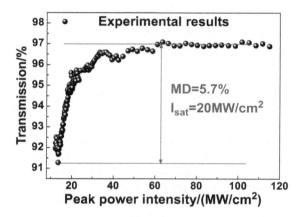

FIGURE 3.4 Nonlinear absorption of SWCNT-D$_2$O-DF SA [86].

density of the SWCNTs on the DF remains unchanged even if some of them are taken off from the DF, and this is useful for ML operation over a long period.

A picosecond pulsed YDF laser (central wavelength: 1053 nm, pulse duration: 10 ps, repetition rate: 26.9 MHz) was used as an illumination source to study the nonlinear saturable absorption of SWCNT-D$_2$O-DF SA. Figure 3.4 shows the nonlinear transmission curve as a function of peak power intensity. It is seen that for the SWCNT-D$_2$O-DF SA, the saturable intensity is 20 MW/cm^2, and the MD is 5.7%, and these values are sufficient for ML [30, 88, 89]. The inherent transmission losses of the DF used here is just 3%. Thus, the nonsaturable losses caused by the D$_2$O solution can be neglected. To the best of our knowledge, the nonsaturable loss of this SA device is 3%, which is the lowest value compared with that of other CNT-SAs.

A schematic diagram of a YDF laser is depicted in Figure 3.5. The ring fiber laser cavity consists of a gain fiber, wavelength division multiplexer (WDM), polarization independent isolator (PI-ISO), optical coupler (OC), polarization controller (PC), and SWCNT-D$_2$O-DF SA. A 20-cm long YDF (LiekkiYb 1200-4/125) with an absorption coefficient of 1200 dB/m at 976 nm was used as the gain medium. YDF was pumped by a 980-nm laser diode (LD). PI-ISO was used to force unidirectional operation in the fiber ring cavity, and was used to achieve different polarization states. The optical coupler was used in the cavity, and a 10% portion of the laser was coupled out from the laser cavity. Because the laser cavity in this case is in a state of all-normal dispersion, a spectral filter is essential for achieving stable ML operation. As a result, a fiber-pigtailed filter centered at 1053 nm with a bandwidth of 8 nm was inserted in the cavity to obtain ML at a wavelength of 1 μm. The total length of the laser oscillator cavity was ~9 m. A power meter, optical spectrum analyzer (YOKOGAWA AQ6370B), autocorrelator (APE Pulse Check SM1200), and digital oscilloscope (Tektronix TDS3024C) were used with a home-made 2.5 GHz photodiode detector to monitor the output properties of the fiber laser.

The YDF laser initiated the continuous wave at a pump power of 30 mW, and the self-started ML operation was obtained at 590 mW with an output power of 9.5 mW. The maximum output power

Novel Mode-Locked Fiber Lasers

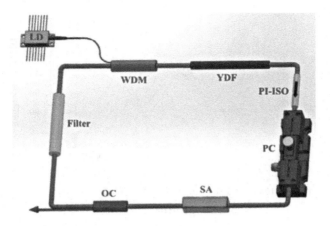

FIGURE 3.5 Yb-doped mode-locked fiber laser setup [86].

reached 15 mW with a pump power of 645 mW, which corresponds to an intracavity power of 150 mW. It has been reported that the efficiency of the fiber laser is relatively lower, and this is primarily attributed to the higher cavity loss in the fiber laser and the lower output ratio. In the experiments, the YDF fiber was LiekkiYb 4/125 with a core diameter of 4 µm, which does not match the HI 1060 fibers (diameter: 6µm). Thus, the cavity loss is relatively larger. In further experiments, enhancing the output ratio is helpful for achieving a high-power output. As shown in Figure 3.6(a), the typical bell-shaped ML optical spectrum profile is centered at 1054.16 nm with a 3 dB spectral bandwidth

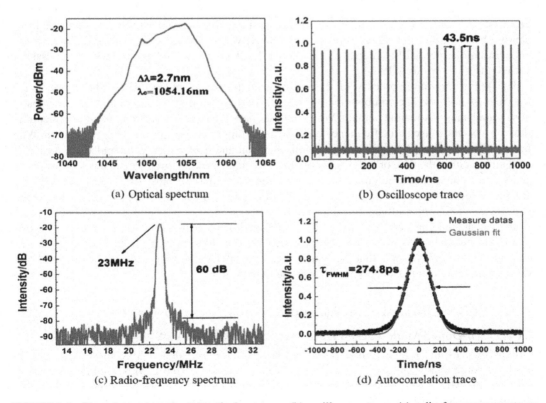

FIGURE 3.6 Experimental results: (a) optical spectrum, (b) oscilloscope trace, (c) radio-frequency spectrum, and (d) autocorrelation trace [86].

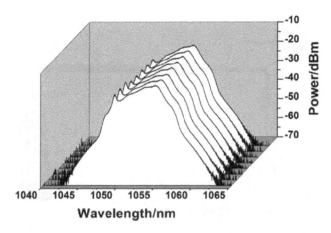

FIGURE 3.7 Long-term optical spectrum measured at a 5-h interval [86].

of 2.7 nm. When the pump power reached 650 mW, the ML operation became unstable, and the fluctuation of the spectrum was obvious. However, stable ML operation was observed again when the pump power was decreased to 590 mW. As illustrated in ref. [19], an intracavity power that exceeds 50 mW can damage CNT-SA. In our experiments, when the intracavity power reached 150 mW, the proposed SWCNT-D2O-DF SA was not destroyed. The corresponding oscilloscope trace is shown in Figure 3.6(b); the pulse trace on the oscilloscope occurs with a relatively uniform intensity with a pulse-to-pulse interval of 43.5 ns that corresponds to the roundtrip time of the cavity. Radio-frequency spectral measurements of the output pulses were conducted, and the results are shown in Figure 3.6(c). A strong signal peak with a fundamental repetition rate of 23 MHz was clearly observed, and the signal-to-noise ratio (SNR) was measured to be ~60 dB. The measured autocorrelation trace is illustrated in Figure 3.6(d). The FWHM is 274.8 ps, and when a Gaussian fit is used, the pulse width is 194 ps.

The pulse formation process of this all-normal dispersion mode-locked fiber laser can be described as follows: The combination of the positive dispersion effect and nonlinear effect lead to pulse broadening in the time domain. When the bandpass filter (BPF) is added, the filter effect of the filter narrows the pulse in the frequency domain and compresses the pulse in the time domain. When the pulse passes through an SA and acts on it, the front and back edges of the pulse are further filtered out, in either the time or frequency domain. The above process circulates in the ring-oscillating cavity until a stable pulse output form. It can be concluded that the pulse of the all-positive dispersion mode-locked fiber laser is formed when the filter effect, saturable absorption effect, dispersion, gain, loss, and nonlinear effect are balanced.

To evaluate the long-term stability of ML, the experiments were performed over 48 h; the power fluctuation percentage is <5%. We also recorded the optical spectrum of the laser every 5 h, as shown in Figure 3.7. We note that the central spectral peak locations, spectral bandwidth, and spectral strength remained reasonably stable over the time period.

3.4.2 Yb-Doped Mode-Locked Fiber Lasers with Bi_2Te_3-based SAs

3.4.2.1 Yb-Doped Mode-Locked Fiber Laser with Bi_2Te_3/PVA SA

Although CNTs and graphene materials have been used as broadband SAs, these two materials still have some disadvantages. When graphene (which has zero band gap) is used as a broadband absorber, the MD is too small [59]. In 2013, topological insulators were proposed as ML materials, mainly including Bi_2Se_3, Bi_2Te_3, and Sb_2Te_3. The researchers found that topological insulator materials have a greater MD: C. Zhao et al. reported that the MDs of Bi_2Se_3 and Bi_2Te_3 measured

Novel Mode-Locked Fiber Lasers

using the Z-scan method are 27% and 28%, respectively [27, 67, 72]. Similar to graphene, topological insulators are also layered materials. Sb_2Te_3 material was prepared via mechanical lift-off, and it was first used in mode-doped fiber lasers [67]. At present, the broadband saturable absorption characteristics of topological insulators have been fully proven, and these insulators are considered to be a potential saturable absorption material.

As an example, Bi_2Te_3 was reported [80]. Under weak light irradiation, when the energy of any photon is greater than 0.15 eV, the valence band electrons can be excited to the corresponding conduction band. However, when the incident light intensity is sufficiently large and the photon energy is greater than 0.15 eV, electrons are continuously excited to the corresponding conduction band. Ultimately, the sub-bands in the conduction band are completely occupied by electrons. Because of the Pauli blocking principle, the material shows photobleaching. At this time, there is no effect of absorbing light; that is, there are no saturated absorber characteristics. Moreover, Bi_2Te_3 also has a band gap-free surface energy state. For microwave photons in the surface metal state absorption band, electrons in the surface energy state play a key role in their absorption. However, under strong microwave radiation, Bi_2Te_3 will exhibit saturation absorption under the limitation of the number of electrons filled in the conduction band of the surface energy state. Therefore, according to the unique electronic properties of topological insulators, there are two different saturation absorption mechanisms that explain the saturation absorption of light and microwave radiation, respectively. The saturation absorption of light comes from the internal insulation state, and the saturation absorption of the microwave band depends on the metal state of the surface.

First, the SAs of Bi_2Te_3/PVA film were prepared, and their optical properties were studied. Then, the optical properties of the SAs of Bi_2Te_3 were fabricated via the pulsed laser deposition (PLD) method, and nonlinear tests of the two kinds of absorbers are carried out. The MDs of the SAs of Bi_2Te_3/PVA film and Bi_2Te_3 were respectively measured to be 8.4% and 10%. The mechanism of the evanescent field increases the interaction length between light and the absorber material, and this is conducive to improving the MD. Moreover, the damage threshold of the Bi_2Te_3 tapered fiber absorber device was greatly improved. The Bi_2Te_3/PVA film was inserted into the Yb-doped all-normal dispersion fiber laser as an SA, and continuous mode-locked operation was achieved. The central wavelength of the mode-locked pulse spectrum was 1052.7 nm, the repetition frequency of the mode-locked pulse sequence was 25.6 MHz, and the pulse width was 417 ps. At the same time, the SA of Bi_2Te_3 was also inserted into the all-normal dispersion YDF laser. Again, continuous mode-locked operation was achieved. The central wavelength of the mode-locked pulse spectrum was 1052.5 nm, the repetition frequency of the mode-locked pulse sequence was 19.7 MHz, and the pulse width was 317 ps.

Bi_2Te_3 nanoplatelets were synthesized via hydrothermal intercalation. Bi_2Te_3 nanoplatelets were dispersed in deionized water. Sodium dodecyl sulfate power was added to the Bi_2Te_3 dispersion as a surfactant. The dispersion was ultrasonically agitated for 6 h. Then, some polyvinyl alcohol (PVA) power was dissolved in deionized water with ultrasonic agitation at 90 °C for 3 h. The Bi_2Te_3 dispersion and PVA solution were uniformly mixed and poured into polystyrene cells. Finally, these cells were placed into an oven for evaporation, and the temperature of the oven was set at 40 °C. About 2 days were needed for complete evaporation. The PVA solution has a high viscosity to polystyrene cells so that it adheres to the walls of the cell. When the cell was dried, the PVA film lost the viscosity in the cell; therefore, the PVA film cell can be easily stripped off the polystyrene cell using tweezers. Bi_2Te_3 was carried by PVA to the surface of the wall and the bottom of the polystyrene cell during the evaporation process. The Bi_2Te_3/PVA film on the wall of the cell was of a higher quality than that on the bottom, and thus, the former can be used as an absorber for ML.

The measured Raman spectrum of the Bi_2Te_3 layer is shown in Figure 3.8. Three typical Raman optical phonon peaks were identified as A_1g_1 at 61 cm^{-1}, Eg_2 at 101 cm^{-1}, and A_1g_2 at 133 cm^{-1}. The linear transmittance of the Bi_2Te_3/PVA film was measured over the range from 900 to 2000 nm, as shown in Figure 3.9(a). The Bi_2Te_3/PVA film has a smooth transmission curve in the near-infrared wavelength band, and this suggests that it is a promising broadband optical material. As seen in the

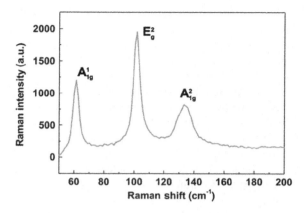

FIGURE 3.8 Raman spectrum of Bi_2Te_3 excited by a 633-nm laser [28].

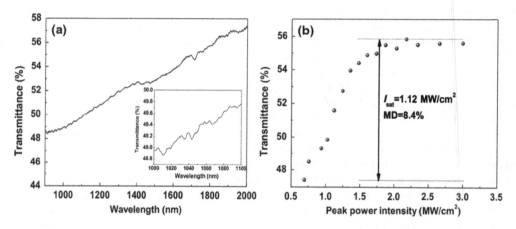

FIGURE 3.9 (a) Linear and (b) nonlinear transmittance of the Bi_2Te_3/PVA film [28].

inset graph of Figure 3.9(a), the transmittance of the Bi_2Te_3/PVA film was 49.4% at 1053 nm. A ps pulse laser centered at 1053 nm was used to measure the nonlinear transmittance. In (b), the MD was measured to be 8.4%. The MD of the Bi_2Te_3/PVA is lower than that of the bulk-structured Bi_2Te_3 that was used for the SPF[90]. The saturable optical intensity of the Bi_2Te_3/PVA film was 1.12 MW/cm², which is much lower than the value of a recently reported Bi_2Te_3 film [91]. It is predicted that the mode-locked fiber laser based on a Bi_2Te_3/PVA film can operate with a relatively low threshold, according to a recent report by Yan et al.[88].

A schematic diagram of the YDF laser, including the laser oscillator and laser amplifier, is depicted in Figure 3.10. A 0.2-m long YDF (LiekkiYb 1200-4/125) with an absorption coefficient of 1200 dB/m at 976 nm was used as the gain medium. YDF was pumped by a 980-nm LD. PI-ISO was used to force unidirectional operation in the fiber ring cavity. PC was engaged to achieve different polarization states, and 10% OC was used. Because the laser oscillator cavity in this case is all-normal dispersion, a spectral filter is essential for stable ML operation. As a result, a fiber-pigtailed filter centered at 1053 nm with a bandwidth of 2 nm was inserted into the cavity to achieve ML at a 1 μm wavelength. The total length of the laser oscillator cavity was about 8 m. A YDF amplifier was used to further amplify the output pulse. A SMF-pigtailed LD with a maximum output power of 700 mW was used as the pump source in the amplification stage, and a 0.3-m long YDF (LiekkiYb 1200-4/125) with an absorption coefficient of 1200 dB/m at 976 nm was used as the gain medium. A power meter, optical spectrum analyzer (YOKOGAWA AQ6370B), autocorrelator (APE Pulse

Novel Mode-Locked Fiber Lasers

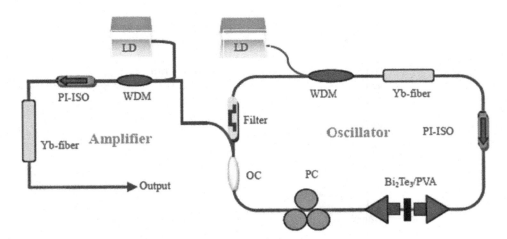

FIGURE 3.10 Yb-doped mode-locked fiber laser setup [28].

Check SM1200), and digital oscilloscope (Tektronix TDS3024C) with a home-made 2.5-GHz photodiode detector were used to monitor the output properties of the laser oscillator and amplifier.

AYDF laser oscillator initiated a continuous wave at the pump of 50 mW, and the ML operation was achieved at 75 mW. In Figure 3.11(a), the optical spectrum of the mode-locked pulses is centered at 1052.7 nm, and the 3-dB spectral width is 0.45 nm. The spectrum has steep edges, displaying the character typical of a dissipative soliton that formed because Bi_2Te_3 was used and also because of the normal dispersion and filtering effect of the laser oscillator [92]. The corresponding oscilloscope trace is depicted in Figure 3.11(b). A pulse was circulating in the cavity at the fundamental cavity repetition rate of 25.6 MHz. The pulse trace exhibits a relatively uniform intensity with a pulse interval of 39 ns, corresponding to the cavity roundtrip time.

The average output power from the oscillator was only 1mW (which is too low to be detected by an autocorrelator), and the measured autocorrelation trace of the YDF amplifier is illustrated in Figure 3.11(c). The FWHM was 587.5 ps, which means that the pulse width was 417 ps when a Gaussian fit was used. The time bandwidth was calculated to be 50.8, which indicates that the pulse was strongly chirped. The pulse width of 417 ps was slightly longer, and this is primarily attributed to the large net positive dispersion resulted by the 5-m-long all-normal dispersion YDF amplifier cavity.

To comprehensively show the performance of the YDF amplifier, the pump power of the YDF oscillator was fixed at 80 mW. The amplified average power was plotted as a function of the amplifying pump power, and the results are shown in Figure 3.12(a). The largest output power was 19.3 mW. As seen in Figure 3.12(b), the spectral intensity increased with an increase in the amplifying pump power, whereas there were no obvious changes in the spectral bandwidth with a different amplifying pump power, and the spectrum remained stable.

To evaluate the ML stability, the optical spectrum of the laser was recorded every 2 h, and the results are shown in Figure 3.13. It is noted that the central spectral peak locations, spectral bandwidth, and spectral strength remained reasonably stable over the time period.

3.4.2.2 Yb-Doped Passive Mode-Locked Fiber Laser with Bi_2Te_3 SA

Bi_2Te_3 was deposited on a tapered fiber using a PLD instrument [88]. PLD is the ideal production technology for growing Bi_2Te_3 because it maintains the consistent composition of the original material [93, 94]. Unlike the high-temperature requirement of CVD, the tapered fiber is deposited under low temperature in PLD; thus, the fiber is not destroyed via thermal softening. The deposited material has good contact with the tapered fiber and does not fall off. To date, the shortest pulse width of 128 fs generated from a mode-locked fiber laser by TI is operated at 1.55 μm[95]. At a 2

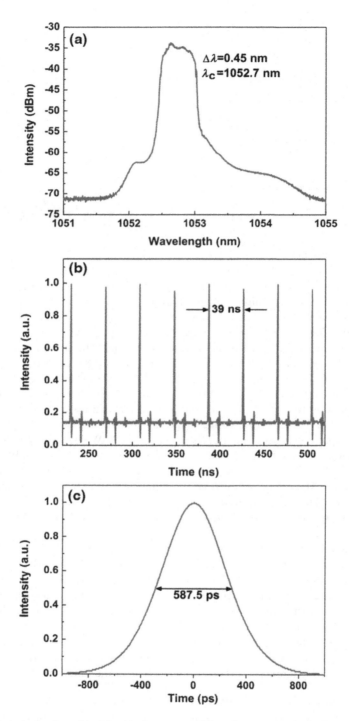

FIGURE 3.11 Experimental results: (a) optical spectrum, (b) oscilloscope trace, and (c) autocorrelation trace [28].

Novel Mode-Locked Fiber Lasers

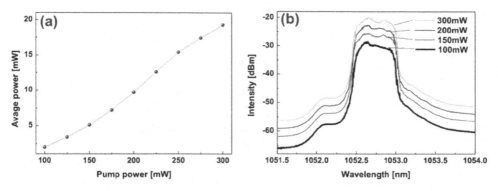

FIGURE 3.12 (a) Amplified average output power versus pump power. (b) Spectra with different amplifying pump power [28].

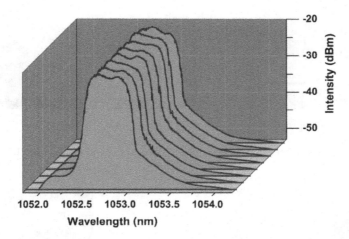

FIGURE 3.13 Long-term optical spectra measured at 2 h intervals [28].

μm waveband, a pulse width of 795 fs was achieved [96]. However, there are a few reports about mode-locked fiber lasers operating near 1μm with TI. Here, aYb-doped mode-locked fiber laser is achieved with a repetition rate of 19.8 MHz when fiber-taper TI-SA Bi_2Te_3 is used. The amplified single pulse energy and pulse duration were 2.8 nJ and 317 ps, respectively.

A schematic diagram of the typical PLD setup is shown in Figure 3.14. In an ultrahigh vacuum (UHV) chamber, targets are illuminated at an angle of 45° by a pulsed and focused laser beam. An Nd:YAG laser (=1064 nm. Mode: SL II-10, Surelite) is employed in the PLD system as an ablation source with a 10-Hz repetition rate. The single pulse energy is 2mJ, and the average power of the laser is 2 W. The target is Bi_2Te_3, and the vacuum pressure of the chamber is fixed at 5×10^{-4} Pa during the deposition procedure. The deposition time is 90 min. Figure 3.15 shows a scanning electron microscope (SEM) image of the tapered fiber and the morphology of the deposited Bi_2Te_3. The length of the taper is about 0.6 mm, and the waist is about 55 μm.

Bi_2Te_3 solution was observed using transmission electron microscopy (TEM) to understand the crystalline orientations of nanosheets (Figure 3.16). Figure 3.16(a) shows a low magnification bright field TEM image of the Bi_2Te_3 nanosheets, exhibiting a hexagonal morphology. Figure 3.16(b) displays a high-resolution TEM image of the same nanosheets and single crystalline quality. The circle indicates the position where a selected-area electron diffraction pattern, which is shown in Figure 3.16(c), was recorded. The selected-area diffraction pattern can be indexed as a 6-fold symmetry zone axis pattern, which is in agreement with the layered structure.

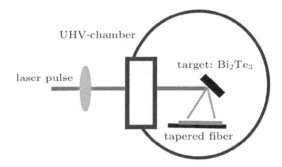

FIGURE 3.14 Schematic diagram of a typical pulsed laser deposition setup [97].

(a) SEM images of tapered fiber   (b) SEM images of the Bi₂Te₃ film

FIGURE 3.15 SEM images of (a) a tapered fiber and (b) morphology of deposited Bi$_2$Te$_3$ [97].

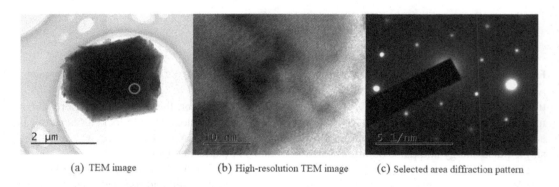

(a) TEM image   (b) High-resolution TEM image   (c) Selected area diffraction pattern

FIGURE 3.16 (a) TEM image of Bi$_2$Te$_3$ nano-sheets, (b) high-resolution TEM image of Bi$_2$Te$_3$ nanosheet, and (c) selected-area diffraction pattern of Bi$_2$Te$_3$ nanosheets [97].

The measured Raman spectrum of the Bi$_2$Te$_3$ layer is shown in Figure 3.17(a). Three typical Raman optical phonon peaks are identified as $A_{1g}1$ at 61 cm^{-1}, Eg2 at 101 cm^{-1}, and $A_{1g}2$ at 133 cm^{-1}. To study the nonlinear characteristics of the fiber-taper TI-SA, a balanced twin-detector measurement system was used, as reported in ref.[98]. to investigate nonlinear absorption. A ps pulsed YDF laser (central wavelength: 1053 nm, pulse duration: 10 ps, and repetition rate: 26.9 MHz) works as an illumination source. Figure 3.17(b) shows the transmission curve of TI-SA as a function of average pump power. As seen here, the MD is 10%. The nonsaturable loss of the fiber-taper TISA is relatively large, and this is mainly a result of scatter losses that arise from the process of the tapered fiber fabrication. To obtain a higher quality fiber-taper TI-SA device, the tapered fiber fabrication technology should be improved and further modified.

Novel Mode-Locked Fiber Lasers 65

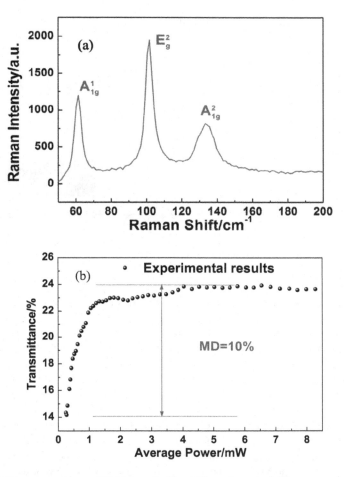

FIGURE 3.17 (a) Raman spectrum of Bi_2Te_3 excited by 633 nm laser and (b) measured nonlinear saturable absorption curve [97].

A schematic diagram of the YDF laser including a laser oscillator and laser amplifier is shown in Figure 3.18. The ring laser oscillator cavity is composed of a gain fiber, WDM, PI-ISO, OC, PC, and fiber-taper TI-SA.

A 20-cm long YDF (LiekkiYb 1200-4/125) with an absorption coefficient of 1200 dB/m at 976 nm was used as a gain medium. YDF was pumped by a 980-nm LD. PI-ISO was used to force unidirectional operation in the fiber ring cavity. PC was used to achieve different polarization sates. An optical coupler was used, and a 10% portion of the laser was coupled out from the oscillator cavity. Because the laser oscillator cavity in this case was an all-normal dispersion cavity, a spectral filter was essential for stable ML operation. As a result, a fiber-pigtailed filter centered at 1053 nm with a bandwidth of 2 nm was inserted in the cavity to achieve ML at 1 μm wavelength. The total length of the laser oscillator cavity was about 10.5 m. A YDF amplifier was used to further amplify the output pulse. An SMF-pigtailed LD with a maximum output power of 700 mW was used as a pump source in the amplification stage, and a 30-cm-long YDF (LiekkiYb 1200-4/125) with an absorption coefficient of 1200 dB/m at 976 nm was used as a gain medium.

The YDF laser oscillator initiated a continuous wave (CW) at a pump power of 60 mW, and the ML operation was achieved at 230 mW. This ML threshold is relatively high, and it is mainly a result of the large scatter losses caused by the fabrication process of the tapered fiber. As seen in the inset

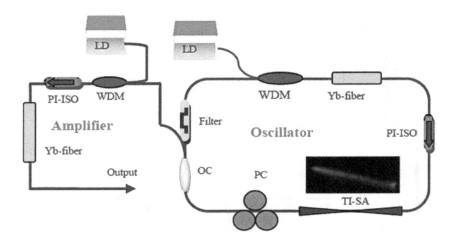

FIGURE 3.18 Yb-doped mode-locked fiber laser setup [97].

of Figure 3.18, an evanescent field around the tapered fiber is easily observed using an infrared night vision viewer while light is confined in other parts of fiber core.

As seen in Figure 3.19(a), the optical spectrum of mode-locked pulses is centered at 1052.5 nm, and the 3-dB spectral width is 1.245 nm. When the pump power reached 290 mW, the ML operation became unstable, and fluctuation of the spectrum was obvious. However, stable ML operation was again observed when the pump power was decreased from 290 mW. This phenomenon indicates that TI-SA was not destroyed by thermal accumulation. Unstable ML may be interpreted as the oversaturation of TI-SA at 290 mW. The corresponding oscilloscope trace is shown in Figure 3.19(b), and the pulse trace on the oscilloscope shows relatively uniform intensity with a pulse-to-pulse interval of 50.75 ns, corresponding to the roundtrip time of the cavity. To evaluate the long-term stability of ML, the laser oscillator worked continuously for 48 h, and the spectrum remained reasonably stable, which indicates that the ML state is very stable. The average output power from the oscillator was only 3mW, which is too low to be detected by an autocorrelator. The measured autocorrelation trace of the YDF amplifier is illustrated in Figure 3.19(c). The FWHM is 447 ps, which means that the pulse width is 317 ps when a Gaussian fit is used. The combination of PLD with a tapered fiber improves the damage threshold of an SA, and this is helpful for the laser working in the high-power regime. However, because the tapered fiber that was used here was fabricated via manual manipulation and the waist of the tapered fiber was not optimized, the laser scatter loss was relatively large, and this reduced the ML output power. To overcome this trouble, we used tapered fiber fabrication equipment. Dispersion management introduces laser losses to some extent. In further experiments, the tapered fiber fabrication issue would be settled to enhance the average output power and then optimize the dispersion compensation work, as shown in ref.[99]. Finally, an ultrashort pulse output with high average power can be achieved.

To show the comprehensive performance of the laser amplifier, we fixed the pump power of the laser oscillator at 250 mW. The amplified average power and single pulse energy with respect to the amplifying pump power were investigated, and the results are shown in Figure 3.20. The largest output power was 55 mW, and the corresponding pulse energy was estimated to be 2.8 nJ. Changes in the spectral bandwidth of the amplified pulse at different pump powers are not obvious, and the spectra remain stable.

In previous content, the fabrication of the Bi_2Te_3/PVA SA and fiber-taper TI-SA Bi_2Te_3 is introduced. Bi_2Te_3/PVA thin film absorbers have the advantages of flexibility, convenience, and ease of manufacturing, but they have the disadvantage of having a low damage threshold. The fiber-taper TI-SA Bi_2Te_3 was made via PLD, which has the advantages of cleanness and high purity. The

Novel Mode-Locked Fiber Lasers

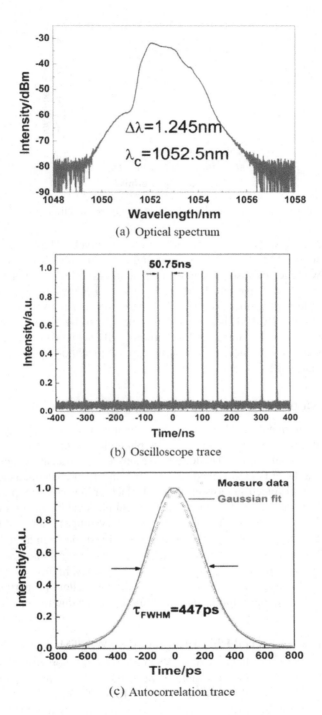

FIGURE 3.19 Experimental results: (a) optical spectrum, (b) oscilloscope trace, and (c) autocorrelation trace [97].

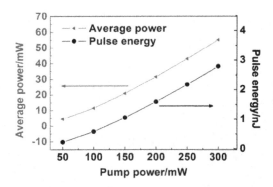

FIGURE 3.20 Amplified average power and single pulse energy versus pump power [97].

MDs of the two kinds of absorbers are 8.4% and 10%, respectively. The mechanism of the evanescent field increases the interaction length between light and the absorber material and effectively improves the MD. Two kinds of absorbers are used to achieve mode-locked operation in all-normal dispersion fiber lasers. The damage threshold of fiber-taper TI-SA Bi_2Te_3 is greatly improved. The pulse width of the mode-locked fiber laser is 417 ps based on Bi_2Te_3/PVA SA, and is 317 ps based on fiber-taper TI-SA Bi_2Te_3 SA. PLD is a good way to make absorber devices, but in the experiment described in this chapter, the advantages of PLD are not fully achieved because the manufacturing process of tapered fiber is not refined. If we can improve the technology of making tapered fibers, we will make high-quality SA devices.

3.4.3 Yb-Doped Mode-Locked Fiber Laser Based on WS_2/FM

In recent years, transition metal sulfide semiconductor materials have gradually become a research hotspot in the field of lasers because of their unique photoelectric properties. Similar to graphene, transition metal sulfides are also layered materials. In a single-layer transition metal sulfide structure, a single transition metal layer is sandwiched between two layers of chalcogen elements. Because of the 2D constraint of specific electron movement and the lack of interlayer coupling perturbation, 2D transition metal sulfides have the following unique optical physical characteristics: (1) The energy band characteristics have adjustable value (usually in 1–2 eV range) related to the number of layers. (2) With a decrease in the number of layers, there is a transition from an indirect band gap to a direct band gap. (3) They have better photoluminescence and electroluminescence properties. (4) They have stable laser characteristics, such as high binding energy, high vibrator strength, and long life. (5) They have ultrafast carrier characteristics. It is these excellent properties that cause transition metal sulfide materials to have application potential in the production of high-performance SA devices.

3.4.3.1 Yb-Doped Mode-Locked Fiber Laser Based on WS_2/FM

Figure 3.21 shows a schematic diagram of a WS_2/FM-based Yb-doped mode-locked fiber laser device. In the experiment, a ring cavity structure was used, and the optical path included the following components: a pump coupler (WDM), YDF, PI-ISO, PC, WS_2/FMSA, OC, and BPF. The semiconductor laser outside the cavity was used as a pump source with a maximum output power of 650 mW.

Among them, the absorption coefficient of YDF at 976 nm was 1200 dB/m, and the length used was 0.5 m. PI-ISO was used to ensure the laser unidirectionality in the resonator. Adjusting the PC enables different polarization states of the fiber laser. The output coupling ratio was set to 30:70, that is, 70% of the ports were connected to the cavity, and 30% of the ports were used as pulse output terminals. The center wavelength of BPF was 1053 nm, and the bandwidth was 2 nm. In the

Novel Mode-Locked Fiber Lasers

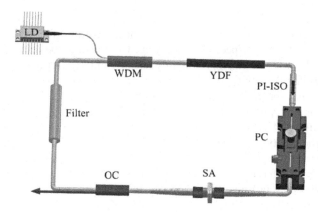

FIGURE 3.21 WS$_2$/FM Yb-doped mode-locked laser device [100].

experiment, WS$_2$/FM was cut to an appropriate size and attached to the ceramic optical fiber head; it was also connected to another ceramic optical fiber head and embedded in the laser cavity through the flange.

Figure 3.22(a) shows the mode-locked spectrum, which has a bell-shaped envelope. As seen in the figure, the central wavelength of the mode-locked spectrum was 1053 nm, and the bandwidth of 3 dB was 0.29 nm. Similar to values reported in the literature [101], the 3-dB bandwidth of the spectrum

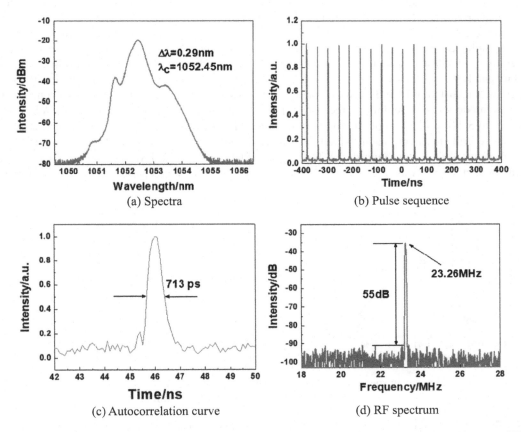

FIGURE 3.22 Experimental results of mode-locked spectrum, pulse sequence, autocorrelation curve, and RF spectrum of WS$_2$/FM Yb-doped mode-locked laser device [100].

was smaller than that of the filter, and this is mainly because the pulse width in an all-normal dispersion fiber laser is usually a few hundred ps. Thus, the nonlinear effect in the cavity is relatively weak, and this is not conducive to spectral broadening. We note that the obtained spectral base was about 2 nm, and it can be inferred that this is mainly because of the spectral filtering effect caused by the insertion of a filter into the cavity. When the pump power reaches to 635 mW, the ML operation begins to be unstable, and spectral defects and jitters begin to appear. ML was still observed when the pump power was again reduced to 550 mW. Figure 3.22(b) shows a diagram of the mode-locked pulse train. The time interval of the two pulses was 43 ns, and this corresponds to the time for the light pulse to run in the cavity for one week. Figure 3.22(c) shows a pulse curve measured using a 6 GHz oscilloscope (Lecroy 8600A) and a 10 GHz domestic detector; the pulse width determined from the figure is 713 ps. Different from the soliton ML at 1.55 m, the pulse width was mostly on the order of femtoseconds, and a pulse width at 1 m is usually several hundred picoseconds. The pulse curve could not be read well because of the limits of the 6 GHz oscilloscope. Figure 3.22(d) shows the radio frequency spectrum. The repetition frequency of the output pulse was 23.26 MHz, and the SNR was as high as 55 dB, indicating that the mode-locked state was very stable. From a comparison to the relevant laser parameters, we find that the 30-mW pulsed laser power that was obtained in our experiment was the highest output power of the transition metal-based sulfide fiber laser. Considering that the output coupling ratio is 30:70, it can be inferred that the power in the cavity reached 100 mW. At this point, the WS_2/FM absorber device was not damaged, and this indicates that using inorganic FM as the substrate greatly improved the damage threshold of the absorber device.

To verify the long-term stability of the operation of the WS_2/FM ytterbium-doped mode-locked fiber laser, we conducted the experiment for 10 days. It was found from the experiment that the power jitter was less than 4.5%, and changes in the spectrum were recorded; the results are shown inFigure 3.23(a). As seen from Figure 3.23(b), the central wavelength of the spectrum, spectral width, and spectral intensity remain basically unchanged for 10 days. To verify whether the ML operation of the laser was really because of the WS_2/FM SA device, we removed WS_2/FM from the cavity and kept the other devices unchanged. Regardless of how the power changed and how the PC angle was adjusted, no mold clamping phenomenon was observed. From this, it can be concluded that the mode-locked operation of the laser is indeed because of the SA characteristics of WS_2/FM.

Figure 3.24is a diagram of an erbium-doped mode-locked fiber laser device based on WS_2/FM. In the experiment, a ring cavity structure was used, and the optical path included the following components: WDM, ytterbium-doped fiber (EDF), PI-ISO, PC, WS_2/FM SA, and OC. A 976 nm semiconductor laser was used outside of the cavity as a pump source, and the maximum output

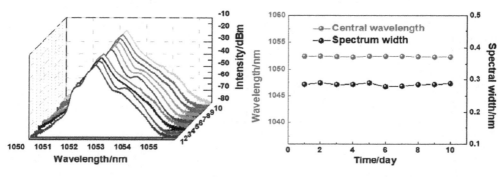

(a) Graph of changes in spectrum measured in 10 days

(b) Graph of changes in center wavelength and spectral width in 10 days

FIGURE 3.23 Stability test results of WS_2/FM Yb-doped mode-locked laser device [100]. (a) Graph of changes in spectrum measured in 10 days. (b) Graph of changes in center wavelength and spectral width in 10 days.

Novel Mode-Locked Fiber Lasers 71

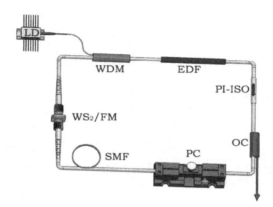

FIGURE 3.24 Schematic diagram of the WS$_2$/FM erbium-doped mode-locked fiber laser device [102].

power was 650 mW. The absorption coefficient of EDF at 976 nm was 3 dB/m, and the length used was 4 m. Other pigtails and compensated fibers were SMFs. PI-ISO was used to ensure resonance of the unidirectional laser in the cavity. PC was adjusted to enable different polarization states of the fiber laser. The output coupling ratio of the OC was 10:90, that is, 90% of the ports were connected to the cavity, and 10% of the ports were used as the pulse output. The total length of the laser ring cavity was 25.5 m. The dispersion parameter D of EDF and SMF at 1550 nm were 16ps/(nm.km) and 17 ps/(nm.km), respectively, and the net dispersion in the cavity was 0.47 ps. The autocorrelation instrument used in the test was an autocorrelation instrument made by Alnair Corporation of Japan; the model was the Alnair HAC 200.

In the experiment, the pump power was gradually adjusted to 4mW, and the WS$_2$/FM erbium-doped fiber laser began to achieve continuous laser output. When the pump power was increased to 30 mW, the WS$_2$/FM erbium-doped fiber laser achieved continuous mode-locked operation; however, at this point, when the pump power was reduced to 23 mW, the mode-locked state can still be maintained, mainly because of hysteresis effect. The maximum output power of ML was 1mW, and the pump power was 41 mW. Figure 3.25(a) shows a typical mode-locked pulse spectrum. As seen in the figure, the center wavelength of the mode-locked pulse spectrum is 1558 nm, and there is a small peak of the DC component at the top of the spectrum. This is mainly a result of the WS$_2$/FM SA. The MD of the device is a bit low [103]. Two pairs of Kelly sidebands are evenly distributed on both sides of the spectrum, and this proves that the mode-locked fiber laser works in the traditional soliton ML interval. Figure 3.25(b) shows a diagram of the mode-locked pulse sequence. The time interval between the two pulses is 121.95 ns, and this corresponds to the time of the optical pulse running in the cavity for one week. Figure 3.25(c) shows the measured autocorrelation curve. The FWHM of the autocorrelation curve is 1.28 ps. If the hyperbolic secant function is used, the pulse width is 830 fs. Figure 3.25(d) shows the radio frequency spectrum. The repetition frequency of the output pulse is 8.2 MHz, and the SNR is as high as 70 dB. The inset in Figure 3.25(d) is the spectrum in the 1 GHz range, and it indicates that the mode-locked state is very stable.

3.4.3.2 Erbium-Doped Mode-Locked Fiber Laser Based on DF-WS$_2$ SA

Figure 3.26 shows a schematic diagram of the erbium-doped mode-locked fiber laser device based on DF-WS$_2$ SA. The annular cavity structure was used in the experiment, and the optical path included the following devices: WDM, EDF, PI-ISO, PC, DF-WS$_2$ SA, and OC. A 976 nm semiconductor laser outside the cavity was used as the pump source with a maximum output power of 650 mW. The absorption coefficient of EDF at 976 nm was 3 dB/m, and the length used was 4 m. The other tail fiber and compensation fiber were SMFs; PI-ISO was used to ensure a unidirectional laser in the resonator. Adjusting PC can result in different polarization states in the fiber laser.

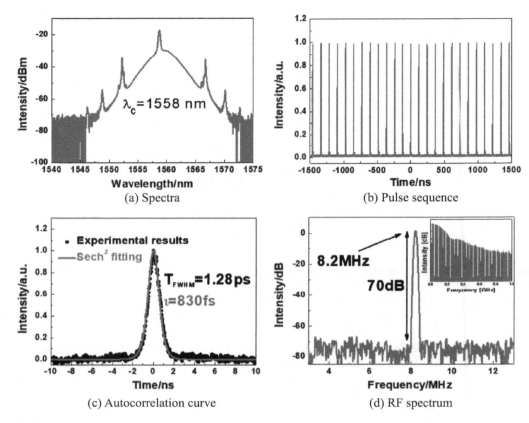

FIGURE 3.25 Experimental results of mode-locking the WS$_2$/FM erbium-doped mode-locked fiber laser device [102].

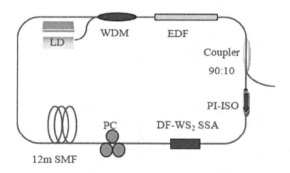

FIGURE 3.26 Erbium-doped mode-locked fiber laser device of DF-WS$_2$ SA [104].

The output coupling ratio of OC was 10:90, that is, 90% of the port was connected in the cavity, and 10% of the port was used as a pulse output terminal. The total length of the laser ring cavity was 20.3 m. The dispersion parameter D of EDF and SMF at 1550 nm were 16ps/(nm.km) and 17 ps/(nm.km), respectively, and the amount of net dispersion in the cavity was 2ps and 0.33 ps for EDF and SMF, respectively.

Figure 3.27(a) shows a typical mode-locked pulse spectrum. As seen in the figure, the center wavelength of the mode-locked pulse spectrum is 1557 nm, and the 3-dB bandwidth is 4 nm. Two pairs of Kelly sidebands are evenly distributed on both sides of the spectrum, and this proves that

Novel Mode-Locked Fiber Lasers 73

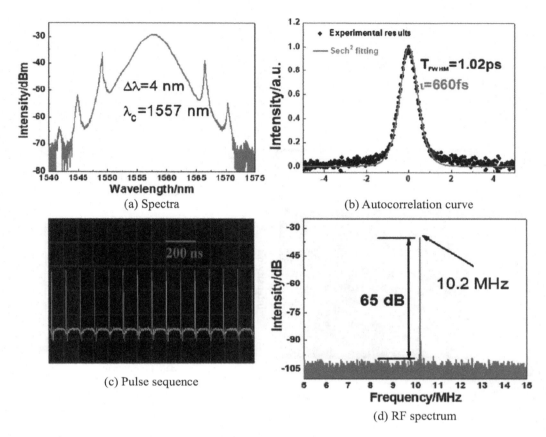

FIGURE 3.27 Experimental results of mode-locking [104].

the mode-locked fiber laser works in the traditional soliton ML interval. Figure 3.27(b) shows the measured autocorrelation curve. The FWHM of the autocorrelation curve is 1.02 ps. If the hyperbolic secant function is used, the pulse width is 660 fs. The time-bandwidth product is 0.327. Figure 3.27(c) shows the mode-locked pulse sequence. The time interval between the two pulses is 98 ns, and this corresponds to the time of the optical pulse running in the cavity for one week. Figure 3.27(d) shows the radio frequency spectrum. The repetition frequency of the output pulse is 10.2 MHz, and the SNR is as high as 65 dB, which shows that the mode-locked state is very stable.

The pump power was further increased to 350 mW, and the PC angle was adjusted; at this point, the laser achieves continuous harmonic ML operation. Figure 3.28(a) is a typical mode-locked pulse spectrum; the center wavelength of the mode-locked pulse spectrum is 1563.7 nm, and the 3-dB bandwidth is 3.9 nm. Figure 3.28(b) shows the measured autocorrelation curve. The FWHM of the autocorrelation curve is 1.1 ps. If the hyperbolic secant function is used, the pulse width is 710 fs. The time-bandwidth product is 0.34. Figure 3.28(c) shows the harmonic mode-locked pulse sequence. The time interval between the two pulses is 2.17 ns, and this corresponds to a repetition frequency of 460.7 MHz.Figure 3.28(d) shows the radio frequency spectrum. The repetition frequency of the output pulse is 460.7 MHz, and the SNR is as high as 66 dB, which shows that the harmonic ML state is very stable.

In this chapter, WS_2/FM and DF-WS_2 SA devices were fabricated. A WS_2/FM SA device formed by depositing a WS_2 layered film on a 20 μm single-layer fluoromica sheet using a cracking method.

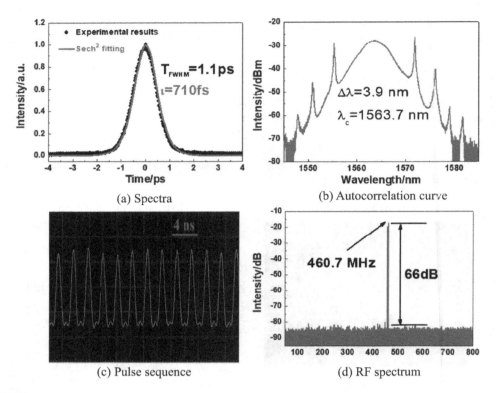

FIGURE 3.28 Harmonic ML experiment results [104].

The advantages of WS$_2$/FM SA devices are that they have good heat dissipation and a high damage threshold. The laser damage threshold of the WS$_2$/FM SA device is 406 MW/cm^2, which is two times higher than that of the WS$_2$/PVA SA. The device was tested nonlinearly in the 1μm band. The measured MD were 5.8%, and the nonsaturable loss was 14.8%; these values are better than those of other types of absorbers. When WS$_2$/FM was inserted into the ytterbium-doped fiber laser cavity continuous mode-locked operation was achieved at 1052 nm with a maximum output power of 30 mW, which is the highest output power of transition metal sulfide fiber lasers. When the WS$_2$/FM absorber device was measured in the 1.55 μm band, the MD was 3.1%, and the nonsaturable loss was 15%; when the device was inserted into the erbium-doped fiber laser cavity, continuous ML was achieved at the center wavelength of 1558 nm. When in operation, the pulse width was 830 fs. These experimental results confirm the broadband SA characteristics of WS$_2$ materials and also prove that WS$_2$ materials can support the output of femtosecond pulses. The DF-WS$_2$ SA device is made of a D-type optical fiber combined with a mixture of alcohol and water, and the damage threshold was greatly improved. When the device was tested nonlinearly in the 1.55 μm band, the MD was 11%, the nonsaturable loss was 18%, and the saturation flux density was 5 MW/cm^2. When the device was used in erbium-doped mode-locked fiber lasers, the fundamental frequency mode-locked operation was achieved at the pump power, and the pulse width is 660 fs. High-order harmonic mode-locked operation of 460.7 MHz was achieved at high pump power with a pulse width of 710 fs and a SNR of up to 66 dB; these values are superior to the harmonic mode-locked experimental results of other 2D materials. These results show that WS$_2$/FM and DF-WS$_2$ SA devices are very practical absorber devices and have wide application prospects in the field of ultrafast lasers.

3.4.4 ERBIUM-DOPED MODE-LOCKED FIBER LASER BASED ON DF-BP SSA

Recently, the rediscovered BP material has joined the large family of 2D absorber materials. The rediscovery of BP is another important area of research progress similar to graphene and transition metal sulfides, adding a new member to the 2D material family. Similar to transition metal sulfides (such as MoS_2 and WS_2) BP is a high-mobility semiconductor material that has properties similar to those of other 2D materials, such as broadband saturable absorption and ultrafast carrier dynamics. The semiconductor energy band structure of BP is a direct energy band structure, regardless of whether it is a single layer, multilayer, or a bulk structure. This feature means that its optical and optoelectronic performances give it a huge advantage over other materials (including silicon and MoS_2). The direct energy gap structure of BP enhances the direct coupling of BP and light, which means BP can be used as a promising candidate material in the field of optoelectronics in the future. Thus far, the saturable absorption characteristics of black phosphor materials in the 0.6–3.0 μm band have been fully verified in passive Q-switched lasers and mode-locked lasers [105–108].

We proposed a new type of BP SSA device and used it in erbium-doped mode-locked fiber lasers to achieve pulse output. This absorber device was prepared by immersing a polished D-type optical fiber in a solution of N-methylpyrrolidone (NMP). This method effectively utilizes the interaction between the evanescent field and the BP material and also avoids direct contact between the BP absorber material and air to prevent oxidation. The nonlinear optical characteristics of the DF-BP SSA device were tested. The modulation depth and nonsaturable loss were 10% and 7.75%, respectively. When the DF-BP SSA device was inserted into an erbium-doped fiber laser, stable mode-locked operation was achieved. The pulse width was 580 fs, and the measured SNR was as high as 65 dB. These results prove that DF-BP SSA is a promising device in the field of ultrafast fiber lasers.

Figure 3.29 shows a diagram of an erbium-doped mode-locked fiber laser device based on DF-BP SSA. In the experiment, a ring cavity structure was used, and the optical path included the following components: WDM, EDF, PI-ISO, PC, and DF-BP SSA, OC. A 976 nm semiconductor laser outside the cavity was used as a pump source with a maximum output power of 650 mW. The absorption coefficient of EDF at 976 nm was 3 dB/m, and the length used was 4 m; the other pigtails and compensated fibers were SMF; PI-ISO was used to ensure that the laser was unidirectional in the cavity. Different polarization states of the fiber laser were achieved when the PC was adjusted; the output coupling ratio of OC was 10:90, that is, 90% of the ports were connected in the cavity, and 10% of the ports were used as pulse output ends. The total length of the laser ring cavity was 13.6 m. The autocorrelation instrument used in the test was an autocorrelation instrument made by Alnair Corporation of Japan, and the model was Alnair HAC 200.

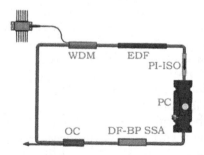

FIGURE 3.29 Erbium-doped mode-locked fiber laser device based on DF-BP SSA [109].

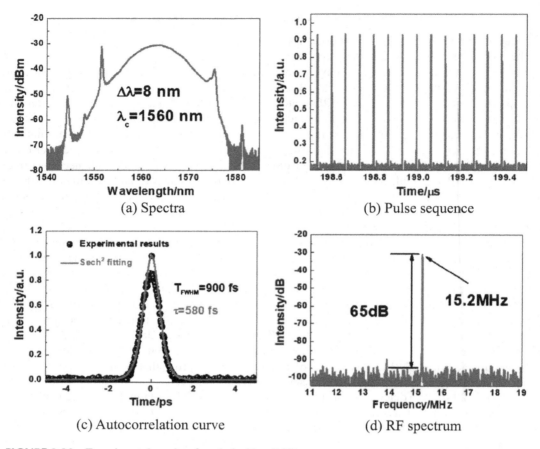

FIGURE 3.30 Experimental results of mode-locking [109].

In the experiment, when the pump power reached 3mW, the erbium-doped fiber laser began to output continuous laser light. When the pump power reached 13.2 mW, the laser achieved continuous mode-locked operation. However, when the pump power was reduced to 9.6 mW, the mode-locked state was still maintained. Figure 3.30(a) shows a typical mode-locked pulse spectrum; as seen in the figure, the center wavelength of the mode-locked pulse spectrum is 1560 nm, and the 3-dB bandwidth is 8 nm. Two pairs of Kelly sidebands are distributed on both sides of the spectrum, and this proves that the mode-locked fiber laser works in the traditional soliton mode-locked interval. Figure 3.30(b) shows a mode-locked pulse sequence. The intensity of the pulse sequence remains the same, and the time interval between the two pulses is 65.8 ns. Figure 3.30(c) shows the measured autocorrelation curve. The FWHM of the autocorrelation curve is 900 fs. If the hyperbolic secant function is used, the pulse width is 580 fs. As reported in the literature [110], the shortest pulse width of the mode-locked fiber laser based on BP material is 272 fs. However, the method they used to make the saturable absorption of BP was mechanical stripping, and then the stripped layered BP was transferred to the end face of the optical fiber. However, the reliability and repeatability of this fabrication method is not high, and this often leads to the appearance of Q-switching operation [78]. As reported in the literature [111], a 280 fs fully negative dispersion erbium-doped fiber laser based on BP SA was fabricated. As previously discussed, the manufacturing method does not avoid oxidation of the BP absorber material. Thus, from a comprehensive comparison point of view, the DF-BP SSA device that we made is a good choice. At the same time, the current method of making the BP absorber is mainly limited to the mechanical peeling method, and thus, it is necessary to

Novel Mode-Locked Fiber Lasers 77

explore a new and feasible manufacturing method. Figure 3.30(d) shows the radio frequency spectrum. The repetition frequency of the output pulse is 15.2 MHz, and the SNR is as high as 65 dB, indicating that the mode-locked state is very stable.

In recent years, BP material has attracted attention as the newest SA material. However, because it has the characteristics of easy oxidation and low damage threshold, further development and application of BP is limited. The DF-BP SSAdevice was made of a D-type optical fiber combined with BP NMP solution. The laser damage threshold was greatly improved, and this method effectively avoided the oxidation of the BP material. The nonlinearity of device was tested in the 1.55 µm, the MD was 7.75%, and the nonsaturable loss was only 10%. 10% is the lowest nonsaturable loss in the results of the currently reported BP materials as SA devices. When the device was applied in erbium-doped fiber lasers, we achieved stable mode-locked operation. The center wavelength of the mode-locked pulse spectrum was 1560 nm, the 3-dB bandwidth was 8 nm, the pulse width was 580 fs, and the measured SNR was as high as 65 dB. These results show that the DF-BP SSA device is a very practical absorber device and has wide application prospects in the field of ultrafast lasers.

3.5 CONCLUSIONS

This chapter focuses on various types of new broadband SAs, and studies the preparation methods that can be used to improve the damage threshold, nonsaturable loss reduction, and mode-locked laser pulse characteristics with new broadband absorber devices in all-fiber mode-locked lasers.

1. The method of making a D-type optical fiber combined with a single-walled CNT in heavy aqueous solution to make CNT SA devices is discussed. The device effectively reduces the nonsaturable loss of single-walled CNT absorber devices, and the obtained 3% nonsaturable loss is much lower than that of similar SA mode-locked devices.

2. PLD technology was used to fabricate a Bi_2Te_3 tapered fiber SA device. The optical characteristics of the Bi_2Te_3/PVA film and Bi_2Te_3 tapered fiber SA were compared. The Bi_2Te_3/PVA film type and Bi_2Te_3 tapered fiber were measured. The two SA devices have MDs of 8.4% and 10%, respectively. It was found that an evanescent field increases the interaction distance between light and the SA material, thereby effectively increasing the modulation depth. For all-positive dispersion mode-locked fiber lasers based on the Bi_2Te_3/PVA film type and Bi_2Te_3 tapered fiber type SA, the pulse widths are 417ps and 317 ps, respectively. Experiments verify that the higher modulation depth of the absorber device reduces the output pulse width.

3. For the first time, a WS_2/FM SA device using single-layer fluoromica as the substrate instead of the traditional polymer was proposed and fabricated. The damage threshold of the device at 406 MW/cm² was two times greater than that of the WS_2/PVA device. WS_2/FM was inserted into the ytterbium-doped fiber laser cavity, and continuous mode-locked operation was achieved at 1052 nm. The average output power of the laser pulse was 30 mW, and this was the highest output power of the transition metal sulfide fiber laser. The WS_2/FM SA device overcomes the shortcomings of the previous absorber device, including the low damage threshold, large insertion loss, and low mechanical performance. It also provides an effective design for the development of high-performance absorber devices.

4. When a D-type optical fiber was combined with BP NMP solution, a BP solution-type SA device based on a D-type optical fiber was fabricated. This method effectively avoids oxidation of the BP material, and the damage threshold was further improved by the effect of the laser evanescent field and the absorber material. The absorber device was used in an erbium-doped mode-locked fiber laser and achieved 580 fs ML laser pulse output. From a comprehensive comparison point of view, the BP SSA devices we produced have greater advantages in similar absorber devices and have broad application prospects in the field of ultrafast lasers.

The above data are summarized in the Table 3.1

TABLE 3.1
The optimized modulation depth, pulse widths, and advantages of SA materials

Name	Modulation depth	Pulse width	Advantage
DF-SWCNT-D_2O SA	5.7%	194 ps	1. Effectively isolate the single-walled carbon nanotube material from direct contact with air to avoid its oxidation; 2. Use the good heat dissipation characteristics of liquid to increase its thermal damage threshold; 3. Low linearity to light in the 1–1.55 μm band Absorption further reduces the unsaturated loss of the absorber device.
Bi_2Te_3/PVASA	8.4%	417 ps	It is flexible, convenient, and easy to manufacture, but its disadvantage is that the absorber has a low damage threshold.
Bi_2Te_3 tapered fiber SA	10%	317 ps	It is made by PLD method, which has the advantages of cleanness and no impurities; higher power range; higher damage threshold.
WS_2/FM	3.1%	830 fs	Excellent electrical insulation and high temperature resistance; low high-frequency dielectric loss, stable dielectric constant, non-aging, and non-fragile; high temperature resistance up to 1200 °C, and no reaction with strong acids and alkalis, pure and transparent; good flatness, no impurities adsorption; strong optical transparency.
DF-WS_2	11%	660 fs	With anti-oxidation, good heat dissipation, high damage threshold.
DF-BP SSA	7.75%	580 fs	The laser damage threshold has been greatly improved. This method can effectively avoid the oxidation of black phosphorous materials; stable mode-locking operation.

The table compared the modulation depths, the pulse widths and advantages of different SA materials in this chapter.

REFERENCES

[1] Maiman T H. Stimulated optical radiation in ruby. *Nature*, 1960, 187(4736): 493–494.

[2] Hargrove L E, Fork R L, Pollack M A. Locking of He-Ne laser modes induced by synchronous intracavity modulation. *Applied Physics Letters*, 1964, 5(1): 4–5.

[3] Hall J L. Optical frequency measurement: 40 years of technology revolutions. *IEEE Journal of Selected Topics in Quantum Electronics*, 2000, 6(6): 1136–1144.

[4] Osellame R, Hoekstra H, Cerullo G, et al. Femtosecond laser microstructuring: an enabling tool for optofluidic lab-on-chips. *Laser & Photonics Reviews*, 2011, 5(3): 442–463.

[5] Wysocki P F, Digonnet M J, Kim B Y. Broad-spectrum, wavelength-swept, erbium-doped fiber laser at 1.55 microm. *Optics Letters*, 1990, 15(16): 879–881.

[6] Zellmer H, Willamowski U, Tunnermann A, et al. High-power cw neodymium-doped fiber laser operating at 9.2 W with high beam quality. *Optics Letters*, 1995, 20(6): 578–580.

[7] Guy M J, Noske D U, Taylor J R. Generation of femtosecond soliton pulses by passive mode locking of an ytterbium-erbium figure-of-eight fiber laser. *Optics Letters*, 1993, 18(17): 1447–1449.

[8] Wu J, Yao Z, Zong J, et al. Highly efficient high-power thulium-doped germanate glass fiber laser. *Optics Letters*, 2007, 32(6): 638–640.

[9] Diddams S A, Hollberg L, Mbele V. Molecular fingerprinting with the resolved modes of a femtosecond laser frequency comb. *Nature*, 2007, 445(7128): 627–630.

[10] Machida S, Yamamoto Y, Itaya Y. Observation of amplitude squeezing in a constant-current – driven semiconductor laser. *Physical Review Letters*, 1987, 58(10): 1000–1003.

[11] Turitsyn S K, Babin S A, Churkin D V, et al. Random distributed feedback fibre lasers. *Physics Reports*, 2014, 542(2): 133–193.

[12] Haus H A. Mode-locking of lasers. *IEEE Journal of Selected Topics in Quantum Electronics*, 2000, 6(6): 1173–1185.

Novel Mode-Locked Fiber Lasers

[13] Klank H, Kutter J P, Geschke O. CO2-laser micromachining and back-end processing for rapid production of PMMA-based microfluidic systems. *Lab on a Chip*, 2002, 2(4): 242–246.

[14] Mogensen F, Olesen H, Jacobsen G. Locking conditions and stability properties for a semiconductor laser with external light injection. *IEEE Journal of Quantum Electronics*, 1985, 21(7): 784–793.

[15] Lang R. Injection locking properties of a semiconductor laser.*IEEE Journal of Quantum Electronics*, 1982, 18(6): 976–983.

[16] Shimony Y, Burshtein Z, Kalisky Y. Cr4+:YAG as passive Q-switch and Brewster plate in a pulsed Nd:YAG laser. *IEEE Journal of Quantum Electronics*, 1995, 31(10): 1738–1741.

[17] Hideur A, Chartier T, Brunel M, et al. Mode-lock, Q-switch and CW operation of an Yb-doped double-clad fiber ring laser. *Optics Communications*, 2001, 198(1): 141–146.

[18] Grelu P, Akhmediev N. Dissipative solitons for mode-locked lasers. *Nature Photonics*, 2012, 6(2): 84–92.

[19] Liu X, Cui Y, Han D, et al. Distributed ultrafast fibre laser. *Scientific Reports*, 2015, 5(1): 9101.

[20] Yan P, Liu A, Chen Y, et al. Passively mode-locked fiber laser by a cell-type WS2 nanosheets saturable absorber. *Scientific Reports*, 2015, 5(1): 12587.

[21] Sun Z, Hasan T, Wang F, et al. Ultrafast stretched-pulse fiber laser mode-locked by carbon nanotubes. *Nano Research*, 2010, 3(6): 404–411.

[22] Qi Y, Danhua L, Jie L, et al. Efficient diode-pumped $Yb:LuY_2SiO_5$ laser mode locked by single-walled carbon nanotube absorber. *Optical Engineering*, 2011, 50(11): 1–5.

[23] Li X, Wang Y, Wang Y, et al. Nonlinear absorption of SWNT film and its effects to the operation state of pulsed fiber laser. *Optics Express*, 2014, 22(14): 17227–17235.

[24] Li X, Wang Y, Wang Y, et al. Yb-doped passively mode-locked fiber laser based on a single wall carbon nanotubes wallpaper absorber. *Optics and Laser Technology*, 2013, 47(144–147.

[25] Song Y-W, Jang S Y, Han W-S, et al. Graphene mode-lockers for fiber lasers functioned with evanescent field interaction. *Applied Physics Letters*, 2010, 96051122.

[26] Li H, Wang Y, Yan P, et al. Passively harmonic mode locking in ytterbium-doped fiber laser with graphene oxide saturable absorber. *Optical Engineering*, 2013, 52(12): 126102.

[27] Zhao C, Zhang H, Qi X, et al. Ultra-short pulse generation by a topological insulator based saturable absorber. *Applied Physics Letters*, 2012, 101(21): 211106.

[28] Li L, Wang Y, Sun H, et al. All-normal dispersion passively mode-locked Yb-doped fiber laser with Bi2Te3 absorber. *Optical Engineering*, 2015, 54(4): 046101.

[29] Liu H, Luo A-P, Wang F- Z, et al. Femtosecond pulse erbium-doped fiber laser by a few-layer MoS2 saturable absorber. *Optics Letters*, 2014, 39(15): 4591–4594.

[30] Mao D, Wang Y, Ma C, et al. WS2 mode-locked ultrafast fiber laser. *Scientific Reports*, 2015, 5(1): 7965.

[31] Renninger W H, Chong A, Wise F W. Dissipative solitons in normal-dispersion fiber lasers. *Physical Review A*, 2008, 77(2): 023814.

[32] Chong A, Buckley J, Renninger W, et al. All-normal-dispersion femtosecond fiber laser. *Optics Express*, 2006, 14(21): 10095–10100.

[33] Zhao L M, Tang D Y, Zhang H, et al. Dissipative soliton operation of an ytterbium-doped fiber laser mode locked with atomic multilayer graphene. *Optics Letters*, 2010, 35(21): 3622–3624.

[34] Im J H, Choi S Y, Rotermund F, et al. All-fiber Er-doped dissipative soliton laser based on evanescent field interaction with carbon nanotube saturable absorber. *Optics Express*, 2010, 18(21): 22141–22146.

[35] Zhang H, Tang D Y, Knize R J, et al. Graphene mode locked, wavelength-tunable, dissipative soliton fiber laser. *Applied Physics Letters*, 2010, 96(11): 111112.

[36] Boguslawski J, Sobon G, Zybala R, et al. Dissipative soliton generation in Er-doped fiber laser mode-locked by Sb2Te3 topological insulator. *Optics Letters*, 2015, 40(12): 2786–2789.

[37] Ortac B, Plotner M, Limpert J, et al. Self-starting passively mode-locked chirped-pulse fiber laser. *Optics Express*, 2007, 15(25): 16794–16799.

[38] Fermann M E, Hartl I. Ultrafast fiber laser technology. *IEEE Journal of Selected Topics in Quantum Electronics*, 2009, 15(1): 191–206.

[39] Reid D T. Ultrafast lasers – technology and applications. *Journal of Microscopy*, 2003, 211(1): 101.

[40] Sotocrespo J M, Akhmediev N, Grelu P, et al. Quantized separations of phase-locked soliton pairs in fiber lasers. *Optics Letters*, 2003, 28(19): 1757–1759.

[41] Haus H A. Applications of nonlinear fiber optics. *Physics Today*, 2002, 55(6): 58–59.

[42] Ortac B, Hideur A, Chartier T, et al. 90-fs stretched-pulse ytterbium-doped double-clad fiber laser. *Optics Letters*, 2003, 28(15): 1305–1307.

[43] Haxsen F, Ruehl A, Engelbrecht M, et al. Stretched-pulse operation of a thulium-doped fiber laser. *Optics Express*, 2008, 16(25): 20471–20476.

[44] Hideur A, Wang H, Tang M, et al. "High-energy Ultrafast Fiber Lasers." in Advanced Photonics, OSA Technical Digest (online) (Optical Society of America, 2014.

[45] Tamura K, Ippen E P, Haus H A, et al. 77-fs pulse generation from a stretched-pulse mode-locked all-fiber ring laser. *Optics Letters*, 1993, 18(13): 1080–1082.

[46] Fermann M E, Kruglov V I, Thomsen B C, et al. Self-similar propagation and amplification of parabolic pulses in optical fibers. *Physical Review Letters*, 2000, 84(26): 6010–6013.

[47] Kruglov V I, Peacock A C, Harvey J D. Exact Self-similar solutions of the generalized nonlinear schrodinger equation with distributed coefficients. *Physical Review Letters*, 2003, 90(11): 113902.

[48] Roser F, Eidam T, Rothhardt J, et al. Millijoule pulse energy high repetition rate femtosecond fiber chirped-pulse amplification system. *Optics Letters*, 2007, 32(24): 3495–3497.

[49] Ilday F O, Buckley J R, Clark W G, et al. Self-similar evolution of parabolic pulses in a laser. *Physical Review Letters*, 2004, 92(21): 213902.

[50] Liu X. Dissipative soliton evolution in ultra-large normal-cavity-dispersion fiber lasers. *Optics Express*, 2009, 17(12): 9549–9557.

[51] Dou Z, Song Y, Tian J, et al. Mode-locked ytterbium-doped fiber laser based on topological insulator: Bi2Se3. *Optics Express*, 2014, 22(20): 24055–24061.

[52] Keller U, Knox W H, Roskos H. Coupled-cavity resonant passive mode-locked Ti:sapphire laser. *Optics Letters*, 1990, 15(23): 1377–1379.

[53] Haiml M, Grange R, Keller U. Optical characterization of semiconductor saturable absorbers. *Applied Physics B*, 2004, 79(3): 331–339.

[54] Okhotnikov O G, Grudinin A B, Pessa M. Ultra-fast fibre laser systems based on SESAM technology: new horizons and applications. *New Journal of Physics*, 2004, 6(1): 177.

[55] Gomes L A, Orsila L, Jouhti T, et al. Picosecond SESAM-based ytterbium mode-locked fiber lasers. *IEEE Journal of Selected Topics in Quantum Electronics*, 2004, 10(1): 129–136.

[56] Saraceno C J, Schriber C, Mangold M, et al. SESAMs for high-power oscillators: design guidelines and damage thresholds. *IEEE Journal of Selected Topics in Quantum Electronics*, 2012, 18(1): 29–41.

[57] Guerreiro P T, Ten S, Borrelli N F, et al. PbS quantum-dot doped glasses as saturable absorbers for mode locking of a Cr: forsterite laser. *Applied Physics Letters*, 1997, 71(12): 1595–1597.

[58] Set S Y, Yaguchi H, Tanaka Y, et al. *Laser mode locking using a saturable absorber incorporating carbon nanotubes*; proceedings of the optical fiber communication conference, F, 2004 [C].

[59] Bao Q, Zhang H, Wang Y, et al. Atomic-layer graphene as a saturable absorber for ultrafast pulsed lasers. *Advanced Functional Materials*, 2009, 19(19): 3077–3083.

[60] Sun Z, Hasan T, Torrisi F, et al. Graphene mode-locked ultrafast laser. *ACS Nano*, 2010, 4(2): 803–810.

[61] Zhang H, Tang D Y, Zhao L M, et al. Large energy mode locking of an erbium-doped fiber laser with atomic layer graphene. *Optics Express*, 2009, 17(20): 17630–17635.

[62] Cizmeciyan M N, Kim J W, Bae S, et al. Graphene mode-locked femtosecond Cr:ZnSe laser at 2500 nm. *Optics Letters*, 2013, 38(3): 341–343.

[63] Wang Q, Chen T, Zhang B, et al. All-fiber passively mode-locked thulium-doped fiber ring laser using optically deposited graphene saturable absorbers. *Applied Physics Letters*, 2013, 102(13): 131117.

[64] Tolstik N, Sorokin E, Sorokina I T. Graphene mode-locked Cr:ZnS laser with 41 fs pulse duration. *Optics Express*, 2014, 22(5): 5564–5571.

[65] Cho W B, Kim J W, Lee H W, et al. High-quality, large-area monolayer graphene for efficient bulk laser mode-locking near 1.25 μm. *Optics Letters*, 2011, 36(20): 4089–4091.

[66] Zhu G, Zhu X, Balakrishnan K, et al. Fe2+:ZnSe and graphene Q-switched singly Ho3+-doped ZBLAN fiber lasers at 3 μm. *Optical Materials Express*, 2013, 3(9): 1365–1377.

[67] Sotor J, Sobon G, Macherzynski W, et al. Mode-locking in Er-doped fiber laser based on mechanically exfoliated Sb2Te3 saturable absorber. *Optical Materials Express*, 2014, 4(1): 1–6.

[68] Wang K, Wang J, Fan J, et al. Ultrafast saturable absorption of two-dimensional MoS2 nanosheets. *ACS Nano*, 2013, 7(10): 9260–9267.

[69] Li L, Yu Y, Ye G J, et al. Black phosphorus field-effect transistors. *Nature Nanotechnology*, 2014, 9(5): 372–377.

[70] Xia F, Wang H, Xiao D, et al. Two-dimensional material nanophotonics. *Nature Photonics*, 2014, 8(12): 899–907.

[71] Chua C K, Pumera M. Chemical reduction of graphene oxide: a synthetic chemistry viewpoint. *Chemical Society Reviews*, 2014, 43(1): 291–312.

[72] Chen S, Zhao C, Li Y, et al. Broadband optical and microwave nonlinear response in topological insulator. *Optical Materials Express*, 2014, 4(4): 587–596.

[73] Rodin A S, Carvalho A, Neto A H C. Strain-induced gap modification in black phosphorus. *Physical Review Letters*, 2014, 112(17): 176801.

[74] Yamashita S, Martinez A, Xu B. Short pulse fiber lasers mode-locked by carbon nanotubes and graphene. *Optical Fiber Technology*, 2014, 20(6): 702–713.

[75] Nicholson J W, Windeler R S, Digiovanni D J. Optically driven deposition of single-walled carbon-nanotube saturable absorbers on optical fiber end-faces. *Optics Express*, 2007, 15(15): 9176–9183.

[76] Hasan T, Sun Z, Wang F, et al.Nanotube–polymer composites for ultrafast photonics. *Advanced Materials*, 2009, 21(3874–3899.

[77] Martinez A, Uchida S, Song Y, et al. Fabrication of carbon nanotube-poly-methyl-methacrylate composites for nonlinear photonic devices. *Optics Express*, 2008, 16(15): 11337–11343.

[78] Chen Y, Jiang G, Chen S, et al. Mechanically exfoliated black phosphorus as a new saturable absorber for both Q-switching and mode-locking laser operation. *Optics Express*, 2015, 23(10): 12823–12833.

[79] Yamashita S, Inoue Y, Maruyama S, et al. Saturable absorbers incorporating carbon nanotubes directly synthesized onto substrates and fibers and their application to mode-locked fiber lasers. *Optics Letters*, 2004, 29(14): 1581–1583.

[80] Martinez A, Sun Z. Nanotube and graphene saturable absorbers for fibre lasers. *Nature Photonics*, 2013, 7(11): 842–845.

[81] Set S Y, Yaguchi H, Tanaka Y, et al. Ultrafast fiber pulsed lasers incorporating carbon nanotubes. *IEEE Journal of Selected Topics in Quantum Electronics*, 2004, 10(1): 137–146.

[82] Luo Z, Wu D, Xu B, et al. Two-dimensional material-based saturable absorbers: towards compact visible-wavelength all-fiber pulsed lasers. *Nanoscale*, 2016, 8(2): 1066–1072.

[83] Martinez A, Fuse K, Yamashita S. Enhanced stability of nitrogen-sealed carbon nanotube saturable absorbers under high-intensity irradiation. *Optics Express*, 2013, 21(4): 4665–4670.

[84] Song Y, Yamashita S, Goh C S, et al. Carbon nanotube mode lockers with enhanced nonlinearity via evanescent field interaction in D-shaped fibers. *Optics Letters*, 2007, 32(2): 148–150.

[85] Luo Z, Wang J, Zhou M, et al. Multiwavelength mode-locked erbium-doped fiber laser based on the interaction of graphene and fiber-taper evanescent field. *Laser Physics Letters*, 2012, 9(3): 229–233.

[86] Li L, Wang Y, Sun H, et al. Single-walled carbon nanotube solution-based saturable absorbers for mode-locked fiber laser.*Optical Engineering*, 2015, 54(8): 086103.

[87] Liu H, Yang Y, Chow K K. Enhancement of thermal damage threshold of carbon-nanotube-based saturable absorber by evanescent-field interaction on fiber end. *Optics Express*, 2013, 21(18975–18982.

[88] Yan P, Lin R, Ruan S, et al. A practical topological insulator saturable absorber for mode-locked fiber laser. *Scientific Reports*, 2015, 5(1): 8690.

[89] Luo Z, Liu M, Liu H, et al. 2 GHz passively harmonic mode-locked fiber laser by a microfiber-based topological insulator saturable absorber. *Optics Letters*, 2013, 38(24): 5212–5215.

[90] Lee J, Koo J, Jhon Y M, et al. A femtosecond pulse erbium fiber laser incorporating a saturable absorber based on bulk-structured Bi2Te3 topological insulator. *Optics Express*, 2014, 22(5): 6165–6173.

[91] Yu Z, Song Y, Tian J, et al. High-repetition-rate Q-switched fiber laser with high quality topological insulator Bi2Se3 film. *Optics Express*, 2014, 22(10): 11508–11515.

[92] Chong A, Renninger W, Wise F. Properties of normal-dispersion femtosecond fiber lasers. *Journal of the Optical Society of America B*, 2008, 25(140–148.

[93] Zhang S X, Yan L, Qi J, et al. Epitaxial thin films of topological insulator Bi2Te3 with two-dimensional weak anti-localization effect grown by pulsed laser deposition. *Thin Solid Films*, 2012, 520(21): 6459–6462.

[94] Yan P, Lin R, Ruan S, et al. A 2.95 GHz, femtosecond passive harmonic mode-locked fiber laser based on evanescent field interaction with topological insulator film. *Optics Express*, 2015, 23(1): 154–164.

[95] Sotor J, Sobon G, Abramski K M. Sub-130 fs mode-locked Er-doped fiber laser based on topological insulator. *Optics Express*, 2014, 22(11): 13244–13249.

[96] Jung M, Lee J, Koo J, et al. A femtosecond pulse fiber laser at 1935 nm using a bulk-structured Bi2Te3 topological insulator. *Optics Express*, 2014, 22(7): 7865–7874.

[97] Lu L, Pei-Guang Y, Yong-Gang W, et al. Yb-doped passively mode-locked fiber laser with Bi2Te3-deposited*. Chinese. *Physics B*, 2015, 24: 124204.

[98] Du J, Wang Q, Jiang G, et al. Ytterbium-doped fiber laser passively mode locked by few-layer Molybdenum Disulfide (MoS2) saturable absorber functioned with evanescent field interaction. *Scientific Reports*, 2015, 4(1): 6346.

[99] Tyszkazawadzka A, Janaszek B, Szczepanski P. Tunable slow light in graphene-based hyperbolic metamaterial waveguide operating in SCLU telecom bands. *Optics Express*, 2017, 25(7): 7263–7272.

[100] Li L, Jiang S, Wang Y, et al. WS2/fluorine mica (FM) saturable absorbers for all-normal-dispersion mode-locked fiber laser. *Optics Express*, 2015, 23(22): 28698–28706.

[101] Lee C C, Miller J, Schibli T R. Doping-induced changes in the saturable absorption of monolayer graphene. *Applied Physics B*, 2012, 108(1): 129–135.

[102] Li L, Wang Y, Wang X, et al. Er-doped mode-locked fiber laser with WS2/fluorine mica (FM) saturable absorber. *Optics & Laser Technology*, 2017, 90(Complete): 109–112.

[103] Zhao C, Zou Y, Chen Y, et al. Wavelength-tunable picosecond soliton fiber laser with Topological Insulator: Bi_2Se_3 as a mode locker. *Optics Express*, 2012, 20(25): 27888.

[104] Li L, Su Y, Wang Y, et al. Femtosecond passively er-doped mode-locked fiber laser with WS2 solution saturable absorber. *IEEE Journal of Selected Topics in Quantum Electronics*, 2016, 23(1): 1–6.

[105] Hisyam M B, Rusdi M F M, Latiff A A, et al. Generation of mode-locked ytterbium doped fiber ring laser using few-layer black phosphorus as a saturable absorber. *IEEE Journal of Selected Topics in Quantum Electronics*, 2017, 23(1): 39–43.

[106] Wang Z, Zhao R, He J, et al. Multi-layered black phosphorus as saturable absorber for pulsed Cr:ZnSe laser at 2.4 μm. *Optics Express*, 2016, 24(2): 1598–1603.

[107] Qin Z, Xie G, Zhang H, et al. Black phosphorus as saturable absorber for the Q-switched Er:ZBLAN fiber laser at 2.8 μm. *Optics Express*, 2015, 23(19): 24713–24718.

[108] Zhang R, Yu H, Zhang H, et al. Broadband black phosphorus optical modulator in visible to mid-infrared spectral range. *arXiv: Materials Science*, 2015, 3(12): 1787–1792.

[109] Li L, Wang Y, Wang X. Ultrafast pulse generation with black phosphorus solution saturable absorber. *Laser Physics*, 2017, 27(8): 085104.

[110] Sotor J, Sobon G, Macherzynski W, et al. Black phosphorus saturable absorber for ultrashort pulse generation. *Applied Physics Letters*, 2015, 107(5): 051108.

[111] Chen Y, Chen S, Liu J, et al. Sub-300 femtosecond soliton tunable fiber laser with all-anomalous dispersion passively mode locked by black phosphorus. *Optics Express*, 2016, 24(12): 13316–13324.

4 Thin Coating Technologies and Applications in High-Temperature Solid Oxide Fuel Cells

San Ping Jiang
Curtin University, Perth, Australia

Shuai He
University of St Andrews, St Andrews, Fife, UK

CONTENTS

4.1 Introduction 84
4.2 Chemical Deposition Methods 85
    4.2.1 Chemical Vapor Deposition and Atomic Layer Deposition 86
    4.2.2 Electrochemical Vapor Deposition (EVD) 87
4.3 Physical Deposition Techniques 88
    4.3.1 Magnetron Sputtering Techniques 88
    4.3.2 Pulse Laser Deposition (PLD) 91
    4.3.3 Plasma Spray Deposition 94
4.4 Colloidal and Ceramic Powder Techniques 96
    4.4.1 Colloidal Techniques 97
        4.4.1.1 Slurry Coating, Spin Coating, and Dip-Coating Methods 97
        4.4.1.2 Sol-Gel Method 100
        4.4.1.3 Screen-Printing Method 101
        4.4.1.4 Electrophoretic Deposition (EPD) 102
    4.4.2 Ceramic Powder Techniques 105
        4.4.2.1 Tape-Casting and Freeze-Tape-Casting Process 105
        4.4.2.2 Tape-Calendering Process 107
        4.4.2.3 Dry-Pressing Method 108
4.5 Flame-Assisted Colloidal Process 109
    4.5.1 Spray Pyrolysis and Flame-Assisted Vapor Deposition 109
    4.5.2 Electrostatic Spray Pyrolysis Deposition 111
4.6 Lithography and Etching Techniques for μ-SOFCs 112
4.7 Concluding Remarks 115
Acknowledgment 116
References 116

4.1 INTRODUCTION

Solid oxide fuel cell (SOFC) is an electrochemical device to convert the chemical energy of fuels, such as hydrogen and hydrocarbons, to electricity, with potential applications in transportation, distributed generation, remote power, defense, and many others (Minh 1993, Singhal 2002, Williams et al. 2004, Singhal 2014). They offer extremely high chemical-to-electrical conversion efficiencies because the efficiency is not limited by the Carnot cycle of heat engines. Further energy efficiency can be achieved when the produced heat is used in combined heat and power, or gas turbine applications. Furthermore, the greenhouse gas emission from SOFC is much lower than that from conventional power generation technologies. Due to its high operating temperature, SOFC has a high tolerance to typical catalyst poisons, produces high-quality heat for reforming of hydrocarbons and offers the possibility of direct utilization of hydrocarbon fuels.

An individual SOFC cell is composed of a porous anode or fuel electrode, a fully dense solid electrolyte, and a porous cathode or air electrode (see Figure 4.1). Driven by the differences in oxygen chemical potential between fuel and air compartments of the cell, oxygen ions migrate through the electrolyte to the anode where they are consumed by oxidation of fuels such as hydrogen, methane, and hydrocarbons (C_nH_{2n+2}). Thus, the electrolyte must be dense in order to separate the air and fuel, must possess high oxygen ionic conductivity but negligible electronic conductivity, and must be chemically and structurally stable over a wide range of partial pressure of oxygen and temperatures (Yokokawa et al. 2005). On the other hand, the cathode and anode must be porous, chemically and thermally compatible with the electrolyte and interconnect, and be electrocatalytic active for the oxygen reduction and fuel oxidation reaction, respectively. In the case of hydrocarbon fuels, the anode must also possess certain tolerance toward sulfur and carbon deposition under SOFC operating conditions. Yttria-stabilized zirconia (YSZ) is the commonly used solid electrolyte, while lanthanum strontium manganite (LSM) and nickel-YSZ cermets often serve as the cathode and anode, respectively. Issues facing the development of electrode materials of SOFC have been reviewed (Jiang 2003, Gorte et al. 2004, Jiang and Chan 2004, Shri Prakash et al. 2014, Mahato et al. 2015).

Traditional SOFCs operate at high temperatures of 900–1000°C because of the low oxygen ion conductivity and high activation energy of oxide electrolytes such as YSZ. However, lowering of the operating temperature of SOFCs brings both dramatic technical and economic benefits. The cost of an SOFC system can be substantially reduced by using less costly metal alloys as interconnect and compliant temperature gaskets (Fergus 2005). Furthermore, as the operation temperature is reduced, thermodynamic efficiency, system reliability, and durability of cell performance increase. This increases the possibility of using SOFCs for a wide variety of applications, including residential,

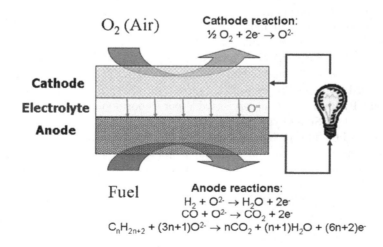

FIGURE 4.1 Schematic diagram of a solid oxide fuel cell.

Thin Coating Technologies

automotive, and portable applications. On the other hand, reduction in operation temperature results in a significant increase in the electrolyte and electrode resistivity and the polarization losses. To compensate for the performance losses associated with a lower operating temperature, the thickness of electrolyte layer has to be reduced in order to lower the ohmic resistance of the cell. Using a thin electrolyte layer, the electrolyte can no longer mechanically support the cell. Thus, anode- or cathode-supported cell structures need to be employed. Anode-supported structure based on Ni/YSZ or Ni/GDC (Gd-doped ceria) cermets is the most popular one for the deposition of a thin electrolyte film on a thick, mechanically strong and porous anode substrate.

The state-of-the-art anode-supported SOFC is based on porous Ni/YSZ cermets as a support. To reduce the electrolyte ohmic resistance and to enhance the cell efficiency, the electrolyte layer deposited should be as thin as possible. As a general role, the film thickness is inversely proportional to the pore size and/or propagated roughness of the surface, which means that the larger the pore size, the more difficult it is to get a thinner electrolyte. Thus, an anode function layer (AFL) is also required to have proper pore structure with less surface roughness. The electrolyte thin layer on the Ni/YSZ anode porous substrate/support should have (i) high ionic conductivity and negligible electrical conductivity, (ii) high density, uniformity, and pinhole-free to prevent the fuel or oxidant crossover, (iii) enough mechanical strength to resist high gas pressure gradient, (iv) good adhesion and contact to the anode as well as cathode, and (v) chemical stability with cell components with no formation of resistive phases at the electrode/electrolyte interface.

Chromia-forming ferric stainless steel has been considered to be the primary candidate as the interconnect materials of SOFCs for the intermediate temperature SOFCs, IT-SOFC, due to the economic and easy processing benefits. However, chromia-forming alloy has a serious oxidation problem because of the rapid growth oxide scale at the SOFC operation temperature. Without effective coating, the vaporization of chromium species poisons the cathode of SOFCs and seriously degrades the cell performance (Jiang et al. 2005, Jiang and Chen 2014). To reduce the vaporization of Cr species, a thin and dense coating is commonly used to deposit on the metallic interconnect. In this case, the thin coating technique should not only deliver an electrolyte and protective film with high density, pore-free or at least no cross-pores but also with the targeted composition and stoichiometry. Various deposition techniques have been applied to fabricate thin and dense electrolyte films and protective coatings for SOFCs (Will et al. 2000).

Microfuel cells are a potential replacement as high efficiency and high specific energy batteries in portable power generation. To date, miniaturized fuel cells utilizing proton exchange membranes and liquid methanol fuels (i.e., direct methanol fuel cells [DMFCs]) have been the primary focus of interests. However, high loading of precious metal catalysts such as Pt and PtRu is required for DMFCs to obtain the beneficial energy output due to the CO poisoning, and under operating conditions methanol crossover is still a serious problem (Wasmus and Kuver 1999, Jiang et al. 2006). Due to the persistent challenges with polymer electrolyte-based fuel cells, there is a growing interest in the development of micro-SOFCs (μ-SOFCs) for portable power sources. With these systems, hydrocarbon fuels, in addition to hydrogen, can be used directly and this reduces the need for the pre-forming of fuels (Shao et al. 2005). To develop μ-SOFCs, the cell configuration design and the deposition techniques for the thin film electrolyte have been the areas of challenge (Beckel et al. 2007). This chapter will focus on the most commonly used thin film techniques for the fabrication and development of thin film components such as electrolyte, electrode, and protective coatings in SOFCs. The techniques are classified into chemical and physical methods based on the nature of the process. Examples are given for the preparation and characterization of electrolyte films as well as electrode and coating for metallic interconnect of SOFCs.

4.2 CHEMICAL DEPOSITION METHODS

Chemical deposition methods can be further divided into chemical vapor deposition (CVD) techniques and liquid precursor techniques. There are two main CVD techniques: the CVD and the electrochemical vapor deposition (EVD). Recent studies have also shown significant development in chemical solution deposition (CSD) technique. These methods make it possible to control chemical composition and to form a dense film. They are also known to be suitable for mass production.

4.2.1 Chemical Vapor Deposition and Atomic Layer Deposition

CVD is a chemical process in which one or more gaseous precursors form a solid material by means of an activation process. CVD has been widely used for fabricating microelectronics. Therefore, the underlying processes are well understood. In recent years, the application of CVD shows promising prospects as an alternative method to solution based fabrication due to its ease in patterning, precise composition control, capability of batch processing, superior material compatibility, and potential in uniform large-area deposition.

A schematic diagram of a typical CVD setup is shown in Figure 4.2 (Mbam et al. 2019). Typically fused-silica glass is used as the substrate material which is heated to deposition temperatures of 600–1200°C depending on the reactivity of the precursors. For the fabricating SOFC components, halogen compounds such as $ZrCl_4$ and YCl_3 (Yamane and Hirai 1987, Yamane and Hirai 1989), metal organic compounds such as metal alkoxides (Chour et al. 1997) or β-diketones (Aizawa et al. 1993) have been used as precursor materials. Growth rates of the film thickness are in the range of 1–10 μm h^{-1}, depending on the evaporation rate and substrate temperature. The chemical and physical structure of the film can be tuned by regulating the chemical reaction and deposition process parameters, for example, temperature, reactor geometry, gas flow rates, and pressure and input concentration. However, the capital cost for the CVD equipment is high (Itoh et al. 1994). In the last decades, the development of CVD has led to the invention and progression of an array of CVD-derived techniques, such as aerosol-assisted CVD (AACVD), hybrid physical-chemical vapor deposition (HPCVD), plasma-enhanced CVD (PECVD), and laser CVD (LCVD). Kim et al. fabricated nanogranular GDC at the cathode/electrolyte boundary by AACVD for high-performance SOFCs working at low temperatures (below 500°C). It was reported that the cells with AACVD-GDC nanogranular surface exhibited improved output performance and cathodic kinetics as compared to the bare cell; samples with smaller grains, which were determined by deposition temperature showed reduced polarization resistance (Kim et al. 2016). Jang et al. successfully coated a dense and uniform lanthanum nickelate (LNO) layer on nanoporous $La_{0.6}Sr_{0.4}Co_{0.2}Fe_{0.8}O_{3-\delta}$ (LSCF) substrate by AACVD (Jang et al. 2019). The LNO nanolayer significantly enhanced the SOFC performance and long-term stability with a peak power density as high as 466 mWcm^{-2} at 600°C obtained, an increase of 60% as compared to the uncoated one. The enhancement was ascribed to the improved cathodic surface kinetics and current collection.

Atomic layer deposition (ALD) is a modified CVD technique, also known as pulsed CVD. In ALD, the substrate surface is exposed alternately to different vaporized precursors. Because gaseous precursors are strictly separated from each other during deposition and the precursors have self-limiting chemistry, one reaction cycle may produce only one atomic layer. For this reason,

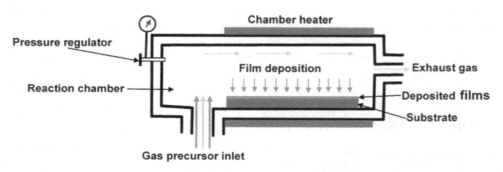

FIGURE 4.2 Schematic diagram of a typical CVD setup (Mbam et al. 2019).

Thin Coating Technologies

ALD can be an ideal technique to grow ultrathin oxide films because composition of ALD films can be altered at each atomic layer with desired ratio. The growth rate for zirconia is 0.1–0.17 nm/cycle (Hausmann et al. 2002). As ALD-deposited films often show distinct advantages in terms of conformality, precise thickness, and composition control over the conventional CVD techniques, ALD has been utilized in a wide range of applications nowadays such as optics, demanding protective coatings, magnetic recording head, and micro-electrochemical systems (Marichy et al. 2012).

Prinz's group (Shim et al. 2007) fabricated free-standing ultrathin YSZ electrolyte films with a target stoichiometry, $(ZrO_2)_{0.92}(Y_2O_3)_{0.08}$, by ALD. For ZrO_2 and Y_2O_3, commercial tetrakis(dimethylamido) zirconium $(Zr(NMe_2)_4)$ and tris(methylcyclopentadienyl)-yttrium $(Y(MeCp)_3)$ were used as precursors with distilled water as oxidant. ALD YSZ films were grown on a Si_3N_4-buffered Si(100) wafer substrate. After ALD deposition, silicon nitride layer was removed by plasma-assisted chemical etching, leaving freestanding YSZ layers. Porous Pt layers were deposited using direct current sputtering as cathode and anode. A maximum power density of 270 mWcm^{-2} was reported for a cell based on an ultrathin YSZ film of 60 nm at 350°C. By using corrugated thin film YSZ electrolyte design, the maximum power density was reported to increase to 677 mWcm^{-2} at 400°C (Su et al. 2008). The high power density output of the cell can be further pushed forward by controlling the three-dimensional geometry of the nanostructured electrode and electrolyte (Chao et al. 2011). For example, a maximum power density of 1.3 Wcm^{-2} at 450°C was acquired through constructing a nanostructured YSZ/yttrium-doped ceria (YDC) bilayered microcell (An et al. 2013). However, the success rate of the cells would be limited by the possible electric shorts between electrodes through the nanoscale electrolyte. The process–microstructure–performance relationship in ALD-engineered SOFCs has been comprehensively reviewed recently by Jihwan's group (Shin et al. 2019).

4.2.2 ELECTROCHEMICAL VAPOR DEPOSITION (EVD)

EVD is a modified CVD process, originally developed by Siemens–Westinghouse for the fabrication of thin YSZ electrolyte layer on tubular SOFCs (Pal and Singhal 1990). EVD is a two-step process. The first step involves the pore closure by a normal CVD type reaction between the steam (or oxygen), metal chloride, and hydrogen through the porous air electrode (phase I). These react to fill the air electrode pores with the YSZ electrolyte according to the following reaction:

$$2MeCl_y + yH_2O = 2MeO_{y/2} + 2yHCl \qquad (31.1)$$

$$4MeCl_y + yO_2 + yH_2 = 4MeO_{y/2} + 4yHCl \qquad (31.2)$$

where Me is the cation species (zirconium and/or yttrium) and y is the valence associated with the cation.

After the pores in the air electrode are closed, film growth then proceeds due to the presence of an electrochemical potential gradient across the deposited film. In this step, oxygen ions formed on the water vapor side of the substrate (i.e., the high oxygen partial pressure side) diffuse through the thin metal oxide layer to the metal chloride side (i.e., the low oxygen partial pressure side). The oxygen ions react with the metal chloride vapor to form the metal oxide products. The solid product or electrolyte is deposited as a thin film spreading over the internal pore surface in a desired region across the air electrode or membrane substrate. This second stage of the reaction is termed the electrochemical vapor deposition or EVD (Phase II). Figure 4.3 shows the schematic diagram of the basic principles of the EVD process (Pal and Singhal 1990).

The growth of YSZ electrolyte film by the EVD process is parabolic with time and the rate-determining step in the EVD process was found to be the electronic transport through the electrolyte film (Pal and Singhal 1990). The EVD techniques were used by others for the fabrication of thin YSZ electrolyte films and thick Ru/YSZ cermet anodes (Sasaki et al. 1994, Suzuki and Kajimura 1997). The dissociated oxygen from metal oxide substrates, such as NiO, was also suggested as an oxygen

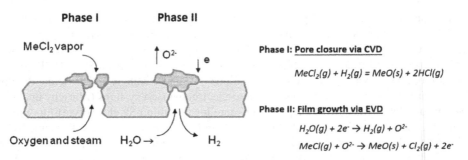

FIGURE 4.3 Schematic diagram of the EVD process, after Pal and Singhal (Pal and Singhal 1990).

source for the reaction instead of gaseous oxygen to form dense YSZ electrolyte films (Ogumi et al. 1995). The effect of NiO content of the substrate on the growth rate of YSZ film was studied in detail by Kikuchi et al. (2006). The high reaction temperature, the presence of corrosive gases and relatively low deposition rates are some of the limiting factors in the application of EVD process in IT-SOFCs. A recent study conducted by Hermawan et al. (2017) used a porous metal support as the substrate to deposit dense and gas-tight YSZ electrolyte layer for metal-supported SOFCs. A combination of atmospheric plasma spray (APS) and EVD techniques were adopted to resolve the slow film growth rate issue in EVD.

4.3 PHYSICAL DEPOSITION TECHNIQUES

Physical deposition techniques reviewed in this chapter such as magnetron sputtering, laser ablation, and plasma spray have the common feature that atoms are brought to the gas phase through a physical process from a solid or molten target. The processes include evaporation, sputtering, laser ablation, and hybrid methods. The deposited films are typically polycrystalline with columnar structure, and the grain size can be tailored by varying the deposition conditions.

4.3.1 Magnetron Sputtering Techniques

Magnetron sputtering is one of the most common physical deposition techniques that is widely used to grow alloy and component films in which one or more of the constituent elements are volatile. Low-defect-density films of high-melting point materials can be grown on unheated substrates because phase formation is mainly governed by kinetics, rather than by thermodynamics.

Radio frequency (RF) magnetron sputtering using an oxide target and DC relative magnetron sputtering using metallic targets have been utilized to produce oxide thin films, for example, YSZ, of high quality. RF sputtering has frequently been utilized to deposit YSZ thin films in part because of the ability to use either metallic or electrically insulating target and the generally high quality of the deposit. In RF-sputtering deposition, an evacuated chamber is filled with the sputtering gas (e.g., Ar). A large negative voltage is applied to the cathode. The sputtering gas forms a self-sustained glow discharge. Physical sputtering of the target occurs when positive ions from the plasma that are accelerated across the space strike the target surface. A metal oxide film is grown by sputtering a metal target in a discharge containing oxygen, usually in conjunction with a noble gas. The metal, metal oxide, and oxygen species that arrive at the substrate are adsorbed and ultimately incorporated into stable nuclei to form a continuous film. Oxygen is often included in the sputtering gas mixture as a means of controlling the metal to oxygen ratio in the target. However, RF sputtering deposition rate decreases significantly with increased oxygen partial pressure (Bae et al. 2000). Oxygen partial pressure has also been reported to impact the electric properties of RF sputtering deposited films (Gong et al. 2010).

DC current magnetron sputtering has also been widely used to deposit thin YSZ electrolyte onto porous supports (Wanzenberg et al. 2003, Hidalgo et al. 2011, Hidalgo et al. 2013). The target can be a single target consisting of an alloy of zirconium and yttrium or can be multiple targets of the pure metal, where the composition of the deposit is controlled by the relative exposed surface area of the targets. Barnett and co-workers described the formation of fully dense YSZ thin films on porous as well as dense electrode substrates by reactive DC current magnetron sputtering with post-deposition annealing temperatures as low as 350°C (Wang and Barnett 1993, Tsai and Barnett 1995). Electrical conductivities of the YSZ deposit were similar to those films formed by other methods. The metal to oxygen ratio in the deposit was found to depend critically on the oxygen partial pressure in the sputter gas, and the crystalline orientation of the deposited film was suggested to directly related to the oxygen partial pressure as well (Karpinski et al. 2012). Similar to RF sputtering, deposition rates are adversely affected by the oxygen partial pressure in the sputtering gas for DC reactive magnetron sputtering of YSZ films. This behavior was shown to be particular apparent for oxygen partial pressures less than 10 mbar, as shown in Figure 4.4 (Fedtke et al. 2004).

Surface morphology of the substrate such as pore size and pore size distribution is critical for the deposition of thin and high-density electrolyte film on the Ni/YSZ anode substrates as the morphology of the deposited film follows that of the substrate surface. Figure 4.5 shows the morphology of the YSZ film deposited on YSZ and Ni/YSZ cermet substrate by magnetron sputtering (Jiang and Chan 2004). Thin YSZ film deposited is characterized by columnar structure and has the same grain and grain boundary pattern as that of the YSZ substrate (Figure 4.5a), indicating that the surface morphology of the YSZ thin film follows closely the morphology of the substrate. The large pores on the substrate surface resulted in open pores on the YSZ electrolyte films deposited (Figure 4.5b and c). Thus, the pore diameter of the substrate surface needs to be in the order of the grain size of the deposited film to achieve dense and uniform YSZ electrolyte film (Figure 4.5d and e). This can be done by using an interlayer or functional layer with fine microstructure and low porosity (Tao and Irvine 2002, Wang, et al. 2011). The function layer is also used to promote the electrode reaction at the electrode/electrolyte interface (Ai et al. 2007, Kim et al. 2007, Solovyev et al. 2017). In addition,

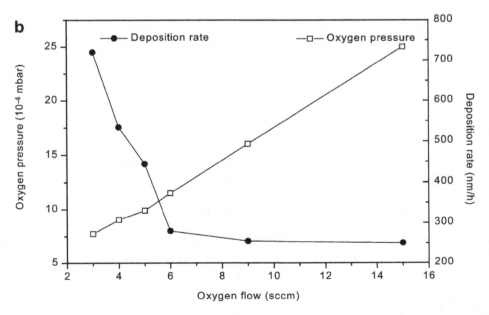

FIGURE 4.4 Relation of YSZ deposition rate by DC reactive magnetron sputtering to oxygen flow rate and oxygen partial pressure, after Fedtke et al. (Fedtke et al. 2004).

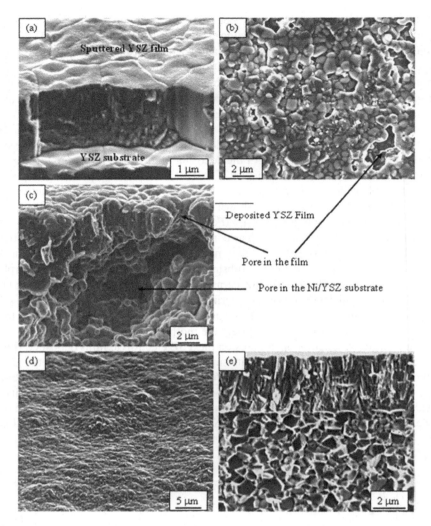

FIGURE 4.5 Morphology of the YSZ thin film deposited on YSZ and NiO/YSZ cermet anode substrate by magnetron sputtering. (a) YSZ film deposited on YSZ substrate; (b) and (c) YSZ film deposited on NiO/YSZ cermet substrate, showing that the defects in the YSZ electrolyte film is due to the large pores on the anode substrate surface; (d) and (e) dense and uniform YSZ film on NiO/YSZ anode substrate with optimized surface morphology, after Jiang and Chan (Jiang and Chan 2004).

successful deposition of uniform, highly porous, columnar Ni/YSZ thin film on dense YSZ substrate using magnetron sputtering was also reported lately (Garcia-Garcia et al. 2015).

Huang et al. (2007) fabricated thin film SOFC structure containing ultrathin YSZ electrolyte membrane 50–150 nm thick using RF sputtering, lithography, and etching. The porous Pt anode and cathode were deposited by dc sputtering at 10 Pa Ar pressure, 100 W, and room temperature. Dense YSZ electrolyte films were deposited at 200°C by RF sputtering. The DC sputtering target was Pt and the RF sputtering target was $Y_{0.16}Zr_{0.84}O_{1.92}$. The maximum power density achieved is 200 mWcm^{-2} at 350°C. DC magnetron sputtering was used by Srivastava et al. (1997) to deposit YSZ electrolyte films (5–16 μm) on Ni/YSZ anode substrates. In order to ensure a dense and impervious electrolyte layer it is desirable to optimize the deposition conditions as such that YSZ films would remain in a state of compressive stress while adhering to Ni/YSZ substrate. This implies that only a narrow range of deposition conditions would be suitable. The pressure in the sputtering chamber,

Thin Coating Technologies   91

target–substrate distance, and the flow of oxygen close to the critical zone during deposition were shown to be important parameter and required careful control (Srivastava et al. 1997, Lamas et al. 2012). Magnetron sputtering was also used to deposit a GDC protecting layer on YSZ electrolyte to prevent Sr^{2+} migration from $(La,Sr)(Co,Fe)O_3$ cathode toward the electrolyte (Uhlenbruck et al. 2007). Solovyev et al. utilized magnetron sputtering to deposit YSZ/GDC bilayer electrolyte, in which YSZ thin film acts as a blocking layer to prevent electrical current leakage in the GDC layer (Solovyev et al. 2016). The cell with YSZ/GDC bilayer electrolyte achieved a maximum power density of 1.25 Wcm^{-2} at 800°C.

Magnetron sputtering techniques combined with photolithography are used to produce unique, patterned electrodes such as LSM, Ni, platinum, and gold. The patterned, thin, dense, and uniformly structured electrodes allow the fundamental aspects of the reaction mechanism to be studied, not possible with conventional, heterogeneous, porous electrode structure. Horita et al. (1998) studied the active site distribution of the O_2 reduction reaction on patterned LSM electrodes and demonstrated that the O_2 reduction reaction occurs primarily at the three phase boundaries (TPB) where LSM cathode, oxygen reactant gas, and YSZ electrolyte meet by the SIMS image technique. The linear relation between the length of TPB and the rate of the reaction was also found for the H_2 oxidation reaction on pattered Ni anodes of SOFCs (Mizusaki et al. 1994). Li et al. employed patterned Ni/YSZ electrode to study carbon deposition and found carbon content near TPB was significantly higher (Li et al. 2015). Sputtering method is also used to prepare protective coating for metallic interconnect. Lee and Bae (2008) applied RF magnetron sputtering technique to deposit $La_{0.6}Sr_{0.4}CrO_3$ (LSCr) and $La_{0.6}Sr_{0.4}CoO_3$ (LSCo) as oxidant-resistance coating on Fe-Cr ferric stainless steel SS430. The oxide layer shows dense structure and good adhesion to the SS430 alloy. LSCr-coated SS430 shows a low electrical resistance as compared to LSCo-coated SS430 probably due to the low diffusivity of manganese or chromium in the LSCr layer. The RF magnetron sputtered lanthanum chromite film on stainless steel substrate formed orthorhombic perovskite structure after annealing at 700°C, and exhibited a dendritic microstructure (Orlovskaya et al. 2004). Additionally, Geng and co-workers fabricated a double-layer oxide structure on stainless steel with sputtered Ni coating in air at 800°C. The outer layer of $NiO/(Ni, Fe, Cr)_3O_4$, formed from the sputter Ni and Fe/Cr, was found to act as a sealing barrier to suppress the evaporation of the inner Cr_2O_3; the double layer structure also lowered the surface oxide area specific resistance (ASR) (Geng et al. 2012).

4.3.2 Pulse Laser Deposition (PLD)

PLD or laser ablation is a physical method of thin film deposition in which a pulsed laser beam, usually of wavelength in the UV range, is employed to ablate a target composed of the desired thin film composition, which is subsequently deposited onto a substrate. The usual range of laser wavelengths for thin film growth by PLD lies between 200 nm and 400 nm for most materials. In PLD, the temperature of the substrate is one of the main parameter affecting atomic surface mobility during the deposition process. PLD enables fabrication of multicomponent stoichiometric films from a single target, and with an appropriate choice of the laser (e.g., Nd:YAG, KrF, XeCl), any material can be ablated and the growth can be carried out in a pressure of any kind of gas, reactive or not. To date, PLD is utilized in the deposition of a wide spectrum of materials, ranging from metals, to polymers, to semiconductors, to insulators, and to biological materials (Huang et al. 2019). Shown in Figure 4.6 is the schematic diagram of a PLD process (Scandurra et al. 2020).

The microstructure of the PLD films depends on the substrate temperature and pressure. Mengucci et al. reported the formation of dense YSZ films by PLD at room temperature and oxygen pressure below 0.05 mbar (Mengucci et al. 2005). Dense amorphous YSZ thin film was also fabricated by Lippert et al. using PLD at ambient temperature, and the crystallization and microstructural evolution of the film can be effectively controlled by low temperature thermal annealing (Heiroth et al. 2011). On the other hand, highly porous YSZ films can be obtained at high pressure of 0.3 mbar (Nair et al. 2005). Infortuna et al. studied in detail, the characteristics of YSZ and GDC thin films

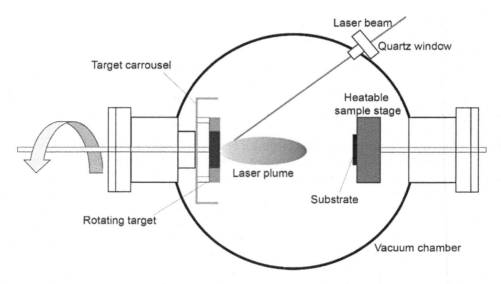

FIGURE 4.6 Schematic diagram of a pulse laser deposition process, after Scandurra et al. (Scandurra et al. 2020).

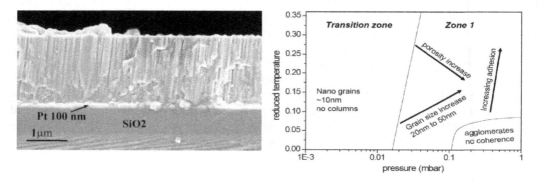

FIGURE 4.7 SEM micrograph of a dense YSZ thin film deposited at 800°C under 0.026 mbar oxygen pressure on a Si/SiO$_2$/TiO$_2$/Pt substrate (left); and structural map for GDC and YSZ thin film deposited by PLD (right). The normalized deposition temperature was calculated from the melting temperature of the materials (2500°C was used for both YSZ and GDC), after Infortuna et al. (Infortuna et al. 2008).

by PLD and the microstructure of GDC and YSZ thin films deposited, ranging from highly porous to dense depending on the substrate temperature and oxygen pressure during the deposition process (Infortuna et al. 2008). Figure 4.7 shows a typical SEM micrograph of a dense YSZ thin film deposited at 800°C and structural map for GDC and YSZ thin films deposited by PLD (Infortuna et al. 2008). By using a split target, Park et al. managed to fabricate YSZ/GDC multilayers of different periodicities and thickness ratio (Park et al. 2014). The electrical conductivity of the PLD films appears to be lower than that of the bulk materials. For example, the conductivity of a thin YSZ film deposited at 400°C in 0.026 mbar is 3×10^{-4} Scm^{-1} at 500°C, compared to ~10^{-3} Scm^{-1} for single YSZ crystal. Low electrical conductivity for YSZ thin films by PLD is also reported by Joo and Choi (2006). One potential problem is that PLD techniques result in crystalline microstructures that usually show columnar grains and texture, resulting in unfavorable anisotropic electrical conductivity (Chen et al. 2003). The surface artifacts of the function layer should also be smaller than the film thickness in order (Li et al. 2013) to avoid the defects on the YSZ film deposited by PLD technique

(Hobein et al. 2002). Nevertheless, it was suggested that by introducing strain confinement to the multilayered YSZ films, a two-order-of-magnitude increase in oxide-ion conductivity compared to bulk YSZ was achieved, reaching 0.01 Scm^{-1} at 475°C (Li et al. 2013).

PLD has been used to deposit multielement films such as LSM, LSCF, and LSGM for SOFC (Zomorrodian et al. 2010, Plonczak et al. 2012, Hwang et al. 2015). Due to its unique method of dislodging atoms from the target, PLD provides some distinct advantages among thin film techniques. As a result of the weak interaction of lasers with gaseous species, ambient atmospheres can be used with little contamination – a significant advantage compared to other deposition techniques based on ions or electrons. Thus, stoichiometric materials transfer from the target to the substrate could be achieved. Koep et al. (2006) studied microstructure and electrochemical properties of LSM and LSC prepared by PLD. Deposition of LSM above 500°C results in thin films in the orthorhombic phase while the low-temperature grown LSM films are amorphous. The stoichiometry of the film is the same as the target composition and the conductivity of the LSM dense film is 21 Scm^{-1} at 700°C. In comparison, epitaxially grown LSCF films exhibited electronic conductivities of 2.3×10^3 Scm^{-1} at 600°C, with an activation energy of 0.09 eV (Zomorrodian et al. 2010). Microfabrication process in combination of photolithographic process is also used to produce well-defined three-dimensional geometries for the investigation of electrode reaction process and reaction sites at electrode materials (Koep et al. 2005, Simrick et al. 2011), and to fabricate µSOFC devices with patterned electrode structure (Yang et al. 2015). PLD is also used to fabricate cobalt and ferrite–based perovskite cathodes such as $La_{0.6}Ca_{0.4}Fe_{0.8}Ni_{0.2}O_{3-x}$ (LSFN) (de Larramendi et al. 2007) and $(La,Sr)CoO_3$ (LSC) (Garbayo et al. 2014) on YSZ electrolyte because the crystalline phase of the perovskite can be formed at substrate temperature of 700°C to avoid the interfacial reaction between YSZ and lanthanum cobaltite-based cathodes. Reasonable ASR of 1.59 Ωcm^2 at 850°C was reported on the LSFN dense cathode layer by PLD (de Larramendi et al. 2007). Whereas a significantly low ASR of 0.274 Ωcm^2 at 700°C was reported for a $SrNb_{0.1}Co_{0.9}O_{3-\delta}$ (SNC) symmetrical cell deposited on YSZ electrolyte (Chen et al. 2015). Due to the dense structure of the PLD films, PLD method would be best suitable for the electrode application with high mixed ionic and electronic conducting oxides. For example, for the LSM cathode films prepared by PLD, the ASR can be as high as 32.1 Ωcm^2 at 850°C (Endo et al. 1996). This is due to the fact that LSM is an electronic conductor with negligible oxygen ion conductivity. PLD is also used to deposit $La_{0.8}Sr_{0.2}Cr_{0.97}V_{0.03}O_3$ and $MnCr_2O_4$ protective film on Crofer 22APU interconnect at a substrate temperature of 750°C (Mikkelsen et al. 2007). The deposited films were dense and significantly reduced the growth of the oxide scale of the alloy. Likely, Brylewski et al. deposited a layer of Co from cobalt metal film on DIN 50049 steel (Kruk et al. 2015). The Co-coated steel not only effectively reduced the formation of volatile Cr species, but also led to a significant reduction in ASR of the steel (5×10^{-6} Ωcm^2 at 1073K).

Recently, there has been significant progress in using PLD to deposit GDC barrier layer between electrode and electrolyte. To investigate cation transport behavior, De Vero et al. deposited a GDC interlayer on single crystal YSZ substrate by PLD, and a LSCF electrode layer was prepared on the GDC layer through the same procedure (De Vero et al. 2018). The Sr diffusion through epitaxial GDC layer is essentially inactive, whereas Zr transport was observed on (100)-oriented GDC interlayer with severe pore formation. The presence of a network of dislocations in PLD deposited GDC interlayer provides a fast diffusion pathway for Zr. Morales et al. performed multiscale analysis on PLD-deposited GDC barrier layer in anode-supported SOFCs, which were industrially fabricated and assembled in stacks (Morales et al. 2017). Cation-interdiffusion behavior at the cathode/barrier layer/electrolyte interfaces was studied under real operation conditions for 3000 h. A gradual loss of Gd and its segregation to grain boundaries for migrating across the CGO barrier layer was revealed and identified as the major cause of the reduction of the layer conductivity at high sintering temperatures. It was concluded that typically employed CGO diffusion barriers present cation interdiffusion problems during the fabrication process. To systematically study the effects of grain boundary (GB) on the oxygen ion transport at cathode/electrolyte interface, Choi et al. deposited GDC layer by PLD on polycrystalline GDC substrate to control its GB density (Choi et al. 2017). Significant improvement in peak power density (three-fold) was obtained on the cell using Pt as electrodes. The reinforced performance by high GB density was attributed to the accumulation of oxygen vacancies, which provided more active sites than grain cores for oxygen ion migration. Moreover, Morales et al. has made great progress in using PLD to deposit large area (~80 cm^2) GDC interlayers, achieving high density and reproducibility (Morales et al. 2018). The cell with deposited GDC layer was later annealed at 1150°C to improve its chemical and microstructural stability, and an excellent power density of 1.25 Wcm^{-2} at 0.7 V and 750°C was achieved. The fuel cell stack was finally operated in realistic conditions for 4500 h and revealed a low degradation rate of 0.5%/1000 h.

Another technique suitable for large-scale production for industrial purpose is electron-beam physical vapor deposition (EB-PVD). EB-PVD is a reliable technique for the deposition of thin film by the vacuum evaporation of a target using an electron beam heating device. This technique has the advantages of the deposited films having an exact target composition, having high deposition rates at low temperatures, and realizing large-scale production for industrial application. The deposition rates of EB-PVD are typically in the range 1–2 μm min^{-1}. In addition, this method can make the deposited film crystallized without additional annealing, thus saving time and energy in the manufacturing process. Another variation of PLD is the large area filtered arc deposition (LAFAD) which has been applied to deposit protective thin coatings on Cr-Fe alloys using CoMn and CrAlY target alloys (Gannon et al. 2008).

Jung et al. (2006) and Mozammel et al. (Tanhaei and Mozammel 2017) investigated the deposition and characteristics of the YSZ thin films deposited on Ni/YSZ anode supports using EB-PVD technique. The YSZ target for the electron beam deposition was prepared by conventional uniaxial pressing with commercial Tosoh TZ8Y powder under 30 MPa followed by sintering in air at 1400°C for 5 h. The grain size of the YSZ film before heat treatment is 0.5–1 μm and then grows to very dense columns with a grain size of 1–2 μm. Heat treatment improves the densification and the adhesion of YSZ film with the anode functional layer (AFL). Highly conductive and dense YSZ film deposited on Ni-YSZ substrate was fabricated by Shin et al., and the quality of the film can be controlled by optimizing the distance between the substrate and target, angle for deposition area, and substrate temperature (Shin et al. 2017). The YSZ electrolyte, prepared by EB-PVD, cell with Ni-YSZ and LSM electrodes showed a power density of 1 Wcm^{-2} at 0.75 V and 900°C. The electrical conductivity of EB-PVD YSZ electrolyte is smaller than that of the bulk YSZ at temperatures higher than 600°C (Gibson et al. 1998). The higher electrical resistance of the YSZ film deposited by EB-PVD could cause additional internal ohmic losses of the cell operating in the intermediate temperatures above 600°C. EB-PVD was also used to deposit GDC thin films on porous NiO/YSZ anode substrates (Laukaitis and Dudonis 2008). The main dominating crystallite orientation of the GDC film repeats the characteristics of the used powder and the crystalline size of the GDC thin film is influenced by the e-beam gun power (Laukaitis and Dudonis 2008). Similarly, the electrical conductivity of GDC films deposited by EB-PVD were reported to be much smaller than the bulk sample, possibly due to the large amount of defects and microstrain in the crystal lattice (Pérez-Coll et al. 2014). EB-PVD technique was also employed to deposit nanocomposite film, for example, NiO-CeO$_2$ thin layer (Kuanr and Krishna Moorthy 2016). The kinetic energy of adatoms, influenced by the electron beam gun power, was found to play a significant role in determining the crystallinity of the deposited films.

4.3.3 PLASMA SPRAY DEPOSITION

The plasma spray process is a high-temperature process (up to 15,000 K for a typical DC torch operating at 40 kW). Figure 4.8 are the schematic diagrams of DC and RF plasma spray processes (Pederson et al. 2006). The plasma spray process is based on the generation of a plasma jet consisting of argon or argon with admixture of H$_2$ and He, which are ionized by a high current arc discharge in a plasma torch. The powders to be sprayed are injected into the plasma where they are accelerated, melted, and finally projected onto a substrate. The coating is formed by solidification and

Thin Coating Technologies

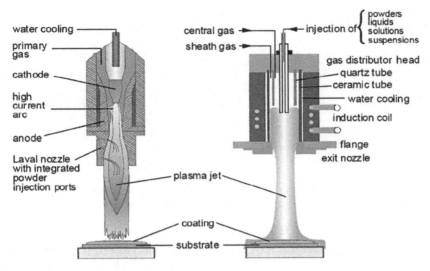

FIGURE 4.8 Schematic diagrams of DC and RF plasma spray processes, after Pederson (Pederson et al. 2006).

flattering of the particles at impact on the substrate. This technique offers the possibility to deposit thin or thick layers in the mm range, which is hardly possible with other physical methods such as magnetron sputtering and PLD. Comparing to DC deposition, RF plasma spray has the ability to use either metallic or electrically insulating and the scalability to large substrates, while its deposition rate is influenced by oxygen partial pressure. With thermal spray technology, cells could be produced without using time and energy consuming sintering steps, reducing the production cost of SOFCs. The deposited layers often show the characteristics of anisotropic microstructure, microcracks due to thermal stress, and interlamellar porosity. To produce coatings with finer microstructure, that is, finer lamellae with grain size in the nanometer range while keeping the high deposition rate, there has been significant development in plasma spray process utilizing liquid suspension or solution as feedstock instead of conventional powder (Vardelle et al. 2014). The films produced by vacuum plasma spray (VPS) typically possess higher density than those produced by APS (Kulkarni et al. 2003). Thus, VPS becomes particularly important in the fabrication of the SOFC electrolyte.

Tai et al. (Tai and Lessing 1991) used plasma spray to prepare porous LSM electrodes. Addition of pore-formers such as carbon is necessary for spraying a porous coating. Coatings with a porosity of ~40% were deposited from LSM with a broad particle size distribution of 53–180 μm and 15wt% solid Carbospheres as a pore-former. The microstructure of the plasma-sprayed LSM coating is characterized by large agglomerates and coarse porous structure. The plasma-sprayed LSM coatings with much finer structure can be achieved by using suspension or solution precursor as feedstock (Wang et al. 2010, Wang and Coyle 2011). Li et al. (2005) studied the effect of spray parameters such as plasma power and spray distance on the electrical conductivity of plasma-sprayed LSM coating and observed lower conductivity compared to that of the sintered LSM. This is explained by the formation of lamella structure of the plasma-sprayed LSM coating. The disadvantage of the method is that rapid heating and quenching involved induces non-stoichiometry and residual strain. Post-treatment, for example, at 1000°C in air for 2 h, was shown to be able to recover the crystallinity and stoichiometry of the plasma-sprayed LSM coating (Lim et al. 2005) and to increase the electrical conductivity (Li et al. 2005). Compressive stress was observed in the YSZ electrolyte film after deposition on cathode support, which gradually decreased when treated at a temperature above 600°C (Huang and Harter 2010). The residue stress was suggested to be indicative of the good bonding and low contact resistance between the film and the substrate. Rambert et al. (1999) prepared LSM/YSZ composite cathodes using VPS technique. The performance of the LSM/YSZ composites

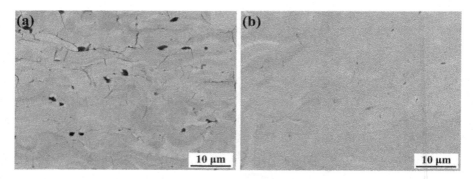

FIGURE 4.9 Polished cross-sectional microstructures of $La_{0.8}Sr_{0.2}CrO_3$ coatings prepared at different deposition temperatures. (a) RT and (b) ~500 °C (Chen et al. 2019).

is affected by the mixing process of the powders. Van Herle et al. (1994) prepared YSZ electrolyte by VPS and observed a strong anisotropy in the ionic conductivity of YSZ electrolytes. The cross-plane conductivities are several times lower than in-plane conductivities. This anisotropy can be eliminated by sintering the deposited YSZ layer at 1500°C for 2h. Mirahmadi et al. reported that the addition of aluminum (~5 wt.%) considerably improved the gas tightness and ionic conductivity of the plasma-sprayed YSZ, increased from 0.88–1.14 Scm^{-1} at 1000 K (Mirahmadi and Pourmalek 2010). The processing parameters such as the heat treatment, melting temperature of the raw materials, and particle size distribution have been shown to have a significant influence in the microstructure and conductivities of the films (Fauchais et al. 2005, Ma et al. 2005, Chen et al. 2019), see Figure 4.9.

There have been attempts to fabricate SOFCs and LSM/YSZ composite cathodes on metallic substrates using APS, taking advantage of the relative simple and low cost of this process (Zheng et al. 2005, White et al. 2007). Vassen et al. applied APS to fabricate porous NiO/YSZ coatings for anode, dense YSZ coatings for electrolyte, and functional coatings as Cr-evaporation layer on interconnect (Vassen et al. 2008). A cell with APS deposited NiO/YSZ anode and YSZ electrolyte and screen-printed LSM cathode produced a power density of 0.8 Wcm^{-2} at 800°C. The influence of feedstock spraying routes, pore former content, and size distribution on the porosity and homogeneity of Ni/YSZ film deposited via APS has been thoroughly reviewed by Gupta et al. (2017). By using high-speed plasma torches and specially designed nozzles, thin and gas-tight yttria- and scandia-stabilized ZrO_2 (YSZ and ScSZ) electrolyte layers (~30 μm in thickness and 1.5–2.5% in porosity) and of porous electrode layers with high material deposition rates were fabricated by VPS techniques. The plasma-sprayed cells showed good electrochemical performance and low internal resistances. Power densities of 300–400 mW/cm² at 750–800°C were reported for plasma-sprayed cells (Schiller et al. 2000, Lang et al. 2001). In addition, the development of low-pressure plasma spraying (LPPS, usually operating at less than 50,000 Pa) has enabled the deposition of dense electrolyte in short time with high quality in the 1–100 micron thickness range with lamellar or columnar microstructure (Smith et al. 2011, Yuan et al. 2017). Wang et al. employed the very low-pressure plasma spraying technique (VLLPS, operating at ~100–1000 Pa) to deposit thin ScSZ electrolyte layer (~50 μm) for a metal-supported cell, and the cell with a configuration of NiO-ScSZ|ScSZ|LSCF exhibited a maximum power density of 1112 $mWcm^{-2}$ at 750°C (Wang et al. 2019).

4.4 COLLOIDAL AND CERAMIC POWDER TECHNIQUES

Colloidal and ceramic powder techniques are simple and cost-effective methods for the manufacturing of thin electrolyte films and are less restricted by the geometric factors. The colloidal and ceramic powder methods include tape-casting (Meier et al. 2004, Song et al. 2008, Nishihora et al. 2018), slurry coating (Cai et al. 2002, Kim et al. 2005), spin coating (Kim et al. 2002, Xu et al.

Thin Coating Technologies

2005, Zhang et al. 2010), dip-coating (Tikkanen et al. 2011), and screen-printing (Chu 1992, Zhang, Huang et al. 2006, Somalu et al. 2017).

4.4.1 Colloidal Techniques

4.4.1.1 Slurry Coating, Spin Coating, and Dip-Coating Methods

In the case of slurry coating, spin coating, and dip-coating, additives such as dispersion and surfactants are added to the electrode or electrolyte powder in order to form a suitable and stable colloidal particle suspension. Moreover, in order to avoid cracks during removal of the organic additives, careful drying and heat treatment procedures have to be adopted. Figure 4.10 shows the schematic diagram for spin coating and dip-coating apparatus. The process is simple and requires little investment. However, the coating process needs to be repeated several times to form a dense and pore-free thin electrolyte layer.

Chen et al. (2007) prepared YSZ slurry mixed with ethyl cellulose and terpineol in a weight ratio of 25:3.4:71.6. The results indicate that heat treatment at 400°C for 10 min for each spin coated layer is effective to remove the organic additives and to form a dense YSZ layer by repeated spin coating process. A single cell with a 14 μm thick YSZ film and Sm-doped ceria (SDC)-impregnated LSM cathode yields a maximum power density of 634 mW cm^{-2} at 700°C in H$_2$/air. A dense bi-layered YSZ/GDC electrolyte membrane with thickness of 2 μm was fabricated directly on NiO-YSZ anode support by spin coating. The GDC layer was spin coated after the completion of YSZ, and the cell with GDC-LSCF composite cathode showed a maximum power density of 200 mW cm^{-2} at 600°C (Kim, Kim et al. 2016). YSZ electrolyte layer as thin as 0.5 μm was prepared by spin coating method using YSZ precursors containing zirconia chloride hydroxide and yttrium chloride hydroxide as the source of Zr and Y with poly-vinylpyrrolidone (PVP) additive (Chen and Wei 2006). Dense and gas-tight YSZ films were obtained after repeating the spin coating process three times (see Figure 4.11). In addition to spin coated YSZ electrolyte, the fabrication of La$_{0.9}$Sr$_{0.1}$Ga$_{0.8}$Mg$_{0.2}$O$_3$ (LSGM) (Sun et al. 2014) and yttrium-doped barium zirconate (BZY) electrolyte (Luisetto et al. 2012) thin film has been reported as well. Spin coating was also used for the preparation of thin porous Sm$_{0.5}$Sr$_{0.5}$CoO$_3$ (SSC) cathode for SOFCs, using a suspension solution made from SSC powder and ethyl cellulose in ethanol (Wang et al. 2008). The suspension was applied to the doped ceria electrolyte surface and spun at 6000 rpm for 30 s to form a uniform layer and to remove the solvent. To form ~20 μm thick SSC cathode, 50 spin coating cycles were performed (Wang, Weng et al. 2008). However, by combination of thick slurry coating and thinning and smoothing with a spinning step, a thin and dense YSZ electrolyte layer can be fabricated by a single coating-drying cycle (Wang et al. 2007).

Aerosol spray deposition was also used to fabricate YSZ thin electrolyte films on metal-supported structure from an isopropanol-based (Matus et al. 2005) or ethanol-based (Łatka et al. 2012) solution. A metal-supported SOFC with thin YSZ electrolyte film achieved a maximum power density

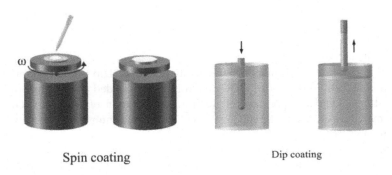

FIGURE 4.10 Schematic diagrams of spin coating and dip coating processes.

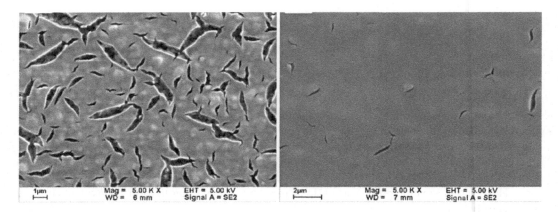

FIGURE 4.11 SEM micrographs of the surface of YSZ-coated samples after heat treatment at 600°C: one spin coating (left) and three spin coatings (right), after Chen and Wei (Chen and Wei 2006).

of 332 mWcm^{-2} with H$_2$/air at 700°C (Tucker et al. 2007). Wang et al. (2001) described a fabrication process for the Ni/YSZ anode-supported thin film YSZ cells based on tape-casting and spray-coating. The Ni/YSZ tape-cast green tapes were cut to desired size and annealed in air at ~1000°C for 1 h to obtain porous substrate with a thickness between 200 and 250 μm. The YSZ thin film electrolyte was then deposited by spray-coating a YSZ-water suspension on substrates under controlled conditions and sintered at 1400°C for 4 h. YSZ electrolyte film as thin as 3 μm can be obtained. The cell shows a very low-ASR, 0.071 Ωcm^2 at 800°C, and has a power output of 0.85 W/cm^2 at 800°C.

Slurry or dip-coating has been used extensively for the fabrication of dense YSZ electrolyte (Will et al. 2000, Torabi et al. 2011). Zhang et al. used the slip-casting method in combination of dip-coating to fabricate anode-supported tubular SOFCs (Zhang et al. 2009). The Ni/YSZ anode tube support was prepared by the slip-casting method and pre-sintered at 1000°C for 2 h. The AFL and YSZ electrolyte thin film were prepared by dip-coating. A suspension of 10wt% NiO/YSZ (NiO:YSZ = 1:1 by weight) in *iso*-propanol was dip-coated onto the anode tubular substrate to form AFL. Mahata et al. successfully developed a co-pressing and co-firing route to prepare YSZ electrolyte dip-coated on the tubular NiO-YSZ anode support (Mahata et al. 2012). The dip-coated YSZ layer along with the anode support was co-pressed by cold isostatic pressing (CIP) and co-fired at 1350°C, and the tubular single cell with dip-coated LSM-YSZ cathode delivered a maximum power density of 62 mW cm^{-2} at 900°C. Similarly, a suspension of 10 wt% YSZ in *iso*-propanol was dip-coated onto the outer surface of the AFL to form an electrolyte layer, followed by drying at room temperature. The process was repeated 10 times in order to prepare a dense and pore-free YSZ thin electrolyte layer. The YSZ-coated NiO/YSZ tubular substrate tubes were sintered at 1380°C for 2h to form a bilayer structure with a porous anode substrate and a dense electrolyte thin film. Multilayered cathodes of LSM, LSM/YSZ, and SSC were deposited to the thin YSZ electrolyte by dip-coating method. Figure 4.12 shows the SEM micrographs of an anode-supported tubular cell with a thin YSZ film and a multilayer cathode prepared by dip-coating method. With impregnation of nanosized GDC particles into both anode-supports and multilayer cathode, the cell achieved a peak power density of 1104 mWcm^{-2} in H$_2$/air and 770 mWcm^{-2} in CH$_4$/air at 800°C. Dip-coating method was also used by Yamaguchi et al. (2007) to deposit GDC electrolyte on extruded anode-supported microtubular cells. A microtubular anode-supported cell with a configuration of NiO-GDC/GDC/LSCF-GDC was prepared by Kituta et al., and the cell performed with a maximum power density of 0.34 W cm^{-2} at 550°C (Usui et al. 2010).

Dip-coating technique is also applied to deposit YSZ or GDC electrolyte film as thin as ~100 nm (Van Gestel et al. 2008). In this case, it is important to control the quality of the microstructure of the substrate as well as the YSZ and GDC sol containing nanoparticles (5–6 nm). For example,

Thin Coating Technologies 99

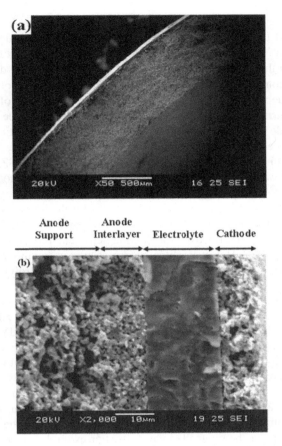

FIGURE 4.12 SEM micrographs of (a) an overview of the tubular cell and (b) the cross section of the tubular cell after testing, showing the thin YSZ electrolyte film and multilayer cathode prepared by dip coating method, after Zhang et al. (Zhang et al. 2009).

nanostructured zirconia sol can be produced by controlled hydrolysis of $Zr(n-O_3H_7)_4$ and $Y(i-O_3H_7)_3$ in the presence of diethanol amine (DEA) as precursor modifier/polymerization inhibitor. DEA is acted as a drying controlling additive and is important for the formation of nanoparticles in the synthesis process. Using dip-coating technique, Van Gestel et al. also fabricated a 1–2 μm thick YSZ layer on an anode-supported cell with GDC as barrier layer and LSCF as cathode, and the cell showed a current density of 1.6 A cm^{-2} at 0.7 V and 650°C. This is an increase of nearly 70% by lowering the electrolyte thickness from 7–10 μm to 1 μm (Van Gestel et al. 2011). In conventional slurry- or dip-coating process, as many as 10 sequential dip–drying–calcination steps are required to obtain sufficiently thick and dense films. These repeated dip–coating–drying–calcination processes make it time-consuming.

Wang et al. (2008) showed that the composition of the dip-coating slurry has a significant influence on the quality of the YSZ thin films. Among the three commonly used composite solvents: trichloroethylene (TCE)/MEK, MEK/EtOH, and EtOH/TCE, YSZ suspension in MEK/EtOH solvents gave the best dispersibility and stability. Also, MEK/EtOH has an evaporation temperature of 74.8°C, low enough for facilitating the formation of YSZ electrolyte thin films. Adding binder (PVB) and plasticizer (PEG and PHT) improves the stability. A thin and dense 16-μm-thick YSZ films were prepared by dip-coating twice and the cell with NiO/YSZ anode and Pt cathode reached a power density of 262 mWcm^{-2} at 800°C. Rather than dip-coating, Liu et al. (2008) drop-coated the YSZ suspension on porous electrode support to forming a thin and dense dual SDC-YSZ electrolyte films. Tao et al. also used drop-coating method to prepare $BaZr_{0.1}Ce_{0.7}Y_{0.2}O_3$ (BZCY) electrolyte with a thickness of 20 μm. The proton conducting electrolyte cell achieved a maximum power density of 377 mW cm^{-2} at 700°C (Tao et al. 2014). Zhang et al. showed that slurry-casting could also be used to deposit thin YSZ electrolyte film on a NiO/YSZ substrate disc assisted by a revolving rod to spread and compact the YSZ layer (Zhang et al. 2008). The revolving or rotating-assisted slurry-casting technique would be difficult to scale up for the fabrication of large area planar cells.

Vacuum slip- or slurry-casting was also used to deposit thin GDC electrolyte layer on LSCF/GDC composite cathode substrates (Serra et al. 2008). The vacuum condition not only improved the bonding strength between the electrode and electrolyte, but also increased the packing density of electrolyte particles (Shi et al. 2017). In addition to the preparation of YSZ electrolyte layer, slurry coating or slurry spray was also used for the deposition of anode and cathode interlayers (Reitz and Xiao 2006). The results show that the inclusion of electrode interlayers significantly improved the performance of the anode-supported cells and the reasons for the improvement could be related to the introduction of a diffuse mixed conduction region associated with the interlayers. The interlayers can also act as barrier layer, effectively preventing detrimental reactions between electrode and electrolyte (Khan et al. 2016). However, slurry or slip-casting is commonly used to fabricate thick NiO/YSZ anode substrates or tubes for the anode-supported thin electrolyte SOFCs.

4.4.1.2 Sol-Gel Method

In sol-gel methods, organo-metallic salts such as metal alkoxides (e.g., zirconium propoixde and yttrium propoxide) are deposited on porous electrode substrate and hydrolyzed under controlled conditions, forming a colloidal sol and a condensation step with organic monomers to form a gel. The deposition by methods such as by spin coating or dip-coating, is followed by a drying and firing process, leading to the formation of a dense electrolyte film. The key feature in this sequence is sol-gel polymerization which can be described by a two-step reaction: initiation via the hydrolysis of alkoxy ligands and polycondensation via an oxylation reaction. The particle concentration, viscosity, concentration, and stability of the sol-gel influence the deposition parameters and film quality and have to be controlled carefully. For example, since the electrode substrate is generally porous, the viscosity of the sol-gel solution has to be optimized to prevent infiltration. Hence organic additives such as PVP are adopted to increase the viscosity of the solution and to make the coating layer more elastic for accommodating stress (Kozuka and Higuchi 2003, Lee et al. 2017). The nature of the porous substrate is also critical in the sol-gel processes. Large pores could lead to pore-induced defects. Thus, the substrate should have a porosity that is both submicron and uniform. Dunn et al. (Dunn et al. 1993) gave an overview and briefly discussed the sol-gel chemistry.

Mehta et al. (1998) used sol-gel method to deposit thin (100–300 nm) YSZ layer on YDC electrolyte to block the electron transfer. An ethyl alcohol solution of YSZ was made by dissolving 11mol% yttrium isopropoxide and 89mol% zirconium isopropoxide in anhydrous isopropanol. The solvent was slightly heated and a small amount of HNO_3 was added to promote dissolution and hydrolysis. The YSZ precursor solution was applied to YDC electrolyte substrate by spin coating. Formation of the cubic YSZ phase was achieved at 600°C in air. A 0.25 M solution used in spin coating results in ~100 nm thick YSZ layer after heat treatment. The open circuit voltage (OCV) of the two-layer YSZ-YDC electrolyte cell is increased by 150–200 mV as compared to an uncoated YDC electrolyte cell in the temperature range of 600–800°C in H_2/O_2. Instead of depositing dense electrolyte film, Choi et al. used GDC sol-gel solution as a densification agent to infiltrate into porous GDC substrate. The cell with densified GDC layer showed highly improved performance with a low degradation rate of 1.72%/1000 h (Choi et al. 2016). The compatibility of the two-layer YSZ-YDC electrolyte cell was also studied by Kim et al. (2002) using a sol-gel spin coating method. The YSZ film deposited by sol-gel spin coating method showed a crack and pinhole free microstructure after sintered at 1400°C. A 2-μm-thick YSZ film on YDC electrolyte can be obtained after six repetitive spin coatings. However, the maximum power density of the two-layer cell is comparable to a YSZ single layer cell with the same thickness at 1000°C. A study conducted by Veldhuis et al. achieved a maximum density of 95% for the sol-gel spin coated YSZ film within 5min at 1000°C. The utilization of a microwave-assisted rapid thermal annealing process resulted in crack-free 70nm thin films (Veldhuis et al. 2015). Besides, it has been reported that the homogeneity of the film can be improved by dispersing the corresponding nanoparticles into the solution. These nanoparticles were suggested to be effective in controlling the stress fields and to reduce cracks in the deposited films (Lee et al. 2017).

Lee et al. (2006) used sol-gel dip-coating method to deposit lanthanum chromite-based perovskite coating on a ferritic stainless steel (SUS444) to reduce the oxidation of the metallic interconnect. Precursor solutions for $(La, Ca)CrO_3$ (LCC) and $(La, Sr)CrO_3$ (LSC) coatings were prepared by adding nitric acid and ethylene glycol into an aqueous solution of lanthanum, strontium (or calcium), and chromium nitrates. Dried LCC and LSC gel films were heat-treated at 400–800°C after dip-coating on the SUS444 substrate. The SEM results show that microstructure of LCC film is denser than that of LSC. The porous structure of LSC layer is considered to be attributed to the formation of the $SrCrO_4$ phase. The presence of LCC and LSC thin film layers depress the oxidation of the SUS444 metallic substrate. Similarly, to reduce the oxidation and evaporation of Cr in the SUS430 ferritic stainless steel substrate, Jalilvand et al. deposited a Fe-doped Ni-Co spinel layer on the substrate using sol-gel dip-coating technique. The Fe doping was found to reduce the oxidation rate and ASR of the substrate, and the $NiCoFeO_4$-coated sample showed a low ASR value of 0.017 Ω cm^2 after 504 h oxidation at 750°C (Jalilvand and Faghihi-Sani 2013).

Sol-gel method possesses the advantages of precise composition control, simple and inexpensive processing procedure, and low processing temperatures. However, thin films based on the sol-gel process also have several demerits. For example, a large shrinkage during heat-treatment and low density from inherent high organic content may cause local defects and large tensile stress in the films, resulting in the formation of cracks and hence serious gas leakage and crossover.

4.4.1.3 Screen-Printing Method

In screen-printing process, a highly viscous paste consisting of a mixture of ceramic powder, organic binder, and plasticizer is forced through the open meshes of a screen using a squeegee. The screen-printed films are dried and sintered at high temperatures. Parameters such as grain size, grain form, slurry viscosity, printing times, and sintering temperature and time are important for the quality and densification of the screen-printed film. Slip casting and screen-printing usually involve large shrinkage associated with the removal of polymeric binders and plasticizers in subsequent sintering and heat treatment stages. This would deteriorate the quality of the thin films. A recent review by Somalu et al. comprehensively discussed the influence of ink rheology and processing conditions on the screen-printed film quality for high-performance SOFC electrode and/or electrolyte (Somalu et al. 2017).

Zhang et al. (2006) used screen-printing technique to fabricate $Sm_{0.2}Ce_{0.8}O_2$ (SDC, 15 μm) single-layer and YSZ (5 μm) + SDC (15μm) bilayer on Ni/YSZ cermet substrate, followed by co-firing. Co-firing at 1400°C led to the formation of Zr-rich microislands, indicating the Zr migration from the NiO/YSZ substrate during the co-firing process. Screen-printing was used to deposit $Sm_{0.2}Ce_{0.8}O_2$ (SDC) electrolyte thin film onto the green NiO/SDC anode substrates using a printing slurry consisting of SDC powder, methylcellulose, terpineol, and ethanol vehicle (Xia et al. 2001). A dense SDC electrolyte layer with thickness of ~30 μm was obtained after co-firing at 1350°C in air for 5 h. The cell with SSC cathode, Ni/GDC anode, and screen-printed SDC electrolyte achieved a power output of 397 mWcm^{-2} in H_2 and 304 mWcm^{-2} in methane at 600°C – an impressive performance at these low temperatures. Screen-printing was also used to deposit protective coating on metallic interconnect for SOFCs. Brylewski et al. screen-printed a layer of $(La,Sr)CrO_3$ on the surface of a DIN 50049 stainless steel (Brylewski et al. 2012). It was found that after long-term thermal oxidation, a Cr-rich spinel multilayer interfacial zones were formed between the coating and the metal, which acted as a barrier layer to suppress the volatilization of Cr species.

1. NiO paste prepared by mixing NiO powder with organic solution and three-roll milled.

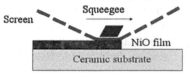

2. NiO paste screen-printed the ceramic substrate.

3. NiO film reduced in H_2 atmosphere at 700~750°C and removed from substrate.

4. Porous Ni film attached to dense Ni plate.

5. GDC, LSC, and Pt films deposited sequentially on porous Ni support.

FIGURE 4.13 Fabrication process for μ-SOFC based on a screen-printed porous Ni substrate, after Joo and Choi (Joo and Choi 2008).

Joo and Choi (2008) fabricated μ-SOFCs based on a porous and thin Ni substrate using the screen-printing technique. Figure 4.13 shows the fabrication process for μ-SOFC using screen-printing techniques. The NiO ink was made of commercial NiO powder mixed with an organic solution of α-terpineol and ethyl cellulose in a weight ratio of 10:1. A phosphate ester-based surfactant and dihydroterpineol acetate were used as dispersants. The Ni film was fabricated by screen-printing a NiO film on a ceramic substrate and subsequently reducing the printed film at 700–750°C in hydrogen. After reduction, a free-standing and porous Ni film was obtained. A thin GDC electrolyte film (~3 μm) was deposited on Ni substrate by PLD method at an oxygen partial pressure of ~30 mm torr, followed by the deposition of a porous LSC cathode for 90 min at room temperature. The deposition at room temperature would produce a porous structure of LSC as required for the cathode. The cell achieved a maximum power output of 26 mWcm^{-2} at 450°C in H_2/air. The μ-SOFC structure was further optimized by Choi et al., in which a dual-layer substrate consisting of porous stainless steel and (La, Sr)(Ti, Ni)O$_3$-YSZ was designed (Kim, Park et al. 2016). The microcell using YSZ as electrolyte and La$_{0.8}$Sr$_{0.2}$CoO$_{3-\delta}$ as cathode delivered a maximum power density of 235 mWcm^{-2} at 450°C in wet H_2. The advantage of the screen-printing is its simplicity and low cost particularly in comparison to the lithography and etching processes commonly used in the fabrication of μ-SOFCs.

4.4.1.4 Electrophoretic Deposition (EPD)

Electrophoretic deposition (EPD) is one of the colloidal processes by which ceramic films are shaped directly onto substrates from an electrostatically stabilized colloidal suspension in a DC electrical field. A DC electrical field causes these charged particles to move forward, and deposit, on an electrode with opposite charge. The EPD process has been used for the fabrication of SOFC components including YSZ (Ishihara et al. 1996, Ishihara et al. 2000, Zou et al. 2011), GDC (Ye et al. 2010, Hu et al. 2017) and (LaSr)(GaMg)O$_3$ electrolyte films (Mathews et al. 2000, Matsuda

Thin Coating Technologies 103

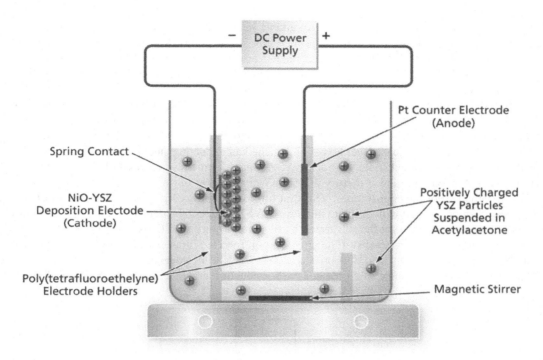

FIGURE 4.14 Schematic diagram of the electrophoretic deposition apparatus. Particles of YSZ are deposited on a NiO-YSZ substrate, after Besra et al. (Besra, Compson et al. 2006).

et al. 2003, Eba et al. 2018). A schematic diagram of the EPD apparatus is shown in Figure 4.14 (Besra, Compson et al. 2006). The EPD process is very simple and has the advantage of uniformity of deposition and high deposition rates. In the case of deposition of YSZ thin electrolyte layer, YSZ nanoparticles are dispersed in organic suspension medium such as ethanol, isopropanol, acetone, acetylacetone or iodine-dissolved acetylacetone, instead of water, to avoid the detrimental effect of water electrolysis on the quality of deposited film. Positive charges are developed on the YSZ particles due to the presence of some residual water. Maleki-Ghaleh et al. found that the EPD deposited YSZ film from acetylacetone suspension was uniform, smooth, and crack-free, whereas the films produced from ethanol or acetone suspension were rough and cracked due to particle agglomeration and low suspension viscosity (Maleki-Ghaleh et al. 2013). A solid concentration of 10 gL^{-1} was found to be suitable for EPD using acetylacetone as solvent (Besra et al. 2007). Iodine-dissolved acetylacetone is an effective solvent for EPD as it has a much lower resistance than ethanol (Ishihara et al. 1996, Chen and Liu 2001). Lower solution resistance reduces the applied voltage, depressing evolution of gases on the negative (i.e., deposited) electrode. Dispersion of YSZ in iodine (I_2) and acetone solution is effective to form charged YSZ particles as the reaction between acetone and iodine leads to the formation of hydrogen iodine (HI) and then produces dissociated protons which are absorbed by the YSZ particles; subsequently, the YSZ particles become positively charged by the addition of I_2 (Peng and Liu 2001). However, at high iodine concentration, current loss will be seen since the excessive amount of free protons with higher mobility would migrate faster than the ceramic particles toward cathode, thus consuming electrons (Zarabian et al. 2013).

EPD process can be generally characterized by two steps (Sarkar et al. 2004). Under the application of an electric field, the charged particles first migrate toward an electrode with opposite charge. The migration depends on the bulk properties of the colloidal dispersion (bath conductivity, viscosity, particle concentration, size distribution, and surface charge density) and the actual field strength in the bath. The other is coagulation of the charged particles at or near the surface of the deposited electrode, forming a solid deposit layer. A suitable heat treatment (firing or sintering) is usually required in order to further densify the deposited film and to eliminate porosity. For a detailed discussion and application of the EPD process, readers are encouraged to read an excellent review by Besra and Liu (2007). In addition, the fundamentals and recent advances of EPD process for SOFCs technology has been reviewed by Pikalova et al. (2019).

One of the prerequisites for EPD is that the substrate should be electrically conductive. Thus, for deposition of YSZ or GDC electrolyte thin films on non-conducting NiO/YSZ anode supports, thin conducting layers such as graphite and Pt are coated onto porous NiO/YSZ composite substrates to facilitate conduction on the surface. Das et al. attached a conducting steel plate on the reverse side of the porous NiO/YSZ substrate, and dense YSZ film (~3 μm) was successfully fabricated (Das et al. 2017). A maximum power density of 756 mWcm^{-2} at 800°C was achieved for the cell using LSM as cathode. For the deposition of YSZ thin films onto porous NiO/YSZ composite substrates that have been pre-coated with graphite thin layers, YSZ layer can be deposited onto the substrates or onto the graphite layers on the substrates, as shown in Figure 4.15 (Hosomi et al. 2007). Figure 4.16 shows the SEM micrographs of the cross section for YSZ films electrophoretically deposited onto NiO/YSZ substrate surface (Hosomi et al. 2007). For YSZ powders deposited onto porous NiO/YSZ substrates whose reverse side is coated with conducting graphite layers, the YSZ powders are

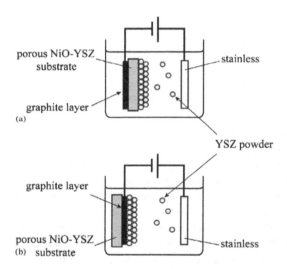

FIGURE 4.15 Experimental setup for (a) EPD of YSZ powders onto porous NiO/YSZ substrates whose reverse side is coated with conducting graphite layers and (b) EPD of YSZ powder onto conducting graphite layers on NiO/YSZ substrates after co-firing, after Hosomi et al. (Hosomi et al. 2007).

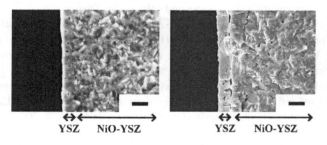

FIGURE 4.16 SEM micrographs of cross-section EPD of YSZ thin films electrophoretically deposited onto porous NiO/YSZ substrates whose reverse side is coated with conducting graphite layers (left) and onto conducting graphite layers on NiO/YSZ substrates (right), after Hosomi et al. (Hosomi et al. 2007). Bars = 5 μm.

subsequently transferred into dense and continuous films with thickness of 3–15µm after co-firing with substrates (left picture, Figure 4.15). For the YSZ films deposited onto graphite layer, an opening is observed at the interface after co-firing (right picture, Figure 4.15), most likely due to the burning and decomposition of the graphite interlayer. Instead of using graphite as the conducting layer, Das et al. directly synthesized a layer of conducting polymer onto the NiO/YSZ substrate by chemical oxidation and polymerization of pyrrole monomer, followed by the deposition of YSZ thin film on top (Das and Basu 2014). The polymer interlayer was then burnt out at high-temperature sintering step, leaving behind a well-adhered dense YSZ film on the substrate. The cell with LSM as cathode delivered a maximum power density of 0.91 Wcm^{-2} at 800°C.

Besra et al. found that in the case of EPD of YSZ thin layers on non-conducting NiO/YSZ substrates with carbon sheet backing, the deposition of YSZ increases with increasing substrate porosity, indicating the presence of a continuous "conducting path" between the electrical contact and the particles through pores in the substrate (Besra et al. 2007). The results indicate the existence of minimum threshold porosity of the NiO/YSZ substrates for EPD to happen, which is 52.5% and 58.5% porosity for 100 V and 25 V applied potentials, respectively. Cherng et al. (2008) showed that Ni/YSZ cermet presintered at 1200°C and reduced at 700°C behaves like a metal electrode and does not require the use of the additional electrical conducting backing layer. Besides, EPD has also been employed to deposit protective coatings on SOFC metallic interconnect (Zhang et al. 2011). Sun et al. deposited a layer of $Cu_{1.3}Mn_{1.7}O_4$ spinel coating on Crofer 22 APU interconnect by EDP process, and it was found that the coating layer has low-ASR and acted as a good barrier layer for oxygen and chromium migration (Sun et al. 2017).

Cherng et al. (2008) studied the EPD of YSZ thin films using aqueous suspension. To prevent the colloids from agglomeration and sedimentation during EPD, negatively charged polyelectrolytes such as ammonium polyacrylate (PAA-NH$_4$) are added as dispersant. PAA-NH$_4$ dissociates in water, forming negatively charged polyanions which would adsorb to YSZ particles to stabilize them electrostatically. Addition of ~0.1 wt% PAA-NH$_4$ is sufficient to stabilize the YSZ slurry. The aqueous EPD is characterized by low currents (<3 mAcm^{-2}) and low voltages (<20 V) on conducting substrates. By performing consecutive aqueous EDPs, Cherng et al. managed to deposit porous NiO/YSZ anode, dense YSZ electrolyte, and porous LSM cathode onto a thin Cu wire electrode, resulting in an anode-supported microtubular SOFC device (Cherng et al. 2013). The tubular cell showed a maximum power density of 363.8 mWcm^{-2} at 800°C. Will et al. (2001) showed that through the adjustment of shrinkage and the shrinkage rate of the deposited zirconia layer on the pre-sintered porous Ni/YSZ substrate, thin, dense layers without cracks can be prepared. The thickness of the deposited YSZ layer could be monitored from the total charge transfer during the EPD process. EPD method was also used to deposit NiO/YSZ anode and YSZ bilayer structure with a good interface bonding between the anode and electrolyte (Besra, Zha et al. 2006) or SDC/YSZ bilayer electrolyte structure (Wang et al. 2015), and to deposit YSZ on porous LSM cathode substrate (Ishihara et al. 2000, Peng and Liu 2001, Kalinina et al. 2014).

The EPD can be used for mass production and has the advantages of short formation times (approximately 1 µm film/min deposition rate), little restriction in the shape of deposition substrates, suitability for mass production, and a simple deposition apparatus. In addition, there is no requirement of binder burnout because the green coating contains little or no organics. However, it has been shown that five or more successive repetitions of the process are necessary to produce a gastight and dense YSZ layer and to achieve a good cell performance (Ishihara et al. 1996, Ishihara et al. 2000).

4.4.2 Ceramic Powder Techniques

4.4.2.1 Tape-Casting and Freeze-Tape-Casting Process

Tape-casting is a commercial processing technology which has been used extensively for the manufacturing of electronic and structural ceramics with thicknesses typically ranging from 25 to 1000 µm and is particularly suitable for making anode-supported planar SOFC. Tape-casting involves spreading of slurries of the ceramic powders and organic ingredients such as the modifier, binders, plasticizer, surfactant, and solvents (ethanol and toluene) onto a flat surface where solvents are allowed to evaporate. After drying, the resulted tape develops a leather-like consistency and can be stripped off from the casting surface. The ceramic powder is the only ingredient that determines the properties of the final tape. Typically, tape-cast anode layers 1–2 mm thick and tape-cast electrolyte layers 40–100 μm thick are produced and these two tapes are laminated and after rolling or calendaring, the green laminates are cut to size and sintered at 1300–1400°C in a specially designed kiln furnace to ensure the parts are all flat. The final anode-supported structures are produced with anode support thicknesses of 500–1000 μm and electrolyte thicknesses of 15–40 μm. Tape-casting technique is also used to produce thick electrolytes of ~100 μm for electrolyte-supported cells. Figure 4.17 shows the typical structure of YSZ electrolyte-supported and Ni/YSZ cermet anode-supported thin YSZ cells. Mismatch in densification of such bilayer structure will cause intolerable bending and/or cracking of the structure during co-firing (Jean et al. 1997, Rubio et al. 2017). Thus, match in the sintering profile of the laminated anode and electrolyte tapes is critical in producing flat anode-supported structures.

Tape-casting is also used to prepare porous anode substrate for anode-supported thin electrolyte cells. Strategies adopted to produce porous substrate usually include the use of pore formers, freeze-casting, phase inversion, and partial sintering. Park et al. (2001) reported the fabrication of anode-supported thin electrolyte SOFCs by using two tape-casting layers of YSZ, one containing a pore former as a porous YSZ anode matrix and one without pore former as the electrolyte layer. The porosity of the anode tape was controlled by the addition of pore formers. Addition of 30% pore former resulted in the formation of ~50% porosity for the YSZ anode substrate tape. The shape of pores in the porous YSZ matrix is somehow related to the shape of the pore formers that are used. After the formation of the composite structure at 1550°C, copper and ceria were added to the porous YSZ layer by wet impregnation. However, the ability to engineer pore structures is limited to the manipulation of thermal fugitive particle orientation and stability during the slurry process. A recent review published by Nishihora et al. presented an overview of the fundamental aspects of the tape-casting technique, and summarized the latest strategies and advancements to manufacture thin ceramic membrane tape with tailored porosity (Nishihora et al. 2018).

A combination of the tape-casting process and freeze-casting process has resulted in a new freeze-tape-casting process that has been developed as a direct means of forming and controlling complex pore structures in green tapes. The freeze-tape-casting process not only allows tailoring of continuously graded pores through the entire cross section but also slows down for long-range

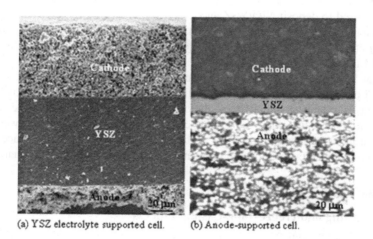

(a) YSZ electrolyte supported cell. (b) Anode-supported cell.

FIGURE 4.17 YSZ electrolyte-supported cell and Ni/YSZ anode-supported cells prepared by tape-casting technique.

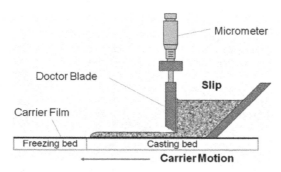

FIGURE 4.18 Freeze-tape-casting apparatus.

alignment of acicular pores from the surface. The final pore morphology can be tuned by the solid particle size and content in the slurry, the freezing conditions (pressure, temperature, time and rate), and types of solvent and additives (Liu et al. 2016). The freeze-tape-casting process starts with the traditional tape-casting process, where an aqueous ceramic slip is cast onto a Mylar™ or Teflon™ carrier film via a doctor blade apparatus. A standard tape-caster is utilized that has been modified only with a thermally isolated freezing bed to allow for unidirectional solidification of the slurry after casting. As with traditional tape-casting, the slip contains sufficient organic binders that make the tape strong and flexible after solvent evaporation for handling and cutting. The freezing of the tape, typically solidified in just several minutes, eliminates particles settling out of the suspension. After the solidification process, the tape is subsequently cut and freeze-dried under vacuum for quick solvent removal through sublimation, where the frozen liquid transforms from a solid to a gas without an intermediate liquid phase. Figure 4.18 shows a typical freeze-tape-casting apparatus. Freeze-casting has several advantages, such as readily controllable membrane porosity, simple sintering process, wide applicability to various materials, and cost-effective production. Sofie (2007) investigated in detail, the fabrication of functionally graded and continuously aligned porous YSZ tapes (500–1000 μm) by a modified freeze-tape-casting process using aqueous and tertiary butyl alcohol (TBA) solvents. The result indicates that the freeze-tape-casting technology can be effective in fabricating fuel cell components. Freeze-casting can be very useful in fabricating electrodes with introduced hierarchical porosity to increase TPB while maintaining adequate fuel flow. Miller et al. studied the effect of alcohol addition on the pore structure of freeze-cast YSZ to evaluate the suitability of the alcohol for use in making freeze-tape-cast SOFC stacks (Miller et al. 2018). The use of alcohol additives in aqueous YSZ slurries was found to result in highly complex pore structures with many ceramic bridges and domains of disordered, non-parallel lamellar pores, which gave qualitatively higher mechanical strength. Panthi et al. successfully employed freeze-casting method to fabricate tubular Ni-YSZ anode supports with radically aligned and graded pore channels for SOFCs, while the YSZ electrolyte and LSM-YSZ cathode layers were coated on the anode support by dip-coating (Panthi et al. 2019). With a mixture of H_2 and N_2 as the fuel and ambient air as the oxidant, the single cell delivered maximum power densities of 0.47, 0.36, and 0.27 W/cm² at 800°C, 750°C, and 700°C, respectively. With a mixture of H_2 and N_2 as the fuel and ambient air as the oxidant, the single cell delivered maximum power densities of 0.47, 0.36, and 0.27 W/cm² at 800°C, 750°C, and 700°C, respectively.

4.4.2.2 Tape-Calendering Process

The tape-calendering process for making thin film SOFCs is based on the progressive rolling of green (unfired) ceramic tapes to produce a thin electrolyte film (typically 0.5–10 μm) on an electrode support (Minh 1988, Singh and Minh 2004). The tape-calendering process (using an anode as the support) and green electrolyte/anode bilayer are shown in Figure 4.19 (Minh 2004, Medvedev et al. 2016). In this fabrication process, electrolyte and anode powders are first mixed with organic

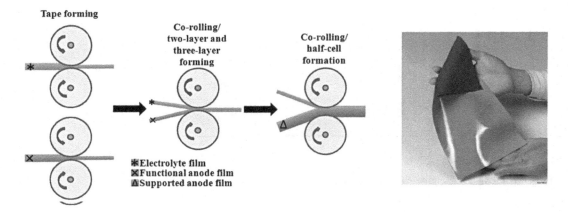

FIGURE 4.19 Tape calendaring process and green electrolyte/anode bilayer, after Medvedev and Minh (Minh 2004, Medvedev et al. 2016).

binders to form ceramic masses. The masses, having a doughy consistency with many of the characteristics of a plastic, are rolled into tapes using a two-roll mill. Electrolyte and anode tapes of certain thickness ratios are laminated and rolled into a bilayer tape. This thin bilayer is then laminated with a thick anode tape, and the lamination is rolled again into a thin bilayer tape. The process can be repeated with different tape thickness ratios until a desired electrolyte film thickness is obtained. Shin et al. reported that by using tape-calendering, the defect generation and residue stress inhomogeneity were significantly reduced in the NiO/YSZ composite tape (13 × 13 cm^2), as compared to a conventional uniaxial press (Shin et al. 2020). In general, the process requires only three rollings to achieve a bilayer with micrometer-thick electrolyte films. The bilayer is fired at elevated temperatures to remove the binders and sinter the ceramics. To form a thin film SOFC single cell, a cathode layer is applied on the electrolyte surface of the sintered bilayer. The process is simple and scalable. Cell size as large as 500 cm^2 can be fabricated by this method. Medvedev et al. managed to fabricated Ba(Ce, Zr)O3-based proton-conducting electrolyte fuel cell using tape-calendering, and the proton ceramic fuel cell (PCFC) delivered an acceptable maximum power density of 308 mWcm^{-2} at 725°C (Medvedev et al. 2016).

4.4.2.3 Dry-Pressing Method

Xia and Liu developed a simple and cost-effective method to fabricate a thin GDC electrolyte film on NiO/GDC anode substrates by dry pressing (Xia and Liu 2001a, 2001b, Xia and Liu 2002). In this method, the green NiO/GDC substrate was formed by uniaxial pressing of NiO/GDC powder under 200 MPa. Highly porous or "foam" GDC powder, synthesized by a glycine nitrate combustion process, was carefully spread on top of the prepressed NiO/GDC substrate, and then compressed at 250 MPa, forming a bilayer structure. Co-firing of the bilayer structure at 1350°C for 5 h yields a dense GDC membrane on a porous NiO/GDC electrode substrate. Care must be taken to ensure the uniform distribution of the GDC powder on the substrate. Figure 4.20 is the SEM micrographs of the as-synthesized GDC powder and the cross sections of a fuel cell consisting of thin GDC electrolyte layer prepared by dry pressing (Xia and Liu 2001b). The GDC powder is highly porous (Figure 4.19a and b). The relative density of the GDC foam powder is 0.84%, substantially smaller than the theoretical density of GDC (7.12 gcm^{-3}). Low relative density indicates the low fill density of the powder. It is suggested that the extremely low fill density of the foam-like GDC is critical for the successful preparation of a thin, dense GDC electrolyte on a porous NiO/GDC substrate. The thickness of the GDC electrolyte is controlled by the amount of GDC powder. By using combined spray-coating and dry pressing method, Choi et al. managed to prepare an anode-supported YSZ (8 μm) /GDC (40 μm) bilayer electrolyte cell. The YSZ slurry was spray-coated on the NiO/GDC

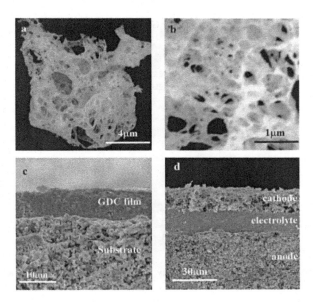

FIGURE 4.20 SEM micrographs of (a) a highly porous and foam-like GDC particle, (b) a portion of the particle shown in (a), (c) cross section of an 8-μm-thick GDC film on a NiO/GDC substrate, and (d) a fuel cell consisting of a 15-μm-thick GDC electrolyte, a Ni/GDC anode and an $Sm_{0.5}Sr_{0.5}CoO_3$ cathode, after Xia and Liu (Xia and Liu 2001a, b).

anode support, and then GDC powder was dry-pressed on top of the YSZ layer, followed by high-temperature sintering at 1400°C (Choi et al. 2014). The final cell with LSCF/GDC cathode showed a peak power density of 218 mWcm^{-2} at 600°C.

The Gd-doped ceria electrolyte on anode substrates can also be formed in situ by solid-state reaction by dry-pressing the stoichiometric Gd_2O_3 and CeO_2 oxides (Leng et al. 2004). The XRD analysis indicates that a single solid solution of Gd-doped CeO_2 is formed after sintering at 1450°C. A maximum power density of 0.58 Wcm^{-2} at 600°C was obtained on a 10-μm-thick GDC film and a LSCF/GDC composite cathode. Duan et al. also reported the successful fabrication of $Ba(Ce, Zr)O_3$-based proton conducting electrolyte fuel cells by dry pressing the anode and electrolyte precursors, followed by high-temperature co-sintering to form dense electrolyte layer (Duan et al. 2015). The PCFC button cells were reported to produce a maximum power density of 455 mWcm^{-2} at 500°C on H_2. The dry pressing technique is simple and cost-effective. However, to form a uniform and thin electrolyte film requires experience and skill particularly for the powder with high fill density. The scale-up of the dry pressing process may also be a problem due to the nature of the manual handing process.

4.5 FLAME-ASSISTED COLLOIDAL PROCESS

4.5.1 Spray Pyrolysis and Flame-Assisted Vapor Deposition

Figure 4.21 shows a typical spray pyrolysis setup (Park et al. 2016). A sufficient force, for example, using a stream of gas at high speed, applied to the surface of a metal salt precursor solution (usually aqueous or alcoholic) or colloidal suspension in the atomizer/nozzle causes the emission of droplets. The solution can also be forced through the nozzle with a syringe pump. Sprayed droplets reaching the substrate surface undergo pyrolytic decomposition. Newly deposited flat droplets, with thickness in the 10–20 nm range, pile up on the previously deposited ones and undergo pyrolytic decomposition as well. This process continues until a film thickness of 100–500 nm is reached (Perednis and

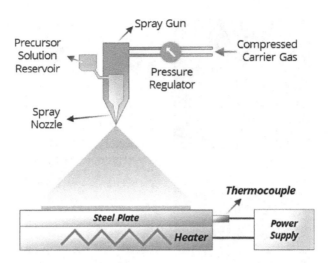

FIGURE 4.21 A schematic diagram of a typical spray pyrolysis setup, after Park et al. (Park et al. 2016).

Gauckler 2005). The degree of decomposition is determined by the relationship between the substrate temperature, the boiling temperature of the solvents, and the melting point of the salts used for the precursor. A compromise has to be found between sufficiently high deposition temperature to achieve complete decomposition if possible and the facility to deposit the droplets still in a wet state on the substrate for a pilling up of the droplets. A wide range of precursors, including metal acetates, chlorides, nitrates, lactate, citrate, and oxalates solution, can be employed to fabricate a variety of materials. The spray pyrolysis synthesis method has been used widely for the preparation of SOFC electrolyte thin films, thin film porous electrodes as well as dense protective coatings for interconnect. Spray pyrolysis method offers high film quality and low processing costs compared to other thin film deposition techniques such as PLD and chemical or physical vapor deposition. The spray pyrolyzed thin films are usually amorphous after deposition (Perednis and Gauckler 2004) and the amorphous state can be converted to a nanocrystalline isotropic microstructure with grain boundaries perpendicular to the film surface (Rupp et al. 2006).

Rupp et al. did a comprehensive characterization of $Gd_{0.2}Ce_{0.8}O_{1.9-x}$ (GDC) thin film deposited by spray pyrolysis technique at 350°C and annealed at 1000°C (Rupp et al. 2007). The film composition was $Gd_{0.23}Ce_{0.77}O_{1.9-x}$, slightly different from the ratio of the components in the precursor. The results also show that the pyrolysis process would lead to thin films that contain trace amounts of carbon and hydroxyl groups from the precursors. However, the residues of the spray pyrolysis film appear to have little effect on the electrical properties of the GDC films. The cracks formed during the thermal decomposition and heat treatment could be eliminated by optimizing the substrate temperature and repeating the film deposition and heating cycle (Setoguchi et al. 1990). In addition, it was reported that highly dense GDC films (~2.5 μm) with low defect concentration can be realized when the deposition was changed from continuous, single step, to various shorter steps (Halmenschlager et al. 2013). Reolon et al. also adopted spray pyrolysis method to deposit dense GDC electrolyte thin film on metal-supported SOFC device (Reolon et al. 2014). The GDC films were found crystalline after deposition without requiring post heat treatment, and the quality of the films depended greatly on the air flow rate. When the film was deposited for 12 cycles (with a final thickness of ~3.30 μm), the maximum power density of the metal-supported cell reached 510 mW cm^{-2} at 650°C.

Spray pyrolysis was also used to deposit porous and catalytic active NiO/SDC composite anode films at substrate temperature 350°C and annealing temperature 500°C (Patil et al. 2008). Using a precursor sol consisting of solution prepared from strontium acetate, lanthanum nitrate, and manganese nitrate dissolved in propane-1,2-diol and LSM powder at a concentration of 0.2 ml/L, a thin porous LSM electrode with distribution of pores between 2 and 3 μm is deposited by spray pyrolysis

Thin Coating Technologies

(Charpentier et al. 2000). Ultrafine, nanostructured Sr-doped $SmCoO_3$ and SDC composite powders were successfully synthesized by Shimada et al. using spray pyrolysis (Shimada et al. 2016). The application of the nanosized composite powder achieved an extremely fine cathode microstructure, with which the anode-supported SOFC delivered a maximum power density of 0.76 Wcm^{-2} at 650°C. Porous coating can also be obtained by powder spray without subsequent decomposition at the substrate (i.e., wet powder spraying). However, the adhesion and uniformity of the porous coating would be difficult to be controlled. Spray pyrolysis is also an effective technique to deposit protective coatings on SOFC interconnects. Korb et al. applied a $La_{0.6}Sr_{0.4}CoO_3$ layer on ferritic stainless steel, and its behavior was investigated after oxidation in air at 800°C (de Angelis Korb et al. 2013). The layer film was uniformly deposited and adhered well to the metallic substrate; the coated steel evidently showed lowered oxidation rates than that of the uncoated counterpart. In addition, Kamecki et al. coated a $MnCo_2O_4$ spinel layer on steel substrates by spray pyrolysis, and managed to lower the deposition temperature down to 390°C which was much lower than previously reported (usually in excess of 900°C). Dense coating with good adhesion and crack-free microstructure was obtained without requiring post high temperature treatment, and the electrical conductivity of the coating was found the same as for bulk spinel sample (Kamecki et al. 2018).

Flame-assisted vapor deposition (FAVD) is a combination of spray pyrolysis and flame synthesis. In this method, an atomized solution is sprayed through a flame in an open atmosphere in which decomposition and combustion reactions occur, resulting in a stable film deposited on a heated substrate. This method requires simple apparatus and is performed at a relatively low temperature and at a high deposition rate. Choy et al. (1997) prepared porous LSM cathodes by FAVD. Ethanol is added to the nitrite precursor solution to make the solution more inflammable. The morphology and porosity of the LSM coating depend strongly on the deposition temperature (i.e., the substrate temperature). An LSM electrode deposited at a deposition temperature of 710°C produced a rather high polarization resistance of 1.34 Ωcm^2 at 900°C. FAVD was also used to fabricate multilayer LSM/doped CeO_2 on YSZ electrolyte (Charojrochkul et al. 1999). Sansernnivet et al. fabricated $La_{0.8}Sr_{0.2}CrO_3$ (LSC) films on a stainless steel substrate using the FAVD technique for SOFC application (Sansernnivet et al. 2010). Both dense and porous LSC films were prepared depending on the various processing parameters such as fuel to water ratio, air pressure, flow rate of a precursor, and the distance between the spray nozzle and the substrate. The porous LSC film showed comparable catalytic activity to metallic catalysts but with less inlet steam required for hydrogen production *via* a steam reforming reaction.

Similar to FAVD, combustion CVD (CCVD) is an open-air, flame-assisted chemical deposition process, capable of producing a wide range of coating morphologies from very dense to highly porous structures. Liu et al. (Liu et al. 2004) successfully employed CCVD to fabricate functionally graded LSM/LSC/GDC cathodes on YSZ electrolyte using nitrate solution precursors. In this method, methane is used as the fuel gas and oxygen as the oxidant for the combustion flame. Grain size as small as ~50 nm can be obtained. By using CCVD method, Dhonge et al. managed to prepare thin zirconia films with monoclinic, tetragonal, and cubic structures depending on the concentration of yttrium-doped (Dhonge et al. 2011). With increased deposition temperature, coagulation of clusters of nanocrystallites (~100 nm) to large grains led to the formation of relatively dense YSZ film.

The spray pyrolysis deposition has the advantages of simple set-up, inexpensive and non-toxic precursors, high deposition efficiency (can be easily scaled up with a productivity of tons per day), and direct deposition under ambient atmosphere.

4.5.2 Electrostatic Spray Pyrolysis Deposition

Spray pyrolysis of aerosol under electrostatic field, electrostatic spray pyrolysis deposition or electrostatic spray deposition (ESD), has been developed and used to prepare thin electrolyte and electrode layers for SOFCs (Fu et al. 2005). In the case of ESD, a high DC voltage (e.g., 6–10 kV) is applied between the nozzle (positive polarity) and the grounded substrate (negative polarity) of a normal spray pyrolysis device of Figure 4.20. The distance between the nozzle and the grounded substrate is in the range of 30–50 mm. ESD makes use of electrostatic charging to disperse the liquid. The advantage of electrostatic dispersion is that the unipolar (usually positive) charge helps to achieve very small drop size. The charge also prevents coalescence of drops, hence agglomeration of particles, during spray. Also, the electric field allows a high degree of control over the direction of flight and the distribution of the rate of deposition over the substrate, by which materials can be prepared with various microstructures, such as porous, dense, and granular. In a typical ESD process, generally five steps are involved: (1) spray formation, (2) droplet transport, evaporation and disruption, (3) droplet landing, (4) droplet spreading and penetration, and (5) decomposition, reaction, and surface diffusion. For fundamental aspects of ESD or electrospraying techniques in general, readers should refer to a detailed review by Jaworek (Jaworek 2007).

Nomura et al. (2005) studied the effect of ESD process on the morphology and density of thin YSZ layers on Ni/YSZ anode substrates using a colloidal suspension of YSZ. Direct use of colloidal suspension would avoid the chemical reactions associated with the precursor solutions during the dispersion and deposition process. Operating parameters such as colloidal concentration, particle size of feed solution, solution medium, flow rate, nozzle tip shape, distance between nozzle and substrate, substrate temperature, and applied voltage were found to be important on the quality and morphology of the deposited YSZ films. For example, the shape of the nozzle tip strongly affects the type of spray. A thin-layer (~3 µm thickness) YSZ electrolyte was deposited on Ni/YSZ anode substrate, achieving an open circuit of 1.06 V at 800°C. Bailly et al. investigated the electrical properties of the ESD deposited YSZ film on Ni/YSZ substrate. The electrical properties of the films (~4 µm) obtained from suspension, determined using an original calibration method of a bulk platinum microelectrode, were found to be identical to that of a bulk reference sample (Bailly et al. 2012).

Fu et al. (2005) used ESD to deposit porous LSCF films on stainless steel and glass substrates using metal nitrate precursor solutions. The as-deposited films are amorphous. After calcination at 750°C, the deposited films crystallize to form perovskite phase. Films obtained at the deposition temperature of 350°C are much more porous than that obtained at 150 and 250°C. The reason is considered due to the preferential landing of aerosol droplets and agglomeration of particles. Princivalle and Djurado (2008) studied the effect of YSZ content and nozzle-to-substrate distance on the morphology of LSM/YSZ composite cathodes on YSZ substrate by ESD. Shown in Figure 4.22 are the SEM micrographs of the LSM/YSZ composite films deposited at 350°C as a function of YSZ content in the composites (Princivalle and Djurado 2008). The nozzle-to-substrate distance was 45 mm. When the YSZ content in the composites is lower than 50wt%, the deposited composite film morphology was found reticulate and highly porous (Figure 4.21(a-c)). In the case of 40%LSM/60%YSZ composite coatings, the ESD LSM/YSZ coating becomes much denser (Figure 4.21d). The simultaneous boiling and drying of the solution with low YSZ content is suggested to be the origin of the formation of reticulate coatings. ESD was also used to deposit a highly porous LSCF thin layer on SDC electrolyte before applying the screen-printed LSCF cathode (Hsu et al. 2008). The LSCF double layer cathode reduces the electrode polarization resistance by 50% as compared to single layered LSCF cathode. Protective coatings can be deposited on metal interconnect for SOFC by ESD technique as well. Conceição et al. applied a dense, crack-free, and thin LSM layer on SS446 alloy interconnect using ESD method (da Conceição et al. 2014). It was found that a dense LSM coating as thin as 300 nm efficiently reduced the oxidation rate of the steel by limiting outward Cr^{3+} diffusion, and the ASR of the coated steel was found very low (30 mΩ cm²) after 200 h oxidation in air at 800°C.

4.6 LITHOGRAPHY AND ETCHING TECHNIQUES FOR µ-SOFCs

Lithography in combination of etching process based on substrates such as Ni or silicon wafers has been used in the design and fabrication of miniature or µ-SOFCs. Chen et al. (2004) fabricated a micro-thin-film SOFC based on thin film deposition and microlithographic processes. The µ-SOFC

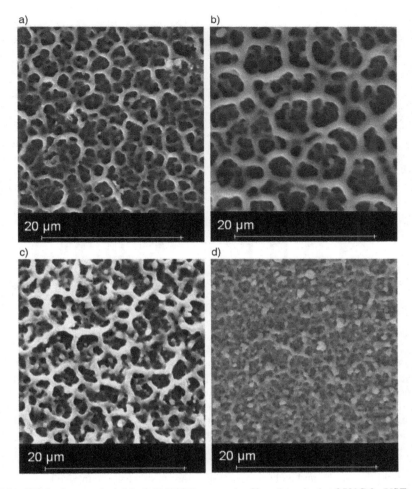

FIGURE 4.22 SEM micrographs of the LSM/YSZ composite films deposited at 350°C for YSZ content of (a) 21%, (b) 43%, (c) 50%, and (d) 60% in the composites, after Princivalle and Djurado (Princivalle and Djurado 2008).

is composed of a thin film electrolyte deposited on a nickel foil substrate by a PLD method. The Ni foil substrate was then processed into a porous anode by photolithographic patterning and wet etching to develop pores for gas transport into the fuel cell. A $La_{0.5}Sr_{0.5}CoO_3$ (LSC) thin film cathode was then deposited on the electrolyte, and a porous NiO-YSZ cermet layer is added to the anode to improve the electrode performance. The μ-SOFC yielded a maximum output power density of 110 mW/cm^2 at 570°C. Huang et al. (2007) fabricated a thin film SOFC containing 50–150 nm thick YSZ or GDC electrolyte and 80 nm porous Pt cathode and anode, using sputtering, lithography, and etching techniques. The peak power densities are 200 and 400 mW/cm^2 at 350 and 400°C, respectively. The high power densities achieved was attributed to the ultrathin electrolyte and the high charge-transfer reaction rates at the interfaces between the nanoporous electrodes (cathode and/or anode) and the nanocrystalline thin electrolyte. The unit cell area is 0.06 mm^2. Increasing the cell area may be difficult due to the inadequate mechanical strength of the free-standing electrolyte films. Symmetrical $La_{0.6}Sr_{0.4}CoO_{3-\delta}$ (LSC)/YSZ/LSC free-standing μ-SOFC membrane supported on silicon substrate was fabricated by Garbayo et al. (2014). The all-PLD deposited symmetrical cell as large as 3.5 × 3.5 μm^2 showed an in-plane conductivity of 300 S cm^{-1} within the temperature range of 450–750°C, and a small ASR of 0.3 Ω cm^2 was demonstrated at 700°C. In order to avoid

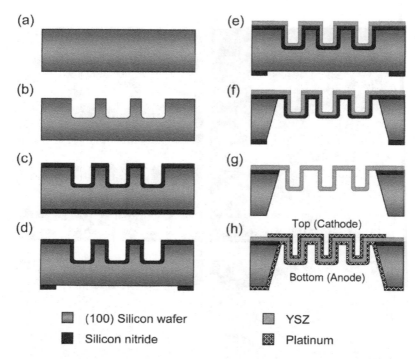

FIGURE 4.23 Process flow diagram for fabrication of the corrugated thin film SOFC. The (100) silicon substrate is etched by DRIE to generate the template for pattern transfer (a,b). A 100 nm thick silicon nitride layer is deposited with LPCVD on both sides of wafer (c). The backside of silicon nitride is patterned with openings (d), followed by ALD deposition of YSZ onto template (e). Silicon template is etched in KOH (f) and silicon nitride etch stop is removed by plasma etching (g). The free-standing corrugated electrolyte is deposited with porous platinum on top (cathode) and bottom (anode) sides acting as both electrode and catalyst, after Prinz et al. (Su et al. 2008).

the complicated lithography and etching processes, porous substrates can also be used as an alternative support. Lee et al. designed a novel metal support composed of nanoporous Ni layer coated on a macroporous ferric stainless steel, and a single cell featuring Ni-GDC/GDC/YSZ/LSC was deposited via PLD method (Lee et al. 2014). The ohmic resistance of the μSOFC deposited on the porous Ni/STS-support is stable and it shows a maximum power density of ~28 mW cm^{-2} after operation for ~112 h at 450°C.

Prinz's group (Su et al. 2008) fabricated free-standing ultrathin corrugated YSZ electrolyte cells using a sequence of MEMS processing steps. The SOFC is based on a corrugated thin film electrolyte, which is generated by a pattern transfer technique. Figure 4.23 shows the process flow for fabrication of the corrugated thin film μ-SOFC. A 350 μm thick, (100) n-type silicon wafer is used to generate the template. One side of the silicon surface is patterned by standard photolithography and deep reactive ion etching (DRIE) to create cup-shaped trenches with smooth cup side wall surface. A 100 nm thick low stress silicon nitride is deposited on both sides of the wafer with low-pressure chemical vapor deposition (LPCVD). On the opposite side to the etched trenches, the silicon nitride layer is patterned to generate square openings for subsequent silicon etching. Next, ALD technique is applied to deposit the ultrathin YSZ electrolyte, forming a corrugated film. The thickness is ~70 nm. To release the YSZ membrane, the silicon substrate is immersed into 30 wt % KOH solution for 5 h at 85°C. Pt anode and cathode are then sputtered on both sides of the free-standing YSZ membrane to form the cell. The maximum power density is 677 mWcm^{-2} at 400°C (Su et al. 2008). In order to improve the elastic and mechanical properties of a free-standing membrane for MEMS integration, Rupp's group designed thin YSZ film architectures with either straight, zigzag or spiral-shaped

Thin Coating Technologies

columnar grain nanostructures (Shi et al. 2019). A free-standing Si_3N_4 thin film template was firstly fabricated through a series of photolithography, reactive ion etching and wet etching processes, and thin YSZ films were deposited by PLD on these substrates. Zigzag and spiral-shaped structures were achieved via rotating the sample holder around the substrate normal during deposition. The variation of morphology from the typically straight columnar to the spiral-type results in nearly a 44% reduction of the elastic modulus; the improved mechanical stability has no negative effect on the ionic transport properties of the YSZ films.

However, photolithography, in combination with the etching process, is complicated and often difficult since the etching process could damage the fuel cell materials. In addition, the resolution of wet etching is generally limited to ~10 µm. Dry etching would allow much finer patterns but is time-consuming for multilayered Ni-based substrates.

4.7 CONCLUDING REMARKS

Many different techniques are available for thin film deposition and fabrication of SOFC components. They can be classified into three main groups: (1) the vapor-phase deposition, such as physical or chemical deposition, spray pyrolysis, plasma spray, etc.; (2) the liquid/colloidal-phase deposition such as EPD, sol-gel, slip casting, and dip-coating methods; and (3) the particle deposition/consolidation methods such as tape-casting and screen-printing. Physical vapor deposition techniques such as PLD and sputtering are generally expensive techniques from the point of view of equipment and have low deposition rates of a few nm per minute. But the quality of the films can be precisely controlled. Similarly, vacuum deposition techniques such as CVD, EVD, and MOCVD also suffer from relatively high equipment cost and complexity, particularly when compared to more conventional particle deposition/consolidation methods of thin film techniques that include tape-casting, EPD, screen-printing, or transfer-printing, among others. Despite these challenges, physical deposition and vacuum methods offer a number of unique advantages. Very thin, fully dense films can be produced on either porous or dense substrates, and can be formed at temperatures much lower than required in traditional ceramic processing, avoiding unwanted interfacial reactions. Physical deposition and vacuum methods are also well suited to the formation of interlayers, where small grain sizes and thin layer thickness are required. Among the most unique aspects of physical and vacuum deposition is the ability to produce unique structures that are not otherwise achievable (Pederson et al. 2006). On the other hand, VPS has the advantages of high deposition rate and the ability to be automated. However, there are two major concerns for the electrolyte deposited by VPS: anisotropic properties and significant variation in film thickness.

In contrast, the particle deposition/consolidation methods such as tape-casting and screen-printing are popular, inexpensive, and are easily scale-up. The main issue is the thickness reduction. The minimum scale of the layers produced by the particle deposition/consolidation methods such as tape-casting is a few hundred of microns in the state-of-the-art. On the other hand, slurry or dip-coating techniques exhibit advantages of low capital cost and simplicity of equipment and can be applied to produce very thin and high-quality electrolyte films through the control of the surface microstructure of the substrate and the sol particle and particle size distribution (Van Gestel et al. 2008). Moreover, slurry or dip-coating can be employed to fabricate thin electrolyte films not only for planar SOFCs, but also tubular SOFCs. The major concerns of the liquid or colloidal based deposition methods are the strain induced during drying or sintering, which may result in some processing defects such as cracks, pores, or delamination. Thus, the most feasible way to overcoming these problems is by a multicoating process. However, the repeated processing approach may not be economically viable for the industrial fabrication and production of the SOFCs components. For the deposition processes involving colloidal solution or liquid slurry, care should also be taken to prevent the solution from filtration into the porous substrates.

The choice of an appropriate thin film deposition technique for SOFC applications is strongly influenced by the material to be deposited, the desired film quality and microstructure, process complexity and scalability, areas of deposition, and the cost of the instrumentation and operation. The existing SOFC technology has demonstrated much higher energy efficiency with extremely low pollutant emission over conventional energy technologies; the cost of the current SOFC systems is still prohibitively high for wide commercial applications. One of the major cost items is the cost of SOFC stack fabrication. The cost of thin electrolyte films, electrode coatings, and protective coatings for metallic interconnect is an important factor in the commercial realization of SOFC technologies. Nevertheless, in cases such as μ-SOFCs, the high film quality and the ability to produce unusual configurations would outweigh other considerations for the selection of thin film deposition techniques.

ACKNOWLEDGMENT

The project was supported by the Australian Research Council under the Discovery Project Scheme (Project Nos. DP180100568 and DP180100731).

REFERENCES

Ai, N., Z. Lu, K. F. Chen, X. Q. Huang, X. B. Du and W. H. Su (2007). "Effects of anode surface modification on the performance of low temperature SOFCs." *Journal of Power Sources* 171(2): 489–494.

Aizawa, M., C. Kobayashi, H. Yamane and T. Hirai (1993). "Preparation of ZrO2-Y2O3 films by CVD using β-diketone metal chelates." *Nippon Seramikkusu Kyokai Gakujutsu Ronbunshi-Journal of the Ceramic Society of Japan* 101(3): 291–294.

An, J., Y. B. Kim, J. Park, T. M. Gur and F. B. Prinz (2013). "Three-dimensional nanostructured bilayer solid oxide fuel cell with 1.3 W/cm(2) at 450 degrees C." *Nano Lett* 13(9): 4551–4555.

Bae, J. W., J. Y. Park, S. W. Hwang, G. Y. Yeom, K. D. Kim, Y. A. Cho, J. S. Jeon and D. Choi (2000). "Characterization of yttria-stabilized zirconia thin films prepared by radio frequency magnetron sputtering for a combustion control oxygen sensor." *Journal of the Electrochemical Society* 147(6): 2380–2384.

Bailly, N., S. Georges and E. Djurado (2012). "Elaboration and electrical characterization of electrosprayed YSZ thin films for intermediate temperature-solid oxide fuel cells (IT-SOFC)." *Solid State Ionics* 222–223: 1–7.

Beckel, D., A. Bieberle-Hutter, A. Harvey, A. Infortuna, U. P. Muecke, M. Prestat, J. L. M. Rupp and L. J. Gauckler (2007). "Thin films for micro solid oxide fuel cells." *Journal of Power Sources* 173(1): 325–345.

Besra, L., C. Compson and M. Liu (2006). "Electrophoretic deposition of YSZ particles on non-conducting porous NiO/YSZ substrates for solid oxide fuel cell applications." *Journal of the American Ceramic Society* 89(10): 3003–3009.

Besra, L., C. Compson and M. L. Liu (2007). "Electrophoretic deposition on non-conducting substrates: The case of YSZ film on NiO-YSZ composite substrates for solid oxide fuel cell application." *Journal of Power Sources* 173(1): 130–136.

Besra, L. and M. Liu (2007). "A review on fundamentals and applications of electrophoretic deposition (EPD)." *Progress in Materials Science* 52(1): 1–61.

Besra, L., S. W. Zha and M. L. Liu (2006). "Preparation of NiO-YSZ/YSZ bi-layers for solid oxide fuel cells by electrophoretic deposition." *Journal of Power Sources* 160(1): 207–214.

Brylewski, T., J. Dabek, K. Przybylski, J. Morgiel and M. Rekas (2012). "Screen-printed (La,Sr)CrO3 coatings on ferritic stainless steel interconnects for solid oxide fuel cells using nanopowders prepared by means of ultrasonic spray pyrolysis." *Journal of Power Sources* 208: 86–95.

Cai, Z., T. N. Lan, S. Wang and M. Dokiya (2002). "Supported Zr(Sc)O-2 SOFCs for reduced temperature prepared by slurry coating and co-firing." *Solid State Ionics* 152: 583–590.

Chao, C. C., C. M. Hsu, Y. Cui and F. B. Prinz (2011). "Improved solid oxide fuel cell performance with nanostructured electrolytes." *ACS Nano* 5(7): 5692–5696.

Charojrochkul, S., K. L. Choy and B. C. H. Steele (1999). "Cathode electrolyte systems for solid oxide fuel cells fabricated using flame assisted vapour deposition technique." *Solid State Ionics* 121(1–4): 107–113.

Charpentier, P., P. Fragnaud, D. M. Schleich and E. Gehain (2000). "Preparation of thin film SOFCs working at reduced temperature." *Solid State Ionics* 135(1–4): 373–380.

Chen, D., C. Chen, Y. Gao, Z. Zhang, Z. Shao and F. Ciucci (2015). "Evaluation of pulsed laser deposited SrNb0.1Co0.9O3–δ thin films as promising cathodes for intermediate-temperature solid oxide fuel cells." *Journal of Power Sources* 295: 117–124.

Chen, F. L. and M. L. Liu (2001). "Preparation of yttria-stabilized zirconia (YSZ) films on La0.85Sr0.15MnO3 (LSM) and LSM-YSZ substrates using an electrophoretic deposition (EPD) process." *Journal of the European Ceramic Society* 21(2): 127–134.

Chen, K. F., Z. Lu, N. Ai, X. Q. Huang, Y. H. Zhang, X. D. Ge, X. S. Xin, X. J. Chen and W. H. Su (2007). "Fabrication and performance of anode-supported YSZ films by slurry spin coating." *Solid State Ionics* 177(39–40): 3455–3460.

Chen, L., C. L. Chen, D. X. Huang, Y. Lin, X. Chen and A. J. Jacobson (2003). *High temperature electrical conductivity of epitaxial Gd-doped CeO2 thin films. 14th International Conference on Solid State Ionics*, Monterey, CA.

Chen, X., N. J. Wu, L. Smith and A. Ignatiev (2004). "Thin-film heterostructure solid oxide fuel cells." *Applied Physics Letters* 84(14): 2700–2702.

Chen, X., S.-L. Zhang, C.-X. Li and C.-J. Li (2019). "Optimization of Plasma-Sprayed Lanthanum Chromite Interconnector Through Powder Design and Critical Process Parameters Control." *Journal of Thermal Spray Technology* 29(1–2): 212–222.

Chen, Y. Y. and W. C. J. Wei (2006). "Processing and characterization of ultra-thin yttria-stabilized zirconia (YSZ) electrolytic films for SOFC." *Solid State Ionics* 177(3–4): 351–357.

Cherng, J. S., J. R. Sau and C. C. Chung (2008). "Aqueous electrophoretic deposition of YSZ electrolyte layers for solid oxide fuel cells." *Journal of Solid State Electrochemistry* 12(7–8): 925–933.

Cherng, J. S., C. C. Wu, F. A. Yu and T. H. Yeh (2013). "Anode morphology and performance of micro-tubular solid oxide fuel cells made by aqueous electrophoretic deposition." *Journal of Power Sources* 232: 353–356.

Choi, H.-J., Y.-H. Na, D.-W. Seo, S.-K. Woo and S.-D. Kim (2016). "Densification of gadolinia-doped ceria diffusion barriers for SOECs and IT-SOFCs by a sol–gel process." *Ceramics International* 42(1): 545–550.

Choi, H., G. Y. Cho and S.-W. Cha (2014). "Fabrication and characterization of anode supported YSZ/GDC bilayer electrolyte SOFC using dry press process." *International Journal of Precision Engineering and Manufacturing-Green Technology* 1(2): 95–99.

Choi, M., J. Y. Koo, M. Ahn and W. Lee (2017). "Effects of grain boundaries at the electrolyte/cathode interfaces on oxygen reduction reaction kinetics of solid oxide fuel cells." *Bulletin of the Korean Chemical Society* 38(4): 423–428.

Chour, K. W., J. Chen and R. Xu (1997). "Metal-organic vapor deposition of YSZ electrolyte layers for solid oxide fuel cell applications." *Thin Solid Films* 304(1–2): 106–112.

Choy, K. L., S. Charojrochkul and B. C. H. Steele (1997). "Fabrication of cathode for solid oxide fuel cells using flame assisted vapour deposition technique." *Solid State Ionics* 96(1–2): 49–54.

Chu, W. F. (1992). "Thin-film and thick-film solid ionic devices" *Solid State Ionics* 52(1–3): 243–248.

da Conceição, L., L. Dessemond, E. Djurado and E. N. S. Muccillo (2014). "La0.7Sr0.3MnO3–δ barrier for Cr2O3-forming SOFC interconnect alloy coated by electrostatic spray deposition." *Surface and Coatings Technology* 254: 157–166.

Das, D., B. Bagchi and R. N. Basu (2017). "Nanostructured zirconia thin film fabricated by electrophoretic deposition technique." *Journal of Alloys and Compounds* 693: 1220–1230.

Das, D. and R. N. Basu (2014). "Electrophoretic Deposition of Zirconia Thin Film on Nonconducting Substrate for Solid Oxide Fuel Cell Application." *Journal of the American Ceramic Society* 97(11): 3452–3457.

de Angelis Korb, M., I. D. Savaris, E. E. Feistauer, L. S. Barreto, N. C. Heck, I. L. Müller and C. de Fraga Malfatti (2013). "Modification of the La0.6Sr0.4CoO3 coating deposited on ferritic stainless steel by spray pyrolysis after oxidation in air at high temperature." *International Journal of Hydrogen Energy* 38(11): 4760–4766.

de Larramendi, I. R., N. Ortiz, R. Lopez-Anton, J. I. R. de Larramendi and T. Rojo (2007). "Structure and impedance spectroscopy of La0.6Ca0.4Fe0.8Ni0.2O3-delta thin films grown by pulsed laser deposition." *Journal of Power Sources* 171(2): 747–753.

De Vero, J. C., K. Develos-Bagarinao, H. Matsuda, H. Kishimoto, T. Ishiyama, K. Yamaji, T. Horita and H. Yokokawa (2018). "Sr and Zr transport in PLD-grown Gd-doped ceria interlayers." *Solid State Ionics* 314: 165–171.

Dhonge, B. P., T. Mathews, S. Rajagopalan, S. Dash, S. Dhara and A. K. Tyagi (2011). *Cubic fluorite yttria stabilized zirconia (YSZ) film synthesis by combustion chemical vapour deposition(C-CVD). International Conference on Nanoscience, Engineering and Technology (ICONSET 2011)*, Chennai, India, IEEE.

Duan, C., J. Tong, M. Shang, S. Nikodemski, M. Sanders, S. Ricote, A. Almansoori and R. O'Hayre (2015). "Readily processed protonic ceramic fuel cells with high performance at low temperatures." *Science* 349(6254): 1321–1326.

Dunn, B., G. C. Farrington and B. Katz (1993). *Sol-Gel Approaches For Solid Electrolytes And Electrode Materials. 9th International Conference on Solid State Ionics*, The Hague, Netherlands.

Eba, H., C. Anzai and S. Ootsuka (2018). "Observation of cation diffusion and phase formation between solid oxide layers of lanthanum gallate-based fuel cells." *Materials Transactions* 59(2): 244–250.

Endo, A., M. Ihara, H. Komiyama and K. Yamada (1996). "Cathodic reaction mechanism for dense Sr-doped lanthanum manganite electrodes." *Solid State Ionics* 86–8: 1191–1195.

Fauchais, P., V. Rat, U. Delbos, J. F. Coudert, T. Chartier and L. Bianchi (2005). "Understanding of suspension DC plasma spraying of finely structured coatings for SOFC." *IEEE Transactions on Plasma Science* 33(2): 920–930.

Fedtke, P., M. Wienecke, M. C. Bunescu, T. Barfels, K. Deistung and M. Pietrzak (2004). "Yttria-stabilized zirconia films deposited by plasma spraying and sputtering." *Journal of Solid State Electrochemistry* 8(9): 626–632.

Fergus, J. W. (2005). "Metallic interconnects for solid oxide fuel cells." *Materials Science and Engineering a-Structural Materials Properties Microstructure and Processing* 397(1–2): 271–283.

Fu, C. Y., C. L. Chang, C. S. Hsu and B. H. Hwang (2005). "Electrostatic spray deposition of La0.8Sr0.2Co0.2Fe0.8O3 films." *Materials Chemistry and Physics* 91(1): 28–35.

Gannon, P., M. Deibert, P. White, R. Smith, H. Chen, W. Priyantha, J. Lucas and V. Gorokhousky (2008). *Advanced PVD protective coatings for SOFC interconnects. Symposium on Materials in Clean Power Systems II held at the 2007 TMS Annual Conference and Exposition*, Orlando, FL.

Garbayo, I., V. Esposito, S. Sanna, A. Morata, D. Pla, L. Fonseca, N. Sabaté and A. Tarancón (2014). "Porous La0.6Sr0.4CoO3–δ thin film cathodes for large area micro solid oxide fuel cell power generators." *Journal of Power Sources* 248: 1042–1049.

Garcia-Garcia, F. J., F. Yubero, A. R. González-Elipe, S. P. Balomenou, D. Tsiplakides, I. Petrakopoulou and R. M. Lambert (2015). "Porous, robust highly conducting Ni-YSZ thin film anodes prepared by magnetron sputtering at oblique angles for application as anodes and buffer layers in solid oxide fuel cells." *International Journal of Hydrogen Energy* 40(23): 7382–7387.

Geng, S., Q. Wang, W. Wang, S. Zhu and F. Wang (2012). "Sputtered Ni coating on ferritic stainless steel for solid oxide fuel cell interconnect application." *International Journal of Hydrogen Energy* 37(1): 916–920.

Gibson, I. R., G. P. Dransfield and J. T. S. Irvine (1998). "Influence of yttria concentration upon electrical properties and susceptibility to ageing of yttria-stabilised zirconias." *Journal of the European Ceramic Society* 18(6): 661–667.

Gong, L., J. Lu and Z. Ye (2010). "Transparent and conductive Ga-doped ZnO films grown by RF magnetron sputtering on polycarbonate substrates." *Solar Energy Materials and Solar Cells* 94(6): 937–941.

Gorte, R. J., J. M. Vohs and S. McIntosh (2004). "Recent developments on anodes for direct fuel utilization in SOFC." *Solid State Ionics* 175(1–4): 1–6.

Gupta, M., A. Weber, N. Markocsan and N. Heiden (2017). "Development of plasma sprayed Ni/YSZ anodes for metal supported solid oxide fuel cells." *Surface and Coatings Technology* 318: 178–189.

Halmenschlager, C. M., R. Neagu, L. Rose, C. F. Malfatti and C. P. Bergmann (2013). "Influence of the process parameters on the spray pyrolysis technique, on the synthesis of gadolinium doped-ceria thin film." *Materials Research Bulletin* 48(2): 207–213.

Hausmann, D. M., E. Kim, J. Becker and R. G. Gordon (2002). "Atomic layer deposition of hafnium and zirconium oxides using metal amide precursors." *Chemistry of Materials* 14(10): 4350–4358.

Heiroth, S., R. Frison, J. L. M. Rupp, T. Lippert, E. J. Barthazy Meier, E. Müller Gubler, M. Döbeli, K. Conder, A. Wokaun and L. J. Gauckler (2011). "Crystallization and grain growth characteristics of yttria-stabilized zirconia thin films grown by pulsed laser deposition." *Solid State Ionics* 191(1): 12–23.

Hermawan, E., G. Sang Lee, G. Sik Kim, H. Chul Ham, J. Han and S. Pil Yoon (2017). "Densification of an YSZ electrolyte layer prepared by chemical/electrochemical vapor deposition for metal-supported solid oxide fuel cells." *Ceramics International* 43(13): 10450–10459.

Hidalgo, H., E. Reguzina, E. Millon, A. L. Thomann, J. Mathias, C. Boulmer-Leborgne, T. Sauvage and P. Brault (2011). "Yttria-stabilized zirconia thin films deposited by pulsed-laser deposition and magnetron sputtering." *Surface and Coatings Technology* 205(19): 4495–4499.

Hidalgo, H., A. L. Thomann, T. Lecas, J. Vulliet, K. Wittmann-Teneze, D. Damiani, E. Millon and P. Brault (2013). "Optimization of DC reactive magnetron sputtering deposition process for efficient ysz electrolyte thin film sofc." *Fuel Cells* 13(2): 279–288.

Hobein, B., F. Tietz, D. Stover and E. W. Kreutz (2002). "Pulsed laser deposition of yttria stabilized zirconia for solid oxide fuel cell applications." *Journal of Power Sources* 105(2): 239–242.

Horita, T., K. Yamaji, M. Ishikawa, N. Sakai, H. Yokokawa, T. Kawada and T. Kato (1998). "Active sites imaging for oxygen reduction at the La0.9Sr0.1MnO3-x/yttria-stabilized zirconia interface by secondary-ion mass spectrometry." *Journal of the Electrochemical Society* 145(9): 3196–3202.

Hosomi, T., M. Matsuda and M. Miyake (2007). "Electrophoretic deposition for fabrication of YSZ electrolyte film on non-conducting porous NiO-YSZ composite substrate for intermediate temperature SOFC." *Journal of the European Ceramic Society* 27(1): 173–178.

Hsu, C. S., B. H. Hwang, Y. Xie and X. Zhang (2008). "Enhancement of Solid Oxide Fuel Cell Performance by La0.6Sr0.4Co0.2Fe0.8O3-delta Double-Layer Cathode." *Journal of the Electrochemical Society* 155(12): B1240–B1243.

Hu, S., W. Li, M. Yao, T. Li and X. Liu (2017). "Electrophoretic deposition of gadolinium-doped ceria as a barrier layer on yttrium-stabilized zirconia electrolyte for solid oxide fuel cells." *Fuel Cells* 17(6): 869–874.

Huang, H., M. Nakamura, P. C. Su, R. Fasching, Y. Saito and F. B. Prinz (2007). "High-performance ultrathin solid oxide fuel cells for low-temperature operation." *Journal of the Electrochemical Society* 154(1): B20–B24.

Huang, K. and H. D. Harter (2010). "Temperature-dependent residual stresses in plasma sprayed electrolyte thin-film on the cathode substrate of a solid oxide fuel cell." *Solid State Ionics* 181(19–20): 943–946.

Huang, Y.-L., H.-J. Liu, C.-H. Ma, P. Yu, Y.-H. Chu and J.-C. Yang (2019). "Pulsed laser deposition of complex oxide heteroepitaxy." *Chinese Journal of Physics* 60: 481–501.

Hwang, J., H. Lee, J.-H. Lee, K. J. Yoon, H. Kim, J. Hong and J.-W. Son (2015). "Specific considerations for obtaining appropriate La1−Sr Ga1−Mg O3− thin films using pulsed-laser deposition and its influence on the performance of solid-oxide fuel cells." *Journal of Power Sources* 274: 41–47.

Infortuna, A., A. S. Harvey and L. J. Gauckler (2008). "Microstructures of CGO and YSZ thin films by pulsed laser deposition." *Advanced Functional Materials* 18(1): 127–135.

Ishihara, T., K. Sato and Y. Takita (1996). "Electrophoretic deposition of Y2O3-stabilized ZrO2 electrolyte films in solid oxide fuel cells." *Journal of the American Ceramic Society* 79(4): 913–919.

Ishihara, T., K. Shimose, T. Kudo, H. Nishiguchi, T. Akbay and Y. Takita (2000). "Preparation of yttria-stabilized zirconia thin films on strontium-doped LaMnO3 cathode substrates via electrophoretic deposition for solid oxide fuel cells." *Journal of the American Ceramic Society* 83(8): 1921–1927.

Itoh, H., M. Mori, N. Mori and T. Abe (1994). "Production cost estimation of solid oxide fuel-cells." *Journal of Power Sources* 49(1–3): 315–332.

Jalilvand, G. and M.-A. Faghihi-Sani (2013). "Fe doped Ni–Co spinel protective coating on ferritic stainless steel for SOFC interconnect application." *International Journal of Hydrogen Energy* 38(27): 12007–12014.

Jang, D. Y., G. D. Han, H. R. Choi, M. S. Kim, H. J. Choi and J. H. Shim (2019). "La0.6Sr0.4Co0.2Fe0.8O3-δ cathode surface-treated with La2NiO4+δ by aerosol-assisted chemical vapor deposition for high performance solid oxide fuel cells." *Ceramics International* 45(9): 12366–12371.

Jaworek, A. (2007). "Electrospray droplet sources for thin film deposition." *Journal of Materials Science* 42(1): 266–297.

Jean, J. H., C. R. Chang and Z. C. Chen (1997). "Effect of densification mismatch on camber development during cofiring of nickel-based multilayer ceramic capacitors." *Journal of the American Ceramic Society* 80(9): 2401–2406.

Jiang, S. P. (2003). "Issues on development of (La,Sr)MnO3 cathode for solid oxide fuel cells." *Journal of Power Sources* 124(2): 390–402.

Jiang, S. P. and S. H. Chan (2004). "A review of anode materials development in solid oxide fuel cells." *Journal of Materials Science* 39(14): 4405–4439.

Jiang, S. P. and X. Chen (2014). "Chromium deposition and poisoning of cathodes of solid oxide fuel cells – A review." *International Journal of Hydrogen Energy* 39(1): 505–531.

Jiang, S. P., Z. C. Liu and Z. Q. Tian (2006). "Layer-by-layer self-assembly of composite polyelectrolyte-nafion membranes for direct methanol fuel cells." *Advanced Materials* 18(8): 1068.

Jiang, S. P., S. Zhang and Y. D. Zhen (2005). "Early interaction between Fe-Cr alloy metallic interconnect and Sr-doped LaMnO3 cathodes of solid oxide fuel cells." *Journal of Materials Research* 20(3): 747–758.

Joo, J. H. and G. M. Choi (2006). "Electrical conductivity of YSZ film grown by pulsed laser deposition." *Solid State Ionics* 177(11–12): 1053–1057.

Joo, J. H. and G. M. Choi (2008). "Simple fabrication of micro-solid oxide fuel cell supported on metal substrate." *Journal of Power Sources* 182(2): 589–593.

Jung, H. Y., K. S. Hong, H. Kim, J. K. Park, J. W. Son, J. Kim, H. W. Lee and J. H. Lee (2006). "Characterization of thin-film YSZ deposited via EB-PVD technique in anode-supported SOFCs." *Journal of the Electrochemical Society* 153(6): A961–A966.

Kalinina, E. G., N. A. Lyutyagina, A. P. Safronov and E. S. Buyanova (2014). "Electrophoretic deposition of Y2O3-stabilized ZrO2 nanoparticles on the surface of dense La0.7Sr0.3MnO3−δ cathodes produced by pyrolysis and solid-state reaction." *Inorganic Materials* 50(2): 184–190.

Kamecki, B., J. Karczewski, T. Miruszewski, G. Jasiński, D. Szymczewska, P. Jasiński and S. Molin (2018). "Low temperature deposition of dense MnCo2O4 protective coatings for steel interconnects of solid oxide cells." *Journal of the European Ceramic Society* 38(13): 4576–4579.

Karpinski, A., A. Ferrec, M. Richard-Plouet, L. Cattin, M. A. Djouadi, L. Brohan and P. Y. Jouan (2012). "Deposition of nickel oxide by direct current reactive sputtering." *Thin Solid Films* 520(9): 3609–3613.

Khan, M. Z., M. T. Mehran, R.-H. Song, J.-W. Lee, S.-B. Lee, T.-H. Lim and S.-J. Park (2016). "Effect of GDC interlayer thickness on durability of solid oxide fuel cell cathode." *Ceramics International* 42(6): 6978–6984.

Kikuchi, K., F. Tamazaki, K. Okada and A. Mineshige (2006). "Yttria-stabilized zirconia thin films deposited on NiO-(Sm2O3)(0.1) (CeO2)(0.8) substrates by chemical vapor infiltration." *Journal of Power Sources* 162(2): 1053–1059.

Kim, H. J., M. Kim, K. C. Neoh, G. D. Han, K. Bae, J. M. Shin, G.-T. Kim and J. H. Shim (2016). "Slurry spin coating of thin film yttria stabilized zirconia/gadolinia doped ceria bi-layer electrolytes for solid oxide fuel cells." *Journal of Power Sources* 327: 401–407.

Kim, J. W., D. Y. Jang, M. Kim, H. J. Choi and J. H. Shim (2016). "Nano-granulization of gadolinia-doped ceria electrolyte surface by aerosol-assisted chemical vapor deposition for low-temperature solid oxide fuel cells." *Journal of Power Sources* 301: 72–77.

Kim, K. J., B. H. Park, S. J. Kim, Y. Lee, H. Bae and G. M. Choi (2016). "Micro solid oxide fuel cell fabricated on porous stainless steel: a new strategy for enhanced thermal cycling ability." *Sci Rep* 6: 22443.

Kim, S. D., S. H. Hyun, J. Moon, J. H. Kim and R. H. Song (2005). "Fabrication and characterization of anode-supported electrolyte thin films for intermediate temperature solid oxide fuel cells." *Journal of Power Sources* 139(1–2): 67–72.

Kim, S. D., J. J. Lee, H. Moon, S. H. Hyun, J. Moon, J. Kim and H. W. Lee (2007). "Effects of anode and electrolyte microstructures on performance of solid oxide fuel cells." *Journal of Power Sources* 169(2): 265–270.

Kim, S. G., S. P. Yoon, S. W. Nam, S. H. Hyun and S. A. Hong (2002). "Fabrication and characterization of a YSZ/YDC composite electrolyte by a sol-gel coating method." *Journal of Power Sources* 110(1): 222–228.

Koep, E., C. Compson, M. L. Liu and Z. P. Zhou (2005). "A photolithographic process for investigation of electrode reaction sites in solid oxide fuel cells." *Solid State Ionics* 176(1–2): 1–8.

Koep, E., C. M. Jin, M. Haluska, R. Das, R. Narayan, K. Sandhage, R. Snyder and M. L. Liu (2006). "Microstructure and electrochemical properties of cathode materials for SOFCs prepared via pulsed laser deposition." *Journal of Power Sources* 161(1): 250–255.

Kozuka, H. and A. Higuchi (2003). "Stabilization of Poly(vinylpyrrolidone)-containing Alkoxide Solutions for Thick Sol-Gel Barium Titanate Films." *Journal of the American Ceramic Society* 86(1): 33–38.

Kruk, A., A. Adamczyk, A. Gil, S. Kąc, J. Dąbek, M. Ziąbka and T. Brylewski (2015). "Effect of Co deposition on oxidation behavior and electrical properties of ferritic steel for solid oxide fuel cell interconnects." *Thin Solid Films* 590: 184–192.

Kuanr, S. K. and S. B. Krishna Moorthy (2016). "Structural and growth aspects of electron beam physical vapor deposited NiO-CeO2nanocomposite films." *Journal of Vacuum Science & Technology A: Vacuum, Surfaces, and Films* 34(2): 021507-1–021507-7.

Kulkarni, A. A., S. Sampath, A. Goland, H. Herman, A. J. Allen, J. Ilavsky, W. Q. Gong and S. Gopalan (2003). *Plasma spray coatings for producing next-generation supported membranes. Symposium on Syntheric Clean Fuels from Natural Gas and Coal-Bed Methane*, New York, NY.

Lamas, J. S., W. P. Leroy and D. Depla (2012). "Influence of target–substrate distance and composition on the preferential orientation of yttria-stabilized zirconia thin films." *Thin Solid Films* 520(14): 4782–4785.

Lang, M., R. Henne, S. Schaper and G. Schiller (2001). "Development and characterization of vacuum plasma sprayed thin film solid oxide fuel cells." *Journal of Thermal Spray Technology* 10(4): 618–625.

Łatka, L., A. Cattini, L. Pawłowski, S. Valette, B. Pateyron, J.-P. Lecompte, R. Kumar and A. Denoirjean (2012). "Thermal diffusivity and conductivity of yttria stabilized zirconia coatings obtained by suspension plasma spraying." *Surface and Coatings Technology* 208: 87–91.

Laukaitis, G. and J. Dudonis (2008). "Microstructure of gadolinium doped ceria oxide thin films formed by electron beam deposition." *Journal of Alloys and Compounds* 459(1–2): 320–327.

Lee, C. and J. Bae (2008). "Oxidation-resistant thin film coating on ferritic stainless steel by sputtering for solid oxide fuel cells." *Thin Solid Films* 516(18): 6432–6437.

Lee, E. A., S. Lee, H. J. Hwang and J. W. Moon (2006). "Sol-gel derived (La0.8M0.2)CrO3 (M=Ca, Sr) coating layer on stainless-steel substrate for use as a separator in intermediate-temperature solid oxide fuel cell." *Journal of Power Sources* 157(2): 709–713.

Lee, K., J. Kang, S. Jin, S. Lee and J. Bae (2017). "A novel sol–gel coating method for fabricating dense layers on porous surfaces particularly for metal-supported SOFC electrolyte." *International Journal of Hydrogen Energy* 42(9): 6220–6230.

Lee, Y., Y. M. Park and G. M. Choi (2014). "Micro-solid oxide fuel cell supported on a porous metallic Ni/stainless-steel bi-layer." *Journal of Power Sources* 249: 79–83.

Leng, Y. J., S. H. Chan, S. P. Jiang and K. A. Khor (2004). "Low-temperature SOFC with thin film GDC electrolyte prepared in situ by solid-state reaction." *Solid State Ionics* 170(1–2): 9–15.

Li, B., J. Zhang, T. Kaspar, V. Shutthanandan, R. C. Ewing and J. Lian (2013). "Multilayered YSZ/GZO films with greatly enhanced ionic conduction for low temperature solid oxide fuel cells." *Phys Chem Chem Phys* 15(4): 1296–1301.

Li, C. J., C. X. Li and M. Wang (2005). "Effect of spray parameters on the electrical conductivity of plasma-sprayed La1-xSrxMnO3 coating for the cathode of SOFCs." *Surface & Coatings Technology* 198(1–3): 278–282.

Li, W., Y. Shi, Y. Luo, Y. Wang and N. Cai (2015). "Carbon deposition on patterned nickel/yttria stabilized zirconia electrodes for solid oxide fuel cell/solid oxide electrolysis cell modes." *Journal of Power Sources* 276: 26–31.

Lim, D. P., D. S. Lim, J. S. Oh and I. W. Lyo (2005). "Influence of post-treatments on the contact resistance of plasma-sprayed La0.8Sr0.2MnO3 coating on SOFC metallic interconnector." *Surface & Coatings Technology* 200(5–6): 1248–1251.

Liu, M. F., D. H. Dong, F. Zhao, J. F. Gao, D. Ding, X. Q. Liu and G. Y. Meng (2008). "High-performance cathode-supported SOFCs prepared by a single-step co-firing process." *Journal of Power Sources* 182(2): 585–588.

Liu, R., T. Xu and C.-a. Wang (2016). "A review of fabrication strategies and applications of porous ceramics prepared by freeze-casting method." *Ceramics International* 42(2): 2907–2925.

Liu, Y., C. Compson and M. L. Liu (2004). "Nanostructured and functionally graded cathodes for intermediate temperature solid oxide fuel cells." *Journal of Power Sources* 138(1–2): 194–198.

Luisetto, I., S. Licoccia, A. D'Epifanio, A. Sanson, E. Mercadelli and E. Di Bartolomeo (2012). "Electrochemical performance of spin coated dense BaZr0.80Y0.16Zn0.04O3-δ membranes." *Journal of Power Sources* 220: 280–285.

Ma, X. Q., H. Zhang, J. Dai, J. Roth, R. Hui, T. D. Xiao and D. E. Reisner (2005). "Intermediate temperature solid oxide fuel cell based on fully integrated plasma-sprayed components." *Journal of Thermal Spray Technology* 14(1): 61–66.

Mahata, T., S. R. Nair, R. K. Lenka and P. K. Sinha (2012). "Fabrication of Ni-YSZ anode supported tubular SOFC through iso-pressing and co-firing route." *International Journal of Hydrogen Energy* 37(4): 3874–3882.

Mahato, N., A. Banerjee, A. Gupta, S. Omar and K. Balani (2015). "Progress in material selection for solid oxide fuel cell technology: A review." *Progress in Materials Science* 72: 141–337.

Maleki-Ghaleh, H., M. Rekabeslami, M. S. Shakeri, M. H. Siadati, M. Javidi, S. H. Talebian and H. Aghajani (2013). "Nano-structured yttria-stabilized zirconia coating by electrophoretic deposition." *Applied Surface Science* 280: 666–672.

Marichy, C., M. Bechelany and N. Pinna (2012). "Atomic layer deposition of nanostructured materials for energy and environmental applications." *Adv Mater* 24(8): 1017–1032.

Mathews, T., N. Rabu, J. R. Sellar and B. C. Muddle (2000). "Fabrication of La1-xSrxGa1-yMgyO3-(x+y)/2 thin films by electrophoretic deposition and its conductivity measurement." *Solid State Ionics* 128(1–4): 111–115.

Matsuda, M., O. Ohara, K. Murata, S. Ohara, T. Fukui and M. Miyake (2003). "Electrophoretic fabrication and cell performance of dense Sr- and Mg-doped LaGaO3-based electrolyte films." *Electrochemical and Solid State Letters* 6(7): A140–A143.

Matus, Y. B., L. C. De Jonghe, C. P. Jacobson and S. J. Visco (2005). "Metal-supported solid oxide fuel cell membranes for rapid thermal cycling." *Solid State Ionics* 176(5–6): 443–449.

Mbam, S. O., S. E. Nwonu, O. A. Orelaja, U. S. Nwigwe and X.-F. Gou (2019). "Thin-film coating; historical evolution, conventional deposition technologies, stress-state micro/nano-level measurement/models and prospects projection: a critical review." *Materials Research Express* 6(12): 122001-1–122001-28.

Medvedev, D., J. Lyagaeva, G. Vdovin, S. Beresnev, A. Demin and P. Tsiakaras (2016). "A tape calendering method as an effective way for the preparation of proton ceramic fuel cells with enhanced performance." *Electrochimica Acta* 210: 681–688.

Mehta, K., R. Xu and A. V. Virkar (1998). "Two-layer fuel cell electrolyte structure by sol-gel processing." *Journal of Sol-Gel Science and Technology* 11(2): 203–207.

Meier, L. P., L. Urech and L. J. Gauckler (2004). "Tape casting of nanocrystalline ceria gadolinia powder." *Journal of the European Ceramic Society* 24(15–16): 3753–3758.

Mengucci, P., G. Barucca, A. P. Caricato, A. Di Cristoforo, G. Leggieri, A. Luches and G. Majnia (2005). "Effects of annealing on the microstructure of yttria-stabilised zirconia thin films deposited by laser ablation." *Thin Solid Films* 478(1–2): 125–131.

Mikkelsen, L., N. Pryds and P. V. Hendriksen (2007). "Preparation of La0.8Sr0.2Cr0.97V0.03O3-delta films for solid oxide fuel cell application." *Thin Solid Films* 515(16): 6537–6540.

Miller, S., X. Xiao, J. Setlock, S. Farmer and K. Faber (2018). "Freeze-cast yttria-stabilized zirconia pore networks: Effects of alcohol additives." *International Journal of Applied Ceramic Technology* 15(2): 296–306.

Minh, N. Q. (1988). "Development of monolithic solid oxide fuel-cells for aerospace applications." *Journal of the Electrochemical Society* 135(8): C344–C344.

Minh, N. Q. (1993). "Ceramic fuel-cells." *Journal of the American Ceramic Society* 76(3): 563–588.

Minh, N. Q. (2004). "Solid oxide fuel cell technology-features and applications." *Solid State Ionics* 174(1–4): 271–277.

Mirahmadi, A. and M. Pourmalek (2010). "Improvement of plasma-sprayed YSZ electrolytes for solid oxide fuel cells by alumina addition." *Ionics* 16(5): 447–453.

Mizusaki, J., H. Tagawa, T. Saito, K. Kamitani, T. Yamamura, K. Hirano, S. Ehara, T. Takagi, T. Hikita, M. Ippommatsu, S. Nakagawa and K. Hashimoto 1994). "Preparation of nickel pattern electrodes on ysz and their electrochemical properties in H_2-H_2O atmospheres." *Journal of the Electrochemical Society* 141(8): 2129–2134.

Morales, M., V. Miguel-Pérez, A. Tarancón, A. Slodczyk, M. Torrell, B. Ballesteros, J. P. Ouweltjes, J. M. Bassat, D. Montinaro and A. Morata (2017). "Multi-scale analysis of the diffusion barrier layer of gadolinia-doped ceria in a solid oxide fuel cell operated in a stack for 3000 h." *Journal of Power Sources* 344: 141–151.

Morales, M., A. Pesce, A. Slodczyk, M. Torrell, P. Piccardo, D. Montinaro, A. Tarancón and A. Morata (2018). "Enhanced performance of gadolinia-doped ceria diffusion barrier layers fabricated by pulsed laser deposition for large-area solid oxide fuel cells." *ACS Applied Energy Materials* 1(5): 1955–1964.

Nair, B. N., T. Suzuki, Y. Yoshino, S. Gopalakrishnan, T. Sugawara, S. Nakao and H. Taguchi (2005). "An oriented nanoporous membrane prepared by pulsed laser deposition." *Advanced Materials* 17(9): 1136.

Nishihora, R. K., P. L. Rachadel, M. G. N. Quadri and D. Hotza (2018). "Manufacturing porous ceramic materials by tape casting—A review." *Journal of the European Ceramic Society* 38(4): 988–1001.

Nomura, H., S. Parekh, J. R. Selman and S. Al-Hallaj (2005). "Fabrication of YSZ electrolyte using electrostatic spray deposition (ESD): I - a comprehensive parametric study." *Journal of Applied Electrochemistry* 35(1): 61–67.

Ogumi, Z., T. Ioroi, Y. Uchimoto, Z. Takehara, T. Ogawa and K. Toyama (1995). "Novel method for preparing nickel/ysz cermet by a vapor-phase process." *Journal of the American Ceramic Society* 78(3): 593–598.

Orlovskaya, N., A. Coratolo, C. Johnson and R. Gemmen (2004). "Structural characterization of lanthanum chromite perovskite coating deposited by magnetron sputtering on an iron-based chromium-containing alloy as a promising interconnect material for SOFCs." *Journal of the American Ceramic Society* 87(10): 1981–1987.

Pal, U. B. and S. C. Singhal (1990). "Electrochemical vapor-deposition of yttria-stabilized zirconia films." *Journal of the Electrochemical Society* 137(9): 2937–2941.

Panthi, D., N. Hedayat, T. Woodson, B. J. Emley and Y. Du (2019). "Tubular solid oxide fuel cells fabricated by a novel freeze casting method." *Journal of the American Ceramic Society* 103(2): 878-888.

Park, J. H., W.-S. Hong, J.-H. Lee, K. J. Yoon, H. Kim, J. Hong, H.-S. Song and J.-W. Son (2014). "Facile fabrication of YSZ/GDC multi-layers by using a split target in pulsed laser deposition and their structural and electrical properties." *Journal of Electroceramics* 33(1–2): 25–30.

Park, S.-I., Y.-J. Quan, S.-H. Kim, H. Kim, S. Kim, D.-M. Chun, C. S. Lee, M. Taya, W.-S. Chu and S.-H. Ahn (2016). "A review on fabrication processes for electrochromic devices." *International Journal of Precision Engineering and Manufacturing-Green Technology* 3(4): 397–421.

Park, S., R. J. Gorte and J. M. Vohs (2001). "Tape cast solid oxide fuel cells for the direct oxidation of hydrocarbons." *Journal of the Electrochemical Society* 148(5): A443–A447.

Patil, B. B., V. Ganesan and S. H. Pawar (2008). "Studies on spray deposited NiO-SDC composite films for solid oxide fuel cells." *Journal of Alloys and Compounds* 460(1–2): 680–687.

Pederson, L. R., P. Singh and X. D. Zhou (2006). "Application of vacuum deposition methods to solid oxide fuel cells." *Vacuum* 80(10): 1066–1083.

Peng, Z. Y. and M. L. Liu (2001). "Preparation of dense platinum-yttria stabilized zirconia and yttria stabilized zirconia films on porous La0.9Sr0.1MnO3 (LSM) substrates." *Journal of the American Ceramic Society* 84(2): 283–288.

Perednis, D. and L. J. Gauckler (2004). "Solid oxide fuel cells with electrolytes prepared via spray pyrolysis." *Solid State Ionics* 166(3–4): 229–239.

Perednis, D. and L. J. Gauckler (2005). "Thin film deposition using spray pyrolysis." *Journal of Electroceramics* 14(2): 103–111.

Pérez-Coll, D., E. Céspedes, A. J. Dos santos-García, G. C. Mather and C. Prieto (2014). "Electrical properties of nanometric CGO-thin films prepared by electron-beam physical vapour deposition." *Journal of Materials Chemistry A* 2(20): 7410–7420.

Pikalova, E. Y. and E. G. Kalinina (2019). "Electrophoretic deposition in the solid oxide fuel cell technology: Fundamentals and recent advances." *Renewable and Sustainable Energy Reviews* 116: 109440-1–109440-23.

Plonczak, P., D. R. Sørensen, M. Søgaard, V. Esposito and P. V. Hendriksen (2012). "Electrochemical properties of dense (La, Sr)MnO3−δ films produced by pulsed laser deposition." *Solid State Ionics* 217: 54–61.

Princivalle, A. and E. Djurado (2008). "Nanostructured LSM/YSZ composite cathodes for IT-SOFC: A comprehensive microstructural study by electrostatic spray deposition." *Solid State Ionics* 179(33–34): 1921–1928.

Rambert, S., A. J. McEvoy and K. Barthel (1999). "Composite ceramic fuel cell fabricated by vacuum plasma spraying." *Journal of the European Ceramic Society* 19(6–7): 921–923.

Reitz, T. L. and H. M. Xiao (2006). "Characterization of electrolyte-electrode interlayers in thin film solid oxide fuel cells." *Journal of Power Sources* 161(1): 437–443.

Reolon, R. P., C. M. Halmenschlager, R. Neagu, C. de Fraga Malfatti and C. P. Bergmann (2014). "Electrochemical performance of gadolinia-doped ceria (CGO) electrolyte thin films for ITSOFC deposited by spray pyrolysis." *Journal of Power Sources* 261: 348–355.

Rubio, D., C. Suciu, I. Waernhus, A. Vik and A. C. Hoffmann (2017). "Tape casting of lanthanum chromite for solid oxide fuel cell interconnects." *Journal of Materials Processing Technology* 250: 270–279.

Rupp, J. L. M., T. Drobek, A. Rossi and L. J. Gauckler (2007). "Chemical analysis of spray pyrolysis gadolinia-doped ceria electrolyte thin films for solid oxide fuel cells." *Chemistry of Materials* 19(5): 1134–1142.

Rupp, J. L. M., A. Infortuna and L. J. Gauckler (2006). "Microstrain and self-limited grain growth in nanocrystalline ceria ceramics." *Acta Materialia* 54(7): 1721–1730.

Sansernnivet, M., N. Laosiripojana, S. Assabumrungrat and S. Charojrochkul (2010). "Fabrication of La0.8Sr0.2CrO3-based Perovskite Film via Flame-Assisted Vapor Deposition for H2 Production by Reforming." *Chemical Vapor Deposition* 16(10–12): 311–321.

Sarkar, P., D. De and H. Rho (2004). "Synthesis and microstructural manipulation of ceramics by electrophoretic deposition." *Journal of Materials Science* 39(3): 819–823.

Sasaki, H., S. Otoshi, M. Suzuki, T. Sogi, A. Kajimura, N. Sugiura and M. Ippommatsu (1994). "Fabrication of high-power density tubular type solid oxide fuel-cells." *Solid State Ionics* 72: 253–256.

Scandurra, R., A. Scotto d'Abusco and G. Longo (2020). "A review of the effect of a nanostructured thin film formed by titanium carbide and titanium oxides clustered around carbon in graphitic form on osseointegration." *Nanomaterials (Basel)* 10(6).

Schiller, G., R. Henne, M. Lang, R. Ruckdaschel and S. Schaper (2000). *Fuel Cells Bulletin* 3: 7.

Serra, J. M., V. B. Vert, O. Buchler, W. A. Meulenberg and H. P. Buchkremer (2008). "IT-SOFC supported on mixed oxygen ionic-electronic conducting composites." *Chemistry of Materials* 20(12): 3867–3875.

Setoguchi, T., M. Sawano, K. Eguchi and H. Arai (1990). "Application of the stabilized zirconia thin-film prepared by spray pyrolysis method to sofc." *Solid State Ionics* 40–1: 502–505.

Shao, Z. P., S. M. Haile, J. Ahn, P. D. Ronney, Z. L. Zhan and S. A. Barnett (2005). "A thermally self-sustained micro solid-oxide fuel-cell stack with high power density." *Nature* 435(7043): 795–798.

Shi, N., S. Yu, S. Chen, L. Ge, H. Chen and L. Guo (2017). "Dense thin YSZ electrolyte films prepared by a vacuum slurry deposition technique for SOFCs." *Ceramics International* 43(1): 182–186.

Shi, Y., A. Fluri, I. Garbayo, J. J. Schwiedrzik, J. Michler, D. Pergolesi, T. Lippert and J. L. M. Rupp (2019). "Zigzag or spiral-shaped nanostructures improve mechanical stability in yttria-stabilized zirconia membranes for micro-energy conversion devices." *Nano Energy* 59: 674–682.

Shim, J. H., C. C. Chao, H. Huang and F. B. Prinz (2007). "Atomic layer deposition of yttria-stabilized zirconia for solid oxide fuel cells." *Chemistry of Materials* 19(15): 3850–3854.

Shimada, H., T. Yamaguchi, T. Suzuki, H. Sumi, K. Hamamoto and Y. Fujishiro (2016). "High power density cell using nanostructured Sr-doped SmCoO3 and Sm-doped CeO2 composite powder synthesized by spray pyrolysis." *Journal of Power Sources* 302: 308–314.

Shin, J., J. H. Park, J. Kim, K. J. Yoon, J.-W. Son, J.-H. Lee, H.-W. Lee and H.-I. Ji (2020). "Suppression of processing defects in large-scale anode of planar solid oxide fuel cell via multi-layer roll calendering." *Journal of Alloys and Compounds* 812: 152113-1–152113-7.

Shin, J. W., D. Go, S. H. Kye, S. Lee and J. An (2019). "Review on process-microstructure-performance relationship in ALD-engineered SOFCs." *Journal of Physics: Energy* 1(4): 042002-1–042002-26.

Shin, T. H., M. Shin, G.-W. Park, S. Lee, S.-K. Woo and J. Yu (2017). "Fabrication and characterization of oxide ion conducting films, Zr1−xMxO2−δ (M = Y, Sc) on porous SOFC anodes, prepared by electron beam physical vapor deposition." *Sustainable Energy & Fuels* 1(1): 103–111.

Shri Prakash, B., S. Senthil Kumar and S. T. Aruna (2014). "Properties and development of Ni/YSZ as an anode material in solid oxide fuel cell: A review." *Renewable and Sustainable Energy Reviews* 36: 149–179.

Simrick, N. J., J. A. Kilner, A. Atkinson, J. L. M. Rupp, T. M. Ryll, A. Bieberle–Hütter, H. Galinski and L. J. Gauckler (2011). "Micro-fabrication of patterned LSCF thin-film cathodes with gold current collectors." *Solid State Ionics* 192(1): 619–626.

Singh, P. and N. Q. Minh (2004). "Solid oxide fuel cells: Technology status." *International Journal of Applied Ceramic Technology* 1(1): 5–15.

Singhal, S. C. (2002). "Solid oxide fuel cells for stationary, mobile, and military applications." *Solid State Ionics* 152: 405–410.

Singhal, S. C. (2014). "Solid oxide fuel cells for power generation." *Wiley Interdisciplinary Reviews: Energy and Environment* 3(2): 179–194.

Smith, M. F., A. C. Hall, J. D. Fleetwood and P. Meyer (2011). "Very low pressure plasma spray—a review of an emerging technology in the thermal spray community." *Coatings* 1(2): 117–132.

Sofie, S. W. (2007). "Fabrication of functionally graded and aligned porosity in thin ceramic substrates with the novel freeze-tape-casting process." *Journal of the American Ceramic Society* 90(7): 2024–2031.

Solovyev, A. A., I. V. Ionov, A. V. Shipilova, A. N. Kovalchuk and M. S. Syrtanov (2017). "Magnetron-sputtered La0.6Sr0.4Co0.2Fe0.8O3 nanocomposite interlayer for solid oxide fuel cells." *Journal of Nanoparticle Research* 19(3): 1–9.

Solovyev, A. A., A. V. Shipilova, I. V. Ionov, A. N. Kovalchuk, S. V. Rabotkin and V. O. Oskirko (2016). "Magnetron-sputtered YSZ and CGO electrolytes for SOFC." *Journal of Electronic Materials* 45(8): 3921–3928.

Somalu, M. R., A. Muchtar, W. R. W. Daud and N. P. Brandon (2017). "Screen-printing inks for the fabrication of solid oxide fuel cell films: A review." *Renewable and Sustainable Energy Reviews* 75: 426–439.

Song, J. H., S. I. Park, J. H. Lee and H. S. Kim (2008). "Fabrication characteristics of an anode-supported thin-film electrolyte fabricated by the tape casting method for IT-SOFC." *Journal of Materials Processing Technology* 198(1-3): 414–418.

Srivastava, P. K., T. Quach, Y. Y. Duan, R. Donelson, S. P. Jiang, F. T. Ciacchi and S. P. S. Badwal (1997). "Electrode supported solid oxide fuel cells: Electrolyte films prepared by DC magnetron sputtering." *Solid State Ionics* 99(3-4): 311–319.

Su, P. C., C. C. Chao, J. H. Shim, R. Fasching and F. B. Prinz (2008). "Solid oxide fuel cell with corrugated thin film electrolyte." *Nano Letters* 8(8): 2289–2292.

Sun, H.-Y., W. Sen, W.-H. Ma, J. Yu and J.-J. Yang (2014). "Fabrication of LSGM thin films on porous anode supports by slurry spin coating for IT-SOFC." *Rare Metals* 34(11): 797–801.

Sun, Z., S. Gopalan, U. B. Pal and S. N. Basu (2017). "Cu1.3Mn1.7O4 spinel coatings deposited by electrophoretic deposition on Crofer 22 APU substrates for solid oxide fuel cell applications." *Surface and Coatings Technology* 323: 49–57.

Suzuki, M. and A. Kajimura (1997). "Geometrical analysis of SOFC anodes fabricated by electrochemical vapor deposition." *Denki Kagaku* 65(10): 859–864.

Tai, L. W. and P. A. Lessing (1991). "Plasma spraying of porous-electrodes for a planar solid oxide fuel-cell." *Journal of the American Ceramic Society* 74(3): 501–504.

Tanhaei, M. and M. Mozammel (2017). "Yttria-stabilized zirconia thin film electrolyte deposited by EB-PVD on porous anode support for SOFC applications." *Ceramics International* 43(3): 3035–3042.

Tao, S. W. and J. T. S. Irvine (2002). "Study on the structural and electrical properties of the double perovskite oxide SrMn0.5Nb0.5O3-delta." *Journal of Materials Chemistry* 12(8): 2356–2360.

Tao, Z., B. Wang, G. Hou and N. Xu (2014). "Preparation of BaZr 0.1 Ce 0.7 Y 0.2 O 3−δ thin membrane based on a novel method-drop coating." *International Journal of Hydrogen Energy* 39(28): 16020–16024.

Tikkanen, H., C. Suciu, I. Wærnhus and A. C. Hoffmann (2011). "Dip-coating of 8YSZ nanopowder for SOFC applications." *Ceramics International* 37(7): 2869–2877.

Torabi, A., T. H. Etsell and P. Sarkar (2011). "Dip coating fabrication process for micro-tubular SOFCs." *Solid State Ionics* 192(1): 372–375.

Tsai, T. and S. A. Barnett (1995). "Bias sputter-deposition of dense yttria-stabilized zirconia films on porous substrates." *Journal of the Electrochemical Society* 142(9): 3084–3087.

Tucker, M. C., G. Y. Lau, C. P. Jacobson, L. C. DeJonghe and S. J. Visco (2007). "Performance of metal-supported SOFCs with infiltrated electrodes." *Journal of Power Sources* 171(2): 477–482.

Uhlenbruck, S., N. Jordan, D. Sebold, H. P. Buchkremer, V. A. C. Haanappel and D. Stover (2007). "Thin film coating technologies of (Ce,Gd)O2-delta interlayers for application in ceramic high-temperature fuel cells." *Thin Solid Films* 515(7–8): 4053–4060.

Usui, T., Y. Ito and K. Kikuta (2010). "Fabrication and characterization of LSCF-GDC/GDC/NiO-GDC microtubular SOFCs prepared by multi-dip coating." *Journal of the Ceramic Society of Japan* 118(1379): 564–567.

Van Gestel, T., F. Han, D. Sebold, H. P. Buchkremer and D. Stöver (2011). "Nano-structured solid oxide fuel cell design with superior power output at high and intermediate operation temperatures." *Microsystem Technologies* 17(2): 233–242.

Van Gestel, T., D. Sebold, W. A. Meulenberg and H. P. Buchkremer (2008). "Development of thin-film nano-structured electrolyte layers for application in anode-supported solid oxide fuel cells." *Solid State Ionics* 179(11-12): 428–437.

Van Herle, J., A. J. McEvoy and K. R. Thampi (1994). "Conductivity measurements of various yttria-stabilized zirconia samples." *Journal of Materials Science* 29(14): 3691–3701.

Vardelle, A., C. Moreau, N. J. Themelis and C. Chazelas (2014). "A perspective on plasma spray technology." *Plasma Chemistry and Plasma Processing* 35(3): 491–509.

Vassen, R., H. Kassner, A. Stuke, F. Hauler, D. Hathiramani and D. Stover (2008). "Advanced thermal spray technologies for applications in energy systems." *Surface & Coatings Technology* 202(18): 4432–4437.

Veldhuis, S. A., P. Brinks and J. E. ten Elshof (2015). "Rapid densification of sol–gel derived yttria-stabilized zirconia thin films." *Thin Solid Films* 589: 503–507.

Wang, C. H., W. L. Worrell, S. Park, J. M. Vohs and R. J. Gorte (2001). "Fabrication and performance of thin-film YSZ solid oxide fuel cells." *Journal of the Electrochemical Society* 148(8): A864–A868.

Wang, H., W. Ji, L. Zhang, Y. Gong, B. Xie, Y. Jiang and Y. Song (2011). "Preparation of YSZ films by magnetron sputtering for anode-supported SOFC." *Solid State Ionics* 192(1): 413–418.

Wang, J. M., Z. Lu, X. Q. Huang, K. F. Chen, N. Ai, J. Y. Hu and W. H. Su (2007). "YSZ films fabricated by a spin smoothing technique and its application in solid oxide fuel cell." *Journal of Power Sources* 163(2): 957–959.

Wang, L. S. and S. A. Barnett (1993). "Sputter-deposited medium-temperature solid oxide fuel-cells with multilayer electrolytes." *Solid State Ionics* 61(4): 273–276.

Wang, X.-M., C.-X. Li, C.-J. Li and G.-J. Yang (2010). "Effect of microstructures on electrochemical behavior of La0.8Sr0.2MnO3 deposited by suspension plasma spraying." *International Journal of Hydrogen Energy* 35(7): 3152–3158.

Wang, Y.-p., J.-t. Gao, W. Chen, C.-x. Li, S.-l. Zhang, G.-j. Yang and C.-j. Li (2019). "Development of ScSZ electrolyte by very low pressure plasma spraying for high-performance metal-supported SOFCs." *Journal of Thermal Spray Technology* 29(1-2): 223–231.

Wang, Y. and T. W. Coyle (2011). "Solution precursor plasma spray of porous La1−xSrxMnO3 perovskite coatings for SOFC cathode application." *Journal of Fuel Cell Science and Technology* 8(2): 021005-1–021005-5.

Wang, Z., X. Huang, Z. Lv, Y. Zhang, B. Wei, X. Zhu, Z. Wang and Z. Liu (2015). "Preparation and performance of solid oxide fuel cells with YSZ/SDC bilayer electrolyte." *Ceramics International* 41(3): 4410–4415.

Wang, Z. C., W. J. Weng, K. Chen, G. Shen, P. Y. Du and G. R. Han (2008). "Preparation and performance of nanostructured porous thin cathode for low-temperature solid oxide fuel cells by spin-coating method." *Journal of Power Sources* 175(1): 430–435.

Wang, Z. H., K. N. Sun, S. Y. Shen, N. Q. Zhang, J. S. Qiao and P. Xu (2008). "Preparation of YSZ thin films for intermediate temperature solid oxide fuel cells by dip-coating method." *Journal of Membrane Science* 320(1-2): 500–504.

Wanzenberg, E., F. Tietz, P. Panjan and D. Stover (2003). "Influence of pre- and post-heat treatment of anode substrates on the properties of DC-sputtered YSZ electrolyte films." *Solid State Ionics* 159(1–2): 1–8.

Wasmus, S. and A. Kuver (1999). "Methanol oxidation and direct methanol fuel cells: a selective review." *Journal of Electroanalytical Chemistry* 461(1–2): 14–31.

White, B. D., Kesler, O., Rose, L. (2007). "Electrochemical characterization of air plasma sprayed LSM/YSZ composite cathodes on metallic interconnects." *ECS Transactions* 7(1): 1107–1114.

Will, J., M. K. M. Hruschka, L. Gubler and L. J. Gauckler (2001). "Electrophoretic deposition of zirconia on porous anodic substrates." *Journal of the American Ceramic Society* 84(2): 328–332.

Will, J., A. Mitterdorfer, C. Kleinlogel, D. Perednis and L. J. Gauckler (2000). "Fabrication of thin electrolytes for second-generation solid oxide fuel cells." *Solid State Ionics* 131(1-–2): 79–96.

Williams, M. C., J. P. Strakey and S. C. Singhal (2004). "US distributed generation fuel cell program." *Journal of Power Sources* 131(1–2): 79–85.

Xia, C. R., F. L. Chen and M. L. Liu (2001). "Reduced-temperature solid oxide fuel cells fabricated by screen printing." *Electrochemical and Solid State Letters* 4(5): A52–A54.

Xia, C. R. and M. L. Liu (2001a). "Low-temperature SOFCs based on Gd0.1Ce0.9O1.95 fabricated by dry pressing." *Solid State Ionics* 144(3-4): 249–255.

Xia, C. R. and M. L. Liu (2001b). "A simple and cost-effective approach to fabrication of dense ceramic membranes on porous substrates." *Journal of the American Ceramic Society* 84(8): 1903–1905.

Xia, C. R. and M. L. Liu (2002). "Microstructures, conductivities, and electrochemical properties of Ce0.9Gd0.1O2 and GDC-Ni anodes for low-temperature SOFCs." *Solid State Ionics* 152: 423–430.

Xu, X. Y., C. R. Xia, S. G. Huang and D. K. Peng (2005). "YSZ thin films deposited by spin-coating for IT-SOFCs." *Ceramics International* 31(8): 1061-1064.

Yamaguchi, T., T. Suzuki, S. Shimizu, Y. Fujishiro and M. Awano (2007). "Examination of wet coating and co-sintering technologies for micro-SOFCs fabrication." *Journal of Membrane Science* 300(1-2): 45–50.

Yamane, H. and T. Hirai (1987). "Preparation of ZRO2-film by oxidation of ZRCL4." *Journal of Materials Science Letters* 6(10): 1229–1230.

Yamane, H. and T. Hirai (1989). "Yttria stabilized zirconia transparent films prepared by chemical vapor-deposition." *Journal of Crystal Growth* 94(4): 880–884.

Yang, M., Z. Xu, S. Desai, D. Kumar and J. Sankar (2015). "Fabrication of micro single chamber solid oxide fuel cell using photolithography and pulsed laser deposition." *Journal of Fuel Cell Science and Technology* 12(2): 021004-1–021004-6.

Ye, F., T. Mori, D. R. Ou, J. Zou, J. Drennan, S. Nakayama and M. Miyayama (2010). "Effect of nickel diffusion on the microstructure of Gd-doped ceria (GDC) electrolyte film supported by Ni–GDC cermet anode." *Solid State Ionics* 181(13–14): 646–652.

Yokokawa, H., N. Sakai, T. Horita, K. Yamaji and M. E. Brito (2005). "Electrolytes for solid-oxide fuel cells." *Materials Research Society Bulletin* 30(8): 591–595.

Yuan, K., J. Zhu, W. Dong, Y. Yu, X. Lu, X. Ji and X. Wang (2017). "Applying Low-Pressure Plasma Spray (LPPS) for coatings in low-temperature SOFC." *International Journal of Hydrogen Energy* 42(34): 22243–22249.

Zarabian, M., A. Y. Yar, S. Vafaeenezhad, M. A. F. Sani and A. Simchi (2013). "Electrophoretic deposition of functionally-graded NiO–YSZ composite films." *Journal of the European Ceramic Society* 33(10): 1815–1823.

Zhang, H., Z. Zhan and X. Liu (2011). "Electrophoretic deposition of (Mn,Co)3O4 spinel coating for solid oxide fuel cell interconnects." *Journal of Power Sources* 196(19): 8041–8047.

Zhang, L., F. Chen and C. Xia (2010). "Spin-coating derived solid oxide fuel cells operated at temperatures of 500 °C and below." *International Journal of Hydrogen Energy* 35(24): 13262–13270.

Zhang, L., H. Q. He, W. R. Kwek, J. Ma, E. H. Tang and S. P. Jiang (2009). "Fabrication and characterization of anode-supported tubular solid oxide fuel cells by sip casting and dip coating techniques." *Journal of the American Ceramic Society* 92(2): 302–310.

Zhang, X. G., M. Robertson, C. Deces-Petit, Y. S. Xie, R. Hui, S. Yick, E. Styles, J. Roller, O. Kesler, R. Maric and D. Ghosh (2006). "NiO-YSZ cermets supported low temperature solid oxide fuel cells." *Journal of Power Sources* 161(1): 301–307.

Zhang, Y. H., X. Q. Huang, Z. Lu, X. D. Ge, J. H. Xu, X. S. Xin, X. Q. Sha and W. H. Su (2006). "Effect of starting powder on screen-printed YSZ films used as electrolyte in SOFCs." *Solid State Ionics* 177(3–4): 281–287.

Zhang, Y. H., J. Liu, X. Q. Huang, Z. Lu and W. H. Su (2008). "Performance evaluation of thin membranes solid oxide fuel cell prepared by pressure-assisted slurry-casting." *International Journal of Hydrogen Energy* 33(2): 775–780.

Zheng, R., X. M. Zhou, S. R. Wang, T. L. Wen and C. X. Ding (2005). "A study of Ni+8YSZ/8YSZ/La0.6Sr0.4CoO3-delta ITSOFC fabricated by atmospheric plasma spraying." *Journal of Power Sources* 140(2): 217–225.

Zomorrodian, A., H. Salamati, Z. Lu, X. Chen, N. Wu and A. Ignatiev (2010). "Electrical conductivity of epitaxial La0.6Sr0.4Co0.2Fe0.8O3−δ thin films grown by pulsed laser deposition." *International Journal of Hydrogen Energy* 35(22): 12443–12448.

Zou, Y., W. Zhou, J. Sunarso, F. Liang and Z. Shao (2011). "Electrophoretic deposition of YSZ thin-film electrolyte for SOFCs utilizing electrostatic-steric stabilized suspensions obtained via high energy ball milling." *International Journal of Hydrogen Energy* 36(15): 9195–9204.

5 Mutually Influenced Stacking and Evolution of Inorganic/Organic Crystals for Piezo-Related Applications

Jr-Jeng Ruan and Kao-Shuo Chang
National Cheng Kung University, Taiwan

CONTENTS

5.1 Introduction: Piezoelectricity 128
5.2 Piezoelectric Nitride Thin Films 128
    5.2.1 $ZnSnN_2$ [4, 12] 128
    5.2.2 $Ti_xAl_{1-x}N$ [5] 129
5.3 Piezoelectric Oxide Thin Film 129
    5.3.1 $ZnSnO_3$ [9, 18] 129
    5.3.2 $BaZnO_2$ [10] 130
    5.3.3 $BiFeO_3$ [11] 130
5.4 Fabrication Method and Morphology Tuning 130
    5.4.1 Combinatorial Reactive Sputtering [4] 130
        5.4.1.1 $Zn–Sn_3N_4$ Composition Spreads 131
        5.4.1.2 AlN-TiN Nanocolumn Composite Composition Spreads [5] 132
    5.4.2 Solution-Based Process 132
        5.4.2.1 $ZnSnO_3$ [18] 132
        5.4.2.2 $BaZnO_2$ [10] 132
        5.4.2.3 $BiFeO_3$ [11] 133
5.5 Characterizations of Various Properties 133
    5.5.1 Piezotronic-Related Property [18] 133
    5.5.2 Schottky Behavior [18] 133
    5.5.3 Schottky Barrier Height (SBH) Variation [18] 134
    5.5.4 Piezophotocatalysis [18] 134
    5.5.5 Piezopotential [47] 134
    5.5.6 Incident Photon-to-Current Efficiency (IPCE) [48] 134
5.6 Nanostructured $ZnSnO_3$ and $BaZnO_2$ and Their Applications 134
    5.6.1 $ZnSnO_3$ [9] 134
    5.6.2 $BaZnO_2$ [10] 138
5.7 Piezoelectric Features of Polymer Materials 139
5.8 Strategies to Enhance the Polarization Levels of Polymer Materials 140
    5.8.1 Polarization of Polymer Phases Via Poling Treatment 140
    5.8.2 Growth of Polymer Ferroelectric Crystalline Phases 141

5.8.3	Growth of Relaxor Ferroelectric Phases	142
5.8.4	Oriented Stacking of Ferroelectric Crystals	143
5.8.5	Maxwell–Wagner–Sillars Interfacial Polarization	143
References		145

Piezoelectric effect is one of the fascinating functional properties of materials and leads to various applications, including power sources, sensors, and actuators. Although the effect was observed in the late 19th century, and various piezoelectric inorganic and polymer crystals have been widely studied since, few studies on the direct coupling through polarization tuning have been reported. In this chapter, inorganic materials are discussed first from Sections 5.1 to 5.6, and then followed by polymers and inorganic/organic composites from Section 5.7 to 5.8. We focus on mutually influenced stacking and evolution of inorganic/organic crystals for enhancing the resulting piezoelectricity and exploring advanced piezo-related applications.

5.1 INTRODUCTION: PIEZOELECTRICITY

Piezoelectric effect develops electrical charges (or deformation) in response to an external stress (or electric field) (Equation 5.1 and 5.2):

$$P = d\sigma \tag{5.1}$$

$$\varepsilon = dE \tag{5.2}$$

where P is polarization (pC/m^2), d is a piezoelectric coefficient (pC/N or m/V), σ is an external stress (N/m^2), ε is a strain, and E is an applied electric field. Commonly used piezoelectric coefficients of d_{33}, d_{31}, and d_{15}, which denote longitudinal, transverse, and shear coefficients of a material, respectively [1].

A piezoelectric material has a non-centrosymmetric center (20 crystal structures) [2]. Ten crystal structures out of them exhibit spontaneous polarization along their polar axes at a given temperature, and they are also coined as pyroelectric materials. Furthermore, ferroelectric materials represent a group under pyroelectric materials, and their spontaneous polarization can be switched using an external electric field.

When a piezoelectric material is operated above Curie temperature (T_c), which can be determined by a plot of dielectric permittivity against temperature [3], its piezoelectricity disappears because of the restoration of a centrosymmetry in a crystal structure.

Numerous remarkable piezoelectric materials, including nitrides (e.g., AlN, GaN, InN, InGaN, ZnGeN$_2$, ZnSnN$_2$, and Ti$_x$Al$_{1-x}$N, [4, 5]), wurtzites (e.g., ZnO, ZnS, and CdSe [6]), and perovskites (e.g., Pb(Zr,Ti)O$_3$ [7]), have been explored. Lead-free perovskite piezoelectric materials, including BaTiO$_3$, K$_{0.5}$Na$_{0.5}$NbO$_3$, Na$_{0.5}$Bi$_{0.5}$TiO$_3$, ZnSnO$_3$, BaZnO$_2$, and BiFeO$_3$, have also been developed [7–11]. Certain nitride- (e.g., ZnSnN$_2$ and Ti$_x$Al$_{1-x}$N) and oxide- (e.g., ZnSnO$_3$, BaZnO$_2$, and BiFeO$_3$) films are reviewed in this chapter.

5.2 PIEZOELECTRIC NITRIDE THIN FILMS

5.2.1 ZnSnN$_2$ [4, 12]

Piezoelectric nitrides exhibit various remarkable properties. Popular materials include wurtzite III-nitrides (e.g., AlN, GaN, and InN). Zn-IV-nitrides are alternatives to explore because of their compatibility with III-nitrides, and they comprise earth-abundant and environmentally friendly elements. Their wide band gap (E_g) variation (approximately 1.7–4.5 eV) enables wide optical (IR to UV devices) and optoelectronic applications [13]. Furthermore, Zn-IV-nitrides are also desirable for

Mutually Influenced Stacking

integrated devices because the strain-induced polarization in the junctions among Zn-IV-nitrides is reduced [4] due to the minor lattice mismatch among them.

ZnSnN$_2$ (ZTN) is favorable among Zn-IV-nitrides. Its cation disorder control can yield superior E$_g$ tunability [14], enabling the use of a visible light region of the solar spectrum. In addition, the low formation energy of donor defects leads to an intrinsic n-type ZTN with high carrier concentrations (up to 10^{20} cm^{-3}) [15]. However, ZTN-related studies still remain insufficient because of the challenging fabrication of ZTN, particularly orthorhombic ZTN [16]. Furthermore, extended piezo-related piezotronic and piezophototronic features are yet to be examined [4].

Hsu et al. [12] fabricated orthorhombic ZTN (Pna2$_1$) through the following novel approaches: 1) the use of combinatorial methodology by means of natural Sn$_3$N$_4$ and Zn thickness gradients to accelerate the formation of ZTN (as discussed in Section 5.4.1), and 2) the approach of Zn–Sn$_3$N$_4$ composition spreads to promote the formation of orthorhombic ZTN (ordered-defect configuration) [4]. The fabricated ZTN exhibited [001]-oriented nanocolumns and enhanced extended piezoelectric performance.

5.2.2 Ti$_x$Al$_{1-x}$N [5]

Various studies have reported the mechanical properties of AlN because it possesses high chemical and thermal stabilities, high electrical resistivity, and a low thermal expansion coefficient [17]. In recent years, theoretical calculations have been performed to study the conductivity, piezoelectricity, and E$_g$ of TiN-doped AlN; however, little experimental research has been conducted. Based on the Ti–Al–N ternary phase diagram, the mutual solubility of TiN and AlN is limited. Because the lattice mismatch between a preferred plane of (111) in TiN (lattice constant $a = 0.0300$ pm) and a plane of (002) in AlN ($a = 0.0311$ pm) was minor (approximately 3.6%), an excellent interface between them was expected to sustain nanocolumn arrays. Thus, Lee et al. [5] employed combinatorial methodology to fabricate AlN–TiN nanocolumn composite composition spreads (libraries) by using reactive sputtering (as discussed in Section 5.4.1). In the study, various amounts of non-piezoelectric TiN were used to distort the lattice of AlN for enhancing its piezoelectricity. A optimal ratio of TiN:AlN was then determined. The intimate coupling among AlN, TiN, and Ti$_x$Al$_{1-x}$N was attributable to the improved piezotronic, piezophototronic, piezophotocatalytic, and piezophotoelectrochemical (PPEC) performance.

5.3 PIEZOELECTRIC OXIDE THIN FILM

5.3.1 ZnSnO$_3$ [9, 18]

(YTG) ZnSnO$_3$ has been reported for various applications in nanogenerators, lithium ion batteries, gas sensors, lead-free ferroelectrics, transparent conductors, and photocatalysis [19–24]. However, no substantial attention for ZnSnO$_3$ in the piezoelectric properties has been received. The first-principle calculation has verified that ZnSnO$_3$ has six possible structures, namely cubic perovskite (Pm-3m), ilmenite (IL)-type (R-3), LiNbO$_3$ (LN)-type (R3C), CdSnO$_3$-type (Pnma), HgSnO$_3$-type (R-3c), and post-perovskite-type (Cmcm) structures [25]. Among them, the piezoelectric property is only exhibited by the LN type. In addition, metastable ZnSnO$_3$ can be stabilized under high pressure [26]. Thus, hydrothermal synthesis [27] is promising for its fabrication. For achieving a desirable morphology of ZnSnO$_3$ through a hydrothermal synthesis, numerous parameters of precursor concentration variations, the types and amount of complex agents, the types of substrates, the reaction temperature and time, and pH values can be readily tailored.

Wang et al. [18] reported using various approaches in hydrothermal reactions to fabricate ZnSnO$_3$ nanowire arrays along the substrate normal (as discussed in Section 5.4.2.1). Piezophotocatalysis of a methylene blue (MB) solution was demonstrated. Furthermore, the associated piezotronic and piezophototronic properties were also determined (as discussed in Section 5.6.1).

5.3.2 BaZnO₂ [10]

Quartz groups represent another category of popular piezoelectric materials [28]. Although low piezoelectricity of α-quartz materials (e.g., GeO_2, $GaAsO_4$, and $AlPO_4$ [8, 29, 30]), they have been extensively studied because of low cost and high Curie temperature [28]. $BaZnO_2$ (space group of $P3_121$) exhibits a structure similar to that of α-quartz. However, only little research has been conducted since 1960 when its crystal structure was determined [31]. Spitsbergen reported that the hexagonal indexing was given to the crystal structure of $BaZnO_2$ by using Hull's graphical method [31]. However, Zheng et al. suggested that $BaZnO_2$ was not a quartz because its general formula of ABO_2 (A = Ba and B = Zn) was different from that of quartz [AO_2 (A = Si, Ge) or ABO_4 (A = Ga, B, Fe, and A; B = As and P)] [31]. The lattice constant of $BaZnO_2$ was 5.890 and 6.781 Å for (a) and (c), respectively, calculated using FHI98PP Troullier–Martins pseudopotentials [31]. Furthermore, $BaZnO_2$ exhibited a direct E_g (approximately 2.2 eV), which was determined by Wang et al. [32] using a density functional theory.

The d_{11} (approximately 12 pC/N) of $BaZnO_2$ is approximately five times greater than that of α-quartz SiO_2 [31], and the electromechanical coupling coefficient (k_{11}, approximately 36%) is approximately three times that of SiO_2 [31]. The k_{11} of $BaZnO_2$ is also higher than that of $GaAsO_4$ (approximately 22%). These highest reported coefficients among the quartz-type group [30, 31] indicate the promise of $BaZnO_2$ for extended piezoelectric applications.

Extremely high temperatures are typically required for the fabrication of $BaZnO_2$, which substantially limits its wide application. Chiang et al. [10] reported a facile solvothermal synthesis to fabricate $BaZnO_2$ nanowires (as discussed in Section 5.4.2.2), and studied their associated piezotronic and piezophototronic features (as discussed in Section 5.6.2).

5.3.3 BiFeO₃ [11]

Multiferroic $BiFeO_3$ (BFO) exhibits large ferroelectric polarization, G-type anti-ferromagnetism, and a high transition temperature [33]. The rhombohedrally distorted perovskite structure (R3c) enables the atomic displacement of Bi ions with reference to FeO_6 octahedrals along the [111] direction, which leads to its spontaneous polarization of approximately 90 $\mu C/cm^2$ [34]. Furthermore, BFO owns a direct and small E_g (approximately 2.2–2.7 eV) and excellent chemical stability [35], which enables it as an efficient visible-light photocatalyst.

Various solution-based processes (e.g., sol-gel coating [36], hydrothermal synthesis [37, 38], chemical solution deposition [39], sol-gel spin coating [40], and colloidal dispersion [41]) have been studied to fabricate various morphologies of BFO powders. Hydrothermal synthesis is one of the most appealing alternatives among these approaches (as discussed in Section 5.4.2). However, high concentrations of KOH solutions [42], high pH values [37], and base-resistant STO substrates [38] were typically required for fabricating BFO powders and thin films. Highly alkaline conditions for using indium tin oxide (ITO) and fluorine-doped tin oxide (FTO) substrates are substantially challenging because of severe etching. When the pH was lowered, unfavorable impurities (e.g., Fe_2O_3, $Bi_2Fe_4O_9$, and $Bi_{25}FeO_{40}$) were formed. Arazas et al. [11] reported the hydrothermal fabrication of pure BFO nanorods on ITO with the assistance of a seed layer through spin coating. The hydrothermal parameters, including the amounts of precursors, adjustment to the pH level, types of complex agents, and reaction times and temperatures, were studied to manipulate the morphology and alignment of BFO (as discussed in Section 5.4.2.3).

5.4 FABRICATION METHOD AND MORPHOLOGY TUNING

5.4.1 Combinatorial Reactive Sputtering [4]

Combinatorial methodology is an efficient approach for fabricating a uniform set of samples on a single substrate (library) in a single process [43]. Developed samples are then rapidly screened systemically to determine desirable compositions for specific applications. This methodology enables

Mutually Influenced Stacking

substantial time-, resource-, and manpower-saving, compared with conventional manufacturing approaches.

For the convenience of various characterizations on a specific single piece (location) across libraries, substrates are cut into numerous equivalent pieces and are subsequently assembled for library deposition.

A high working pressure and a long working distance (approximately 7 cm) between a gun and a substrate enable the deposition of low-mobility species, thus leading to the formation of well-separated 1D structures aligned normal to a substrate [44]. Furthermore, RF powers, flow ratios of Ar:N$_2$, and deposition temperatures are also crucial for obtaining desirable-crystallinity materials.

5.4.1.1 Zn–Sn$_3$N$_4$ Composition Spreads
5.4.1.1.1 Moving Shutter [4]

Kuo et al. [4] employed a stainless steel moving shutter involved in combinatorial reactive sputtering for synthesizing Zn–Sn$_3$N$_4$ composition spreads on FTO/glass substrates, which were synthesized by coupling natural Sn$_3$N$_4$ (Figure 5.1(a)) and artificial Zn (Figure 5.1(b)) thickness gradients. A natural Sn$_3$N$_4$ gradient was optimized by adjusting an angle of the Sn-target sputtering gun, and a reactive RF mode (50 W) and a working pressure (20 mTorr) were applied. Various gas ratios of Ar:N$_2$ (e.g., 24:3, 27:3, 28:3, 29:3, 33:3, and 35:3) and substrate temperatures (400–500°C) were tailored to optimize the crystallinity of Sn$_3$N$_4$. Furthermore, a Zn thickness gradient was obtained using the following conditions: pure Ar plasma, DC 50 W, 20 mTorr, and 500°C.

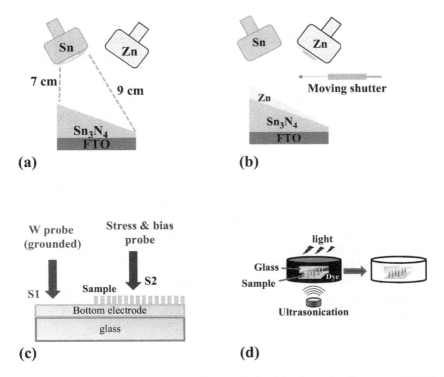

FIGURE 5.1 Fabrication of Zn–Sn$_3$N$_4$ composition spreads: (a) schematic of a natural Sn$_3$N$_4$ thickness gradient and (b) schematic of an artificial Zn thickness gradient. (c) Piezotronic measurement configuration. (d) Schematic of piezophotocatalysis. [Adapted from *Crystal Growth & Design* 17, 4694 (2017); *Journal of The American Ceramic Society* 103, 4129 (2020); *Journal of The American Ceramic Society* 99 [8], 2593 (2016).]

5.4.1.1.2 No Moving Shutter [12]

Another set of experiment to fabricate $Zn–Sn_3N_4$ composition spreads was also studied by Hsu et al. [12], in which no moving shutter was used to avoid mechanical malfunction problems and minimize the complexity of the system. Sn_3N_4 was also first deposited on the substrate, followed by a reversed deposition of Zn (substrate indexed 180°) on top of Sn_3N_4. The total number of layers of Sn_3N_4 and Zn and the respective deposition times were tailored to tune the morphology and crystallinity of $Zn–Sn_3N_4$ composition spreads.

The working pressures (20–35 mTorr), RF powers (40–75 W), and flow ratios of $Ar:N_2$ (38:2, 42:2, 42:3, and 48:3) were studied to optimize Sn_3N_4 nanocolumns at 450 °C. For optimizing Zn crystallinity using a Zn target at a flow rate of Ar (20 sccm) and 450 °C, the working pressures (20–50 mTorr) and DC powers (90 and 150 W) were tuned.

5.4.1.2 AlN-TiN Nanocolumn Composite Composition Spreads [5]

Lee et al. [5] reported a synthesis of AlN–TiN nanocolumn libraries (AlN thickness gradient coupled with reversed TiN thickness gradient) on FTO/glass substrates (24 mm × 24 mm) using combinatorial reactive sputtering. Operation power and pressure, $Ar–N_2$ ratios, and deposition temperature, were investigated to pursue crystalline AlN and TiN nanocolumns. The following conditions: 450-W RF power, 10 mTorr, and $Ar:N_2$ of 10:2 sccm, and 450 °C were used to obtain the required crystallinity of AlN nanocolumn arrays. Similarly, RF of 350 W, 10 mTorr, $Ar:N_2$ of 10:1 sccm, and 450 °C were used to obtain the reversed natural TiN thickness gradient. Seven layers were deposited to achieve the required sample thickness.

5.4.2 SOLUTION-BASED PROCESS

Hydrothermal (solvothermal) reactions are considered one of the sustainable alternatives to replace traditional high-vacuum techniques because of their simplicity of experimental design, sustainability, low cost and pollution, low-temperature synthesis, potential wide-scale throughputs, capability for maneuvering various morphologies, sizes, and orientations of nanostructures [45].

These processes are performed by transferring prepared precursor solutions and substrates into a Teflon-lined stainless steel autoclave and then placed into a box furnace operated at certain temperature for certain time. Substrates are then removed, washed using distilled (DI) water, and then dried to obtain desirable films.

5.4.2.1 ZnSnO₃ [18]

Wang et al. [18] employed a two-step hydrothermal reaction to fabricate $Zn_{1-x}SnO_3$ nanowire arrays. For Step 1 (seed layer preparation), $ZnCl_2$ and $SnCl_4·5H_2O$ precursors were dissolved in DI water, and then $NaOH_{(aq)}$ was used to tune the pH ≈ 7. A complex agent of glucose was subsequently added, and the pH ≈ 11.5 was tuned using $NaOH_{(aq)}$. The total solution volume was sustained approximately 50 mL. Afterward, the solution with an ITO/glass substrate was hydrothermally heated at 160 °C for 1.5 h to grow a $ZnSnO_3$ seed layer. For Step 2, the hydrothermal precursor solution was prepared almost the same as the aforementioned procedure, except using another complex agent of polyethylene glycol (molecular weight (MW): 200). The solution and the seed layer substrate were hydrothermally heated at 200 °C for 4, 6, 8, and 12 h to obtain various $Zn_{1-x}SnO_3$ nanowire arrays.

5.4.2.2 BaZnO₂ [10]

Chiang et al. [10] solvothermally grew $Ba_{1-x}ZnO_2$ nanowires on FTO/glass substrates, in which $ZnCl_2$ and $BaCl_2 • 2H_2O$ precursors were used. $BaCl_2 • 2H_2O$ was first dehydrated before dissolved in methanol. The pH value was tuned using NaOH tablets. The prepared solution and an FTO substrate were subsequently solvothermally heated at 200 °C for 4 h to obtain $Ba_{1-x}ZnO_2$ nanowires.

Mutually Influenced Stacking

5.4.2.3 BiFeO$_3$ [11]

Arazas et al. [11] developed a two-step approach (spin coating and hydrothermal synthesis) to fabricate BiFeO$_3$ nanorods. A seed layer was first spin-coated on an ITO/glass substrate. The spin-coating solution was prepared as follows. 8-g Bi(NO$_3$)$_3$·5H$_2$O and 6.06-g Fe(NO$_3$)$_3$·9H$_2$O precursors were mixed in 2-mL 2-methoxyethanol (2-ME) solvent under vigorous stirring until they were completely dissolved. 10-mL Acetic acid and 0.1-mL ethanolamine were then added to the above solution under continuous stirring. Additional 2-ME was poured into the prepared solution until 20 mL was reached. The solution was spin-coated on an ITO/glass substrate at 6000 rpm for 40 s, dried at 80 °C for 30 min, and finally annealed at 500°C for 1 h under O$_2$ gas to obtain a robust seed layer.

BFO nanorods were then hydrothermally fabricated on the seed-layered substrate. The precursor solutions were prepared by mixing 3.63-g Bi(NO$_3$)$_3$·5H$_2$O, 3.03-g Fe(NO$_3$)$_3$·9H$_2$O, and 1.5-g polyethylene glycol (PEG, MW: 10 k) in 2-mL HNO$_3$ and 20-mL DI water under vigorous stirring for 20 min. 200 mL of DI water was then added to the above solution, and the pH $\approx$ 10–11 was tuned using KOH tablets under constant stirring for 1 h. The precipitates were collected by filtration and washed using DI water to remove NO$_3^-$ and K$^+$ ions. Finally, the precipitates, 1.5-mL PEG (MW: 200 k), and 1.403-g KOH were mixed in 40-mL DI water under vigorous stirring. The prepared solution and seed-layered substrate were then hydrothermally heated at 200 °C for various times to obtain BFO nanorods.

5.5 CHARACTERIZATIONS OF VARIOUS PROPERTIES

5.5.1 Piezotronic-Related Property [18]

A special probe (Sadhu Design Co., Taiwan) is designed to characterize the piezotronic properties of samples [18]. The system involves an analyzer (HP-4145B) and two tungsten (W) electrical probes. One probe (S2) contacts the top of materials under study, which is integrated with a sensitive stress reader and capable of supplying voltage and stress to materials. The other probe (S1) is grounded and contacts a bottom electrode. When samples are additionally illuminated, the piezophototronic effect is measured (Figure 5.1(c)).

5.5.2 Schottky Behavior [18]

According to the thermionic emission-diffusion theory [46], current densities (J$_D$) for a Schottky diode is expressed as follows:

$$J = J_s \left(e^{\frac{qV}{kT}} - 1 \right) = A^{**}T^2 e^{\frac{-q\phi_{Bn}}{kT}} \left(e^{\frac{qV}{kT}} - 1 \right) = J_s \left(e^{\frac{qV}{kT}} - 1 \right) \tag{5.3}$$

where J$_S$ is the saturation current; A** represents effective Richardson constant; T is the absolute temperature; q denotes the electronic charge; ϕ_{Bn} is the barrier height; k represents Boltzmann's constant; and V is the applied bias. When V$_r$ (reversed bias) > 3kT/q, J$_r$ is deduced according to Equation (5.3):

$$J_r \cong J_s = A^{**}T^2 e^{\frac{-q(\phi_{B0}-\Delta\phi)}{kT}} = A^{**}T^2 e^{\frac{-q\phi_{B0}}{kT}} e^{\left(\frac{q\sqrt{\frac{qE}{4\pi\varepsilon_s}}}{kT} \right)} \left(\because e^{\frac{qV}{kT}} \to 0 \right) \tag{5.4}$$

$$\text{where } \phi_{Bn} = \phi_{B0} - \Delta\phi; \Delta\phi = \sqrt{\frac{qE}{4\pi\varepsilon_s}}; E = \sqrt{\frac{2qN_D\left(V + V_{bi} - \frac{kT}{q}\right)}{\varepsilon_s}} \tag{5.5}$$

When $V_r \gg V_{bi}$ & kT/q, $E \propto V^{1/2}$ is obtained.

$$\therefore J_r \propto e^{V^{\frac{1}{4}}} \therefore \ln J_r \propto V^{\frac{1}{4}}$$

Thus, a linear relationship between $\ln(J_r)$ and $V^{1/4}$ under various pressures at a range of bias indicates a Schottky behavior at contacts (e.g., S1 and S2 in Figure 5.1(c)).

5.5.3 Schottky Barrier Height (SBH) Variation [18]

Schottky barrier height variation ($\Delta\phi$) at S2 and S1 under various pressures can be determined from Equation (5.4), which is simplified into Equation (5.6) when a sample is under two different pressures (p1 and p2) because $A^{**}T^2 e^{\frac{-q\phi_{Bo}}{kT}}$ is a constant.

$$\Delta\phi_{p2-p1} = \Delta\phi_{p2} - \Delta\phi_{p1} = \frac{kT}{q} \ln\left(\frac{J_{r_{p2}}}{Jr_{p1}}\right) \tag{5.6}$$

For an easy calculation, p1 is assigned a reference pressure and p2 is varied. Thus, $\Delta\phi_{p2-p1}$ at S1 and S2 under a bias is determined.

5.5.4 Piezophotocatalysis [18]

For photocatalytic activity, sample solutions are maintained in the dark for certain time to achieve an absorption–desorption equilibrium prior to measurement. A photocatalytic efficiency of samples is then measured by photodegrading a dye solution under irradiation. A UV-vis spectrometer is utilized to characterize residual-dye concentrations. For piezophotocatalytic activity, external stress is additionally applied using a piece of transparent glass and ultrasonication operated at a power of approximately 0.2 W and at a frequency of 40 kHz (Figure 5.1(d)).

5.5.5 Piezopotential [47]

Stress-induced piezopotential measurement presents a direct evidence of the piezoelectricity exhibited by a material. The piezopotential can be measured using two W probes, a semiconductor parameter analyzer (HP–4145B), and a current–voltage (I–V) probe system integrated with an oscilloscope.

5.5.6 Incident Photon-to-Current Efficiency (IPCE) [48]

IPCE is measured through a photoelectrochemical (PEC) reaction, in which a sample is used as a working electrode and a platinum film as a counter electrode, and both electrodes are immersed in a suitable electrolyte solution without any applied DC bias. A PEC cell is fabricated by heating two L-shaped polymers (Surlyn) to approximately 120 °C for several minutes and then are attached to the two electrodes. The resulting reservoir is filled with electrolyte solutions, and the cell is then illuminated with a light source (wavelength ranging from 400 to 700 nm).

5.6 NANOSTRUCTURED ZnSnO$_3$ AND BaZnO$_2$ AND THEIR APPLICATIONS

5.6.1 ZnSnO$_3$ [9]

ZnSnO$_3$ nanowire arrays were fabricated using 0.438-mmol (1.75-g) PEG (4000) in precursor solutions (inset, Figure 5.2(a)). The density and surface area were approximately $81/\mu m^2$ and 1.9×10^4 nm^2 in a 1-μm^2 area, respectively. These features were ideal for piezo-related applications.

Mutually Influenced Stacking

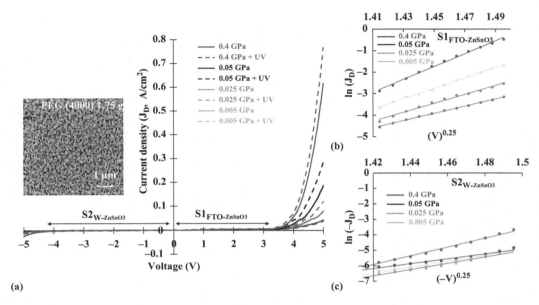

FIGURE 5.2 (a) I-V characteristics of ZnSnO₃ nanowire arrays under various stresses with (dashed line) and without (solid line) UV irradiation. The inset shows a tilted SEM image of ZnSnO₃ nanowire samples fabricated using 0.438-mmol (1.75-g) PEG (4 k). (b) Plot of ln (J_D) as a function of $(V)^{1/4}$ at S1$_{FTO-ZnSnO3}$. (c) Plot of ln ($-J_D$) as a function of $(-V)^{1/4}$ at S2$_{W-ZnSnO3}$. [Adapted from *Journal of The American Ceramic Society* 103, 4129 (2020).]

The piezotronic effect of the ZnSnO₃ nanowire arrays was characterized using a probe design (as discussed in Section 5.5.1). The effect was indicated by asymmetric I–V characteristics under various stress levels (e.g., 0.005, 0.025, 0.05, and 0.4 GPa) (solid line, Figure 5.2(a)). When a positive bias was applied at the interface between W and ZnSnO₃ (S2$_{W-ZnSnO3}$) (Figure 5.1(c)), electrons flowed from the interface between FTO and ZnSnO₃ (S1$_{FTO-ZnSnO3}$) to S2$_{W-ZnSnO3}$ (Figure 5.3(c)). A decrease in SBH at S1$_{FTO-ZnSnO3}$ leads to the improved J_D for each stress level because of the development of a positive piezopotential at S1$_{FTO-ZnSnO3}$ (bottom of ZnSnO₃ nanowires). Meanwhile, when a negative bias was applied at S2$_{W-ZnSnO3}$, the development of a negative piezopotential at S2$_{W-ZnSnO3}$ (top of ZnSnO₃ nanowires) led to an increase in SBH. Thus, electron flow from S2$_{W-ZnSnO3}$ to S1$_{FTO-ZnSnO3}$ was reduced (reduced J_D). Piezophototronic characteristics were determined when external illumination was applied to the ZnSnO₃ nanowire arrays (dashed line, Figure 5.2(a)), in which the same J_D variation trend as the piezotronic effect was observed.

The Schottky behavior of the arrays was studied through I–V characteristics (as discussed in Section 5.5.2) [13]. A plot of ln (J_D) against $(V)^{1/4}$ ranging from approximately 4–5 V at S1$_{FTO-ZnSnO3}$ under various stress levels is illustrated in Figure 5.2(b). Similarly, another plot of ln ($-J_D$) against $(-V)^{1/4}$ ranging from approximately −4 to −5 V at S2$_{W-ZnSnO3}$ under various stress levels is illustrated in Figure 5.2(c). Schottky contacts were then validated from the observation of a linear relationship at S1$_{FTO-ZnSnO3}$ and S2$_{W-ZnSnO3}$.

Although a nominal SBH at S1$_{FTO-ZnSnO3}$ and S2$_{W-ZnSnO3}$ was indicated by the threshold voltage measured from the I–V characteristics (double-headed arrow, Figure 5.2(a)), the SBH variation at S1$_{FTO-ZnSnO3}$ ($\Delta\Phi_{(p-0.005, S1)}$; p denotes various stresses) and S2$_{W-ZnSnO3}$ ($\Delta\Phi_{(p-0.005, S2)}$) was crucial to understand the J_D behavior in Figure 5.2(a). The details are discussed in Section 5.5.3. A decreasing trend as a function of pressure at a bias of 4.9 V in $\Delta\Phi_{(p-0.005, S1)}$ (Figure 5.3(a)) indicated the proportional generation of positive piezopotential at the bottom of the ZnSnO₃ nanowires. By contrast, no substantial variation in $\Delta\Phi_{(p-0.005, S2)}$ as a function of stress was observed (increase of

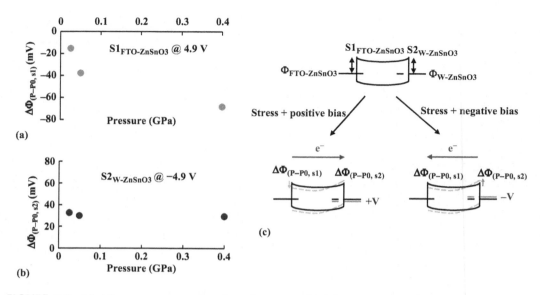

FIGURE 5.3 (a) $\Delta\Phi_{S1(FTO-ZnSnO3)}$ as a function of pressure. (b) $\Delta\Phi_{S2(W-ZnSnO3)}$ as a function of pressure. (c) Energy band diagram evolution under stress and bias. [Adapted from *Journal of The American Ceramic Society* 103, 4129 (2020).]

approximately 30 mV at 0.4 GPa (Figure 5.3(b)), indicating no effective accumulation of negative piezopotential on top of the ZnSnO₃ nanowires. The associated energy band evolution is illustrated in Figure 5.3(c).

The photocatalytic and piezophotocatalytic activities of the ZnSnO₃ nanowire arrays for MB solutions were illustrated in Figure 5.4(a). Random ZnSnO₃ nanowires (without PEG (4000) involvement, brown) and self-degradation of MB (blue) were also measured for comparison. The measurement details are discussed in Section 5.5.4. In addition, the activity for nanowires from our previous study (light blue) [18] was also plotted as a reference. The result indicated that the ZnSnO₃ nanowire arrays (green) outperformed all other samples because of the following: 1) the high densities and surface areas, which enabled strong light absorption and induced substantial amounts of electron–hole (e⁻-h⁺) pairs, 2) enhanced alignment and crystallinity of nanowires, which enabled easy transport of e⁻–h⁺ pairs to the sample surface to react with O_2 and H_2O for the formation of superoxide ($\cdot O_2^-$) and hydroxyl ($\cdot OH$) radicals, respectively, and 3) favorable energy band positions of the sample [9].

The piezophotocatalytic activity for the ZnSnO₃ nanowire arrays was also evaluated. A substantial improvement was observed when two-piece glass and ultrasonication were applied (red), particularly in the first 120 min (red), compared with the condition without any external stresses (green) and with the involvement of one-piece glass and ultrasonication (purple). In addition to the piezopotential buildup, the improvement was also associated with the band bending of the ZnSnO₃ nanowire arrays. Both inhibited the recombination of photogenerated e⁻–h⁺ pairs. A plot of $\ln(C_0/C)$ (C_0 and C represent the original and residual concentrations of MB, respectively) against irradiation time was employed to calculate the degradation rate constant (k) for the samples (inset, Figure 5.4(a)). The highest k (approximately 17.6×10^{-3} min⁻¹) was observed for the ZnSnO₃ nanowire array sample.

A three-run cycling test was conducted to determine the piezophotodegradation reliability of the ZnSnO₃ nanowire array sample. The condition of applying one-piece glass and ultrasonication (lower stress) (Figure 5.4(b)) revealed more reusable piezophotodegradation than that of applying two-piece glass and ultrasonication (higher stress) (Figure 5.4(c)) because the third run started to

Mutually Influenced Stacking 137

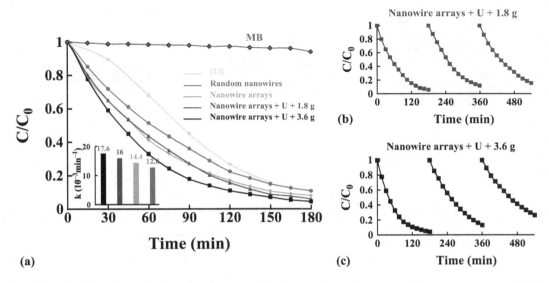

FIGURE 5.4 (a) Photocatalytic and piezophotocatalytic activities of ZnSnO$_3$ nanowire arrays. Photocatalytic activities of random ZnSnO$_3$ nanowires are presented for comparison. The degradation rate constant k is displayed in the inset. (b) and (c) Piezophotocatalytic cycling test of ZnSnO$_3$ nanowire arrays using a one-piece and two-piece glass with ultrasonication, respectively. [Adapted from *Journal of The American Ceramic Society* 103, 4129 (2020).]

deteriorate for the latter condition, which damaged the durability of the ZnSnO$_3$ nanowires because of the long-term application (> 6 h) of overloading stress.

PEC and PPEC water splitting were another promising application of the ZnSnO$_3$ nanowire arrays. The performance was indicated by the induced photocurrent density (J_{ph}), which was then used to calculate the associated applied bias photon-to-current efficiency (ABPE) (Figure 5.5(a)). The PPEC reaction (under both illumination and stress, red) was substantially more enhanced than the PEC reaction (under illumination only, green). The reliability of PEC and PPEC reactions was evaluated through a five-cycle test, which was conducted at a bias of approximately 0.5 V with a 40-s light-on and 40-s light-off cycles (Figure 5.5(b)). Reliable J_{ph} values were observed for both reactions, and the PPEC reaction considerably outperformed the PEC reaction. The J_{ph} pattern for

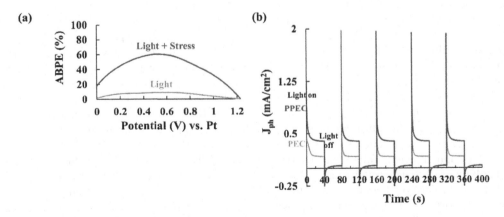

FIGURE 5.5 PEC and PPEC results. (a) ABPE values and (b) cycling test. [Adapted from *Journal of The American Ceramic Society* 103, 4129 (2020).]

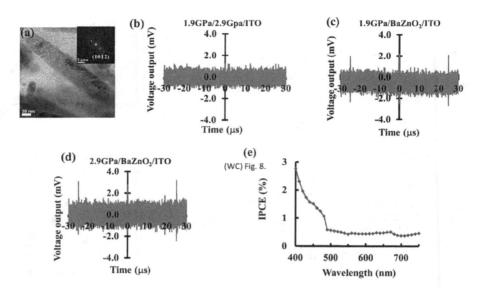

FIGURE 5.6 (a) TEM image of single Ba$_{1-x}$ZnO$_2$ nanowire. The inset shows an SAED pattern taken from the red circle marked in (a). Piezopotential characterizations of various samples at various stresses: (b) ITO at approximately 1.9 and 2.9 GPa; (c) and (d) BaZnO$_2$ films on an ITO/glass substrate at approximately 1.9 and 2.9 GPa, respectively. (e) IPCE results. [Adapted from *Applied Surface Science* 445, 71 (2018); *Materials Chemistry and Physics* 247, 122880 (2020).]

each cycle was attributable to the rapid generation of e$^-$–h$^+$ pairs under illumination in the initial stage and then the J_{ph} gradually stabilized after equilibrium was attained. In summary, the results indicate the potential of the sample for photocatalytic-related applications.

5.6.2 BaZnO$_2$ [10]

To study the growth orientation of the solvothermally grown Ba$_{1-x}$ZnO$_2$ nanowires, a long single Ba$_{1-x}$ZnO$_2$ nanowire was probed (Figure 5.6(a)). An SAED pattern (inset, Figure 5.6(a)) indicated the plane indexed at (10$\bar{1}$2) (d-spacing of approximately 0.27 nm) in Ba$_{1-x}$ZnO$_2$.

The measured piezopotential against time under two different alternating pressures of approximately 1.9 and 2.9 GPa (40 kHz) is illustrated in Figure 5.6(b)–(d) for the BaZnO$_2$ films deposited on an ITO/glass substrate. The measurement details are discussed in Section 5.5.5. A bare ITO substrate was also measured for comparison; non-piezoelectricity was indicated because no piezopotentials were observed at the two pressures (Figure 5.6(b)). However, when a pressure of approximately 1.9 and 2.9 GPa was applied to the sample, a piezopotential of approximately 2 (Figure 5.6(c)) and 3 (Figure 5.6(d)) mV was observed, respectively. These results indicated the piezoelectric property of the BaZnO$_2$ films.

IPCE is another critical indicator for quantifying the conversion efficiency of solar power to hydrogen harvesting under zero bias (Figure 5.6(e)). The IPCE value was approximately 2.8% at a wavelength of 400 nm, indicating the promise of this sample in PEC applications.

Although various aforementioned inorganic materials have been studied, there still exist opportunities for improvement in their piezoelectric performance. Thus, a novel idea by coupling inorganic and polymer materials through mutual stacking is discussed. Piezoelectric inorganic materials typically exhibit robust polarization under stress, which is excellent to induce improved polarization in polymers. Furthermore, easy tuning of polymer crystal orientations may be also excellent to guide the alignment of inorganic crystals. A synergistic piezoelectric performance can then be expected by the mutual influence.

Mutually Influenced Stacking

5.7 PIEZOELECTRIC FEATURES OF POLYMER MATERIALS

For the materials piezoelectricity, it is related to the alteration of materials polarization proportional to either deformation or applied mechanical stresses. The change of polarization ΔP, related to the degree of piezoelectricity, is described by the following equation [49–50].

$$\Delta P = d_{ij} X_j \tag{5.7}$$

$$\Delta P = g_{ij} S_j \tag{5.8}$$

The X and S in above equations denote the applied stress and caused strain respectively. The d_{ij} and g_{ij} are two piezoelectric coefficients related to the relationships between the degree of polarization and stress or strain respectively. Because of the anisotropy of crystal structure, the polarization or piezoelectricity P is described as a vector that has three orthogonal components in the 1, 2, and 3 directions. The first subscript i is related to the direction of electric field or displacement, and the second subscript j denotes the direction of mechanical stress or charge deformation. Compared with piezoelectric inorganics, piezoelectric polymers possess lower transverse piezoelectric coefficient (d_{31}) (or described as piezoelectric strain constant), and nevertheless much higher piezoelectric stress constants (g_{31}). Therefore, piezoelectric polymers are more sensitive to instigated strains. Also, with low density and elastic stiffness, amorphous polymer materials commonly illustrate high voltage sensitivity. Accordingly, piezoelectric polymers are better suitable for the use in sensors as the deformation modulus is rather lower than that of ceramics. Furthermore, typically, the insulating feature makes polymers also possess high dielectric breakdown strength, which renders better voltage endurance.

The above relationships are derived mainly based on the piezoelectric behaviors of crystals. Nevertheless, different from piezoelectric inorganics, both amorphous and crystalline phases of polymer materials are able to yield piezoelectric responses, and however, different mechanisms are involved. The piezoelectricity of polymer single crystals also originates from the polarization change upon the elastic deformation of crystal lattices, similar to the piezoelectric behaviors of ceramics. For polar amorphous of polymers or small organic molecules, molecules are randomly dispersed, and nevertheless the overall polarization across top and bottom surfaces are also adjustable based on the instigated elastic or permanent deformations.

Within amorphous regions of polymer materials, unperturbed molecular chains prefer to curl themselves into coil-like objects as a thermodynamic favorable status. Under applied mechanical stresses, molecular coils can be forced to have either permanent translational displacement, or elastic stretching/compression deformations, which are basically the origin of viscous and elastic properties of polymer amorphous phases respectively. Related molecular weight, backbone rigidity, size of lateral attached groups, broad ranges of viscoelastic properties are available. With the dispersion of entangled or crosslinked long-chain molecules, the elastic feature is to be enhanced, which makes amorphous polymer materials behave better as elastomers. Hence, amorphous polymer materials are able to behave as an elastomer with adjustable and much greater levels of reversible stretching/compression, and these elastic deformations might come with a significant change of molecular polarization as the spatial arrangement of polar groups are modified accordingly.

For a polarizable elastomer film sandwiched between two electrodes, it is basically recognized electrically as a capacitor. The elastic deformation of the sandwiched polymer film caused by input mechanical energy is therefore accompanied with measurable alteration of dielectric properties, provided that the synchronous extension and shrinkage of deposited electrodes are feasible. The accumulated electric charges on electrodes are able to be further augmented or lessened during mechanical deformation as being associated with the capacitance alteration. Accordingly, mechanical energy applied to deform polymer dielectrics is converted to electric energy. On the other hand, as the applied electric field is able to cause elastic deformation of polarized elastomers, input electric energy is converted to stored mechanical energy. Since the dielectric elastomers are able to produce charges on top and bottom electrodes due to mechanical stress, they also can be considered as piezoelectric materials. Accordingly, as being polarizable, amorphous polymer elastomers are able to behave as a type of piezoelectric materials. However, the efficiency of above working principle of polymer elastomers is critically dependent on the available degrees of polarization, which nevertheless are much less satisfied normally. The low degree of polarization is also viewed as the bottleneck for the use of semicrystalline polymer piezoelectric materials. Therefore, after the introduction of basic polarization mechanism, several strategies able to enhance the degree of polymerization of polymer materials are discussed in the following part of this chapter.

Main polarization mechanisms of polymeric materials. For the concerned polarization of polymer materials, possible polarization mechanisms include electronic, vibrational (or atomic), dipolar, ionic, and interfacial polarization [51–59]. Nevertheless, not all these mechanisms are able to yield non-trivial contribution to the polarization of polymeric materials. For chemically saturated insulating polymers, all the available outer shell electrons are tightly bound in the σ-bonds, and there is no free electron able to be delocalized and contribute to an electrical current, rendering the electronic polarization less feasible. For the semiconductive conjugated polymers, the overlapping of P_z orbitals of each carbon is able to cause π-electrons to be delocalized and move along polymer backbones. However, this overlapping of P_z orbitals critically depends on the maintenance of coplanar molecular conformation, which nevertheless is easily to be seriously lessened by thermal agitation and surrounding interactions as molecular chains are in the amorphous state. For the vibrational polarization, it is merely about 10–50% of electronic polarization commonly [60]. The attachment of polar side groups and the replacement of carbons atoms on the molecular backbone with elements of group 14 on periodic table might help to increase the contribution of both electronic and vibrational polarization [56, 57]. However, the chemical stability of molecular backbones and the great challenge of synthesizing such inorganic polymers has established the barriers of these approaches. According to above discussion, vibrational and ionic polarizations are less likely to yield meaningful contribution to the piezoelectric or dielectric. Therefore, most of the strategies adopted for the innovation of piezoelectric polymer materials are related to the development of dipolar and interfacial polarization.

5.8 STRATEGIES TO ENHANCE THE POLARIZATION LEVELS OF POLYMER MATERIALS

5.8.1 POLARIZATION OF POLYMER PHASES VIA POLING TREATMENT

Compared to both the electric and vibrational polarization, the dipolar polarization is theoretically more promising to be largely enhanced and contribute to the piezoelectric performances. With a sinusoidal half-wave direct electric field waveform in the desired period of times, the electric dipoles, either in amorphous or crystalline phases, are likely to rotate accordingly depending on the strength of outer fields and encountered kinetic barriers of dipole rotation. Hence, the poling treatment has been an intensely studied approach for directional dipolar polarization of polymer amorphous phases, as well as polar polymer crystals [61–64].

However, for the amorphous phases of long-chain molecules, they can be at either solid glass state or liquid molten state at working temperatures. Especially for polymer amorphous solids, more than one energy status could exist, which are mainly related to the deviation from unperturbed molecular conformations and available segmental thermal motions, and thus the feasibility of dipole rotation and dielectric relaxation. In addition to intrinsic rotational rigidity imposed by molecular structure, there are many other additional factors able to influence the dipole rotation within polymer amorphous phases under the applied electric field, which includes molecular entanglement, interfacial effects, available space, residual stresses, and interactions with foreign ingredients. Principally, any factor able to influence molecular conformation is able to alter the spatially dispersion of polar groups. As there are many competitive factors able to adjust molecular conformation, the degree of polarization of amorphous phase can be subtly manipulated theoretically. Among all these possible influential factors, mechanical stretching and electrical poling are the two mostly adopted approaches, as the they can be easily adjusted and quantified. Nevertheless, it should be noticed that, as being driven by entropy restoration, molecular relaxations to original unperturbed molecular conformation are thermodynamically favorable. As cooperative segmental thermal motions are available in amorphous polymers at temperatures above Tg, molecular relaxations are able to proceed forward via various types of available kinetic paths, and a significant fraction of molecules are able to efficiently relax back to the status with isotropic orientation of polar group after the input factors have been removed. From this point of view, the true challenge of enhancing the polarization of polymer amorphous phases lies on the preservation of oriented alignment of polar groups, instead of the strategies needed for the alignment of polar groups.

In addition to amorphous phase, as the rotation of structural units within crystal lattices is feasible, the interactions between electric dipoles and outer electric fields are also able to cause the directional polarization of crystal lattices [65–68]. As a result of dipole alignment upon poling treatment, molecular stems are forced to adopt the all-trans conformation in both crystalline and amorphous phases. The all-tans zigzag conformation is principally the most extended form of molecular chains, and is also the molecular conformation able to yield the utmost degree of electric dipoles of molecular segments. Therefore, the main approach to enhance the polarization of polymer materials is universally to create a significant fraction of all-trans molecular segments. However, as molecular segments are organized with crystalline lattices, this modification of molecular conformation upon poling treatment is favorably accompanied with the change of lattice-packing scheme, and ferroelectric crystals are likely to evolve accordingly. In order to facilitate accompanied phase evolution, the poling treatment is usually integrated with other processes like stretching and thermal treatment [69].

5.8.2 Growth of Polymer Ferroelectric Crystalline Phases

For a ferroelectric crystal, the dipolar polarization can be reoriented by an outer electric field, and a certain fraction of electric dipole remain unchanged after the remove of outer electric field. The intrinsic polarization and non-linear polarization reversal with remanent polarization at zero intensity of outer electric field are acknowledged as the proof and features of ferroelectric crystals. The polarization extent of polymer ferroelectric crystals is generally modifiable by applied mechanical stresses, viewed as the piezoelectric feature. Hence, basically, all the ferroelectric materials are able to yield piezoelectric response, but not every piezoelectric material is able to yield ferroelectricity. In view of the increase of available change of thin film polarization, the growth of ferroelectric crystals has been intensely pursued as an approach for polymer films to yield stronger and more sensitive piezoelectric and also dielectric responses.

For the crystallization of a selected polymer to a crystalline form, a specific molecular conformation is generally adopted for molecular stems to reach the same thermodynamic status and repeatedly establish inter- and intramolecular interactions through lattice packing. However, partially because more than one kind of molecular conformation are able to evolve upon the competition between intermolecular and intramolecular interactions, polymorphism is one of the common characteristics of polymers, and nevertheless, only some specific crystalline forms possess desired dipolar polarization. For example, PVDF-based fluoropolymers are the most-known ferroelectric and piezoelectric polymer materials. Depending on adopted conformations of molecular stems, these fluoropolymers are able to crystallize into more than four kinds of crystalline forms, including paraelectric high-temperature phases, ferroelectric low-temperature phase, and intermediate phases [70–75].

For a crystalline phase able to yield promising dipolar polarization, the essential criteria is the lattice packing of molecular stems with all-trans conformation. This all-trans conformation arranges more electronegative fluorine atoms on one side of packed molecular stems and less electronegative hydrogen atoms on the other side, therefore resulting in electric dipoles across packed molecular stems. The lattice packing of all-trans molecular stems has been acknowledged as the most feasible and well-known mechanism able to originate the dipolar polarization and thus ferroelectric features of polymer crystals. Nevertheless, as fluorine atoms are arranged on the same side of molecular stems upon the adoption of all-trans conformation, the mutual repulsion between neighboring fluorine atoms is unavoidable. Accordingly, the all-trans conformation is a less stable molecular conformation, which renders the growth of ferroelectric crystals a thermodynamically less favorable event. As the growth and dispersion of ferroelectric crystals with the lattice packing of all-trans conformer is generally considered able to make the major contribution to dielectric response of polymer materials, most of research strategies reported in literature are focused on the creation and stabilization of lattice packing of molecular stems with all-trans conformation, which includes the application of outer electric field, and mechanical stretching.

Upon uniaxial/biaxial stretching, molecular chains, either in crystalline lattices or amorphous regions, are forced to extend and frequently adopt the all-trans conformation along the stretching directions. As a result, non-polar crystalline form is driven to transform to polar crystals based on the initiated packing of all-trans conformers [76–80]. However, the conducted stretching is also able to initiate the secondary stage of crystallization, and therefore the non-polar phases are likely to further grow as well during stretching process, instead of being transformed to polar phases. The involved competition of phase evolutions is critically dependent on the encountered kinetic barriers and thus selected temperatures. In addition, the increases of stretching rate and ratio also contribute to enhancing the growth of ferroelectric phases as molecular relaxations are better prevented [79].

The fraction of polar phases basically increases with the increase of stretch ratio and stretching rate, especially the stretching ratio [81]. Compared to uniaxial stretching, biaxial stretching has the added advantages of bidirectional enhancement. Also, the piezoelectric coefficient of biaxial stretched films is found to be larger than uniaxially stretched samples [82]. However, the stretching treatment is also able to break polymer crystals, and cause significant decrease of crystallinity. As the crystallinity is also critical for the net polarization of polymer materials, the stretching ratio needs to be limited below a threshold level in order to prevent the occurrence of crystal breaking [83].

5.8.3 Growth of Relaxor Ferroelectric Phases

For the crystallization of homopolymers composed of only one kind of structural unit, the regularity of a specific molecular conformation is generally required for lattice packing. Differently, molecular chains of copolymers are composed of more than one kind of structural units, and the composition of constituent structural units is quite influential to crystallization behavior. As the crystallization of copolymer allows certain variation of composition, the conformation of molecular stems within copolymer crystals might be not completely regular and likely undergo fruitful relaxation processes during heating. Recently, a new composition dependence intermediate phase has been identified, which is called morphotropic phase boundary (MPB). Actually for piezoelectric ceramics, the formation of MPB is a well-established physical concept and has been widely studied. Surprisingly, according to recent experimental and theoretical results, the growth of MPB is also feasible within organic materials. The ferroelectric copolymer poly(vinylidenefluoride-co-trifluoroethylene) (PVDF-TrFE) is found to have MPB regions [84–87]. Specifically, when the vinylidene fluoride (VDF) content decreases, a structural evolution from the all-*trans* conformation to the relaxor phase occurs. In the narrow transition range between 49 mol % $\leq$ VDF $\leq$ 55 mol %, the interconvention between competing ferroelectric phase and relaxor phase appears feasible. The relaxor phase can be viewed as the packing of molecular chains partially adopting both the all-*trans* and 3/1 helical conformations, therefore making the packing of molecular stems less compact. The looser lattice packing in MPB enables easy rotation of polarization, which explains the remarkably enhanced dielectric and piezoelectric properties near this region. The achieved dielectric constant

Mutually Influenced Stacking

can be higher than that of dipolar polymer glass and paraelectric polymers. Therefore, the discovery of MPB not only opens the new science interests but also offers a new route for developing the piezoelectric materials features.

5.8.4 ORIENTED STACKING OF FERROELECTRIC CRYSTALS

Normally within semicrystalline polymer materials, board-like lamellar crystals are randomly dispersed. Regarding this polycrystalline nature, dense stacking of numerous lamellar crystals with similar orientations of dipolar polarization has been viewed critical for collective contribution of dispersed polar crystals [88–90]. For the piezoelectric performance of polycrystalline thin films, the oriented crystal stacking is also required to avoid the mutual cancellation of piezoelectric responses of dispersed crystals.

Different from the self-assembly behaviors of organic molecules, the assembly of crystallites is much less feasible as both the driving forces and possible mechanisms are not well understood yet. With the two-phase nature of polymer solids, the dangling chains of polymer lamellar crystals are frequently entangled with surrounding molecular segments remaining at the amorphous state. Therefore, it is generally believed that self-assembly and coalescence of dispersed polymer crystals are less likely to occur as a mechanism to reach oriented crystal stacking. Nevertheless, the participation of entangled molecular segments into the growth of neighboring crystals likely causes stretching effects to residual amorphous segments being mutually entangled. As dispersed crystals evolves concurrently, in-between amorphous regions are stretched along multiple directions, which results in mutual drawing among evolving crystals. With this mutual drawing phenomenon, the growth and assembly of crystals are coordinated therefore. As a result, several-micrometer spread of oriented crystal stacking arrays has been experimentally observed, similar to a guided self-assembly process. As a rarely recognized means of phase interactions, the mutual drawing among dispersed crystals is conceivably related to the fraction of molecular chains being effectively entangled together, density of lamellar crystals, and the density of dispersed crystals. Hence, several influential factors are available for the initiation of this guided assembly behavior of dispersed crystals, which appears as a new route to engineer crystal stacking within polymer materials.

As the arrayed stacking of ferroelectric crystals evolves everywhere within thin films, a significant fraction of horizontally stretched amorphous molecular segments are present in between stacked crystals. Compared to molecular stems packed inside crystal lattices, these stretched amorphous segments are less constrained kinetically, and therefore can be directionally polarized by the poling treatment. Accordingly, the ferroelectric nature becomes attainable for stretched amorphous polymer phases. With the contribution of both polar phases, ferroelectric crystals and in-between stretched amorphous regions, the collective dipolar polarization of prepared films is able to yield the dielectric constant above 70 at room temperature. Thus, this approach is potentially able to significantly result in directional dipolar polarization of polymer amorphous phases.

5.8.5 MAXWELL–WAGNER–SILLARS INTERFACIAL POLARIZATION

The dispersion of nanofillers with high permittivity constant has been realized as a way to enhance energy storage in polymer films, which involves interfacial polarization and the enhancement of local electric fields. With a large permittivity contrast between nanofillers and polymer matrix, an outer electric field is able to result in dipolar interfacial charges, which results in the accumulation of surface charges on two opposite poles of nanofillers along the electric field. Due to the existence of these interfacial charges, this permittivity contrast results in the enhancement of local electric fields within polymer matrix, which is estimated via the equations below [91, 92].

$$E_p = E_0 \left[f_p \left(1 - \varepsilon_{r,p} / \varepsilon_{r,F} \right) + \varepsilon_{r,p} / \varepsilon_{r,F} \right]^{-1} \tag{5.9}$$

$$E_F = E_0 \left[f_p \left(\varepsilon_{r,F} / \varepsilon_{r,p} - 1 \right) + 1 \right]^{-1} \qquad (5.10)$$

In the above equations, E_p and E_F are the local electric fields in the polymer matrix and fillers respectively, and f_p is polymer volume fraction, and $\varepsilon_{r,p}$ and $\varepsilon_{r,F}$ are the dielectric constant of polymer matrix and nanofillers, respectively. Furthermore, as nanofillers are densely distributed and highly polarized by outer electric fields, the local fields of nearby nanofillers is likely to be superimposed, and therefore further enhanced by each other. The enhancement of local electric field is also likely to stimulate the accumulation of space changes along with the field direction. Thus, in addition to the permittivity contrast, more factors are there for the enhancement of local electric fields in polymer matrix. As a result, the actually reached strength of local electric fields can be much higher than the prediction of the Equation (5.9). Accordingly, the piezoelectric performances of surrounding polymer crystals are to be enhanced.

The above discussions basically evaluate Maxwell–Wagner–Sillars (MWS) interfacial polarization, which is related to enhanced local electric field, and also the presence of effective electron traps. These accumulated charge carriers at interfaces might take a long time to discharge as charge mobility is quite low within nanofillers. However, accumulated positive and negative charges might be driven to combine by the switch of outer electric field, which thus reduces the polarization and creates an internal AC current within nanoparticles.

For optimizing interfacial areas and therefore interfacial polarization, dense and uniform dispersion of nanofillers with a reduced nanoparticle size are desired. As a matter of fact, due to the strong aggregation tendency and poor chemical affinity between inorganic, the homogeneous mixing of inorganic fillers into polymer matrix is frequently not attainable. Filler-rich and filler-poor domains are likely to be created as a result of heterogeneous mixing and limitation of miscibility. The agglomeration is able to largely reduce the ratio of interfacial area to total weight of added nanofillers, and lessen available interfacial areas. However, via chemical and physical modification of nanoparticles, the interactions between polymers and nanoparticles can be largely improved.

The dispersion of nanoparticles creates lots of interfacial regions, and the involved interfacial interactions, either attractive or repulsive, are capable of largely modifying interfacial properties, including crystallinity, free volume, available segmental motions and molecular conformation, and fraction of molecular segments being effectively entangled. With the variation of energy status and the heterogeneous local structure, these interfacial areas of polymer matrix yield a high density of effective traps able to reduce the energy of charge carriers and thus slow down the carrier transportation. The caused hindrance of accumulation and transportation of charge carriers upon the dispersion of nanoscale fillers may serve as the primary mechanism for the enhancement of voltage endurance, provided that the percolation threshold of nanofillers is not reached. Therefore, the dispersion of nanofillers with high dielectric constant has been identified as an approach to enhance both the polarization of polymer matrix and voltage endurance.

Accordingly, the enhancement of energy harvesting, and actuation and piezoelectric responses of piezoelectric polymers via the dispersion of inorganic/organic fillers have been widely recognized [93–99]. However, for the practical uses of piezocomposite consisting of inorganic piezoelectric fillers in a polymer matrix, there is still much room for improvement. The interfacial effect on the enhancement of piezoelectric constant (d_{33}) is analyzed to decay with the increase of distance away from dispersed inorganic crystals, in addition to interfacial areas and properties. Therefore, the homogeneous dispersion of nanocrystals with high permittivity is needed for uniform piezoelectric response of prepared hybrid materials. Furthermore, in order to optimize the contribution of interfacial polarization, oriented close stacking of constituent polymer/inorganic crystals with mutual alignment of polarization axes through the whole sample is desired. The possible approaches of crystal engineering to evolve favorite stacking of inorganic/organic crystals within materials therefore should be widely explored in order to develop novel piezoelectric polycrystalline materials.

REFERENCES

[1] H. Wei, H. Wang, Y. Xia, D. Cui, Y. Shi, M. Dong, C. Liu, T. Ding, J. Zhang, Y. Ma, N. Wang, Z. Wang, Y. Sun, R. Wei, and Z. Guo, An overview of lead-free piezoelectric materials and devices, *Journal of Materials Chemistry C* 6(46), 12446–12467 (2018).

[2] M. S. Vijaya, *Piezoelectric Materials and Devices: Applications in Engineering and Medical Sciences*, Taylor & Francis (2012).

[3] T. Stevenson, D. G. Martin, P. I. Cowin, A. Blumfield, A. J. Bell, T. P. Comyn, and P. M. Weaver, Piezoelectric materials for high temperature transducers and actuators, *Journal of Materials Science: Materials in Electronics* 26(12), 9256–9267 (2015).

[4] C. H. Kuo and K.-S. Chang, Piezotronic and piezophototronic properties of orthorhombic $ZnSnN_2$ fabricated using $Zn-Sn_3N_4$ composition spreads through combinatorial reactive sputtering, *Crystal Growth & Design* 17, 4694–4702 (2017).

[5] H.-Y. Lee, S.-Y. Lee, and K.-S. Chang, Enhancement of piezo-related properties of AlN through combinatorial AlN-TiN nanocolumn composite composition spread, *Ceramics International* 45, 22744 (2019).

[6] Z. L. Wang, Zinc oxide nanostructures: growth, properties and applications, *Journal of Physics: Condensed Matter* 16, R829–R858 (2004).

[7] P. K. Panda, Review: environmental friendly lead-free piezoelectric materials, *Journal of Materials Science* 44, 5049–5062 (2009).

[8] P. Gillet, J. Badro, B. Varrel, and P. F. McMillan, High-pressure behavior in a-$AlPO_4$: Amorphization and the memory-glass effect, *Physical Review B* 51, 11262–11270 (1995).

[9] C.-H. Chou, S.-Y. Li, and K.-S. Chang, High density $ZnSnO_3$ nanowire arrays fabricated using single-step hydrothermal synthesis, *Journal of the American Ceramic Society* 103, 4129 (2020).

[10] Y.-C. Chiang, and K.-S. Chang, Fabrication of $Ba_{1-x}ZnO_2$ Nanowires Using Solvothermal Synthesis: Piezotronic and Piezophototronic Performance Characterization, *Applied Surface Science* 445, 71 (2018).

[11] A. P. R. Arazas, C.-C. Wu, and K.-S. Chang, Hydrothermal fabrication and analysis of piezotronic-related properties of $BiFeO_3$ nanorods, *Ceramics International* 44, 14158 (2018).

[12] A.-J. Hsu and K.-S. Chang, Physical, photochemical, and extended piezoelectric studies of orthorhombic $ZnSnN_2$ nanocolumn arrays, *Applied Surface Science* 470, 19 (2019).

[13] A. Punya and W. R. L. Lambrecht, Quasiparticle band structure of $Zn-IV-N_2$ compounds, *Physical Review B* 84, 165204–165213 (2011).

[14] T. D. Veal, N. Feldberg, N. F. Quackenbush, W. M. Linhart, D. O. Scanlon, L. F. J. Piper, and S. M. Durbin, Band gap dependence on cation disorder in $ZnSnN_2$ solar absorber, *Advanced Energy Materials* 5, 1501462–1501466 (2015).

[15] R. Qin, H. Cao, L. Liang, Y. Xie, F. Zhuge, H. Zhang, J. Gao, K. Javaid, C. Liu, and W. Sun, Semiconducting $ZnSnN_2$ thin films for Si/$ZnSnN_2$ p-n junctions, *Applied Physics Letters* 108, 142104–142108 (2016).

[16] A. N. Fioretti, A. Zakutayev, H. Moutinho, C. Melamed, J. D. Perkins, A. G. Norman, M. Al-Jassim, E. S. Toberer, and A. C. Tamboli, Combinatorial insights into doping control and transport properties of zinc tin nitride, *Journal of Materials Chemistry C* 3, 11017–11028 (2015).

[17] K. Tonisch, V. Cimalla, Ch. Foerster, H. Romanus, O. Ambacher, and D. Dontsov, Piezoelectric properties of polycrystalline AlN thin films for MEMS application, *Sensors and Actuators, A: Physical* 132 658–663 (2006).

[18] Y.-T. Wang and K.-S. Chang, Piezopotential-induced schottky behavior of $Zn_{1-x}SnO_3$ nanowire arrays and piezophotocatalytic applications, *Journal of the American Ceramic Society* 99 (8), 2593 (2016).

[19] J. M. Wu, C. Xu, Y. Zhang, Y. Yang, Y. Zhou, and Z. L. Wang, Flexible and transparent nanogenerators based on a composite of lead-free $ZnSnO_3$ triangular-belts, *Advanced Materials* 24, 6094–6099 (2012).

[20] F. Han, W. C. Li, C. Lei, B. He, K. Oshida, and A. H. Lu, Selective formation of carbon-coated, metastable amorphous $ZnSnO_3$ nanocubes containing mesopores for use as high-capacity lithium-ion battery, *Small* 10, 2637–2644 (2014).

[21] Y. Bing, Y. Zeng, C. Liu, L. Qiao, Y. Sui, B. Zou, W. Zheng, and G. Zou, Assembly of hierarchical $ZnSnO_3$ hollow microspheres from ultra-thin nanorods and the enhanced ethanol-sensing performances, *Sensors and Actuators B: Chemical* 190, 370–377 (2014).

[22] D. Mukherjee, A. Datta, C. Kons, M. Hordagoda, S. Witanachchi, and P. Mukherjee, Intrinsic anomalous ferroelectricity in vertically aligned $LiNbO_3$-type $ZnSnO_3$ hybrid nanoparticle-nanowire arrays, *Applied Physics Letters* 105, 212903–212907 (2014).

[23] J. D. Perkins, J. A. del Cueto, J. L. Alleman, C. Warmsingh, B. M. Keyes, L. M. Gedvilas, P. A. Parilla, B. To, D. W. Readey, and D. S. Ginley, Combinatorial studies of Zn-Al-O and Zn-Sn-O transparent conducting oxide thin films, *Thin Solid Films* 411, 152–160 (2002).

[24] C. Fang, B. Geng, J. Liu, and F. Zhan, D-fructose molecule template route to ultra-thin $ZnSnO_3$ nanowire architectures and their application as efficient photocatalyst, *Chemical Communications* 2350–2352 (2009).

[25] H. Gou, J. Zhang, Z. Li, G. Wang, F. Gao, R. C. Ewing, and J. Lian, A polar oxide $ZnSnO_3$ with a $LiNbO_3$-type structure, *Applied Physics Letters* 98, 091914–091916 (2011).

[26] Y. Inaguma, A. Aimi, Y. Shirako, D. Sakurai, D. Mori, H. Kojitani, M. Akaogi, and M. Nakayama, High-pressure synthesis, crystal structure, and phase stability relations of a $LiNbO_3$-type polar titanate $ZnTiO_3$ and its reinforced polarity by the second-order jahn–teller effect, *Journal of the American Chemical Society* 136, 2748–2756 (2014).

[27] S. H. Ko, D. Lee, H. W. Kang, K. H. Nam, J. Y. Yeo, S. J. Hong, C. P. Grigoropoulos, and H. J. Sung, Nanoforest of hydrothermally grown hierarchical zno nanowires for a high efficiency dye-sensitized solar cell, *Nano Letters* 11, 666–671 (2011).

[28] Y. Zeng, Y. Zheng, J. Xin, and E. Shi, First-principle study the piezoelectricity of a new quartz-type crystal $BaZnO_2$, *Computational Materials Science* 56, 169–171 (2012).

[29] G.S. Smith and P.B. Isaacs, The crystal structure of quartz-like GeO_2, *Acta Cryst* 17, 842–846 (1963).

[30] O. Cambon, P. Yot, S. Rul, J. Haines, and E. Philippot, Growth and dielectric characterization of large single crystals of $GaAsO_4$, a novel piezoelectric material, *Solid State Sciences* 5, 469–472 (2003).

[31] U. Spitsbergen, The crystal structures of $BaZnO_2$, $BaCoO_2$ and $BaMnO_2$, *Acta Cryst* 13, 197–198 (1960).

[32] Y. X. Wang, C. E. Hu, Y. M. Chen, Y. Cheng, and G. F. Ji, First-principles study of the structural, optical, dynamical and thermodynamic properties of $BaZnO_2$ under pressure, *International Journal of Thermophysics* 37, 106–113 (2016).

[33] R. Ramesh and N. A. Spaldin, Multiferroics: progress and prospects in thin films, *Nature Materials* 6, 21–29 (2007).

[34] J. Wang, J.B. Neaton, H. Zheng, V. Nagarajan, S. B. Ogale, B. Liu, D. Viehland, V. Vaithyanathan, D. G. Schlom, U. V. Waghmare, N. A. Spaldin, K. M. Rabe, M. Wuttig, and R. Ramesh, Epitaxial $BiFeO_3$ multiferroic thin film heterostructures, *Science* 299, 1719–1722 (2003).

[35] S. Li, Y.-H. Lin, B.-P. Zhang, Y. Wang, C.-W. Nan, Controlled fabrication of $BiFeO_3$ uniform microcrystals and their magnetic and photocatalytic behaviors, *The Journal of Physical Chemistry C* 114, 2903–2908 (2010).

[36] T.-J. Park, G. C. Papaefthymiou, A. J. Viescas, A. R. Moodenbaugh, and S. S. Wong, Size-dependent magnetic properties of single-crystalline multiferroic $BiFeO_3$ nanoparticles, *Nano Letters* 7, 766–772 (2007).

[37] J.-T. Han, Y.-H. Huang, X.-J. Wu, C.-L. Wu, W. Wei, B. Peng, W. Huang, and J. B. Goodenough, Tunable synthesis of bismuth ferrites with various morphologies, *Advanced Materials* 18, 2145–2148 (2006).

[38] H.-Y. Si, W.-L. Lu, J.-S. Chen, G.-M. Chow, X. Sun, and J. Zhao, Hydrothermal epitaxial multiferroic $BiFeO_3$ thick film by addition of the PVA, *Journal of Alloys and Compounds* 577, 44–48 (2013).

[39] S. Gupta, M. Tomar, and V. Gupta, Ferroelectric photovoltaic properties of Ce and Mn codoped $BiFeO_3$ thin film, *Journal of Applied Physics* 115, 014102 (2014).

[40] K. Prashanthi, R. Gaikwad, and T. Thundat, Surface dominant photoresponse of multiferroic $BiFeO_3$ nanowires under sub-bandgap illumination, *Nanotechnology* 24, 505710 (2013).

[41] G. S. Lotey and N. K. Verma, Synthesis and characterization of $BiFeO_3$ nanowires and their applications in dye-sensitized solar cells, *Materials Science in Semiconductor Processing* 21, 206–211 (2014).

[42] S. H. Han, K. S. Kim, H. G. Kim, H.-G. Lee, H.-W. Kang, J. S. Kim, and C. I. Cheon, Synthesis and characterization of multiferroic $BiFeO_3$ powders fabricated by hydrothermal method, *Ceramics International* 36, 1365–1372 (2010).

[43] K.-S. Chang, W.-C. Lu, C.-Y. Wu, and H.-C. Feng, High-throughput identification of higher-k dielectrics from an amorphous N_2-doped HfO_2–TiO_2 library, *Journal of Alloys and Compounds* 615, 386–389 (2014).

[44] Z.-A. Lin, W.-C. Lu, C.-Y. Wu, and K.-S. Chang, Facile fabrication and tuning of TiO_2 nanoarchitectured morphology using magnetron sputtering and its applications to photocatalysis, *Ceramics International* 40, 15523–15529 (2014).

[45] W.-C. Lu, H.-D. Nguyen, C.-Y. Wu, K.-S. Chang, and M. Yoshimura, Modulation of physical and photocatalytic properties of (Cr,N) codoped TiO_2 nanorods using soft solution processing, *Journal of Applied Physics* 115, 144305 (2014).

[46] S. M. Sze, *Physics of Semiconductor Devices* 2nd edition John Wiley & Sons, Inc. p. 279 (1985).

[47] W.-C. Lu, C. M. Quezada, and K.-S. Chang, Two-step solvothermal synthesis of $BaZnO_2$ films on indium tin oxide substrates and their piezo-related and photoelectrochemical performance, *Materials Chemistry and Physics* 247, 122880 (2020).

[48] Y.-J. Peng, S.-Y. Lee, and K.-S. Chang, Facile fabrication of a photocatalyst of Ta_4N_5 nanocolumn arrays by using reactive sputtering, *Journal of The Electrochemical Society* 162, H371 (2015).

[49] J. S. Harrison and Z. Ounaies, Piezoelectric Polymers. In *Encyclopedia of Polymer Science and Technology*, (Ed.). doi:10.1002/0471440264.pst427 (2002).

[50] K. S. Ramadan, D. Sameoto, and S. Evoy, *Smart Material, Structure* 23, 033001 (2014).

[51] L. Zhu and Q. Wang, Novel ferroelectric polymers for high energy density and low loss dielectrics, *Macromolecules* 45, 2937 (2012).

[52] Q. M. Zhang, C. Huang, F. Xiaand, and J. Su, Electric EAP. In *Electroactive Polymer (EAP) Actuators as Artificial Muscles: Reality, Potential, and Challenges*, 2nd ed.; Y. Bar-Cohen, Ed.; SPIE Press: Bellingham, WA; PM136, 89 (2004).

[53] F. Carpi, D. De Rossi, R. Kornbluh, R. E. Pelrine, and P. Sommer-Larsen, *Dielectric Elastomers as Electromechanical Transducers: Fundamentals, Materials, Devices, Models and Applications of an Emerging Electroactive Polymer Technology*; Elsevier: Boston, MA, (2008).

[54] P. Brochu and Q. Pei, Advances in dielectric elastomers for actuators and artificial muscles, *Macromolecular Rapid Communications* 31, 10 (2010).

[55] R. Pelrine, R. Kornbluh, J. Joseph, R. Heydt, Q. B. Pei, and S. Chiba, High-field deformation of elastomeric dielectrics for actuators, *Materials Science and Engineering: C* 11, 89 (2000).

[56] C. C. Wang, G. Pilania, S. A. Boggs, S. Kumar, C. Breneman, and R. Ramprasad, Computational strategies for polymer dielectrics design, *Polymer* 55, 979 (2014).

[57] M. Lienhard, I. Rushkin, G. Verdecia, C. Wiegand, T. Apple, and L. V. Interrante, Synthesis and characterization of the new fluoropolymer poly(difluorosilylenemethylene); An analogue of poly-(vinylidene fluoride), *Journal of the American Chemical Society* 119, 12020 (1997).

[58] A. R. Blythe and D. Bloor, *Electrical Properties of Polymers*, 2nd ed.; Cambridge University Press: Cambridge; New York (2005).

[59] K.-C. Kao, *Dielectric Phenomena in Solids: with Emphasis on Physical Concepts of Electronic Processes*; Elsevier Academic Press: Boston, MA (2004).

[60] C. C. Wang, G. Pilania, S. A. Boggs, S. Kumar, C. Breneman, and R. Ramprasad, Computational strategies for polymer dielectrics design, *Polymer* 55, 979 (2014).

[61] V. G. Artemov and A. A. Volkov, Water and ice dielectric spectra scaling at 0 °C, *Ferroelectrics* 466, 158 (2014).

[62] K. Sasaki, R. Kita, N. Shinyashiki, and S. Yagihara, Glass transition of partially crystallized gelatin–water mixtures studied by broadband dielectric spectroscopy, *The Journal of Chemical Physics* 140, 124506 (2014).

[63] K. O. Izutsu, *Electrochemistry in Nonaqueous Solutions*; Wiley-VCH: Weinheim, Germany (2002).

[64] G. R. Leader and J. F. Gormile, The dielectric constant of N-methylamides, *Journal of the American Chemical Society* 73, 5731 (1951).

[65] M. V. Kakade, S. Givens, K. Gardner, K. H. Lee, D. B. Chase, and J. F. Rabolt, Electric field induced orientation of polymer chains in macroscopically aligned electrospun polymer nanofibers, *Journal of the American Chemical Society* 129(10), 2777 (2007).

[66] Y. Huan, Y. Liu, and Y. Yang, Simultaneous stretching and static electric field poling of poly (vinylidene fluoride-hexafluoropropylene) copolymer films, Polymer Engineering & *Science* 47(10), 1630 (2007).

[67] F. Guan, J. Pan, J. Wang, Q. Wang, and L. Zhu, Crystal orientation effect on electric energy storage in poly (vinylidene fluoride-co-hexafluoropropylene) copolymers, *Macromolecules* 43(1), 384 (2010).

[68] H. Von Seggern and S. N. Fedosov, Conductivity-induced polarization buildup in poly (vinylidene fluoride), *Applied Physics Letters* 81(15), 2830 (2002).

[69] J. Kułek, C. Pawlaczyk, and E. Markiewicz, Influence of poling and ageing on high-frequency dielectric and piezoelectric response of PVDF-type polymer foils, *Journal of Electrostatics* 56(2), 135 (2002).

[70] C. Wan and C. R. Bowen, Multiscale-structuring of polyvinylidene fluoride for energy harvesting: the impact of molecular-, micro-and macro-structure, *Journal of Materials Chemistry A* 5, 3091 (2017).

[71] M. Benz and W. B. Euler, Determination of the crystalline phases of poly (vinylidene fluoride) under different preparation conditions using differential scanning calorimetry and infrared spectroscopy, *Journal of Applied Polymer Science* 89, 1093 (2003).

[72] A. J. Lovinger, Polymorphic transformations in ferroelectric copolymers of vinylidene fluoride induced by electron irradiation, *Macromolecules* 18, 910 (1985).

[73] K. Tashiro and R. Tanaka, Structural correlation between crystal lattice and lamellar morphology in the ferroelectric phase transition of vinylidene fluoride–trifluoroethylene copolymers as revealed by the simultaneous measurements of wide-angle and small-angle X-ray scatterings, *Polymer* 47, 5433 (2006).

[74] E. Bellet-Amalric and J. F. Legrand, Crystalline structures and phase transition of the ferroelectric P (VDF-TrFE) copolymers, a neutron diffraction study, *The European Physical Journal B-Condensed Matter and Complex Systems* 3, 225 (1998).

[75] K. Tashiro, K. Takano, M. Kobayashi, Y. Chatani, and H. Tadokoro, Structure and ferroelectric phase transition of vinylidene fluoride-trifluoroethylene copolymers: 2. VDF 55% copolymer, *Polymer* 25, 195 (1984).

[76] A. Jain, J. Kumar, D. R. Mahapatra, and H. H. Kumar, Detailed studies on the formation of piezoelectric β-phase of PVDF at different hot-stretching conditions. In *Sensors and Smart Structures Technologies for Civil, Mechanical, and Aerospace Systems, International Society for Optics and Photonics* 7647, 76472C (2010).

[77] V. Sencadas, Jr, R. Gregorio, and S. Lanceros-Méndez, α to β phase transformation and microestructural changes of PVDF films induced by uniaxial stretch, *Journal of Macromolecular Science* 48(3), 514 (2009).

[78] R.P. Vijayakumar, D. V. Khakhar, and A. Misra, Studies on α to β phase transformations in mechanically deformed PVDF films, *Journal of Applied Polymer Science* 117(6), 3491 (2010).

[79] C. H. Du, B. K. Zhu, and Y. Y. Xu, Effects of stretching on crystalline phase structure and morphology of hard elastic PVDF fibers, *Journal of Applied Polymer Science* 104(4), 2254 (2007).

[80] L. Li, M. Zhang, M. Rong, and W. Ruan, Studies on the transformation process of PVDF from α to β phase by stretching, *RSC Advances* 4(8), 3938 (2014).

[81] B. Mohammadi, A. A. Yousefi, and S. M. Bellah, Effect of tensile strain rate and elongation on crystalline structure and piezoelectric properties of PVDF thin films, *Polymer Testing* 26(1), 42 (2007).

[82] G.T. Davis, in: T.T. Wang, J.M. Herbert, A.M. Glass (Eds.), *Production of Ferroelectric Polymer Films. The Applications of Ferroelectric Polymers*, Blackie and Sons Ltd, Glasgow, USA, (Chapter 3) (1998).

[83] H. Lu and L. Li, Crystalline structure, dielectric, and mechanical properties of simultaneously biaxially stretched polyvinylidene fluoride film, *Polymers for Advanced Technologies* 29(12), 3056 (2018).

[84] Y. Liu, H. Aziguli, B. Zhang, W. Xu, W. Lu, J. Bernholc, and Q. Wang, Ferroelectric polymers exhibiting behaviour reminiscent of a morphotropic phase boundary, *Nature* 562(7725), 96 (2018).

[85] M. Ahart, M. Somayazulu, R. E. Cohen, P. Ganesh, P. Dera, H. K. Mao, R. J. Hemley, Y. Ren, P. Liermann, and Z. Wu, Origin of morphotropic phase boundaries in ferroelectrics, *Nature* 451(7178), 545 (2008).

[86] Y. Liu, B. Zhang, A. Haibibu, W. Xu, Z. Han, W. Lu, J. Bernholc, and Q. Wang, Insights into the morphotropic phase boundary in ferroelectric polymers from the molecular perspective, *The Journal of Physical Chemistry C* 123(14), 8727 (2019).

[87] Y. Liu, Z. Han, W. Xu, A. Haibibu, and Q. Wang, Composition-dependent dielectric properties of poly(vinylidene fluoride-trifluoroethylene)s near the morphotropic phase boundary, *Macromolecules* 52(17), 6741 (2019).

[88] W.-H. Liew, M. S. Mirshekarloo, S. Chen, K. Yao, and F. E. H. Tay, Nanoconfinement induced crystal orientation and large piezoelectric coefficient in vertically aligned P (VDF-TrFE) nanotube array, *Scientific Reports* 5, 09790 (2015).

[89] Y. J. Park, S. J. Kang, B. Lotz, M. Brinkmann, A. Thierry, K. J. Kim, and C. Park, Ordered ferroelectric PVDF–TrFE thin films by high throughput epitaxy for nonvolatile polymer memory, *Macromolecules* 41, 8648 (2008).

[90] K.-L. Kim, W. Lee, S.-K. Hwang, S.-H. Joo, S.-M. Cho, G. Song and Y.-J. Yu, Epitaxial growth of thin ferroelectric polymer films on graphene layer for fully transparent and flexible nonvolatile memory, *Nano Letters* 16, 334 (2015).

Mutually Influenced Stacking

[91] J. Wang, F. Guan, J. Pan, Q. Wang, L. Zhu, Achieving high electric energy storage in a polymer nanocomposite at low filling ratios using a highly polarizable phthalocyanine interphase, *Journal of Polymer Science Part B: Polymer Physics* 52(24), 1669 (2014).

[92] E. Kuffel, W. S. Zaengl, and J. Kuffel, *High Voltage Engineering: Fundamentals*, 2nd ed., Butterworth-Heinemann: Boston, MA (2000).

[93] S. Deshmukh and Z. Ounaies, Active single walled carbon nanotube–polymer composites IUTAM Symp, On Multi-Functional Material Structures and Systems. Berlin: Springer 103 (2010).

[94] H. M. Matt and F. L. di Scalea, Macrofiber composite piezoelectric rosettes for acoustic source location in complex structures, *Smart Materials Structure* 16, 1489 (2007).

[95] Y. Yang, L. Tang, and H. Li, Vibration energy harvesting using macrofiber composites, Smart Mater. *Structure* 18, 115025 (2009).

[96] C. Park, Z. Ounaies, K. E. Wise, and J. S. Harrison, In situ poling and imidization of amorphous piezoelectric polyimides, *Polymer* 45, 5417 (2004).

[97] G. M. Atkinson, R. E. Pearson, Z. Ounaies, C. Park, J. S. Harrison, S. Dogan, and J. A. Midkiff, *Novel piezoelectric polyimide MEMS* 12th Int. Conf. on *TRANSDUCERS, Solid State Sensors, Actuators and Microsystems*, 782 (2003).

[98] K. Park, et al, Flexible nanocomposite generator made of $BaTiO_3$ nanoparticles and graphitic carbons, *Advanced Materials* 24, 2999(2012).

[99] K. Prashanthi, N. Miriyala, R. D. Gaikwad, W. Moussa, V. R. Rao, and T. Thundat, Vibtrational energy harvesting using photopatternable piezoelectric nanocomposite cantilevers, *Nano Energy* 2 923 (2013).

6 Single-Atom Catalysts on Nanostructure from Science to Applications

Yi-Sheng Lai, Anggrahini Arum Nurpratiwi, and Yen-Hsun Su
National Cheng Kung University, Taiwan

CONTENTS

6.1 Concept of Single-Atom Catalysts 151
    6.1.1 Single-Atom Catalysts 151
    6.1.2 Synthesis of SACs 153
        6.1.2.1 Mass-Selected Soft-Landing Method 153
        6.1.2.2 Wet-Chemistry Synthetic Method.. 154
6.2 Single-Atom on Nanostructure 156
    6.2.1 Single-Atom on Metal Oxide... 156
    6.2.2 Single-Atom on Metal 157
    6.2.3 Single-Atom on 2D Materials.. 157
6.3 Application of Single-Atom on Nanostructure... 158
    6.3.1 Electrocatalytic Oxygen Reduction Reaction (ORR)..................................... 159
    6.3.2 Electrocatalytic Hydrogen Evolution Reaction (HER).................................. 160
    6.3.3 Electrocatalytic Carbon Dioxide Reduction Reaction (CO_2RR) 161
References.. 162

6.1 CONCEPT OF SINGLE-ATOM CATALYSTS

Single-atom catalysts (SACs) have been attracting many researchers in recent years. It is considered an attractive and powerful technique that can enhance catalytic activity and selectivity. SACs contain isolated individual atoms dispersed on the surface of appropriated supports. Decreasing the size of a metal particle to a single metal atom can maximize the atomic efficiency of metals as well as increase its specific activity performance. This subchapter will explain the concept of SACs as well as the strategy in preparing the SACs. The most recent development in preparing SACs focusing on the high activity and stability of the single-atoms will also be discussed.

6.1.1 SINGLE-ATOM CATALYSTS

Low-coordinated metal atoms are usually used as the active sites of the catalysts, whereas the other-sizes are either inert or can be a trigger to underused side reaction. Decreasing the size of the metal particle can contribute to the increase of the specific activity per metal atoms. Therefore, the size of the metal particle has an important role in determining the performance of catalysts as traditional heterogeneous catalysts usually contain a mixture of a broadside distribution of metal particles (Chen et al. 2018, Yang et al. 2013). So, decreasing the size of the metal particles to a single-atom can result in highly desirable catalytic activity.

A catalyst that contains only isolated single-atom dispersed in an appropriate supports, widely known as SACs, has recently attracted research attention as a new frontier in catalysis science. In 2011, Qiao et al. (2011), for the first time, implemented this term using isolated single Pt atoms anchored to the surface of iron oxide Pt/FeO$_x$) nanocrystallites. The research confirmed the excellent activity and stability of this Pt SAC for both CO oxidation and preferential oxidation of CO in H$_2$. (Liu 2017) in their research also stated that the most efficient option to gain the use of metal atom of the supported metal catalyst is to reduce the size of the metal to well-defined, automatically dispersed active centers, which are SACs (Figure 6.1.).

Single metal atoms have a huge surface free energy. Hence, appropriate support can effectively anchor the single metal atoms since due to their huge free energy, the deposited atoms of a single metal atom are considered too mobile to agglomerate and form large clusters. Therefore, extensive attention was given to the search for an appropriate support for SACs. A series of SACs have been successfully fabricated, characterized, and examined as the characterization and computational modeling was advancing (Figure 6.2.).

Many researchers have experimentally and theoretically demonstrated SACs on many supports to achieve higher activity and selectivity. Metal oxides, such as FeO$_x$, CeO$_2$ (Aranifard, Ammal, and Heyden 2014, Ding et al. 2014), Al$_2$O$_3$ (Hackett et al. 2007, Gao 2016), and ZnO (Gu et al. 2014) are considered as a very important support for SACs. SACs can also be stabilized on the surface of another metal such as Pd single-atoms on (Pei et al. 2017), Ag (Aich et al. 2015), or Au (Kesavan et al. 2011) surfaces. The interaction between the SACs and metal will be obtained in the alloy form. Special attention is also been given to two-dimensional (2D) materials. 2D materials are also considered as good support for SACs due to their excellent physical and chemical properties. Graphene (Yan et al. 2015) and its family such as graphyne (Ma et al. 2015), graphitic (Zhang et al. 2013), and graphitic carbon nitride (g-C$_3$N$_4$) (Gao et al. 2016) are the most favorite 2D materials used in the SACs. Besides the graphene family, other 2D materials such as N-doped carbon nanofibers (N-CNFs) (Bulushev et al. 2016), single-layer hexagonal BN (h-BN) (Mao et al. 2014), C$_2$N (Ma et al. 2016), MoS$_2$ (Du et al. 2015), SrTiO$_3$ (Wang et al. 2014), Ti$_2$CO$_2$ (Zhang et al. 2016), and transition metal phthalocyanine monolayer sheets (TM-Pc) (Wang et al. 2015) have been studied as supports for the SACs.

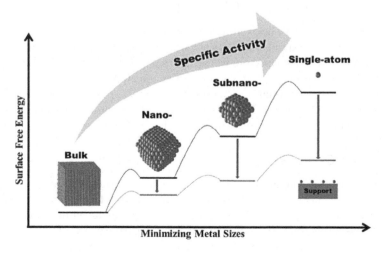

FIGURE 6.1 Schematic illustration of the changes in surface free energy and specific activity per metal atom over metal particle size and the effects of the support in stabilizing single-atom.

Single-Atom Catalysts on Nanostructure 153

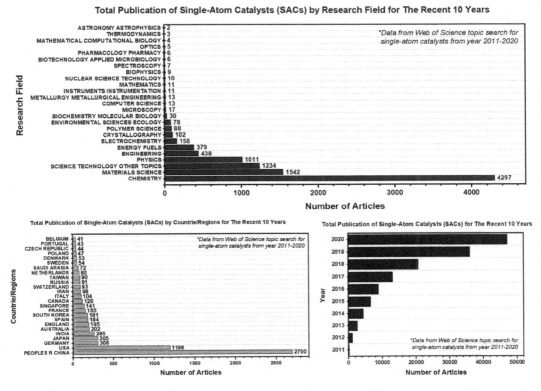

FIGURE 6.2 Total publication of single-atom catalysts for the last 10 years and the distribution by research fields and countries/regions.

6.1.2 Synthesis of SACs

Exploration and improvement for the SACs synthesis method have emerged as one of the most important focuses in the SACs field. Improving the synthesis method of SACs is considered a major challenge because the single-atom tends to agglomerate. The experimental attainability, catalytic activity, and chemical stability of SACs are also unclear. However, the recent development of technologies has made the preparation as well as the characterization of SACs possible.

A variety of synthesis approaches had been suggested for the fabrication of SACs due to the demand for precise design and synthesis of SACs as a highly efficient catalyst. SACs tend to agglomerate the single metal atoms either during the synthesis processes or the subsequent treatments (Liu 2017). Therefore, research about the fabrication of SACs has been developed and several strategies have been proposed such as mass-selected soft-landing and wet chemistry synthetic methods (Figure 6.3.).

6.1.2.1 Mass-Selected Soft-Landing Method

The mass-selected soft-landing method is considered a powerful technique in preparing SACs due to its ability to control the size of metal species by using a mass-selected molecular/atom beam with high precision. This technique can afford an excellent model catalyst for fundamental studies on the atomic level to the cluster size effect.

A wide variety of both experimental and/or theoretical studies of SACs using a mass-selected soft-landing method have been done to obtain the catalytic properties of SACs on support (Heiz et al. 1999, Kaden et al. 2009, Abbet et al. 2000). However, this technique is expensive and has a

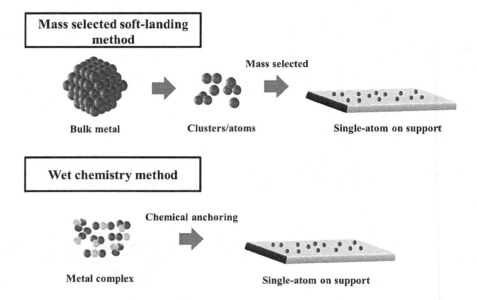

FIGURE 6.3 Method for preparing SACs; mass-selected soft-landing method and wet chemistry method.

low-yield fabrication method, which limits its application. Hence, this method is not considered suitable for industrial applications for heterogeneous catalysts.

6.1.2.2 Wet-Chemistry Synthetic Method

The wet-chemistry method uses a precursor material that contains the single-atom species. The reason for using such precursors is to anchor the metal single-atom on the surface of the support through a chemical reaction. It is also can avoid the agglomeration of single-atoms during post-treatment processes. Pre-treatment of the catalyst was required to remove some useless or even poisonous ligands as metal active sites with high accessibility are needed for activating the reactants. As a result, the possibility of the metal single-atom to agglomerate to the larger particle is higher (Zhang, Shi, and Xu 2005, Kim et al. 2006). Hence, using strong metal support is the key factor in preventing the agglomeration of single-atom on the surface of the support.

6.1.2.2.1 Defect Engineering Strategy

Defects in the nanosheets can shift the electronic structure and coordination number in the surrounding environment. Vacancies and unsaturated coordination sites lead by this defect can serve as "traps". The metal precursors as well as the anchor metal atoms would be captured and further bonded with the atom surrounding the vacancy. Zhang et al. (2018b) have successfully synthesized Pt SACs with high loading, of up to 2.3 wt%.

Optimizing the construction condition of defect engineering is a key factor when we apply the defect engineering strategy. The objective is to ensure the formation of uniform defective centers as anchoring sites as well as stabilizing the single-atom homogeneously. Characterizing the type of defect precisely is also important as it can be useful for selecting a suitable metal species and avoiding the disappearance of another atom resulting in the transformation of other defects during the post-treatment process (Figure 6.4.).

The defect in the form of a vacancy in the defect engineering strategy is considered as an advantage in comparison with the common synthesis process of SACs. The defect concentration can also be controlled by this method which means the metal loading content of SACs can be tuned.

Single-Atom Catalysts on Nanostructure 155

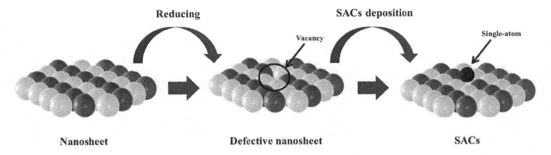

FIGURE 6.4 Defect engineering strategy for preparing SACs.

Moreover, the presence of defect itself can enhance the optical and electronic properties of support, which also contributes to broadening its application in photocatalysis and electrocatalysis.

6.1.2.2.2 Spatial Confinement Strategy

Spatial confinement strategy can be obtained by two steps: (1) separating and enveloping suitable mononuclear metal precursors with the help of porous materials obtaining a high spatial distribution and atomic dispersion of metal; (2) post-treatment method which will remove the ligands of precursors and forming the single metal atoms. The carbonized skeleton support has contribution in stabilizing the single-atom. Spatial confinement strategy, essentially, is a method when the single-atoms are spatially confined into molecular-scale cages to prevent them from shifting to another location. The key factor in achieving excellent SACs which has a stable and uniform spatial confinement metal species in the spatial confinement strategy is selecting suitable mononuclear metal precursors based on the molecular size and charge state (Figure 6.5.).

Metal species tend to migrate and agglomerate. However, during the synthesis process, the spatial confinement strategy will restrict the single metal atom from shifting to another location and also agglomeration. Zhang et al. (2014) have observed that uniform single Mo sites were confined by the robust coordination between molybdenyl acetylacetonate ($MoO_2(acac)_2$) and benzoyl salicylal hydrazone ligands in a COF. Furthermore, the spatial confinement strategy shows some good potential for improving catalyst as well as broadening the application of SACs.

6.1.2.2.3 Coordination Design Strategy

Controlling the uniform coordination sites of the support is important when synthesizing SACs using the coordination design strategy. The objective is to integrate single metal atoms for homogeneous active center formation. Mononuclear metal precursor selection is important and the strong

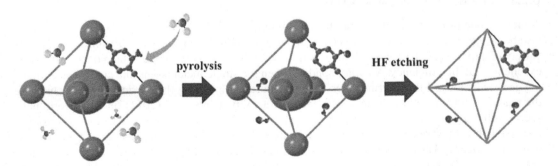

FIGURE 6.5 Formation procedure of single-atoms using immobilized on the carbonized skeleton using spatial confinement strategy for SACs synthesis.

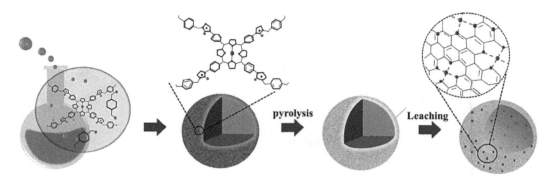

FIGURE 6.6 Illustration of the synthesis of a single-atoms using coordination design strategy.

interaction with coordination sites is a key factor for selecting a suitable mononuclear metal precursor. Coordination sites, during the synthesis process using coordination design strategy, will bind the adsorbed mononuclear metal precursors, which can prevent their shift and agglomeration (Figure 6.6).

The as-formed single metal atoms prepared by coordination strategy are shown to have higher stability compared to the other strategies in the wet-chemistry method. The reason is that this approach makes single metal atoms have a strong interaction with the coordination atoms on the supports. Furthermore, surrounding coordination atoms can adjust the electronic structure of the SACs active centers, which can enhance the catalytic performance of SACs (Chen et al. 2018).

6.2 SINGLE-ATOM ON NANOSTRUCTURE

Minimizing the size of metals particles can lead to the increasing of free surface energy. Single metal atoms are the smallest size of metals, which means that single metal atoms have huge free surface energy. However, due to its huge surface energy, single-atoms are too mobile and considered to be easy to agglomerate and forming large clusters. Appropriate supports are needed to achieve a high stability single-atom and preventing the agglomeration of the single-atoms. The high stability of single-atoms on supports can enhance the catalytic performance of the SACs. Many researchers have been working on discovering appropriate supports. Hence, several supports such as metal oxides, metals, and 2D materials have been experimentally and/or theoretically demonstrated. Hence, in this subchapter, series of the development of single-atom on the nanostructure are discussed.

6.2.1 Single-Atom on Metal Oxide

A metal oxide protection surface may be substituted by a single metal atom. The metal atom will also be firmly anchored due to its contact with the surrounding oxygen anions (Bruix et al. 2014). A variety of experiments have been done to add a single-atom to metal oxide, which demonstrates its excellent catalytic efficiency. A strong bond between the single metal atoms and coordinated lattice oxygen stems toward the intermediates has contributed to lowering the barrier reaction, changing the reaction pathway, and enhancing the catalytic activity. Qiao et al. (2011), in their work, have successfully added a single Pt atom on a metal oxide. The single Pt atoms, which are uniformly dispersed on FeO_x support with high surface area, show an extremely high catalytic activity for CO oxidation reaction. The d orbital vacancy in the single Pt atoms, due to the electron transfer from Pt atoms to the FeO_x surface, can lead to a strong binding and stabilization of single Pt atoms which increases its catalytic activity. The observation of single-atom Pt catalysts has excellent potential in reducing the high cost of commercial noble-metal catalysts as well as in proving the concept of single-atom heterogeneous catalysts. Research on FeO$_x$ as a support has been continued as Li et al. (2015) also examined various single metal atoms on FeO$_x$ catalysts M$_1$/FeO$_x$ (M = Au, Rh, Pd, Co, Cu, Ru, Ni, Ti). Their systems show strong binding energy and a high diffusion barrier of the single metal atoms resulting in higher stability compared to Pt/FeO$_x$.

Studies also confirmed the stabilization of single-atoms on another metal oxide support B$_1$/CeO$_2$ (B = Au, Ru, Pt). Dvorak et al. (2016), in his study, confirmed the stabilization of monodispersed Pt on CeO$_2$. It is observed to be effective in merging the excess oxygen even in unfavorable conditions. The Pt single-atoms were stabilized by the defects on all over solid surfaces of CeO$_2$. The separation of Pt at the monoatomic step edges results in stabilizing the dispersion of the single Pt^{2+} ions and contributing to the oxygen storage of the CeO$_2$. Other metal oxide supports for SACs were Pt$_1$/Au$_1$/ZnO (Gu et al. 2014) and Pd$_1$/TiO$_2$ (Kaden et al. 2009, Sakthivel et al. 2004).

6.2.2 Single-Atom on Metal

A catalytically active metal is atomically dispersed in the surface layer of a more inert host metal. In the surface layer of the host, the metal presents two more active components. When surrounded by the host metal at a much reduced level, the atoms of the more active agent are thermodynamically more stable. Therefore, a single-atom can be stabilized by anchoring it on the surface of another metal in the form of an alloy (Liu 2017). The single-atom alloys suggest a way to weaken the reactivity of a highly reactive element and to design a catalyst with a few amount of the metal. Various researches have been done to design a catalyst that results in their great catalytic activity. Aich et al. (2015) were studying the Pd-Ag alloy catalysts with a small amount of Pd. Their research demonstrates the improved activity of acrolein for hydrogenation due to the presence of the isolated Pd atoms. The site formation when isolating Pd atoms contributes to the decrease of the activation energy for H$_2$ separation, which is explained in Figures 6.7. and 6.8.

6.2.3 Single-Atom on 2D Materials

Major attention has been devoted to single-atoms supports aside from metal oxides and metals. Series of researches were directed toward 2D materials as SACs supports, such as graphene (Yan et al. 2015) and its family which are graphyne (Ma et al. 2015), graphitic (Zhang et al. 2013), and g-C$_3$N$_4$ (Ma et al. 2012); MoS$_2$ (Du et al. 2015), SrTiO$_3$ (Wang et al. 2014), and Ti$_2$CO$_2$ (Zhang et al. 2016). These 2D materials have been used as promising catalyst supports due to their wide specific surface areas, electronic properties, and thermal properties. These special properties of 2D materials play an important role in enhancing the activity of the catalysts.

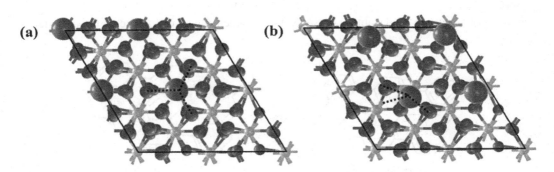

FIGURE 6.7 Top views of single-metal atom anchored on (a) vacancy free and (b) oxygen defective iron-oxide surfaces.

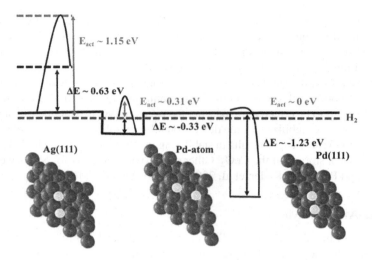

FIGURE 6.8 Potential energy diagram showing how the Pd single-atom alloy surface has an effect on the energies compared to the Ag (111) and Pd (111) surface.

The activity enhancement of the SACs supported on the defective graphene has been demonstrated. Furthermore, Yan et al. (2015) reported that single-atom Pd_1/graphene catalysts showed approximately 100% selectivity to butene. Excellent durability toward deactivation via both agglomeration of metal atoms and carbonaceous during 100 h total reaction time was achieved (Figure 6.9).

Aside from graphene and its families, the other 2D materials such as MoS_2 monolayers show a uniform distribution of anchoring sites for single-atoms. MoS_2 can play as defect-free supporting materials that provide a wide surface-to-volume ratio (Du et al. 2015). Another attractive support for SACs is perovskite oxides, such as $SrTiO_3$. Research on $SrTiO_3$ has received a lot of interest in recent years due to the high thermal stability and excellent dispersibility of the Sr adatoms. Thus, Sr adatoms serve as a nucleation center and lead the growth of the noble-metal nanostructures. Zhang et al. (2013) have successfully applied Ni single-atom on $SrTiO_3$, which can be considered as a system designed to explore SACs on 2D materials support.

6.3 APPLICATION OF SINGLE-ATOM ON NANOSTRUCTURE

Global energy demand and the scarcity of non-renewable energy have enforced a lot of researches to develop green renewable energies. The development of electrochemical energy conversion through oxygen reduction reaction (ORR), hydrogen evolution reaction (HER), and carbon dioxide reduction (CO_2RR) have been challenging in recent years. Many advanced studies have revealed excellent activity, selectivity, and stability of SACs. In this subchapter, we will discuss the importance of

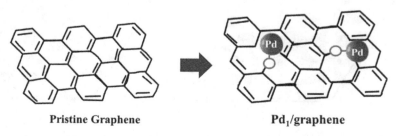

FIGURE 6.9 Schematic illustration of single-atom Pd_1/graphene catalysts.

Single-Atom Catalysts on Nanostructure

SACs on nanostructure for their application as a key factor for enhancing the catalytic performance of energy conversion reaction.

6.3.1 Electrocatalytic Oxygen Reduction Reaction (ORR)

ORR is an important process in life as oxygen (O_2) is the most abundant element on the earth. ORR can be preceded by two kinds of pathways: the four-electron pathway and two-electron pathway. The four-electron pathway converts O_2 directly into H_2O, meanwhile, the two-electron pathway converts O_2 to H_2O by producing H_2O_2 in the process.

Four-electron pathway

$$O_2 + 4H^+ + 4e = 2H_2 \tag{6.1}$$

Two-electron pathway

$$O_2 + 2H^+ + 2e = H_2O_2 \tag{6.2}$$

Polymer electrolyte membrane fuel cells (PEMFC) are one of the most promising electrochemical energy devices which have excellent efficiency and reliability. However, a large amount of Pt metals was used in the catalysts for both anode and cathode, which leads to the significant reduction of Pt metals. The instability of Pt at the cathodes is also considered a challenging obstacle in the commercialization of PEMFC. Moreover, the high cost of Pt metals due to its scarcity is also one of the reasons for researchers to develop a high activity, high stability, and low-cost catalyst for ORR electrocatalysts (Figure 6.10).

A series of single Pt atoms have been reported to show the ORR performances. Appropriate supports also play an important factor in catalytic performance. The preferred way for Pt-based

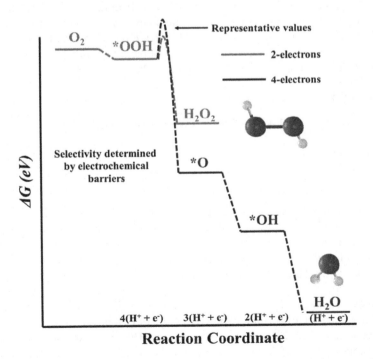

FIGURE 6.10 Illustration of a free energy diagram for four-electron and two-electron oxidation-reduction reaction.

catalysts is the four-electron pathway. However, Yang et al. (2016) report that the Pt single-atom does not show good ORR activity. Despite that, Liu et al. (2016) reported excellent ORR activity of Pt single-atom on N-doped carbon black which had higher utilization efficiency than Pt/C. Furthermore, other studies of non-Pt SACs application on ORR have been reported. Zhang et al. (2013) have successfully trapped single niobium atoms on graphitic layers which demonstrated excellent ORR performance in the alkaline media. In their reports, they explain that the trapped single Ni atoms can suppress the chemical/thermal coarsening in the active particle, which leads to the enhancement of overall conductivity. Moreover, redistribution of d-band electrons produced by Ni single-atoms exhibits high stability and activity for O_2 adsorption.

6.3.2 Electrocatalytic Hydrogen Evolution Reaction (HER)

HER is considered the most efficient process to produce high-quality molecular hydrogen. Electrochemical water splitting suggests a certain solution for a clean and renewable system of H_2 production. Catalysts play an important factor in HER in determining the performance. Based on the theory, HER proceeds through two kinds of steps: the Volmer–Heyrovsky step and the Volmer–Tafel step, wherein this mechanism * denotes an active site on the surface of electrocatalysts. HER activity can be described by the adsorption free energy of H* (ΔG_H), and the favorable value is close to zero. The hydrogen production rate can be calculated by the raw electrode current, H_2 Faraday efficiency, the number of electrochemically active sites from the voltammetric quantification, and experimental temperature and pressure environment. The H_2 production rate is calculated as shown in Equation (6.3):

$$H_2 = \frac{J_{op}RT}{nFP}$$

(6.3)

where J_{op} is the optimum current from the determined H_2 Faraday efficiency, R is the gas constant (0.082 L atm mol^{-1} K^{-1}), T represents the experimental environment temperature, the value n represents the number of electron in the reaction ($n = 2$ for H_2), F represents the Faraday constant (96,485 C mol^{-1} e^{-1}), and P represents the experimental environment pressure. Unit analysis providing for the H_2 production rate is reported as Ls^{-1}m^{-2} (Figure 6.11).

Volmer–Heyrovsky step

$$H^+ + e^- + * \rightarrow H*; H * + H^+ + e^- \rightarrow H_2 + *$$

(6.4)

Volmer–Tafel step

$$H^+ + e^- + * \rightarrow H*; 2H \rightarrow H_2 + 2*$$

(6.5)

The benchmark catalyst for HER is Pt-based catalysts which are very costly and also rare. Developing active, stable, and low-cost electrocatalysts for water splitting became an important issue. Hence, Sun et al. (2013) suggested using a method to produce isolated single Pt atoms on N-doped graphene using atomic layer deposition (ALD). They reported that the single Pt atom catalysts can enhance the HER catalytic activities and stability of up to 37 times compared to the commercial Pt/C catalysts. Another research for single Pt atoms on N-doped graphene support using ALD techniques was also conducted by Chang et al. (2014). Their research reported that HER performance was increased up to 13 times compared to the Pt/C catalysts.

Other than Pt, an excellent catalytic performance for HER was also shown by a single Co atom anchored on N-doped graphene. This observation makes Co SACs a promising candidate in water splitting applications replacing Pt (Fei et al. 2015, Stephens and Chorkendorff 2011). Single-atom Ni also showed that the comparable catalytic performance to Pt/C with the overpotential to achieve a

Single-Atom Catalysts on Nanostructure   161

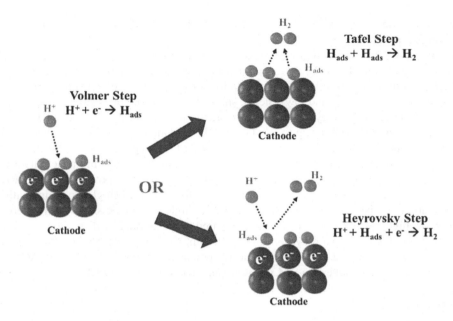

FIGURE 6.11 Schematic illustration presenting possible routes for HER on the cathode.

current density of 10 mA/cm^{-2} was as low as 70 mV for Ni SACs anchored in the defective graphene supports (Zhang et al. 2018a). Density functional theory (DFT) calculation demonstrates that the defects in the graphene and the different coordination configuration of Ni atoms facilitate the high catalytic performance of HER.

6.3.3 Electrocatalytic Carbon Dioxide Reduction Reaction (CO_2RR)

Recent developments on CO_2RR have been done to convert CO_2 into other carbon compounds. CO_2 is a massive air pollutant in the atmosphere, which gradually increases from time to time. Hence, converting CO_2 to other carbon compounds is considered important. However, CO_2 is a stable gas which can only be transformed into other carbon compounds through harsh reaction conditions. CO_2RR is an effective process to convert CO_2 due to the reaction rate and selectivity of their product. The performance of CO_2RR has a high dependence on its catalysts. However, the low specific surface area of the currently developed catalysts, makes the photocatalytic efficiency of these catalysts still far from satisfactory. In recent years, an extremely high activity for CO_2 reduction to CO has been found in SACs. The uniformity of SACs active sites and its geometric structure can provide equal spatial and electronic interactions with substrates. These interactions are beneficial for the enhancement of the catalytic selectivity. Thus, SACs are promising in the CO_2RR process to achieve an excellent activity and high efficiency. The CO production is reported by the reaction turnover frequency (TOF). TOFs can be measured by measuring the raw electrode current, CO Faraday efficiency, and the number of electrochemically active sites voltammetric quantification. The instantaneous TOF for CO production can be calculated from Equation (6.5):

$$TOF = \frac{i_{CO}}{nF\alpha} \quad (6.6)$$

where i_{CO} is the CO partial current obtained from the raw electrode current that determined the CO Faraday efficiency (C s^{-1}), the value n represents the number of electron in the reaction (for CO production, $n = 2$), F represents the Faraday constant (96,485 C mol^{-1} e^{-1}), and α

represents the moles of electrochemically active sites. Unit analysis provided for the TOF for CO production is reported as s^{-1}.

Zhao et al. (2017) were the first to suggest SACs for CO_2RR with Ni single-atom anchored on N-doped porous carbon. These Ni SACs can achieve a high turnover frequency (TOF) of 5273 h^{-1} for CO_2RR selectivity and excellent Faraday efficiency of over 71.9% for CO production. Yang et al. (2018) applied a high-density and low-valence single Ni atomic site on N-doped graphene supports for achieving CO_2RR efficiency. Ni SACs were shown to have an excellent CO_2RR efficiency of 97% with an exceptional TOF of up to 14,800 h^{-1}. This excellent CO_2RR performance of Ni SACs was also contributed by the unique electronic structure of Ni single-atoms. Furthermore, these Ni SACs appear to have long-term durability, which can maintain 98% of its initial activity after 100 h.

Ni SACs, single Co atomic sites were also reported to have great potential for CO_2RR performance enhancement. Wang et al. (2018) have successfully fabricated single Co atoms with a different coordination number of nitrogen. They found that single Co atomic sites with two-coordinated N atoms show higher activity and selectivity compared to the atomically dispersed Co with another coordination number of N. Moreover, in their research, Co SACs exhibit CO efficiency of up to 94%, and a current density of 18.1 mA/cm^{-2} at overpotentials as low as 0.52 V. Co SACs also show excellent stability which makes Co SACs a promising active site for CO_2RR.

REFERENCES

Abbet, S., A. Sanchez, U. Heiz, W. D. Schneider, A. M. Ferrari, G. Pacchioni, and N. Rosch. 2000. "Acetylene cyclotrimerization on supported size-selected Pd-n clusters (1 <= n <= 30): one atom is enough!" *Journal of the American Chemical Society* 122 (14):3453–3457. doi: 10.1021/ja9922476.

Aich, P., H. J. Wei, B. Basan, A. J. Kropf, N. M. Schweitzer, C. L. Marshall, J. T. Miller, and R. Meyer. 2015. "Single-atom alloy Pd-Ag catalyst for selective hydrogenation of acrolein." *Journal of Physical Chemistry C* 119 (32):18140–18148. doi: 10.1021/acs.jpcc.5b01357.

Aranifard, S., S. C. Ammal, and A. Heyden. 2014. "On the importance of metal-oxide interface sites for the water-gas shift reaction over Pt/CeO2 catalysts." *Journal of Catalysis* 309:314–324. doi: 10.1016/j.jcat.2013.10.012.

Bruix, A., Y. Lykhach, I. Matolinova, A. Neitzel, T. Skala, N. Tsud, M. Vorokhta, V. Stetsovych, K. Sevcikova, J. Myslivecek, R. Fiala, M. Vaclavu, K. C. Prince, S. Bruyere, V. Potin, F. Illas, V. Matolin, J. Libuda, and K. M. Neyman. 2014. "Maximum noble-metal efficiency in catalytic materials: atomically dispersed surface platinum." *Angewandte Chemie-International Edition* 53 (39):10525–10530. doi: 10.1002/anie.201402342.

Bulushev, D. A., M. Zacharska, A. S. Lisitsyn, O. Y. Podyacheva, F. S. Hage, Q. M. Ramasse, U. Bangert, and L. G. Bulusheva. 2016. "Single atoms of pt-group metals stabilized by N-doped carbon nanofibers for efficient hydrogen production from formic acid." *Acs Catalysis* 6 (6):3442–3451. doi: 10.1021/acscatal.6b00476.

Chang, H., M. Saito, T. Nagai, Y. Y. Liang, Y. Kawazoe, Z. C. Wang, H. Wu, K. Kimoto, and Y. Ikuhara. 2014. "Single adatom dynamics at monatomic steps of free-standing layer reduced graphene." *Scientific Reports* 4. doi: ARTN 603710.1038/srep06037.

Chen, Y. J., S. F. Ji, C. Chen, Q. Peng, D. S. Wang, and Y. D. Li. 2018. "Single-atom catalysts: synthetic strategies and electrochemical applications." *Joule* 2 (7):1242–1264. doi: 10.1016/j.joule.2018.06.019.

Ding, W. C., X. K. Gu, H. Y. Su, and W. X. Li. 2014. "Single Pd atom embedded in CeO2(111) for NO reduction with CO: a first-principles study." *Journal of Physical Chemistry C* 118 (23):12216–12223. doi: 10.1021/jp503745c.

Du, C. M., H. P. Lin, B. Lin, Z. Y. Ma, T. J. Hou, J. X. Tang, and Y. Y. Li. 2015. "MoS2 supported single platinum atoms and their superior catalytic activity for CO oxidation: a density functional theory study." *Journal of Materials Chemistry A* 3 (46): 23113–23119. doi: 10.1039/c5ta05084g.

Dvorak, F., M. F. Camellone, A. Tovt, N. D. Tran, F. R. Negreiros, M. Vorokhta, T. Skala, I. Matolinova, J. Myslivecek, V. Matolin, and S. Fabris. 2016. "Creating single-atom Pt-ceria catalysts by surface step decoration." *Nature Communications* 7. doi: ARTN 1080110.1038/ncomms10801.

Fei, H. L., J. C. Dong, M. J. Arellano-Jimenez, G. L. Ye, N. D. Kim, E. L. G. Samuel, Z. W. Peng, Z. Zhu, F. Qin, J. M. Bao, M. J. Yacaman, P. M. Ajayan, D. L. Chen, and J. M. Tour. 2015. "Atomic cobalt on nitrogen-doped graphene for hydrogen generation." *Nature Communications* 6. doi: ARTN 866810.1038/ncomms9668.

Single-Atom Catalysts on Nanostructure

Gao, H. W. 2016. "CO oxidation mechanism on the gamma-Al2O3 supported single Pt atom: First principle study." *Applied Surface Science* 379:347–357. doi: 10.1016/j.apsusc.2016.04.009.

Gao, G., Y. Jiao, E. R. Waclawik, and A. Du, 2016. "Single atom (Pd/Pt) supported on graphitic carbon nitride as an efficient photocatalyst for visible-light reduction of carbon dioxide." *Journal of the American Chemical Society* 138 (19): 6292–6297. doi: 10.1021/jacs.6b02692

Gu, X. K., B. T. Qiao, C. Q. Huang, W. C. Ding, K. J. Sun, E. S. Zhan, T. Zhang, J. Y. Liu, and W. X. Li. 2014. "Supported Single Pt-1/Au-1 Atoms for Methanol Steam Reforming." *Acs Catalysis* 4 (11):3886–3890. doi: 10.1021/cs500740u.

Hackett, S. E. J., R. M. Brydson, M. H. Gass, I. Harvey, A. D. Newman, K. Wilson, and A. F. Lee. 2007. "High-activity, single-site mesoporous Pd/Al2O3 catalysts for selective aerobic oxidation of allylic alcohols." *Angewandte Chemie-International Edition* 46 (45): 8593–8596. doi: 10.1002/anie.200702534.

Heiz, U., A. Sanchez, S. Abbet, and W. D. Schneider. 1999. "Catalytic oxidation of carbon monoxide on monodispersed platinum clusters: each atom counts." *Journal of the American Chemical Society* 121 (13):3214–3217. doi: 10.1021/ja983616l.

Kaden, W. E., T. P. Wu, W. A. Kunkel, and S. L. Anderson. 2009. "Electronic structure controls reactivity of size-selected Pd clusters adsorbed on TiO2 Surfaces." *Science* 326 (5954): 826–829. doi: 10.1126/science.1180297.

Kesavan, L., R. Tiruvalam, M. H. Ab Rahim, M. I. bin Saiman, D. I. Enache, R. L. Jenkins, N. Dimitratos, J. A. Lopez-Sanchez, S. H. Taylor, D. W. Knight, C. J. Kiely, and G. J. Hutchings. 2011. "Solvent-free oxidation of primary carbon-hydrogen bonds in toluene using Au-Pd alloy nanoparticles." *Science* 331 (6014):195–199. doi: 10.1126/science.1198458.

Kim, Y. T., K. Ohshima, K. Higashimine, T. Uruga, M. Takata, H. Suematsu, and T. Mitani. 2006. "Fine size control of platinum on carbon nanotubes: from single atoms to clusters." *Angewandte Chemie-International Edition* 45 (3): 407–411. doi: 10.1002/anie.200501792.

Li, F. Y., Y. F. Li, X. C. Zeng, and Z. F. Chen. 2015. "Exploration of high-performance single-atom catalysts on support M-1/FeOx for CO oxidation via computational study." *Acs Catalysis* 5 (2): 544–552. doi: 10.1021/cs501790v.

Liu, J. Y. 2017. "Catalysis by supported single metal atoms." *Acs Catalysis* 7 (1):34–59. doi: 10.1021/acscatal.6b01534.

Liu, P. X., Y. Zhao, R. X. Qin, S. G. Mo, G. X. Chen, L. Gu, D. M. Chevrier, P. Zhang, Q. Guo, D. D. Zang, B. H. Wu, G. Fu, and N. F. Zheng. 2016. "Photochemical route for synthesizing atomically dispersed palladium catalysts." *Science* 352 (6287): 797–801. doi: 10.1126/science.aaf5251.

Ma, D. W., T. X. Li, Q. G. Wang, G. Yang, C. Z. He, B. Y. Ma, and Z. S. Lu. 2015. "Graphyne as a promising substrate for the noble-metal single-atom catalysts." *Carbon* 95: 756–765. doi: 10.1016/j.carbon.2015.09.008.

Ma, D. W., Q. G. Wang, X. W. Yan, X. W. Zhang, C. Z. He, D. W. Zhou, Y. A. Tang, Z. S. Lu, and Z. X. Yang. 2016. "3d transition metal embedded C2N monolayers as promising single-atom catalysts: A first-principles study." *Carbon* 105: 463–473. doi: 10.1016/j.carbon.2016.04.059.

Ma, X. G., Y. H. Lv, J. Xu, Y. F. Liu, R. Q. Zhang, and Y. F. Zhu. 2012. "A strategy of enhancing the photoactivity of g-c3n4 via doping of nonmetal elements: a first-principles study." *Journal of Physical Chemistry C* 116 (44): 23485–23493.

Mao, K. K., L. Li, W. H. Zhang, Y. Pei, X. C. Zeng, X. J. Wu, and J. L. Yang. 2014. "A theoretical study of single-atom catalysis of CO oxidation using Au embedded 2D h-BN monolayer: a CO-promoted O-2 activation." *Scientific Reports* 4. doi: ARTN 544110.1038/srep05441.

Pei, G. X., X. Y. Liu, X. F. Yang, L. L. Zhang, A. Q. Wang, L. Li, H. Wang, X. D. Wang, and T. Zhang. 2017. "Performance of Cu-Alloyed Pd Single-Atom Catalyst for Semihydrogenation of Acetylene under Simulated Front-End Conditions." *Acs Catalysis* 7 (2):1491–1500. doi: 10.1021/acscatal.6b03293.

Qiao, B. T., A. Q. Wang, X. F. Yang, L. F. Allard, Z. Jiang, Y. T. Cui, J. Y. Liu, J. Li, and T. Zhang. 2011. "Single-atom catalysis of CO oxidation using Pt-1/FeOx." *Nature Chemistry* 3 (8): 634–641. doi: 10.1038/Nchem.1095.

Sakthivel, S., M. V. Shankar, M. Palanichamy, B. Arabindoo, D. W. Bahnemann, and V. Murugesan. 2004. "Enhancement of photocatalytic activity by metal deposition: characterization and photonic efficiency of Pt, Au, and Pd deposited on TiO2 catalyst." *Water Research* 38 (13):3001–3008. doi: 10.1016/j.watres.2004.04.046.

Stephens, I. E. L., and I. Chorkendorff. 2011. "Minimizing the use of platinum in hydrogen-evolving electrodes." *Angewandte Chemie-International Edition* 50 (7):1476. doi: 10.1002/anie.201005921.

Sun, S. H., G. X. Zhang, N. Gauquelin, N. Chen, J. G. Zhou, S. L. Yang, W. F. Chen, X. B. Meng, D. S. Geng, M. N. Banis, R. Y. Li, S. Y. Ye, S. Knights, G. A. Botton, T. K. Sham, and X. L. Sun. 2013. "Single-atom catalysis using Pt/graphene achieved through atomic layer deposition." *Scientific Reports* 3. doi: ARTN 177510.1038/srep01775.

Wang, W. L., E. J. G. Santos, B. Jiang, E. D. Cubuk, C. Ophus, A. Centeno, A. Pesquera, A. Zurutuza, J. Ciston, R. Westervelt, and E. Kaxiras. 2014. "Direct observation of a long-lived single-atom catalyst chiseling atomic structures in graphene." *Nano Letters* 14 (2): 450–455. doi: 10.1021/nl403327u.

Wang, X. Q., Z. Chen, X. Y. Zhao, T. Yao, W. X. Chen, R. You, C. M. Zhao, G. Wu, J. Wang, W. X. Huang, J. L. Yang, X. Hong, S. Q. Wei, Y. Wu, and Y. D. Li. 2018. "Regulation of coordination number over single co sites: triggering the efficient electroreduction of CO2." *Angewandte Chemie-International Edition* 57 (7):1944–1948. doi: 10.1002/anie.201712451.

Wang, Y., H. Yuan, Y. F. Li, and Z. F. Chen. 2015. "Two-dimensional iron-phthalocyanine (Fe-Pc) monolayer as a promising single-atom-catalyst for oxygen reduction reaction: a computational study." *Nanoscale* 7 (27):11633–11641. doi: 10.1039/c5nr00302d.

Yan, H., H. Cheng, H. Yi, Y. Lin, T. Yao, C. L. Wang, J. J. Li, S. Q. Wei, and J. L. Lu. 2015. "Single-atom Pd-1/graphene catalyst achieved by atomic layer deposition: remarkable performance in selective hydrogenation of 1,3-butadiene." *Journal of the American Chemical Society* 137 (33):10484–10487.

Yang, H. B., S. F. Hung, S. Liu, K. D. Yuan, S. Miao, L. P. Zhang, X. Huang, H. Y. Wang, W. Z. Cai, R. Chen, J. J. Gao, X. F. Yang, W. Chen, Y. Q. Huang, H. M. Chen, C. M. Li, T. Zhang, and B. Liu. 2018. "Atomically dispersed Ni(i) as the active site for electrochemical CO2 reduction." *Nature Energy* 3 (2):140–147. doi: 10.1038/s41560-017-0078-8.

Yang, S., J. Kim, Y. J. Tak, A. Soon, and H. Lee. 2016. "Single-atom catalyst of platinum supported on titanium nitride for selective electrochemical reactions." *Angewandte Chemie-International Edition* 55 (6): 2058–2062. doi: 10.1002/anie.201509241.

Yang, X. F., A. Q. Wang, B. T. Qiao, J. Li, J. Y. Liu, and T. Zhang. 2013. "Single-atom catalysts: a new frontier in heterogeneous catalysis." *Accounts of Chemical Research* 46 (8):1740–1748. doi: 10.1021/ar300361m.

Zhang, J., X. Wu, W. C. Cheong, W. X. Chen, R. Lin, J. Li, L. R. Zheng, W. S. Yan, L. Gu, C. Chen, Q. Peng, D. S. Wang, and Y. D. Li. 2018b. "Cation vacancy stabilization of single-atomic-site Pt-1/Ni(OH)(x) catalyst for diboration of alkynes and alkenes." *Nature Communications* 9. doi: ARTN 100210.1038/s41467-018-03380-z.

Zhang, L. Z., Y. Jia, G. P. Gao, X. C. Yan, N. Chen, J. Chen, M. T. Soo, B. Wood, D. J. Yang, A. J. Du, and X. D. Yao. 2018a. "Graphene defects trap atomic Ni species for hydrogen and oxygen evolution reactions." *Chem* 4 (2): 285–297. doi: 10.1016/j.chempr.2017.12.005.

Zhang, X. F., J. J. Guo, P. F. Guan, C. J. Liu, H. Huang, F. H. Xue, X. L. Dong, S. J. Pennycook, and M. F. Chisholm. 2013. "Catalytically active single-atom niobium in graphitic layers." *Nature Communications* 4. doi: ARTN 192410.1038/ncomms2929.

Zhang, X., J. C. Lei, D. H. Wu, X. D. Zhao, Y. Jing, and Z. Zhou. 2016. "A Ti-anchored Ti2CO2 monolayer (MXene) as a single-atom catalyst for CO oxidation." *Journal of Materials Chemistry A* 4 (13): 4871–4876. doi: 10.1039/c6ta00554c.

Zhang, X., H. Shi, and B. Q. Xu. 2005. "Catalysis by gold: isolated surface Au3+ ions are active sites for selective hydrogenation of 1,3-butadiene over Au/ZrO2 catalysts." *Angewandte Chemie-International Edition* 44 (43): 7132–7135. doi: 10.1002/anie.200502101.

Zhang, Y. X., Z. Y. Zhao, D. Fracasso, and R. C. Chiechi. 2014. "Bottom-up molecular tunneling junctions formed by self-assembly." *Israel Journal of Chemistry* 54 (5–6): 513–533. doi: 10.1002/ijch.201400033.

Zhao, C. M., X. Y. Dai, T. Yao, W. X. Chen, X. Q. Wang, J. Wang, J. Yang, S. Q. Wei, Y. E. Wu, and Y. D. Li. 2017. "Ionic exchange of metal-organic frameworks to access single nickel sites for efficient electroreduction of CO2." *Journal of the American Chemical Society* 139 (24): 8078–8081. doi: 10.1021/jacs.7b02736.

7 Growth, Characteristics and Application of Nanoporous Anodic Aluminum Oxide Synthesized at Relatively High Temperature

Chen-Kuei Chung
National Cheng Kung University, Taiwan

Ming-Wei Liao
National Tsing Hua University, Taiwan

Chin-An Ku
National Cheng Kung University, Taiwan

CONTENTS

7.1 Introduction: The Brief Review of Growth and Characteristics of Nanoporous Anodic Aluminum Oxide (AAO) and its Applications 166
7.2 A Literature Survey of AAO Formed from the Low-Purity, High-Purity Al Foils and the Al Films on Si Substrates 166
7.3 Hybrid Pulse Anodization (HPA) of Bulk Al Foils at Low-to-High Temperatures 169
    7.3.1 Experiments of HPA from Al Foils 169
    7.3.2 Two-Step Anodization of Bulk Al Foils 169
    7.3.3 Single-Step Hybrid Pulse Anodization 173
7.4 Hybrid Pulse Anodization of Al Films on Si Substrates at Relatively High Temperature 178
    7.4.1 Experiments of Deposition and Anodization of Al Films on Si Substrates 178
    7.4.2 The Microstructure and Morphology of As-deposited Al Films 178
    7.4.3 The Pores Growth and Distribution Uniformity of AAO 181
7.5 Applications of HPA AAO in the Effective Nanoporous Membranes Fabrication and Photocatalysis 184
    7.5.1 A Literature Review of AAO Membrane Detachment Method 184
    7.5.2 One-time Potentiostatic Method for AAO Membrane Detachment 184
    7.5.3 A Literature Review of $P25-TiO_2$ Photocatalysis 186
    7.5.4 Larger Specific Surface Area of 3D AAO for Enhancing $P25-TiO_2$ Photocatalysis 187
    7.5.5 Photocatalytic Performance of TiO_2 P25 on Over-etched 3D AAO 187
7.6 Conclusions 190
References 191

7.1 INTRODUCTION: THE BRIEF REVIEW OF GROWTH AND CHARACTERISTICS OF NANOPOROUS ANODIC ALUMINUM OXIDE (AAO) AND ITS APPLICATIONS

Anodic aluminum oxide (AAO) is a well-known nanostructured template in industry and science research. Compared with other polymer nanomaterial template, that is, polycarbonate or polystyrene [1–3], AAO has advantages of nanoscale pores, self-organization, controllable pore size, good biocompatibility and thermal stability [4, 5]. Besides, aluminum is one of the most commonly used metals in the industry, which makes aluminum surface treatment highly compatible with production lines. These characteristics make AAO an object worth studying. AAO can be classified into two types (i.e., barrier or porous) depending on the structure. Barrier type AAO, with a thin and compact packed structure, has been widely used in the past for protection and recently with lubricant-infused surfaces for antifouling, anti-icing and anti-corrosion applications [6]. However, the porous type AAO, with a highly ordered nanopore arrangement, has received great attention in recent decades as a template for fabricating nanowires or other nanomaterials for wide applications [7–11]. In addition, there are thousands of AAO papers published in international journals for the various applications including nanomaterial synthesis or nano-pattern transfer, bio-applications, humidity sensors, structure color, photocatalysis, surface-enhanced Raman scattering (SERS) and other fields [10, 12–20]. In industry, the characteristic of AAO is utilized on surface colorization for value-added, increasing surface hardness and anti-corrosion for products. In general, a self-ordered AAO can be prepared by performing the two-step anodization of Al foil in an appropriate electrolyte at low temperature [21]. Due to the different degrees of dissociation of H^+ in different electrolytes, the rates of oxidation and dissolution balance under different anodizing potentials, for example, 25 V for sulfuric acid, 40 V for oxalic acid and 195 V for phosphoric acid [22–24]. If a potential higher than this value is applied, the higher current density and joules heat generation cause damage of the pore structure. In addition, it shows that interpore distance of AAO is related to the applied potential with a proportionality of 2.5 nm/V under mild anodization conditions [25]. Thus, the maximum pore size is limited by the optimum anodizing potential of the chosen electrolyte. For example, the maximum pore size of AAO anodized in oxalic acid (anodizing potential 40 V) is around 40 nm. For nanowire applications such as magnetic recording [26], super capacitors [27] or field emission devices [28], the wire diameter (determined by the AAO pore size) has a crucial effect on the device performance. Therefore, the problem of controlling the AAO pore size has attracted significant attention in the literature. For example, Lee et al. [24] proposed a hard anodization process at extremely high potentials of 110 ~ 140 V in oxalic acid for fabricating high aspect ratio pore channels. It was shown that potentials in this range result in a significant increase in the pore distance and oxide growth rate. Bai et al. [29] used a fractional factorial design method to prepare ordered AAO in sulfuric acid at relatively high potentials and electrolyte concentrations. Zaraska et al. [30] obtained different pore diameters by adjusting the post-pore widening time in phosphoric acid immersion. In brief, the growth, morphology and characteristics of nanoporous AAO and its wide applications have attracted great attentions in the traditional and advanced industry as well as scientific research.

7.2 A LITERATURE SURVEY OF AAO FORMED FROM THE LOW-PURITY, HIGH-PURITY AL FOILS AND THE AL FILMS ON SI SUBSTRATES

In the past few decades, most of the AAO process is still based on high-purity aluminum (> 99.99%) and two-stage low temperature (0 ~ 10°C). Many related studies still use this method as a guideline. However, the cost of high-purity aluminum is more than 1,000 times that of low-purity aluminum or alloy. Therefore, some scholars have tried to replace the traditional high-purity aluminum with low-purity aluminum for experiments. Although the cost has been reduced, the regularity of the holes has been sacrificed. However, some scholars have suggested that the regularity of holes has little effect on some applications. For instance, Choi et al. proposed the application of low-purity aluminum

AAO in structural color and explained that the wavelength of color has a large relationship with the porosity, and the regularity of pores has little effect [31]. Chen et al. proposed the application of AAO from 6061 alloy in filtration [32]. In addition, Cho et al. also studied the roughness of AAO grown in phosphoric acid from 6061 aluminum alloy under different polishing conditions [23]. These studies have shown that the method of making AAO with low-purity aluminum substrates has received more and more attention, and related applications have also been more developed.

As introducing other materials into the AAO template, we must remove the left Al and open bottom barrier to obtain through pore structure. Nevertheless, the removing process of Al foil is difficult due to thickness variation. As the AAO template was separated from Al foil successfully, the following encountering problem is how to attach AAO with another substrate. For improving the adhesion problem, more and more studies have directly anodized Al thin films which were pre-coated on silicon (Si) substrates [33–35]. However, it reveals different AAO results between the Al film on Si and the bulk Al foil on the same anodization condition. One of the important reasons is the limited thickness of film (usually less than 1 μm) which restricts to rather short anodizing time. On this condition, some experiments reported that the intrinsic material properties of Al film may directly influence the self-organized pore structure. The internal defects like grain boundaries of Al could lead to disordered pore formation [36]. The small grain size of Al film also limited the pore distance and diameter of AAO [37]. In addition, oxidation of Si after complete anodization of Al induced a mechanical stress and then led to structural change of AAO [38]. Moreover, it was reported that addition of Ti for corrosion protection also influences the porous structures [39, 40].

The porous AAO film with highly ordered nanopore arrangement was considered as one of the most prominent method for template-assisted growth of nanomaterials due to the advantages of controllable diameter, high aspect ratio and economical way in producing. In general, the nanoporous AAO was prepared by performing the two-step anodization of an electrochemically polished bulk Al foil. It encountered an attachment problem when combining AAO onto the Si substrate for depositing nanomaterial [41]. Therefore, numerous researchers commenced directly anodizing Al thin film on Si for growing nano-rod or nanowire of gold, carbon, platinum, zinc oxide, copper and so on [42–52]. Most of investigations were focused on unique material properties or characteristic structure rather than effects of parameters on AAO template. However, the AAO template quality could influence the nanostructured material synthesis. Moreover, the thicknesses of Al films were usually controlled to be over 1 μm [42, 44–46, 49] in order to extend total anodizing time or perform multi-step process [45]. For successfully fabricating well-ordered porous AAO, the thickness of Al film should be over 10 μm [34]. On the other hand, the pre-texture method on Al film surface by using lithography and imprinting have been developed to obtain high-quality pore arrangement [53–55]. However, either depositing such a thick Al film or pre-patterning by lithography requires lots of time, high cost and energy consumption. Until now, the fabrication of AAO template by direct anodizing Al films on Si substrates remains a major challenge, especially in submicron thickness. In comparison with Al foil, some intrinsic material properties of Al film have also been experimentally observed to influence the self-organized pore structure. The internal defects like grain boundary of Al were discussed to lead to disordered pore formation [36] and the small grain size of Al film also limited the pore distance and diameter of AAO [37]. Nevertheless, the effect of surface morphology on pore structure has not been reported despite the fact that it is a significant factor in anodization of Al foil. The electrochemical polish which is a necessary process before anodizing Al foil was reported to influence the pore size distribution and order [56]. In the case of Al film, it is impossible to perform electrochemical polish due to limited thickness. The original surface morphology of as-deposited film plays an important role in pore shape and arrangement. In addition, the above mentioned works on anodization of Al film are conventional two-step process. Few studies devoted to discuss one-step anodization of Al film. In order to deposit nanomaterial on Si substrate directly, anodization of Al film was firstly proposed by Crouse et al. [33]. Following the investigation, the combination of AAO and Si substrate has received much attention. For example, the AAO containing well-ordered straight pore channels can be used as a mask for patterning or transferring nanoporous structures into Si wafer [57]. Pu et al. performed anodization of Al film on Si and post-etching for fabricating nanoscale Si tips [58]. Oide et al. also fabricate the nanostructures of hole and column arrays on Si by the particular interface of porous AAO and Si and chemical etching [59]. Kokonou et al. fabricated high-density stoichiometric SiO_2 dot arrays on Si through an anodic porous alumina template [48]. Sai et al. fabricated a subwavelength structured (SWS) surfaces by porous AAO mask [60]. Due to the different anodization behavior of bulk Al and Al film, there are a few literatures about the relationship between Al film deposition and AAO pore structure. Rabin et al. deposited a thick Al film for electrochemical polish and processing two-step anodization in various electrolytes [61]. Choi et al. demonstrated a well-ordered AAO porous structure using the pre-texture method of nanoindentation [62]. The formation of pores follows the original indentation. Myung et al. also deposited Al film with thickness of 5.5 μm for two-step anodization and examined pore formation in various applied potentials and mixed H_2SO_4 and oxalic acid [44]. Mei et al. analyzed both anodizing I-t curve and found ordering of nanopore arrangement affects the thickness of the deposited Al film [63]. These literatures confirmed that the subsequent AAO formation is closely related to deposition of Al film.

The intrinsic material properties of Al film may directly influence the self-organization of pore structure. The internal defects like grain boundaries of Al could lead to disordered pore formation [36]. The small grain size of Al film also limits the pore distance and diameter of AAO [37]. In addition, oxidation of Si after complete anodization of Al induces a mechanical stress and then leads to structural change of AAO [38]. Moreover, it was reported that addition of Ti for corrosion protection also influences the porous structures [39, 40]. Although the field-assisted dissolution is widely accepted as mechanism of pore formation in recent works for discussing AAO growth, researchers claimed that plastic deformation due to compressive oxide growth stresses dominates pore growth whether in experiment or theorem. Thompson and co-workers used the tungsten tracer layer in Al films to confirm that plastic flow occurred from pore bottom upward to cell wall [64]. Houser and Hebert simulated the behavior of stress-driven material flow by continuum theory which revealed different flow velocity linked to levels of compressive stress [65, 66]. Sulka et al. demonstrated that external tensile stress could affect the pore size and arrangement in anodization of bulk Al [67]. These results suggest that mechanical stress plays an important role in AAO pore structure formation. However, the relationship between residual stresses and pore characterization has not been investigated in anodization of Al film. In fact, the growth of Al film can be understood by the Volmer–Weber (three-dimensional) mode and commonly results in intrinsic stress [68]. In general, the compressive stress during porous alumina growth is about 400 MPa concerned with anodizing condition [69]. The magnitude of residual stress in Al films can reach several tens to hundreds of MPa which is expected to significant affect porous AAO formation. The effect of residual stress in films on pore generation is interesting as well the intrinsic stress commonly occurs in the prepared Al films. Liao and Chung [70] reported the role and effect of the residual stress on the evolution of AAO pore structure. The residual stress in Al films interacting with oxide growth stress during anodization leads to change of pore formation in AAO. The tensile residual stress releases the compressive oxide growth stress of AAO during plastic deformation for higher pore density; that is, the increase of effective flow stress in the films with extra tensile stress reduces the interpore distance of AAO film on Si for higher density of pores. It offers new foundations for realizing AAO films on Si.

Recently, the fabrication, structure and properties of nanoporous AAO formed from the various aluminum metal and alloy are continuously and widely studied in the traditional and advanced industry as well as scientific research due to its wide applications [6–20, 23]. For example, Wu et al. reported the durable lubricant-infused anodic aluminum oxide surfaces with high aspect ratio nano-channels for resistance to environment [6]. Jeong et al. claimed a massive, eco-friendly and facile fabrication of multi-functional anodic aluminum oxides applied to nanoporous templates and sensing platforms [7]. Sulka et al. reviewed the AAO templates with different patterns and channel shapes for various applications [8]. Process-dependent pore structures were also controlled by anodizing parameters, electrolyte concentration and additives, and material composition [8–13]. Wu et al. reported the light-induced nanowetting: erasable and rewritable polymer nanoarrays via solid-to-liquid transitions [9]. Zhou and Nonnenmann reviewed the recent progress in nanoporous templates beyond anodic aluminum oxide and toward functional complex materials [11]. The optical and/or color properties were studied by the dyeing, structure periodicity and electrolyte additives [3, 6, 12, 14, 18]. The humidity sensor, SERS and photocatalysis are also crucial for our environment monitoring, safety, health, process control, food reservation and so on [3, 6, 15, 16, 19, 20]. However, the conventional AAO templates were synthesized using two-step potentiostatic method of direct current anodization (DCA) at low temperature (0–10 °C) to avoid dissolution effect. Here, we propose the patented hybrid pulse anodization (HPA) with normal-positive and small negative voltages for AAO fabrication at relatively high temperatures of 15–25°C for promoting AAO structure and discuss in the following sections.

7.3 HYBRID PULSE ANODIZATION (HPA) OF BULK AL FOILS AT LOW-TO-HIGH TEMPERATURES

7.3.1 EXPERIMENTS OF HPA FROM AL FOILS

High-purity and low-purity aluminum foil (99.997% and 99%, Alfa Aesar, USA) was used for two-step and one-step AAO anodization process. The foil was electropolished in a solution of $HClO_4$: C_2H_5OH = 1 : 4 at 20 V at room temperature. The first step of AAO anodization process was performed in 0.3 or 0.5 M oxalic acid at a temperature of 5–30 °C. In the DCA process, anodization was performed at voltage of 30 or 40 V. Meanwhile in the HPA process, the anodization was performed by a hybrid pulse comprising a positive square wave with an amplitude of 30 V or 40 V followed by a negative square wave with an equal duration, but an amplitude of −2 V. The period of the hybrid pulse was specified 2–10 s. The experiment of AAO formation was performed by means of the potentiostat (Jiehan 5000, Taiwan) and the three-electrode electrochemical cell with a platinum mesh as the counter electrode, the aluminum foil as the working electrode and Ag/AgCl/3M KCl as the reference electrode. For two-step anodization, the specimens were immersed in 5 wt% phosphoric acid at 50 °C for 60 minutes in order to remove all porous alumina structure then second anodization process was conducted.

7.3.2 TWO-STEP ANODIZATION OF BULK AL FOILS

In order to produce a highly ordered nanopore structure, a lot of researches had been studied in terms of process parameters [71–73]. In 1970s–1980s, a new approach of pulse anodization by alternating two different positive potentials had been used in Al industry to produce alumina films for improving hardness, corrosion resistance and thickness uniformity by the multilayer oxide with sandwiched large and small cells [74, 75]. The pulse reverse hard anodization with an equal value of positive and negative currents (/potential) but different duration was used for high-speed anodic oxidation on Al [76, 77]. However, the above pulse reverse hard pulse anodization was avoided for current nanotechnology because of non-uniform and disordered nanostructure [78]. Currently, synthesis of AAO with single, double and even multi-modulated pore structures was performed by cyclic pulse anodization combining conventional so-called mild (or low-field) anodization (MA, 1–4 μm h^{-1}) and hard (or high-field) anodization (HA, 50–100 μm h^{-1}) [78–80]. But, the strict process conditions were required at low temperatures of −1 °C~10 °C from high-purity (>99.997%) Al foils. Moreover, Chung et al. [81–84] had demonstrated the feasibility of AAO production from low-purity Al foil at RT (20°C −30°C) using HPA method with a small negative potential difference, which could overcome the limitation of conventional low-temperature AAO process to form nanopores by suppressing joules heat during reaction. It is impossible to apply conventional DCA to AAO production at severe conditions from low-purity Al and RT due to the joules heat-enhanced dissolution. The limitation can be overcome by applying HPA method in which the pulse potential difference can suppress the joules heat in electronics circuit effectively and even has better cooling capability in the aqueous electrochemical cell [81, 82]. The heat generation is harmful to nanostructures because it accumulates randomly all over the AAO, which would cause the non-uniform distribution of AAO and ruin the structure.

The continuous current-time curve in DCA is shown in Figure 7.1(a), which leads to heat accumulation for the thermally enhanced joules heating dissolution effect with reaction time to destroy the pore structure with increasing temperature from 20°C to 30°C (Figure 7.2(a)–(c)). In contrast, well-ordered AAO nanostructures (Figure 7.2(d)–(f)) are achieved by HPA process at RT because of the HPA-induced alternating current-time curve (Figure 7.1(b)) as well as effective cooling at negative potential (t- duration) to diminish the dissolution. The pore diameter of the AAO nanostructures becomes a little larger at 25–30 °C than at 20 °C due to the temperature-dissolution-widen effect. In comparison with the current-time curve by conventional pulse anodization (PA, Figure 7.1(c)) with HPA process, the distinct negative current is observed during PA while nearly zero negative current occurs during HPA. In the t_{off} duration of PA with 0 V, it behaves as a discharging capacitor and all the charges accumulated in t_{on} duration flow in opposite direction to result in the negative (cathodic) current. The appearance of random reverse current may be attributed to the impurities of Al foil and the quick change of polarization between two electrodes (anode/cathode) in a very short time. Regarding the t- duration of HPA of −2 V, the discharging effect can be overcome by the energy barrier height of cathodic reaction. Therefore, nearly zero negative current occurs accordingly.

The different current-time (I-t) curves between DCA and HPA (Figure 7.1) correspond to the different anodization mechanism. Although the similar trend of current increment with time shows the

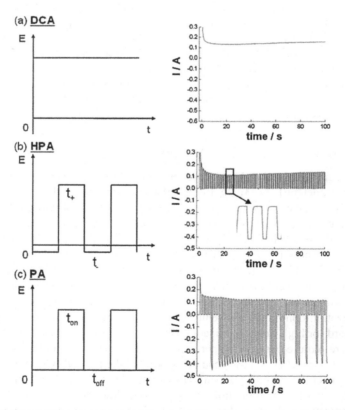

FIGURE 7.1 Real-time monitoring of current-time evolution: Comparison of the relationship between the applied potential difference (V) and the corresponding current (I) as a function of time (t).

Growth, Characteristics and Application of Nanoporous AAO 171

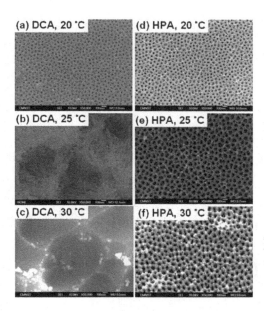

FIGURE 7.2 DC anodization (DCA) vs hybrid pulse anodization (HPA): SEM micrographs of AAO formed by DCA from low-purity 99% Al foil for 1 h at temperatures of: (a) 20 °C, (b) 25 °C and (c) 30 °C while those by HPA for 1 h at temperatures of: (d) 20 °C, (e) 25 °C and (f) 30 °C correspondingly.

similar anodized oxide formation and dissolution for AAO, the variation of HPA at the t-period with a near zero cathodic current exhibits an effective cooling for significantly reducing joules heat to make anodization possible from low-purity Al and relatively high temperature [81, 82]. The I-t curves of DCA and HPA (Figure 7.1(a) and (b)) were further examined in Figure 7.3(a) and (b), respectively.

Four periods I, II, III and IV are defined for the transition of I-t curves during DCA and HPA. Each period in both DCA and HPA is quite similar to the classical AAO I-t curve (Figure 7.3(c)): the initial stage of period I (Figure 7.3(c)) shows the formation of the initiate oxide. The current at the beginning is high due to low resistance of the metallic aluminum. Then the current decreases because of a high-resistance thin barrier oxide formation. Increasing thickness raises resistance for a further decrease of current in period II. Moreover, the current to turn upward with time in period II-III transition is due to small imperfections, for example, roughness in the compact oxide layer. These small imperfections in areas with thinner oxide lead to the current concentration than the rest of surface with thicker oxide. These areas with lower resistance for higher current are the places which the initial formation of pores. As the dissolution reaction starts at a single point, the barrier oxide thickness is reduced to nanoscale size at the bottom of pore. At this time, the oxide layer resistance is also reduced due to the barrier thickness reduction and the current simultaneously increases to repair the damage with oxidation (period III). It will increase the electrolyte temperature for enhancing the dissolution rate. Fortunately, pulse current in HPA can suppress the heat generation at RT with nearly zero anodization current and liquid electrolyte cooling. In period IV, the current will reach a stable value where the oxide dissolution rate and formation rate reach a steady-state level. In pulse current, the average current density can be regarded as I_{avg} = peak current × duty cycle. So, the current efficiency in HPA strongly depends on the duty cycle, which is smaller than that in DCA. Therefore, lower joules heat are generated in HPA compared to DCA. Also, the joules heat in HPA can be dissipated during the negative pulse period with nearly zero negative current and liquid electrolyte cooling. So, the HPA process at RT benefits the AAO pores distribution uniformity at relatively severe conditions and is good for the low-cost low-purity Al foils without any heat treatment.

Figure 7.4(a)–(b) show the schematic mechanisms of joules heat generation by DCA and HPA process in terms of the applied potential (V) waveform and the corresponding elevated temperatures

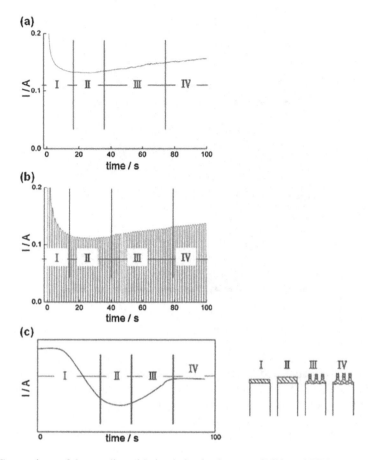

FIGURE 7.3 Comparison of the anodic oxidation behavior between DCA and HPA.

as a function of time (t), respectively. The dash-line is the applied DC potential difference and the solid-line represents the HPA waveforms with the potential differences and times. The elevated temperature with joules heat under DCA process rises initially due to heat accumulation and tends to be saturated at higher temperature than that under HPA. Regarding effective cooling in HPA, the aluminum substrate has very high thermal conductivity around 236 W/mK and aluminum oxide also has good thermal conductivity around 35 W/mK) [85]. The anions-incorporated aluminum oxide in AAO is believed to have even a higher conductivity. If joules heat are generated at the aluminum substrate and the interface between the substrate and alumina, the heat will conduct to AAO rapidly to raise reaction temperature. In chemical reactions, raising temperature hastens reactions due to the energy of molecules for more collisions per unit time. Therefore, the dissolving reaction will be much enhanced. The joules heat-enhanced dissolution reaction can be suppressed by means of HPA (Figure 7.1(b)). The pulse potential difference was used for suppressing the raising temperature of electronic devices in air. Here, the cooling of AAO formation in aqueous electrochemical cell is better in air, because the thermal conductivity of electrolyte is much higher than air. Therefore, the joules heat generated during pulse positive on period can be eliminated during negative period with nearly zero current.

If joules heat from DCA are added to higher process temperature of 25–30 °C, it will enhance the dissolution and damage nanopore structures. In contrast, the joules heat generated during HPA t_+ period rises first and then drops to zero during t_- period. Therefore, the temperature increment in HPA is unobvious. The quantitative discussion is done on the heat generation estimated from the anodizing current through the barrier layers (Figure 7.1 and Figure 7.3). The power generated in the

Growth, Characteristics and Application of Nanoporous AAO

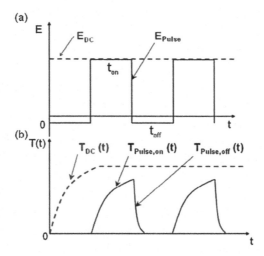

FIGURE 7.4 Schematic mechanisms of joules heat under DCA and HPA conditions: (a) the applied potential difference (V) waveform and (b) the corresponding temperatures as a function of time (t) by DCA and HPA, respectively.

process is related to the current flowing through the barrier layer, that is, $P_{joule} = I^2R$ where P_{joule} is measured in watts, I in amperes, R in ohms. It is noticed that R is usually measured as an amount of resistance of the template which depends on its length, cross-sectional area and material resistivity, which can be expressed as $R = \rho L / A$. However, the resistance of such tiny template can't be accurately calculated since the resistance from electrolyte and Al foil is difficult to locate. Alternatively, we can apply the formula, $P_{joule}(t) = I(t) \cdot V$ to be the generated joules heat per time and the total energy expended over a period of time is computed as $P_{joule}(t) = \int I(t) V \, dt$.

Figure 7.5 depicts the accumulated joules heat generation from above methods in the process of 100 seconds. The data are computed by MATLAB R2008a V.7.6, in which we can see that HPA successfully produces the lowest joules heat for AAO during the experimental process by effective "cooling" mechanisms from the nearly zero current and high thermal conductivity liquid electrolyte to diminish the dissolution reactions for the promote of pore uniformity.

7.3.3 Single-Step Hybrid Pulse Anodization

A traditionally electrochemical method for producing AAO is two-step anodization process proposed by Masuda and Fukuda [21]. The two-step anodization on Al foil at constant voltage can result in a well-ordered AAO configuration compared with one-step anodization. However, the electrolyte temperature usually maintains at 0–5 °C to reduce the dissolution and growth rate of alumina during anodization. Thus, there are some researchers devoted to produce AAO at relatively high temperatures [86, 87]. On the other hand, we had demonstrated that the HPA could reduce the resistive heating effect of anodizing current and form nanopores at room temperatures [81]. The heat generation is harmful to nanostructures due to accumulation randomly all over the AAO nanostructure especially the discontinuous geometry to enhance the dissolution effect.

Most of the investigations focused on the high and low positive voltage pulsing instead of positive and negative voltage. One of the reasons is that the negative voltage in AC pulse anodizing leads to excess hydrogen ions which will harm the porous AAO structure [88]. HPA with high positive voltage and low negative voltage can solve this problem. Moreover, pulse reverse with an equal value of long positive and short negative voltages has been studied to increase the uniformity of AAO [77]. Thus, it attracts our interest to investigate more details about HPA on different anodization conditions, especially for one-step process. It has been used to form the porous AAO nanostructure on

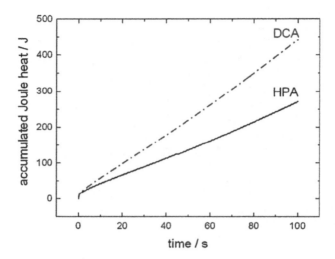

FIGURE 7.5 The comparison of time-evolution of accumulated joules heat for HPA and DCA.

Si-based substrates because the aluminum film deposited on Si wafer is too thin for two-step anodization [35, 41]. For industrial applications, the advantages of one-step anodization is low cost and timesaving. It is noted that little attention has been paid to study the behavior of one-step anodized Al foil because of non-uniform and disordered pore structure [89]. However, some applications have no obvious relationship with AAO pore regularity, so the one-step process is still worth investigating for improving AAO process. In this section, we have investigated the one-step anodization of Al foil in 0.5 M oxalic acid by HPA and DCA at environment temperature of 5–15 °C. The film thickness and nanopore size distribution was analyzed to understand the effect of temperature and voltage mode on AAO between HPA and DCA.

Figure 7.6 (a) and (b) show the SEM micrographs of AAO performed by one-step DCA and HPA from 99.997 % Al foil for 1 h at low temperature of 5 °C and then immersed in phosphoric acid 5 wt% for 30 minutes, respectively. The porous AAO nanostructure is clearly seen and obtained for both cases. Moreover, the morphology of AAO by HPA method exhibits highly ordered pore arrangements, more uniform pore size and higher circularity than that by DCA. The pore distribution of AAO films can be further quantitatively analyzed by ImageJ software of SEM images. Figure 7.7 shows the relationship between the amount of pores per μm² and pore diameters of AAO by one-step DCA (dash-line, triangles) and HPA (solid-line, squares) in Figure 7.6. The range of 40±5 nm pores occupies the main distribution in DCA while the range of 35±5 nm is observed in HPA. It indicated that the pore diameter of AAO by DCA is larger than HPA which may be attributed to heat-dissolution-widening effect. The continuous heat accumulation in DCA leads to higher joules heat than that in HPA. Therefore, higher heat-dissolution-widening effect in DCA results in the larger size of pores than HPA. On the other hand, the distribution uniformity can be calculated by the ratio of the pores located at these main peaks to the whole amount of pores. In the case of HPA, there are around 92.6% pores in the range of 35±5 nm and 82.9% pores are achieved by DCA. It implies that the lower Joule-heat dissolution in HPA is beneficial for the distribution of uniformity of pores, that is, higher quality of AAO films.

Pore circularity is another important factor to determine the quality of porous AAO besides the pore size. The definition of circularity can be expressed as [90]:

$$\text{Circularity} = \left(4\pi \frac{A}{S^2} \right)$$

where A and S represent the area and perimeter of each pore of AAO, respectively. A circularity value of 1.0 indicates a perfect circle. As the value approaches 0, it indicates an increasingly elongated polygon. Figure 7.8 (a) and (b) show the pore circularity distribution of AAO film formed by

Growth, Characteristics and Application of Nanoporous AAO   175

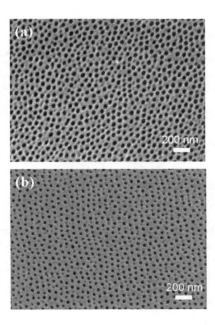

FIGURE 7.6 SEM micrographs of AAO formed by one-step (a) DCA and (b) HPA from 99.997 % Al foil in 0.5M oxalic acid at 5 °C for 1 h and then immersed in phosphoric acid 5 wt% for 30 minutes.

DCA and HPA, respectively, corresponding to Figure 7.6. It shows that the value of pore circularity of AAO produced by HPA is much higher than AAO produced by DCA. The main fraction of pore circularity of AAO by HPA ranges from 0.7 to 0.8 which is consistent with the SEM image (Figure 7.6(b)). In contrast, the pore circularity of AAO by DCA is lower than 0.6 because of more irregular pores. Indeed, most of pore arrangements in the one-step anodization were not well except for long anodizing time of several dozens of hours [91]. The result evidences that HPA is beneficial for improving the distribution and circularity of nanopores by one-step anodization process.

Figure 7.9(a) shows the SEM micrographs of top and cross-section view of AAO formed by one-step HPA from 99.997% Al foil for 1 h at relatively high temperature of 15 °C. The AAO performed by one-step HPA are still good at the elevated temperature, but the AAO nanostructures were ruined by DCA at 15 °C due to raising joules heat. In our experiments, the unexpected rising anodizing current was easily occurred especially in DCA. Thisunexpected rising current could be attributed to some defects of high-purity aluminum foil like grain boundaries or dislocations. Then, the drastically local increasing temperature increases dissolution rate. Finally, the unbalance of formation and dissolution rate of alumina causes the electric breakdown and ruins the whole AAO nanostructure. However, this phenomenon can be suppressed by HPA because the negative period provides sufficient cooling effect. The morphology of barrier layer is shown in cross-section of the SEM micrographs in Figure 7.9(a). It is noted that the well-formed semicircle structure without cracks or defects at the interface between AAO and Al foil is obtained. It is the advantage for one-step HPA-AAO formation at the elevated temperature compared to DCA.

Figure 7.9(b) shows the analysis of the pore size distribution as a function of diameter and the pore circularity of AAO films by one-step HPA at 15 °C for 1 h. The main fraction of pore diameter of AAO by one-step HPA at 15 °C is in the rage of 45±5 nm and around 88.6 % pores are formed within the range of diameter. Compared with the results of HPA at 5 °C (35±5 nm, Figure 7.7), the pore size increases with increasing environment temperatures but the distribution uniformity decreases. The temperature-enhanced dissolution widens the pore but reduces the distribution uniformity. However, it is noted that not only pore diameter but also the uniformity of one-step HPA performed at 15 °C is higher than one-step DCA performed at 5 °C (40±5 nm and 82.6 %). In addition, the main fraction of pore circularity by one-step HPA ranging from 0.5 to 0.6 is higher than one-step DCA. Although

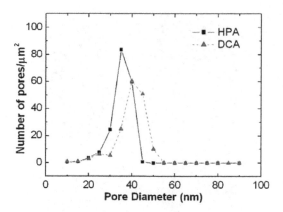

FIGURE 7.7 The relationship between the amount of pores per μm² area and pore diameters of AAO formed by one-step DCA (dash-line, triangles) and HPA (solid-line, squares) from 99.997 % Al foil in 0.5 M oxalic acid at 5 °C for 1 h.

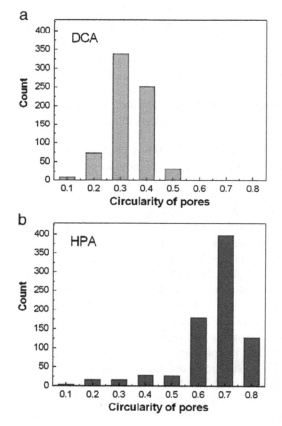

FIGURE 7.8 Pore circularity of AAO formed by (a) DCA and (b) HPA from 99.997 % Al foil in 0.5M oxalic acid for 1 h at 5 °C.

the pore uniformity and pore circularity slightly decrease from the elevated temperature, the quality of porous AAO films is still better than DCA at low temperature. Figure 7.10(a) and (b) show the SEM images of the whole cross-section of AAO formed by the same one-step HPA condition at 5 °C and 15 °C, respectively. The thickness of AAO measured by ImageJ software is 1.2 μm for 5 °C

Growth, Characteristics and Application of Nanoporous AAO 177

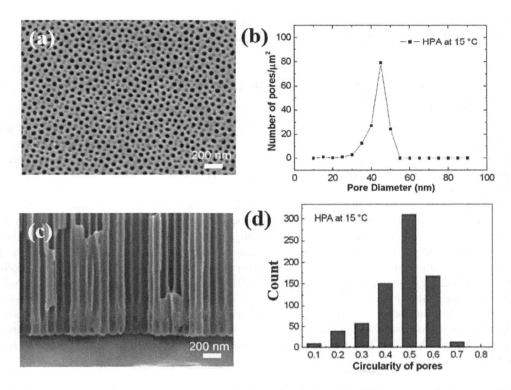

FIGURE 7.9 (a) SEM micrograph, (b) pore diameter distribution, (c) SEM cross-section view and (d) circularity of AAO formed by HPA from 99.997 % Al foil in 0.5M oxalic acid at 15 °C for 1 h.

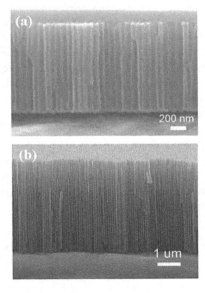

FIGURE 7.10 Cross-section of AAO formed by HPA from 99.997 % Al foil in 0.5M oxalic acid for 1 h at (a) 5 °C and (b) 15 °C.

and greatly increased to 4.1 μm for 15 °C, respectively. From industrial point of view, it is more economic and efficient for producing AAO by HPA at relatively high temperature.

7.4 HYBRID PULSE ANODIZATION OF AL FILMS ON SI SUBSTRATES AT RELATIVELY HIGH TEMPERATURE

Compared to bulk Al anodization, the Al films are experimentally studied for the formation of pore structure. Some conventional works on anodizing Al films are two-step process with polishing for thick Al films of several up to 10 μm. It takes long time to deposit the Al films at high cost for preparation. However, it is difficult to carry out electrochemical polish and two-step anodization for the Al thickness less than 1 μm. Therefore, the original surface morphology of as-deposited film plays an important role in pore shape and arrangement. The one-step anodization of Al films on Si substrate for porous AAO as well as pore widening is important for academic and practical application. For industry, the one-step anodization offers the advantages of low-cost and time-saving process. Using HPA can decrease the probability of failure in single-step anodization of Al film. Also, the HPA can enhance distribution of uniformity and pore circularity in short period anodization due to cooling effect and hydrogen ions attracted in negative period. In this section, the Al thin films were deposited by magnetron sputtering with different depositing parameters of target power and substrate bias in order to control the different morphology of films with thickness less than 700 nm. Then one-step HPA is used to fabricate the AAO templates. The evolution of morphology and microstructure of the sputtered Al thin films was studied and correlated to the sputtering parameter and synthesized AAO characteristics.

7.4.1 EXPERIMENTS OF DEPOSITION AND ANODIZATION OF AL FILMS ON SI SUBSTRATES

The boron-doped p-Si (100) substrates were initially cleaned in a solution of H_2SO_4 and H_2O_2 for the deposition of Al thin films. The Al films were deposited on p-Si (100) by magnetron sputtering in an argon atmosphere at room temperature. The power of 5 cm aluminum target (99.99% purity) was controlled at 50~150 W by DC power supply without or with substrate bias of 0~200 V during deposition. The argon flow rate was adjusted to 50~100 sccm by mass flow controller for maintain constant working pressure. The distance between the substrate and target was 100 mm. The base and working pressure were 0.27 mPa and 0.25 Pa, respectively. The deposition time was adjusted for 1, 2 and 3 hr to study its effect on Al thin films at room temperature. The thicknesses of all as-deposited Al thin films were measured by α-step profiler. The single-step anodization of Al thin films were performed in 0.3 M oxalic acid at 15 °C for 5 min. The applied normal potential was 40 V. In order to reduce joule heat, the hybrid pulse technology was applied for anodization process which was made of a normal positive voltage of 40 V followed by small negative voltage of -2 V with the duty ratio of 1s:1s, that is, the period of 2 s. Formation of AAO was performed by means of the potentiostat (Jiehan 5000, Taiwan) using the three-electrode electrochemical cell consisting of the platinum mesh as the counter electrode, the Al thin film as the working electrode and Ag/AgCl/3 M KCl as the reference electrode. After anodization, the specimens were immersed in 5 wt% phosphoric acid at RT for 40 minutes for the pore widening.

7.4.2 THE MICROSTRUCTURE AND MORPHOLOGY OF AS-DEPOSITED AL FILMS

In GIXRD result, the Al film of about 184 nm deposited at 50 W for 60 min is X-ray amorphous since no diffraction peak is detected (not shown in figure). It is also difficult to perform anodization for 5 min in this ultra-thin Al film. Therefore, the studies are focused on the other four Al films in Table 7.1. Figure 7.11 shows GIXRD patterns of four as-deposited Al thin films formed at the 100~150 W Al target power and 0~200 V substrate bias. Two distinct crystalline Al peaks denoted as

TABLE 7.1
A list of thickness, grain size and roughness (Ra) of all as-deposited Al thin films as a function of the Al target power and substrate bias

Al power (W)	Bias (V)	Deposited Time (min)	Thickness (nm)	Grain Size* (nm)	Ra (nm)
50	0	60	184	N/A**	5.67
100	0	60	552	21.0	10.65
100	100	60	519	20.7	6.00
100	200	60	446	20.1	5.27
150	0	60	694	25.6	12.52

* Estimated by GIXRD
** show amorphous in XRD

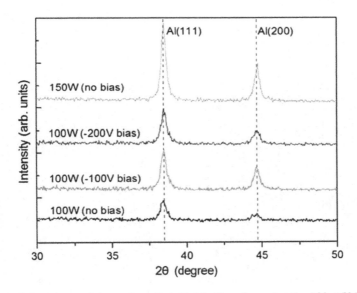

FIGURE 7.11 GIXRD patterns of the as-deposited Al thin films formed at the 100~150 W Al target power and 0~200 V substrate bias.

Al (111) at 38.4° and Al (200) at 44.7° are found for all the as-deposited Al thin film. It is noted that the peak of Al film deposited at 100 W with no bias shows slightly shift toward small angle. It may be attributed to residual stress of film which causes increase in atomic planar distance. Nevertheless, this little shift is unobvious and with limited effect in AAO formation. In addition, the intensity of GIXRD peaks is proportional to film thickness, crystallinity and grain size. The increase of peak intensity from 100 W to 150 W can be attributed to the increase of film thickness. But the Al film with bias added from 0 to -100 V shows slightly decrease in thickness but stronger diffraction intensity. When the bias is applied during sputtering process, it results in ion bombardment and increased energy or momentum of Al atoms. Therefore, suitable bias can enhance more Al crystals formation during the sputtering process. However, excess bias may also reduce film thickness due to Ar^+ bombardment which is known as resputtering effect [92]. In the case of -200 V bias, it shows slight decrease in diffraction peak intensity due to the reduced film thickness. The mean grain size of four crystalline Al films can be estimated according to the well-known Scherrer formula ($0.9\lambda/\beta cos\theta_b$), where β represents the full width at half maximum of XRD peak in radian, θ_b is 2θ of the peak in degrees, and λ is the wavelength of X-ray in Angstrom or nm) listed in table. 1. No obvious peak shift reveals the limited effect of residual stress on the evolution of grain size. As we keep the target power constant at 100 W and increase bias from 0 to 200 V, the mean grain size of Al films slightly decreases from 21.0 to 20.1 nm. It is reported that the grain size is roughly proportional to film thickness [93] so this unclear change may be induced by thickness reduction. On the other hand, the increased target power can substantially increase not only thickness from 552 to 694 nm but also the mean grain size from 21.0 to 25.6 nm.

Figure 7.12 shows two-dimensional AFM topography images of as-deposited Al films revealing the varied surface morphology on different sputtering conditions. The average roughness (Ra) of four Al thin films measured by AFM is in the range of several to ten nanometers as listed in Table 7.1. It indicates increase of bias is beneficial to deposit Al film with low roughness. In addition, higher power created much rough surface due to high deposition rate. It is noted that Al films deposited at 100 W and 200 V bias exhibit less fluctuation and the lowest Ra. It is probably caused by Ar+ bombardment. Also the significantly varied surface morphology is seen from SEM images.

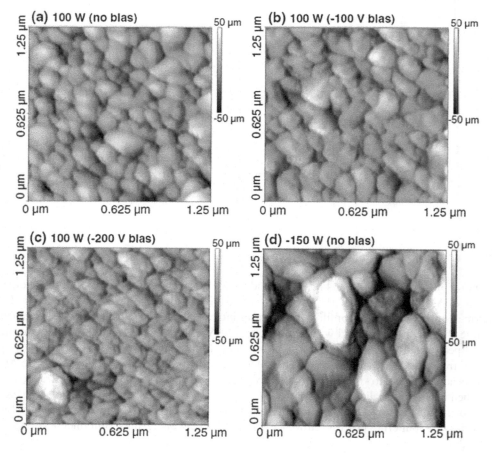

FIGURE 7.12 Two-dimensional AFM topography images of the Al thin film of the as-deposited Al thin films: (a) 100 W target power (no bias), (b) 100 W target power and −100 V bias, (c) 100 W target power and −200 V bias, (d) 150 W (no bias).

Growth, Characteristics and Application of Nanoporous AAO

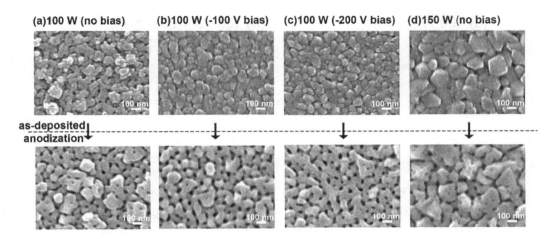

FIGURE 7.13 Top-view SEM micrographs of the as-deposited Al thin films and formed AAO structure: (a) 100 W target power (no bias), (b) 100 W target power and −100 V bias, (c) 100 W target power and −200 V bias, (d) 150 W (no bias).

7.4.3 The Pores Growth and Distribution Uniformity of AAO

Figures 7.13(a)–(d) show SEM micrographs for comparison of before and after anodization of four Al thin films deposited at various sputtering parameters. The formed nanoporous structures after anodizing Al thin films can be closely linked to the surface morphology of the as-deposited Al thin films. The Al thin film deposited at 100 W without substrate bias shows rough surface and many Al particles together with some extrusive ones on surface (Figure 7.13(a)). Consequently, it shows relatively disordered AAO pore arrangement after anodization. In contrast, the Al thin films at the same target power but with the applied bias of −100 V (Figure 7.13(b)) and −200 V (Figure 7.13(c)) show much smooth surface with dense compact particles on the top with much less extrusion. It is attributed to the ion bombardment effect for smoothening surface with reduced roughness. The AAO structure obtained from these two samples shows better pore arrangement and more uniform pore size. In the case of the Al thin film deposited at 150 W without substrate bias, the size of particles apparently increases with target power (Figure 7.13(d)) and the light-dark contrast clearly exists in the edge of many particles. The contrast evidences the large surface fluctuation and relatively high surface roughness. In brief, the particle size increases with increasing target power and reduces with substrate bias. The film with larger size of particles exhibits higher roughness. It is noted that the particle size is approximately proportional to grain size and the fine-grained film generally has smoother surface. After anodization, it reveals that the relatively high plane of particles on the surface of Al film shows very small pore or even no pore formation on it. In order to further investigate the effect of surface morphology on AAO structure, the diameter and amount of pores were calculated.

Figures 7.14(a)–(d) show the histograms of pore size distribution of AAO structures formed by anodizing Al thin films deposited at various sputtering parameters of 100 W/no bias, 100 W/100 V bias, 100 W/200 V bias and 150 W/no bias, respectively. The pore size distribution is manifestly related to surface morphology induced by target power and substrate bias. In the Al thin film deposited at 100 W/no bias, the pores of 20±5 nm in size occupy the main distribution and the calculated average pore size is 20.8 nm (Figure 7.14(a)). When the 100 V substrate bias is applied, the range of pores is shifted to 25±5 nm and the statistic pore size distribution is more uniform (from 20.80±6.82 nm to 24.70±6.76 nm), as shown in Figure 7.14(b). When the substrate bias is further increased to 200 V (Figure 7.14(c)), it shows the similar main distribution of 20~30 nm compared to that at 100

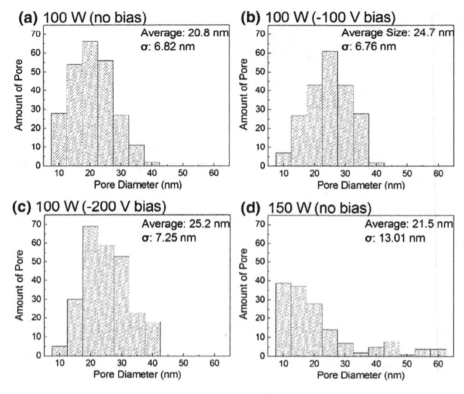

FIGURE 7.14 Pore size distribution of AAO structures formed by anodizing four Al thin films deposited at: (a) 100 W target power (no bias), (b) 100 W target power and 100 V bias, (c) 100 W target power and 200 V bias, (d) 150 W (no bias).

W/100 V bias, but increases the number of pores with diameters of 30~40 nm. The statistic pore distribution is 25.20±7.25 nm. The increase of pore size is probably attributed to local smooth area due to the lowest average roughness Ra of the as-deposited Al film. However, it results in a little increase of standard deviation σ. Furthermore, the Al thin film deposited at 150 W/no bias shows the increase of particle size but the non-uniform pore size distribution (21.50±13.01 nm with the largest standard deviation σ) caused by larger surface fluctuation (Figure 7.14(d)). It is noted that the enhanced pore distribution uniformity can be attributed to improvement of surface fluctuation or morphology. Increasing bias from 100 V to 200 V may increase pore size due to the lower roughness together with the decreased particle size. If the particle size decreases, more particle boundaries will form. Therefore, decreasing particle size in a specific range may be advantageous for pore arrangement and enlarge the pore. It is noted that the irregular-shape pores of anodized bulk Al foil occur in grain boundary. In this case, the small particle size with narrow particle boundary induces more uniform pore distribution together with the decreased irregular-shape pores.

The local uneven electric field distribution due to surface fluctuations is an important factor of affecting AAO synthesis during anodization. Figure 7.15(a) shows the typical Al film with smooth surface deposited by ion beam sputtering (IBS) in ultra-high vacuum and its porous structure after anodization. The surface morphology stays very smooth without any particles and fluctuation due to the vacuum condition. In this highly smooth Al film, the similar nanoporous AAO structure was formed as the anodization of well-polished bulk Al. Therefore, the AAO structure shows larger average pore size and good pore size distribution as 31.06±6.34 nm (standard deviation σ=6.34 nm) even

Growth, Characteristics and Application of Nanoporous AAO 183

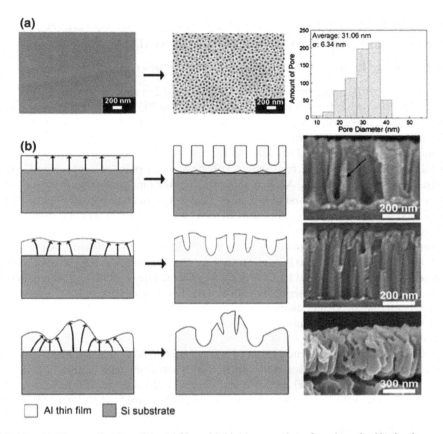

FIGURE 7.15 (a) The anodization of the Al film with highly smooth surface deposited by ion beam sputtering and its pore size distribution, (b) the schematic diagrams for explaining the AAO formation at Al thin films with various surface morphology. The black arrows in the scheme represent the direction of electric field.

in short anodization time. It reveals that surface fluctuations play an important role during anodization. Figure 7.15(b) shows the schematic diagrams and cross-sectional SEM images for explaining the AAO formation from Al thin films with various surface morphologies. The arrow stands for electric field direction. If the Al film is near completely flat, the electric field will be uniformly distributed on surface as potential difference is applied. Consequently, the oxidation and dissolution during AAO process is stable and then forms high uniform pores and straight nano-channels (as the top cross-section SEM image of AAO from the IBS Al film). If the Al films have surface fluctuations, the electrical field strength is concerned with the curvature of locations which may enhance chemical or field dissolution. This non-uniform electrical field causes the smaller pore size and uniformity in the limited anodization time. It is also found nano-channels are not well-perpendicular to the surface (as the middle cross-section SEM image of AAO from the Al film deposited at 100 W and 200 V bias). In a highly rough surface, the strongest electrical field occurs in the valley or cusp of surface fluctuations with high curvature. Thus the pores formed in the area could be enlarged due to stronger electric field. Meanwhile the pore size in top position decreases due to relatively weaker electric field. Also the curved nano-channels are formed due to the direction of electrical field affected by surface curvature (as the bottom cross-section SEM image of AAO from the Al film deposited at 150 W).

7.5 APPLICATIONS OF HPA AAO IN THE EFFECTIVE NANOPOROUS MEMBRANES FABRICATION AND PHOTOCATALYSIS

7.5.1 A LITERATURE REVIEW OF AAO MEMBRANE DETACHMENT METHOD

The AAO membrane with regular nanoporous array can be used in nanomaterial synthesis [7], filters [94] and as substrates for surface-enhanced Raman scattering [95]. Traditionally, AAO membrane separation mostly uses solution containing Hg^{2+} or Cu^{2+} to obtain AAO membranes by etching the aluminum substrate, which is the most common and earliest method. However, in order to avoid environmental pollution and material waste, many researchers proposed the improved methods, including voltage reduction method, reverse bias voltage method, pulse voltage method and double-layer anodization method to peel the AAO membrane. The AAO membrane detachment technology for non-heavy metal solutions was first proposed by Rigby et al. in Nature in 1989 [96]. The voltage reduction method was used to destroy the AAO barrier layer through the gradual drop in potential, which further promotes the AAO membrane detachment from the aluminum substrate. This research was very forward-looking when discussing the structure and principle of AAO, but it is relatively wasteful, and each step lacks a rule to follow. The reverse bias method [97–100] is a method of applying a negative voltage to achieve AAO membrane detachment. The H_2 bubbles generate between the aluminum substrate and the AAO barrier layer so that the hydrogen gas pushes to obtain a complete membrane. A complete AAO membrane can be acquired on the same substrate by this method. However, it still has several shortcomings. First, it is relatively time-consuming with the gradual increase of negative voltage during whole process, and the thickness must exceed a certain degree. In addition, because bubbles are generated between the barrier layer and the aluminum substrate, additional etching is required to achieve the through-hole structure. The double-layer anodization method is to grow a sacrificial layer of AAO and then etch the sacrificial layer with phosphoric acid or phosphochromic acid to obtain an AAO film. In 2015, Masuda et al. [101] proposed to use concentrated sulfuric acid as a sacrificial layer, so it can achieve the membrane detachment process. The method proposed by Masuda et al. has the advantage of being able to separate the complete film repeatedly. However, it takes more time of one more anodization and etching step and concentrated sulfuric acid may also be harmful to the environment. Among these methods, pulse voltage detachment [102–105] is the most efficient method for AAO membrane detachment. It can peel the thin film fast in one simple process. The $HClO_4$-C_2H_5OH solution used for detachment process is the same as the pre-treatment of AAO which can avoid the waste of the solution. The shortcoming of this method is the integrity of membrane and the repetition of process. Here, we will propose an improved voltage detachment method from AA1050 to separate the membrane completely.

7.5.2 ONE-TIME POTENTIOSTATIC METHOD FOR AAO MEMBRANE DETACHMENT

Figure 7.16 shows an optical photograph of detached AAO and Al substrate. The time required for the detachment process by DC 50 V in $HClO_4$-C_2H_5OH solution is just 20s. Compared with

FIGURE 7.16 Complete AAO membrane obtained by our improved voltage detachment method.

Growth, Characteristics and Application of Nanoporous AAO

TABLE 7.2
Comparison of AAO membrane detachment methods and results

Ref.	Method	Al purity	Detachment time	Photographs of complete membrane	Repetitions of AAO membrane
[99]	Reverse bias voltage method	99.999%	20 min	Yes	Yes
[98]	Reverse bias voltage method	99.999%	13 min	NA	NA
[100]	Reverse bias voltage method	99.999%	30–90 s	NA	Yes
[104]	Pulse voltage method	99.999%	3 s	NA	NA
[103]	Pulse voltage method	99.99%	3 s	NA	NA
[105]	Pulse voltage method	99.999%	3–60 s	NA	NA
[101]	Two-layer anodization method	99.999%	15 min	Yes	Yes
[106]	Two-layer anodization method	99.999%	75 min	Yes	Yes
Ours	One time potentiostatic method	Al 1050 alloy (~99.5%)	20 s	Yes	Yes

traditional pulse voltage method by 55 V for 3 s, we tried to lower the voltage to 50 V and prolonged the process time to 20 s, which can reduce the stress during detachment process. It can be seen that the membrane with a diameter of 2 cm is completely separated without defects or cracks.

Table 7.2 lists the comparison of the AAO membrane peeling reported in several journals and ours. The methods used can be divided into three categories: reverse bias voltage method, pulse voltage method and two-layer anodization method. It is noted that only our work successfully produced through-hole AAO membranes from low-purity aluminum. We proposed a fast, low-cost and green process to detach complete through-hole AAO membranes in one-step process repeatedly.

Figure 7.17 shows the SEM images of AAO with the top and bottom view of detached AAO membrane. The through-hole structure can be achieved by a single one-step process; additional etching process is not required. In order to achieve the multi-detachment of AAO membrane, electrochemical polishing can be applied to reduce the surface roughness of Al substrate to achieve multi-detachment, as shown in Figure 7.18.

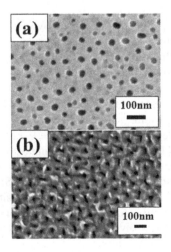

FIGURE 7.17 (a) The top view and (b) the bottom view of detached AAO membrane.

FIGURE 7.18 The 2nd repetition of complete AAO membrane is achieved by electrochemical polishing for reducing Al surface roughness.

7.5.3 A Literature Review of P25-TiO$_2$ Photocatalysis

Titanium dioxide material has intensively been applied in photocatalyst since the Honda–Fujishima effect was discovered in 1972 [107]. Furthermore, O'Regan and Grätzel proposed a high-efficiency dye-sensitized solar cell (DSSC) by using nanosized TiO$_2$ particles as oxide semiconductor [108]. Following these discoveries, titanium dioxide has attracted significant attention in recent decades. Numerous literatures have studied the microstructure of titanium dioxide for enhancing performance of photocatalyst or DSSC. For TiO$_2$ crystal, rutile phase is the most stable structure and is used as a dielectric gate material for MOSFET devices due to the high dielectric constant [109]. While a metastable structure of anatase phase was investigated as a superior structure for higher performance of photocatalysis and photovoltaics [110, 111]. In literatures, one of the accepted views is mixture of major anatase and minor rutile phase can further enhance the photocatalytic activity due to the charge separation by electron transfer from rutile to anatase [112, 113]. However, it is difficult to control specific phase composition by annealing due to the fact that the transition temperature from rutile to anatase is in a range of 700 to 900 °C. Degussa P-25 (a commercial TiO$_2$ powders) containing mixed phase with 78% anatase, 14% rutile and 8% amorphous [114] provides one of a feasible solutions for high photocatalytic activity. While the P-25 TiO$_2$ powder has been commonly used in various photocatalytic processes like water pollution purification, it still has the problem of non-reusable and requirement of additional separation process [115]. In general, the solid TiO$_2$ film coating on a substrate is suitable for reusable photocatalyst and can be fabricated by various technologies, such as sol–gel deposition, spray pyrolysis, pulsed laser deposition or sputtering deposition. However, coating P-25 TiO$_2$ powder encountered adhesion problem due to its nano-powder structure. To solve this problem, some inorganic binder should be added for immobilizing P25 nanoparticles to the substrate during deposition. For example, Chen et al. [116] synthesized macroporous TiO$_2$-P25 composite films by PEG modified sol–gel method and suggested that 500 °C is the calcination temperature for excellent adhesion between the films and stainless steel substrate. On the other hand, some substrates containing particular structure were used as catalyst supports for increasing specific surface area since the photocatalytic action only occurs at the surface of TiO$_2$ films. Yu et al. [117] prepared various nanostructured TiO$_2$ using sol–gel method and CNTs as a support. Rodriguez et al. [118] deposited the P-25 TiO$_2$ on a rough stainless steel created by heat and chemical surface treatment and a porous β-SiC for large surface area. Vargová et al. [119] used macroporous reticulated alumina with pore density of 10 and 15 pores per inch for dip coating P25 TiO$_2$ powder. Recently, AAO containing nanoporous structures can be obtained by anodizing Al in an appropriate electrolyte at low temperature has received increasing attention in recent decades for nanotechnology development. The pore diameter and distance of the porous AAO can be easily adjusted by applied potential during anodization. For photocatalysis application,

Growth, Characteristics and Application of Nanoporous AAO

the nanoporous AAO was used as a template for TiO$_2$ nanotube or nanowire fabrication [120, 121]. However, while these investigations have demonstrated the enhancement of photocatalysis performance by nanostructured TiO$_2$, less study has attempted to directly combine P25 TiO$_2$ powder and nanoporous AAO. In this section, the P25 TiO$_2$ powders is deposited on a nanoporous alumina by sol–gel method for enhancing photocatalytic performance.

7.5.4 Larger Specific Surface Area of 3D AAO for Enhancing P25-TiO$_2$ Photocatalysis

Rising target power from 50 W to 185 W during sputtering deposition results in an exponential increase of surface roughness due to stronger grain orientation of (111). Therefore, the growth of AAO pores on the extremely rough Al film creates a 3D AAO nanostructure. Figure 7.19 shows the side view SEM image with 45° oblique angle, it is clearly seen that the specific 3D nanoporous feature is caused by the extremely rough Al surface containing numerous hillocks. The overall AAO pore number and specific surface area gradually increase by the 3D surface.

According to literature, the etching rate of AAO in 5 wt% phosphoric acid at room temperature for pore widening is approximately 8 nm per hour [33]. In our experience, the pore widening time of 40 min is suitable for one-step anodization of Al film to obtain an intact AAO pore feature without any crack on the cell wall. However, intact AAO pores are not essential for catalysis application. It is worth to further increase pore widening time for creating a much rough surface. Figure 7.20 shows the comparison of 3D nanoporous AAO films with various following pore widening time. As rising pore widening from 40 to 60 min, an evident increase is seen in AAO pore size. Moreover, there are some visibly damaged cell walls in certain regions, as shown in Figure 7.20(b). In fact, the AAO structures by one-step anodization of Al film reveal less ordered than typical two-step anodization of Al foil. The disordered pore arrangement results in the variation in cell wall thickness and possibility of fracture in certain regions after pore widening. As increasing the pore widening time to 80 min, it reveals the much larger pore but cracked cell walls in all regions due to an over etching effect, as shown in Figure 7.20(c). The 80 min is the limited pore widening time from completely removed nanoporous AAO structure in this experiment condition. Comparing to pore widening time of 40 min, the over-etched porous AAO by 80 min pore widening has much larger surface area despite the cracked cell walls.

7.5.5 Photocatalytic Performance of TiO$_2$ P25 on Over-etched 3D AAO

Figure 7.21 shows the SEM images of AAO with planar, 3D, over-etched 3D nanoporous structure for deposition of TiO$_2$ P25 nanoparticles. In the case of planar porous AAO, TiO$_2$ particles are adsorbed on cell walls of each pore due to the van der Waals force. However, it is seen that less P25 TiO$_2$ nanoparticles are introduced to the pore channels, as shown in Figure 7.21(a). Notice

FIGURE 7.19 3D nanoporous AAO film (side view, 45° oblique angle). Each sample is performed following pore widening process for 40 min.

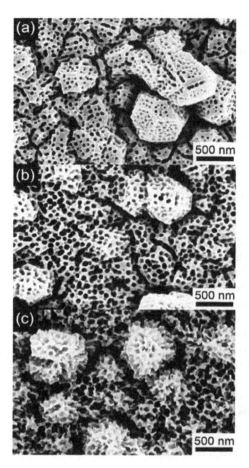

FIGURE 7.20 3D nanoporous AAO films with various following pore widening time: (a) 40 min. (b) 60 min and (c) 80 min.

that the nanoparticles in the solution during deposition are considered as Brownian motion. The random motion leads to difficulty in penetration of P25 TiO_2 nanoparticles in the single-aspect pore channel of planar AAO. By contrast, the 3D porous structure exhibits more TiO_2 particles adsorption not only on the surface but also in pore channels, as shown in Figure 7.21(b). The hillocks in surface containing lots of non-perpendicular pores increase the probability of introduction of P25 TiO_2 nanoparticles in Brownian motion. Moreover, the valleys in the 3D AAO structure are helpful to aggregate more TiO_2 nanoparticles compared to planar structure. Figure 7.21(c) shows the P25 TiO_2 nanoparticles deposited on over-etched porous AAO by 80 min pore widening. It is observed the most TiO_2 nanoparticles in this extremely rough substrate, as our expectancy. Comparing intact 3D AAO structure, the over-etched porous AAO has much larger pore size and surface area to accommodate and adsorb more P25 TiO_2 nanoparticles, respectively. It is believed that the P25 TiO_2 nanoparticles depositedin the nanoporous pore channels are much stable than on the surface.

Figure 7.22(a) demonstrates the adhesion improvement of P25 TiO_2 by using over-etched 3D AAO as support material. In the case of as-deposition, the phtocatalysis results show similar decomposition trend whether in pure Si wafer or AAO (over-etched 3D). Each sample exhibits good photocatalytic performance by completely decomposing MB after 12-hour UV irradiation.

Growth, Characteristics and Application of Nanoporous AAO 189

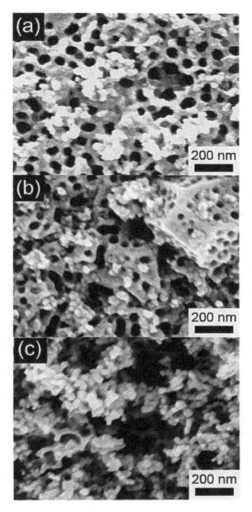

FIGURE 7.21 Deposition of P25 TiO2 nanoparticles on (a) planar nanoporous AAO film, (b) 3D nanoporous AAO film and (c) over-etched 3D nanoporous AAO film.

However, it is seen that the P25 TiO$_2$ deposited on pure Si wafer shows dramaticdecrease in photocatalytic performance after water flushing. After 20-hour UV irradiation, the concentration of MB maintains in 82.3%. It indicates that desorption of P25 TiO$_2$ from supporting Si wafer occurred due to the poor adhesion. By contrast, the concentration of MB decreases to 22.6% after 20-hour UV irradiation. This result confirms the adhesion of P25 TiO$_2$ and AAO. Figure 7.22(b) shows comparison of photocatalytic performance of TiO$_2$ P25 on various supporting materials including Si wafer, 2D AAO, 3D AAO and over-etched 3D AAO after water flushing. The results demonstrate that the over-etched 3D AAO has the best photocatalytic performance toward degradation of MB due to the largest surface area for catalytic reaction and adsorption of TiO$_2$. In addition, the 2D AAO exhibits medium photocatalytic performance with MB concentration of 55.7% after 20-hour photodegradation. This implies that the P25 TiO$_2$ which adsorbed on 2D AAO surface slightly desorbed after to water flushing. However, the porous structure can keep most of P25 TiO$_2$ in pore comparing to flat surface of Si wafer.

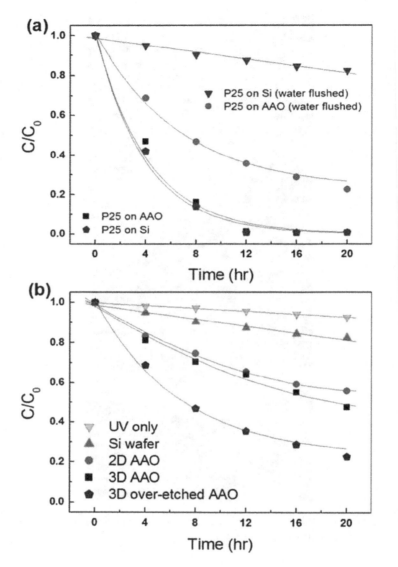

FIGURE 7.22 (a) Comparison of TiO2 P25 nanoparticles on AAO (over-etched 3D structure) and Si for photocatalytic performance and adhesion. The adhesion problem was examined by water flushing. (b) Photocatalytic performance of TiO2 P25 nanoparticles on various supporting materials after water flushing.

7.6 CONCLUSIONS

In this chapter, the HPA with normal-positive and small negative voltages is proposed for AAO fabrication at relatively high temperatures of 15–25°C for promoting AAO structure in comparison with conventional AAO synthesized using two-step potentiostatic method of DCA at low temperature (0–10 °C) to avoid dissolution effect. The growth of nanoporous AAO in bulk Al foils and Al films containing different microstructure surface morphology are discussed. By comparing DCA and HPA of bulk Al, the negative period with nearly zero cathodic current can provide effective cooling for promoting AAO quality. The positive hydrogen ions attracted to the surface lead to the dissolution of the top of AAO. It can improve the rough and distorted AAO structure by one-step

Growth, Characteristics and Application of Nanoporous AAO   191

DCA. The advantages of effective cooling and extra dissolution in HPA are helpful to fabricate AAO by one-step anodization of Al film at room temperature. The effect of surface morphology and microstructure of the sputtered Al thin films on Si substrate for pore growth and uniformity of AAO nanostructure is studied. The surface morphology is greatly concerned with the target power and substrate bias. Increasing target power from 100 to 150 W at no bias leads to the increased thickness of Al film from 552 to 694 nm and the mean grain size of films increases from 21.0 to 25.6 nm. In addition, increasing substrate bias from 0 to 200 V at the constant 100 W power may reduce the thickness from 552 to 446 nm and decrease the mean grain size from 21.0 to 20.1 nm due to ion bombardment effect. The variation in mean grain size of Al films induces change of particle size and surface morphology which affect the pore arrangement and distribution of AAO nanostructure. The investigation is beneficial for improving pore structure regularity and performance enhancement in one-step anodization of Al films on the Si substrates in future applications.

A single one-step DCA method improved from voltage pulse method to detach the AAO film from AA1050 is proposed in this book. Compared with other AAO membrane peeling methods, the process is simple, fast and low cost. Furthermore, it achieves the repetition of the AAO membrane detachment on the same substrate by electropolishing process for reducing surface roughness and defects. By depositing P25 TiO_2 powders on an over-etched 3D nanoporous AAO, the common adhesion problem can be improved. After water flushing, the over-etched 3D nanoporous AAO maintains the ~80% photocatalytic performance of as-deposited sample. Moreover, the over-etched 3D nanoporous AAO shows the largest surface area for catalytic reaction and adsorption of TiO_2 comparing to intact 2D and 3D AAO.

REFERENCES

[1] M. Long, S. Peng, J. Chen, X. Yang and W. Deng, "A new replication method for fabricating hierarchical polymer surfaces with robust superhydrophobicity and highly improved oleophobicity", *Colloids and Surfaces A: Physicochemical and Engineering Aspects* 507, pp. 7–17, 2016.

[2] B. S. Kim, H. J. Kim, S. An, S. Chi, J. Kim, J. Lee, "Micro- and nano-porous surface patterns prepared by surface-confined directional melt crystallization of solvent", *Journal of Crystal Growth* 469, pp. 184–190, 2017.

[3] N. A. Ogurtsov, Y. V. Noskov, K. Y. Fatyeyeva, V. G. Ilyin, G. V. Dudarenko, and A. A. Pud, "Deep impact of the template on molecular weight, structure, and oxidation state of the formed polyaniline", *The Journal of Physical Chemistry B* 117, pp. 5306–5314, 2013.

[4] X. Huang, H. Mutlu and P. Théato, "The toolbox of porous anodic aluminum oxide-based nanocomposites: from preparation to application", *Colloid and Polymer Science*, doi:10.1007/s00396-020-04734-0.

[5] C. L. Liu and H. L. Chen, "Crystal orientation of PEO confined within the nanorod templated by AAO nanochannels", *Soft Matter* 14, 5461, 2018.

[6] D. Wu, D. Zhang, Y. Ye, L. Ma, B. Minhas, B. Liu, H. A. Terryn, J. M. C. Mol, X. Li, "Durable lubricant-infused anodic aluminum oxide surfaces with high-aspect ratio nanochannels", *Chemical Engineering Journal* 368, pp. 138–147, 2019.

[7] S. H. Jeong, H. L. Im, S. Hong, H. Park, J. Baek, D. H. Park, S. Kim & Y. K. Hong, "Massive, eco-friendly, and facile fabrication of multi-functional anodic aluminum oxides: application to nanoporous templates and sensing platforms," *RSC Advances* 7, pp. 4518–4530, 2017.

[8] G. D. Sulka, A. Brzózka, L. Zaraska, E. Wierzbicka, & A. Brudzisz, "AAO templates with different patterns and channel shapes," *Submicron Porous Materials*, P. Bettotti (ed.), Springer, Italy, ch. 5, pp. 107–156, 2017.

[9] K. T. Lin, Y. J. Chen, M. R. Huang, V. K. Karapala, J. H. Ho and J. T. Chen, "Light-induced nanowetting: erasable and rewritable polymer nanoarrays via solid-to-liquid transitions", *Nano Letters* 20, pp. 5853–5859, 2020.

[10] M. Norek, M. Wlodarski, "Morphological and chemical characterization of highly ordered conical-pore anodic alumina prepared by multistep citric acid anodizing and chemical etching process," *Journal of Porous Materials* 25, pp. 45–53, 2018.

[11] Z. Zhou and S. S. Nonnenmann, "Progress in nanoporous templates: beyond anodic aluminum oxide and towards functional complex materials", *Materials* 12, 2535, 2019.

[12] C. C. Chang, F. C. Chiang, S. M. Chen, K. Thangavelu, and H. J. Yang, "Studies on electrochemical oxidation of aluminum and dyeing in various additives towards industrial applications," *International Journal of Electrochemical Science* 11, pp. 2142–2152, 2016.

[13] C. K. Chung, D. Dhandapani, C. J. Syu, M. W. Liao, B. Y. Chu, E. H. Kuo, "Role of oxalate anions on the evolution of widened pore diameter and characteristics of room-temperature anodic aluminum oxide," *Journal of the Electrochemical Society* 166, pp. 121–127, 2019.

[14] X. Zheng, Q. Wang, J. J. Luan, Y. Li, N. Wang, R. Zhang, "Angle-dependent structural colors in a nanoscale-grating photonic crystal fabricated by reverse nanoimprint technology," *Beilstein Journal of Nanotechnology* 10, 1211–1216, 2019.

[15] S. W. Chen, O. K. Khor, M. W. Liao and C. K. Chung, "Sensitivity evolution and enhancement mechanism of porous anodic aluminum oxide humidity sensor using magnetic field", *Sensors and Actuators B: Chemical* 199, pp. 384–388, 2014.

[16] C. K. Chung, O. K. Khor, E. H. Kuo and C. A. Ku, "Total effective surface area principle for enhancement of capacitive humidity sensor of thick-film nanoporous alumina", *Materials Letters*, 260, 126921, 2020.

[17] N. T. Nesbitt, M. J. Burns, & M. J. Naughton, "Facile fabrication and formation mechanism of aluminum nanowire arrays", *Nanotechnology* 31, 095301, 2019.

[18] S. Y. Zhang, Q. Xu, Z. J. Wang, Y. X. Ren, R. J. Yan, W. J. Ma, & J. L. Zhu, "The effect of propylene glycol on the optical properties of iridescent porous anodic alumina films," *Journal of Porous Materials* 25, pp. 1213–1217, 2018.

[19] C. K. Chung, M. W. Liao, E. H. Kuo and Z. W. Wang, "Enhancement of photocatalytic performance of P25-TiO$_2$ nanoparticles by 3D nanoporous anodic alumina at room temperature", *International Journal of Applied Ceramic Technology* 15, pp. 438–447, 2018.

[20] K. T. Tu and C. K. Chung, "Enhancement of surface Raman spectroscopy performance by silver nanoparticles on resin nanorods arrays from anodic aluminum oxide template", *Journal of the Electrochemical Society* 164, B3081–B3086, 2017.

[21] H. Masuda and K. Fukuda, "Ordered metal nanohole arrays made by a two-step replication of honeycomb structures of anodic alumina," *Science* 268, pp. 1466–1468, 1995.

[22] A. P. Li, F. Muller, A. Birner, K. Nielsch, and U. Gosele, "Hexagonal pore arrays with a 50-420 nm interpore distance formed by self-organization in anodic alumina," *Journal of Applied Physics*, vol. 84, pp. 6023–6026, 1998.

[23] K. B. Kim, B. C. Kim, S. J. Ha, & M. W. Cho, "Effect of pre-treatment polishing on fabrication of anodic aluminum oxide using commercial aluminum alloy," *Journal of Mechanical Science and Technology*, vol. 31, pp. 4387–4393, 2017.

[24] W. Lee, R. Ji, U. Goesele, and K. Nielsch, "Fast fabrication of long-range ordered porous alumina membranes by hard anodization," *Nature Materials*, vol. 5, pp. 741–747, 2006.

[25] J. P. O'Sullivan and G. C. Wood, "The morphology and mechanism of formation of porous anodic films on aluminum," *Proceedings of the Royal Society of London. Series A, Mathematical and Physical Sciences*, vol. 317, pp. 511–543, 1970.

[26] S. G. Yang, H. Zhu, D. L. Yu, Z. Q. Jin, S. L. Tang, and Y. W. Du, "Preparation and magnetic property of Fe nanowire array," *Journal of Magnetism and Magnetic Materials*, vol. 222, pp. 97–100, 2000.

[27] X. Y. Wang, X. Y. Wang, W. G. Huang, P. J. Sebastian, and S. Gamboa, "Sol-gel template synthesis of highly ordered MnO2 nanowire arrays," *Journal of Power Sources*, vol. 140, pp. 211–215, 2005.

[28] S. Rahman and H. Yang, "Nanopillar arrays of glassy carbon by anodic aluminum oxide nanoporous templates," *Nano Letters*, vol. 3, pp. 439–442, 2003.

[29] A. Bai, C. Hu, Y. Yang, and C. Lin, "Pore diameter control of anodic aluminum oxide with ordered array of nanopores," *Electrochimica Acta*, vol. 53, pp. 2258–2264, 2008.

[30] L. Zaraska, G. D. Sulka, J. Szeremeta, and M. Jaskula, "Porous anodic alumina formed by anodization of aluminum alloy (AA1050) and high purity aluminum," *Electrochimica Acta*, vol. 55, pp. 4377–4386, 2010.

[31] Y. Bae, J. Yu, Y. Jung, D. Lee, & D. Choi, "Cost-effective and high-throughput plasmonic interference coupled nanostructures by using quasi-uniform anodic aluminum oxide," *Coatings*, vol. 9, pp. 420, 2019.

[32] C. Hun, Y. J. Chiu, Z. Luo, C. Chen, & S. Chen, "A new technique for batch production of tubular anodic aluminum oxide films for filtering applications," *Applied Sciences*, vol. 8, pp. 1055, 2018.

Growth, Characteristics and Application of Nanoporous AAO

[33] D. Crouse, Y. H. Lo, A. E. Miller, and M. Crouse, "Self-ordered pore structure of anodized aluminum on silicon and pattern transfer," *Applied Physics Letters*, vol. 76, pp. 49–51, 2000.

[34] A. L. Cai, H. Y. Zhang, H. Hua, and Z. B. Zhang, "Direct formation of self-assembled nanoporous aluminium oxide on SiO2 and Si substrates," *Nanotechnology*, vol. 13, pp. 627–630, 2002.

[35] S. Inoue, S. Z. Chu, K. Wada, D. Li, and H. Haneda, "New roots to formation of nanostructures on glass surface through anodic oxidation of sputtered aluminum," *Science and Technology of Advanced Materials*, vol. 4, pp. 269–276, 2003.

[36] W. H. Yu, G. T. Fei, X. M. Chen, F. H. Xue, and X. J. Xu, "Influence of defects on the ordering degree of nanopores made from anodic aluminum oxide," *Physics Letters A*, vol. 350, pp. 392–395, 2006.

[37] A. F. Feil, M. V. da Costa, L. Amaral, S. R. Teixeira, P. Migowski, J. Dupont, G. Machado, and S. B. Peripolli, "The influence of aluminum grain size on alumina nanoporous structure," *Journal of Applied Physics*, vol. 107, p. 026103, 2010.

[38] A. F. Feil, P. Migowski, J. Dupont, L. Amaral, and S. R. Teixeira,"Nanoporous aluminum oxide thin films on si substrate: structural changes as a function of interfacial stress," *Journal of Physical Chemistry C*, vol. 115, pp. 7621–7627, 2011.

[39] V. C. Nettikaden, A. Baron-Wiechec, P. Bailey, T. C. Q. Noakes, P. Skeldon, and G. E. Thompson, "Formation of barrier-type anodic films on sputtering-deposited Al-Ti alloys," *Corrosion Science*, vol. 52, pp. 3717–3724, 2010.

[40] V. C. Nettikaden, H. Liu, P. Skeldon, and G. E. Thompson, "Porous anodic film formation on Al-Ti alloys in sulphuric acid," *Corrosion Science*, vol. 57, pp. 49–55, 2012.

[41] M. T. Wu, I. C. Leu, and M. H. Hon, "Anodization behavior of Al film on Si substrate with different interlayers for preparing Si-based nanoporous alumina template," *Journal of Materials Research*, vol. 19, pp. 888–895, 2004.

[42] M. S. Sander and L. S. Tan, "Nanoparticle arrays on surfaces fabricated using anodic alumina films as templates," *Advanced Functional Materials*, vol. 13, pp. 393–397, 2003.

[43] M. J. Kim, J. H. Choi, J. B. Park, S. K. Kim, J. B. Yoo, and C. Y. Park, "Growth characteristics of carbon nanotubes via aluminum nanopore template on Si substrate using PECVD," *Thin Solid Films*, vol. 435, pp. 312–317, 2003.

[44] N. V. Myung, J. Lim, J. P. Fleurial, M. Yun, W. West, and D. Choi, "Alumina nanotemplate fabrication on silicon substrate," *Nanotechnology*, vol. 15, pp. 833–838, 2004.

[45] S. K. Hwang, J. Lee, S. H. Jeong, P. S. Lee, and K. H. Lee, "Fabrication of carbon nanotube emitters in an anodic aluminium oxide nanotemplate on a Si wafer by multi-step anodization," *Nanotechnology*, vol. 16, pp. 850–858, 2005.

[46] G.-Y. Zhao, C.-L. Xu, D.-J. Guo, H. Li, and H.-L. Li, "Template preparation of Pt-Ru and Pt nanowire array electrodes on a Ti/Si substrate for methanol electro-oxidation," *Journal of Power Sources*, vol. 162, pp. 492–496, 2006.

[47] T. Shimizu, M. Nagayanagi, T. Ishida, O. Sakata, T. Oku, H. Sakaue, T. Takahagi, and S. Shingubara, "Epitaxial growth of Cu nanodot arrays using an AAO template on a Si substrate," *Electrochemical and Solid-State Letters*, vol. 9, pp. J13–J16, 2006.

[48] M. Kokonou, A. G. Nassiopoulou, K. P. Giannakopoulos, A. Travlos, T. Stoica, and S. Kennou, "Growth and characterization of high density stoichiometric SiO2 dot arrays on Si through an anodic porous alumina template," *Nanotechnology*, vol. 17, pp. 2146–2151, 2006.

[49] C.-J. Yang, S.-M. Wang, S.-W. Liang, Y.-H. Chang, C. Chen, and J.-M. Shieh, "Low-temperature growth of ZnO nanorods in anodic aluminum oxide on Si substrate by atomic layer deposition," *Applied Physics Letters*, vol. 90, p. 033104, 2007.

[50] N. Tasaltin, S. Ozturk, N. Kilinc, H. Yuezer, and Z. Z. Ozturk, "Simple fabrication of hexagonally well-ordered AAO template on silicon substrate in two dimensions," *Applied Physics A: Materials Science and Processing*, vol. 95, pp. 781–787, 2009.

[51] C.-H. Lai, C. W. Chang, and T. Y. Tseng, "Size-dependent field-emission characteristics of ZnO nanowires grown by porous anodic aluminum oxide templates assistance," *Thin Solid Films*, vol. 518, pp. 7283–7286, 2010.

[52] X. Jin, Y. Hu, Y. Wang, R. Shen, Y. Ye, L. Wu, and S. Wang, "Template-based synthesis of Ni nanorods on silicon substrate," *Applied Surface Science*, vol. 258, pp. 2977–2981, 2012.

[53] T. S. Kustandi, W. W. Loh, H. Gao, and H. Y. Low, "Wafer-scale near-perfect ordered porous alumina on substrates by step and flash imprint lithography," *ACS Nano*, vol. 4, pp. 2561–2568, 2010.

[54] A. S. M. Chong, L. K. Tan, J. Deng, and H. Gao, "Soft imprinting: creating highly ordered porous anodic alumina templates on substrates for nanofabrication," *Advanced Functional Materials*, vol. 17, pp. 1629–1635, 2007.

[55] H. Oshima, H. Kikuchi, H. Nakao, K.-i. Itoh, T. Kamimura, T. Morikawa, K. Matsumoto, T. Umada, H. Tamura, K. Nishio, and H. Masuda, "Detecting dynamic signals of ideally ordered nanohole patterned disk media fabricated using nanoimprint lithography," *Applied Physics Letters*, vol. 91, p. 022508, 2007.

[56] D. C. Leitao, A. Apolinario, C. T. Sousa, J. Ventura, J. B. Sousa, M. Vazquez, and J. P. Araujo, "Nanoscale topography: a tool to enhance pore order and pore size distribution in anodic aluminum oxide," *Journal of Physical Chemistry C*, vol. 115, pp. 8567–8572, 2011.

[57] H. Asoh, M. Matsuo, M. Yoshihama, and S. Ono, "Transfer of nanoporous pattern of anodic porous alumina into Si substrate," *Applied Physics Letters*, vol. 83, pp. 4408–4410, 2003.

[58] L. Pu, Y. Shi, J. M. Zhu, X. M. Bao, R. Zhang, and Y. D. Zheng, "Electrochemical lithography: fabrication of nanoscale Si tips by porous anodization of Al/Si wafer," *Chemical Communications*, pp. 942–943, 2004.

[59] A. Oide, H. Asoh, and S. Ono, "Natural lithography of si surfaces using localized anodization and subsequent chemical etching," *Electrochemical and Solid-State Letters*, vol. 8, pp. G172–G175, 2005.

[60] H. Sai, H. Fujii, K. Arafune, Y. Ohshita, M. Yamaguchi, Y. Kanamori, and H. Yugami, "Antireflective subwavelength structures on crystalline Si fabricated using directly formed anodic porous alumina masks," *Applied Physics Letters*, vol. 88, p. 201116, 2006.

[61] O. Rabin, P. R. Herz, Y. M. Lin, A. I. Akinwande, S. B. Cronin, and M. S. Dresselhaus, "Formation of thick porous anodic alumina films and nanowire arrays on silicon wafers and glass," *Advanced Functional Materials*, vol. 13, pp. 631–638, 2003.

[62] J. S. Choi, G. Sauer, P. Goring, K. Nielsch, R. B. Wehrspohn, and U. Gosele, "Monodisperse metal nanowire arrays on Si by integration of template synthesis with silicon technology," *Journal of Materials Chemistry*, vol. 13, pp. 1100–1103, 2003.

[63] Y. F. Mei, X. L. Wu, T. Qiu, X. F. Shao, G. G. Siu, and P. K. Chu, "Anodizing process of Al films on Si substrates for forming alumina templates with short-distance ordered 25 nm nanopores," *Thin Solid Films*, vol. 492, pp. 66–70, 2005.

[64] P. Skeldon, G. E. Thompson, S. J. Garcia-Vergara, L. Iglesias-Rubianes, and C. E. Blanco-Pinzon, "A tracer study of porous anodic alumina," *Electrochemical and Solid-State Letters*, vol. 9, pp. B47–B51, 2006.

[65] K. R. Hebert and J. E. Houser, "A model for coupled electrical migration and stress-driven transport in anodic oxide films," *Journal of The Electrochemical Society*, vol. 156, pp. C275–C281, 2009.

[66] J. E. Houser and K. R. Hebert, "The role of viscous flow of oxide in the growth of self-ordered porous anodic alumina films," *Nature Materials*, vol. 8, pp. 415–420, 2009.

[67] G. D. Sulka, S. Stroobants, V. V. Moshchalkov, G. Borghs, and J. P. Celis, "Effect of tensile stress on growth of self-organized nanostructures on anodized aluminum," *Journal of The Electrochemical Society*, vol. 151, pp. B260–B264, 2004.

[68] R. Koch, "The intrinsic stress of polycrystalline and epitaxial thin metal-films," *Journal of Physics. Condensed Matter*, vol. 6, pp. 9519–9550, 1994.

[69] S. J. Garcia-Vergara, P. Skeldon, G. E. Thompson, and H. Habazaki, "Stress generated porosity in anodic alumina formed in sulphuric acid electrolyte," *Corrosion Science*, vol. 49, pp. 3772–3782, 2007.

[70] M. W. Liao and C. K. Chung, "The role and effect of residual stress on pore generation during anodization of aluminium thin films", *Corrosion Science* 74, pp. 232–239, 2013.

[71] I. Vrublevsky, V. Parkoun, J. Schreckenbach, and W. A. Goedel, "Dissolution behaviour of the barrier layer of porous oxide films on aluminum formed in phosphoric acid studied by a re-anodizing technique," *Applied Surface Science*, vol. 252, pp. 5100–5108, 2006.

[72] A. Han and Y. Qiao, "Effects of nanopore size on properties of modified inner surfaces," *Langmuir*, vol. 23, pp. 11396–11398, 2007.

[73] I. De Graeve, H. Terryn, and G. E. Thompson, "Influence of local heat development on film thickness for anodizing aluminum in sulfuric acid," *Journal of The Electrochemical Society*, vol. 150, pp. B158–B165, 2003.

[74] H. Takahashi, M. Nagayama, and H. Alahori and A. Kitahara, "Electron-microscopy of porous anodic oxide films on Al by ultrathin sectioning technique", *Journal of Electron Microscopy* 22, pp. 149–157, 1973.

[75] K. Yokoyama, H. Konno, H. Takahashi, and M. Nagayama, "Advantages of pulse anodizing," *Plating and Surface Finishing*, vol. 69, pp. 62–65, 1982.

[76] K. Okubo, D. Toba, and Y. Sakura, "Galvanostatic method with pulse reverse current for high-temperature high speed anodic oxidation", *Japan Journal of Metal Finishing Society*, vol. 39, pp. 512–516, 1988.

[77] K. Okubo, S. Suyama, and Y. Sakura, "Studies of microstructure of anodic oxide films on aluminum by pulse current with a negative component", *Japan Journal of Metal Finishing Society*, vol. 40, pp. 1366–1371, 1989.

[78] W. Lee, R. Scholz, and U. Gösele, "A continuous process for structurally well-defined Al2O3 Nanotubes based on pulse anodization of aluminum," *Nano Letters*, vol. 8, pp. 2155–2160, 2008.

[79] W. Lee and J.-C. Kim, "Highly ordered porous alumina with tailor-made pore structures fabricated by pulse anodization," *Nanotechnology*, vol. 21, p. 485304, 2010.

[80] D. Losic and M. Lillo, "Porous alumina with shaped pore geometries and complex pore architectures fabricated by cyclic anodization," *Small*, vol. 5, pp. 1392–1397, 2009.

[81] C. K. Chung, R. X. Zhou, T. Y. Liu, & W. T. Chang, "Hybrid pulse anodization for the fabrication of porous anodic alumina films from commercial purity (99%) aluminum at room temperature, "*Nanotechnology*, vol. 20, pp. 055301, 2009.

[82] C. K. Chung, W. T. Chang, M. W. Liao, H. C. Chang, & C. T. Lee, "Fabrication of enhanced anodic aluminum oxide performance at room temperatures using hybrid pulse anodization with effective cooling, "*Electrochimica Acta*, vol. 56, pp. 6489–6497, 2011.

[83] C. K. Chung, W. T. Chang, M. W. Liao, & H. C. Chang, "Effect of pulse voltage and aluminum purity on the characteristics of anodic aluminum oxide using hybrid pulse anodization at room temperature, "*Thin Solid Films*, vol. 519, pp. 4754–4758, 2011.

[84] C. K. Chung, M. W. Liao, H. C. Chang, & C. T. Lee, "Effects of temperature and voltage mode on nanoporous anodic aluminum oxide films by one-step anodization, "*Thin Solid Films*, vol. 520, pp. 1554–1558, 2011.

[85] H. B. Michaelson, *CRC Handbook of Chemistry and Physics*, D. R. Lide, Ed, 73 ed., CRC Press, Boca Raton, 1992.

[86] M. A. Kashi and A. Ramazani, "The effect of temperature and concentration on the self-organized pore formation in anodic alumina," *Journal of Physics D: Applied Physics*, vol. 38, pp. 2396–2399, 2005.

[87] G. D. Sulka and W. J. Stępniowski, "Structural features of self-organized nanopore arrays formed by anodization of aluminum in oxalic acid at relatively high temperatures," *Electrochimica Acta*, vol. 54, pp. 3683–3691, 2009.

[88] S. Wernick, R. Pinner, and P. G. Sheasby, *The Surface Treatment and Finishing of Aluminum and its Alloys*, 5 ed. Finshing Publications, United States, 1987.

[89] A. O. Araoyinbo, M. N. A. Fauzi, S. Sreekantan, and A. Aziz, "One-step anodization of aluminum at room temperature," *Sains Malaysiana*, vol. 38, pp. 521–524, 2009.

[90] G. D. Sulka, A. Brzózka, L. Zaraska, and M. Jaskuła, "Through-hole membranes of nanoporous alumina formed by anodizing in oxalic acid and their applications in fabrication of nanowire arrays," *Electrochimica Acta*, vol. 55, pp. 4368–4376, 2010.

[91] Y. Wen Bin and T. Xiao Hong, "One-step anodization preparation and photoluminescence property of anodic aluminum oxide with nanopore arrays," *Materials Science Forum*, vol. 663, pp. 272–275, 2010.

[92] S. I. Shah and P. F. Carcia, "Superconductivity and resputtering effects in rf-sputtered YBa2Cu3O7-x thin-films," *Applied Physics Letters*, vol. 51, pp. 2146–2148, 1987.

[93] Y. Choi and S. Suresh, "Size effects on the mechanical properties of thin polycrystalline metal films on substrates," *Acta Materialia*, vol. 50, pp. 1881–1893, 2002.

[94] F. Y. Wen, P. S. Chen, T. W. Liao, & Y. J. Juang, "Microwell-assisted filtration with anodic aluminum oxide membrane for Raman analysis of algal cells," *Algal Research*, vol. 33, pp. 412–418, 2018.

[95] C. H. Huang, H. Y. Lin, S. Chen, C. Y. Liu, H. C. Chui, & Y. Tzeng, "Electrochemically fabricated self-aligned 2-D silver/alumina arrays as reliable SERS sensors," *Optics Express* vol. 19, pp. 11441–11450, 2011.

[96] R. C. Furneaux, W. R. Rigby, & A. P. Davidson, "The formation of controlled-porosity membranes from anodically oxidized aluminium.," *Nature*, vol. 337, pp. 147, 1989.

[97] J. J. Schneider, J. Engstler, K. P. Budna, C. Teichert, & S. Franzka, Freestanding, highly flexible, large area, nanoporous alumina membranes with complete through-hole pore morphology," *European Journal of Inorganic Chemistry*, vol. 2005, pp. 2352–2359, 2005.

[98] M. Tian, S. Xu, J. Wang, N. Kumar, E. Wertz, Q. Li, P. M. Campbell, M. H. W. Chan & T. E. Mallouk, "Penetrating the oxide barrier in situ and separating freestanding porous anodic alumina films in one step," *Nano Letters*, vol. 5, pp. 697–703, 2005.

[99] Y. K. Hong, B. H. Kim, D. I. Kim, D. H. Park, & J. Joo, "High-yield and environment-minded fabrication of nanoporous anodic aluminum oxide templates," *RSC Advances*, vol. 5, pp. 26872–26877, 2015.

[100] E. Choudhary, & V. Szalai, "Two-step cycle for producing multiple anodic aluminum oxide (AAO) films with increasing long-range order," *RSC Advances*, vol. 6, pp. 67992–67996, 2016.

[101] T. Yanagishita, & H. Masuda, "High-throughput fabrication process for highly ordered through-hole porous alumina membranes using two-layer anodization," *Electrochimica Acta*, vol. 184, pp. 80–85, 2015.

[102] H. D. L. Lira, & R. Paterson, "New and modified anodic alumina membranes: part III. Preparation and characterisation by gas diffusion of 5 nm pore size anodic alumina membranes," *Journal of Membrane Science*, vol. 206, pp. 375–387, 2002.

[103] J. H. Yuan, W. Chen, R. J. Hui, Y. L. Hu, & X. H. Xia, "Mechanism of one-step voltage pulse detachment of porous anodic alumina membranes," *Electrochimica Acta*, vol. 51, pp. 4589–4595, 2006.

[104] S. Zhao, K. Chan, A. Yelon, & T. Veres, "Preparation of open-through anodized aluminium oxide films with a clean method," *Nanotechnology*, vol. 18, pp. 245304, 2007.

[105] A. Brudzisz, A. Brzózka, & G. D. Sulka, "Effect of processing parameters on pore opening and mechanism of voltage pulse detachment of nanoporous anodic alumina," *Electrochimica Acta*, vol. 178, pp. 374–384, 2015.

[106] T. Zhang, Z. Ling, Y. Li, & X. Hu, "A new method for highly efficient fabrication of through-hole porous anodic alumina membranes," *ECS Journal of Solid State Science and Technology*, vol. 6, pp. 862–865, 2017.

[107] A. Fujishima and K. Honda, "Electrochemical photolysis of water at a semiconductor electrode," *Nature*, vol. 238, pp. 37, 1972.

[108] B. Oregan and M. Gratzel, "A low-cost, high-efficiency solar-cell based on dye-sensitized colloidal $TiO2$ films," *Nature*, vol. 353, pp. 737–740, 1991.

[109] O. Carp, C. L. Huisman, and A. Reller, "Photoinduced reactivity of titanium dioxide," *Progress in Solid State Chemistry*, vol. 32, pp. 33–177, 2004.

[110] A. L. Linsebigler, G. Q. Lu, and J. T. Yates, "Photocatalysis on Tio2 surfaces - principles, mechanisms, and selected results," *Chemical Reviews*, vol. 95, pp. 735–758, 1995.

[111] N. G. Park, J. van de Lagemaat, and A. J. Frank, "Comparison of dye-sensitized rutile- and anatase-based $TiO2$ solar cells," *Journal of Physical Chemistry B*, vol. 104, pp. 8989–8994, 2000.

[112] D. Gong, C. A. Grimes, O. K. Varghese, W. C. Hu, R. S. Singh, Z. Chen, and E. C. Dickey, "Titanium oxide nanotube arrays prepared by anodic oxidation," *Journal of Materials Research*, vol. 16, pp. 3331–3334, 2001.

[113] D. C. Hurum, A. G. Agrios, K. A. Gray, T. Rajh, and M. C. Thurnauer, "Explaining the enhanced photocatalytic activity of Degussa P25 mixed-phase $TiO2$ using EPR," *Journal of Physical Chemistry B*, vol. 107, pp. 4545–4549, 2003.

[114] B. Ohtani, O. O. Prieto-Mahaney, D. Li, and R. Abe, "What is Degussa (Evonik) P25? Crystalline composition analysis, reconstruction from isolated pure particles and photocatalytic activity test," *Journal of Photochemistry and Photobiology a-Chemistry*, vol. 216, pp. 179–182, 2010.

[115] S. R. Patil, B. H. Hameed, A. S. Skapin, and U. L. Stangar, "Alternate coating and porosity as dependent factors for the photocatalytic activity of sol-gel derived $TiO2$ films," *Chemical Engineering Journal*, vol. 174, pp. 190–198, 2011.

[116] Y. Chen and D. D. Dionysiou, "A comparative study on physicochemical properties and photocatalytic behavior of macroporous $TiO2$-P25 composite films and macroporous $TiO2$ films coated on stainless steel substrate," *Applied Catalysis A: General*, vol. 317, pp. 129–137, 2007.

[117] Y. Yu, J. C. Yu, J. G. Yu, Y. C. Kwok, Y. K. Che, J. C. Zhao, L. Ding, W. K. Ge, and P. K. Wong, "Enhancement of photocatalytic activity of mesoporous $TiO2$ by using carbon nanotubes," *Applied Catalysis A: General*, vol. 289, pp. 186–196, 2005.

[118] P. Rodriguez, V. Meille, S. Pallier, and M. A. Al Sawah, "Deposition and characterisation of $TiO2$ coatings on various supports for structured (photo)catalytic reactors," *Applied Catalysis A: General*, vol. 360, pp. 154–162, 2009.

[119] M. Vargova, G. Plesch, U. F. Vogt, M. Zahoran, M. Gorbar, and K. Jesenak, "TiO2 thick films supported on reticulated macroporous Al2O3 foams and their photoactivity in phenol mineralization," *Applied Surface Science*, vol. 257, pp. 4678–4684, 2011.

[120] H. Imai, Y. Takei, K. Shimizu, M. Matsuda, and H. Hirashima, "Direct preparation of anatase TiO2 nanotubes in porous alumina membranes," *Journal of Materials Chemistry*, vol. 9, pp. 2971–2972, 1999.

[121] Y. Lin, "Photocatalytic activity of TiO2 nanowire arrays," *Materials Letters*, vol. 62, pp. 1246–1248, 2008.

8 Multifunctional Superhydrophobic Nanocomposite Surface
Theory, Design, and Applications

Yongquan Qing and Changsheng Liu
Northeastern University, China

CONTENTS

8.1 Introduction...200
8.2 Theoretical Model of Superhydrophobic Nanocomposite Surface................200
    8.2.1 Young Equation 200
    8.2.2 Wenzel Model 202
    8.2.3 Cassie-Baxter Model 203
    8.2.4 Cassie-Wenzel Wetting Regime Transition...204
    8.2.5 Superhydrophobic Stability of Hierarchical Structure 205
8.3 Design of Functional Superhydrophobic Nanocomposite Surface................207
    8.3.1 Zero-dimensional Structural Unit to Construct Two-dimensional Superhydrophobic Nanocomposite Surface................207
    8.3.2 Zero-dimensional Structural Unit to Construct Three-dimensional Superhydrophobic Nanocomposite Surface 208
    8.3.3 One-dimensional Structural Unit to Construct Two-dimensional Superhydrophobic Nanocomposite Surface................209
    8.3.4 One-dimensional Structural Unit to Construct Three-dimensional Superhydrophobic Nanocomposite Surface................210
    8.3.5 Porous Bulk Structural Unit to Construct Three-dimensional Superhydrophobic Nanocomposite Surface................210
8.4 Application of Functional Superhydrophobic Nanocomposite Surface 212
    8.4.1 Anti-Icing..212
    8.4.2 Condensation 213
    8.4.3 Oil/Water Separation...213
    8.4.4 Superamphiphobicity...214
    8.4.5 Corrosion Resistance 215
    8.4.6 Wettability Conversion 215
    8.4.7 Other Application...216
8.5 Summary...217
Acknowledgments...217
References...218

8.1 INTRODUCTION

Tens of thousands of organisms, after several hundred million years of evolution, have demonstrated the incomparable excellent characteristics while adapting to the ecological environment. For example, lotus leaves come out of "rising unsullied from mud", geckos walk on walls, pitcher plants prey on insects, spider silk collects water, butterfly wing induces anisotropic of water droplets, and so on [1–7]. It is expected to prepare new materials with special functions by imitating the unique structure and functional principle of biology, which can make up for the shortcomings of traditional materials and promote the continuous progress of human society. Among them, superhydrophobic surfaces have experienced rapid development for more than 20 years (1996-), due to offer exciting promise for self-cleaning, anti-icing, anti-fog, anti-contamination, anti-sticking, corrosion resistance, drag reduction, and oil–water separation, etc., and are widely used in defense, aviation, industry, agriculture, and daily life. For example, superhydrophobic surfaces are used for ship shells and fuel storage tanks to achieve anti-fouling and anti-corrosion effects [8, 9]; used for outdoor antennas and aircraft surfaces with anti-ice/snow, ensuring high-quality signal reception and reducing energy consumption and cost [10, 11]; used in microfluidic pipelines, reducing fluid transport resistance and fluid viscosity loss [12, 13]; used in water vehicles or underwater nuclear submarines, reducing water resistance, improving travel speed, and saving energy [14, 15]; used to modify textiles, making into special clothing for waterproof and anti-fouling [16, 17].

To date, the theoretical model and preparation of superhydrophobic surfaces have obtained great progress, and a series of biomimetic superhydrophobic surfaces have been developed. Scientists have established basic wetting models and equations through unremitting exploration of the relationship between the structure of the animals and plants surface and the special wettability. With the extension of application domain, it is no longer limited to the preparation of a single superhydrophobic surface, but more hoped to prepare a functional superhydrophobic surface by constructing a specific wetting model and synthesizing new surface modifiers. Currently, the research of functionalized superhydrophobic surface is in full swing, showing unlimited charm and business opportunities in design, synthesis, and engineering.

8.2 THEORETICAL MODEL OF SUPERHYDROPHOBIC NANOCOMPOSITE SURFACE

Wettability is a very important feature of solid surface, which refers to the ability of liquid to replace gas to spread on solid surface [18]. Generally, the surface wettability is mainly determined by its free energy and roughness [19, 20]. Its wettability can be artificially controlled by changing and adjusting the chemical composition, micro or macro geometric structure of solid surface. Surface free energy (surface tension, γ) shows a physical quantity between the surface state of solids and liquids and is the theoretical basis for studying surface wettability [21]. γ refers to the pulling force that causes the molecules on the surface to contract toward the interior due to the interaction between the internal molecules of a solid or liquid, which is equal to the work done per unit area under constant temperature and pressure. γ of water is 0.0727 N/m; γ greater than 0.1 mN/m is defined as high-energy surface. Normally, inorganic solid (such as metal, ceramic, etc.) surface easily wetted by water belongs to high-energy surface, while the solid surface with not easily wetted and its γ less than 0.1 mN/m is defined as low-energy surface (such as organic solid, polymer surface, etc.). Theoretically, it can be judged whether the wetting process is spontaneous according to the data of γ, but not that can be measured directly in experiments. In practical applications, static contact angle (CA) is usually used to describe wetting state of liquid on solid surfaces.

8.2.1 Young Equation

Due to the influence of the surface tension of the droplet, the angle between the horizontal line of the solid–liquid contact interface and the tangent line of the liquid–gas interface on the droplet is

Multifunctional Superhydrophobic

the static CA, which is mainly caused by the difference of the interfacial tension among the solid, liquid, and gas phases. Generally, according to the CA of water droplets on different solid surfaces, the surface wettability can be divided into the following four categories: when the CA is less than 90°, the water droplets can infiltrate the solid, which is called hydrophilic surface; when the CA is less than 10°, the water droplets spread easily on the solid surface, which is called superhydrophilic surface; when the CA is greater than 90°, the water droplet is not easy to infiltrate the solid, which is called hydrophobic surface; when the CA is greater than 150°, the water droplet is perfectly spherical shape on the solid surface, which is called superhydrophobic surface.

However, the dynamic process of water droplets should also be considered in accurately judging the hydrophobic effect of a surface, which can be measured by sliding angle (SA); that is, only when the water CA greater than 150° and SA is less than 10°, it can be called a superhydrophobic surface in the real sense. Besides, the SA and CA hysteresis describe the wetting behavior of a droplet when placed on an inclined solid surface. SA refers to the critical inclination angle when the droplet is on the inclined surface and is about to roll down from the inclined surface under the action of gravity.

If a droplet is placed still on the solid surface and the droplet is continuously added with a dropper, when the contact line of the droplet on the solid surface just begins to move outward, the CA of the droplet is called the advancing angle (θ_a). On the contrary, the static droplet is drawn continuously, and when the contact line of the droplet on the surface begins to recede inward, the CA of the droplet is called the receding angle (θ_r). The difference of the hysteresis CA of the droplet ($\Delta\theta = \theta_a - \theta_r$) is called CA hysteresis [22]. The hysteresis of a droplet is explained from the view-point of thermodynamics: each state has a corresponding CA when a droplet is still on the solid surface and exists in many metastable or stable states. Among them, for the stable-state CA, the θ_a and θ_r are the maximum and minimum energy metastable state, respectively, and the difference between the two energies is called the energy resistance. For the droplet to slide, the metastable energy barrier must be overcome.

Furmidge et al. [23] proposed the relation equation between the SA (α) and the CA hysteresis when the droplet moves on the surface:

$$mg \cdot \sin\alpha = w \cdot \gamma_{lv}\left(\cos\theta_r - \cos\theta_a\right) \tag{8.1}$$

where m and w are the mass and diameter of the droplet, respectively; and γ_{lv} is the liquid–gas interfacial tension. Hereinafter, the SA can be used to measure the hysteresis of the CA. The smaller the SA is, the stronger the ability of the liquid to roll off the solid surface is and the smaller the hysteresis is. The equation shows that the smaller the hysteresis of CA on the solid surface is, the smaller the SA is. The result is that the smaller the CA hysteresis and SA, the smaller the adhesion of the solid surface to the liquid or the stronger the ability of the liquid to roll off such surface, which is more conducive to the self-cleaning performance.

The superhydrophobic function of the substrate surface is caused by the gas film formed by the rough surface immersed in air. Scientists have proposed the following theories to explain how roughness should be calculated and expressed. The essence of wetting phenomenon is that a certain liquid wets the solid surface, which is caused by the interaction between the interfacial tension of solid, liquid, and gas phases. For a flat, uniform, and smooth solid surface, when the liquid, solid, and gas phases reach the equilibrium state, $dU_\delta = 0$.

$$\delta U_\sigma = \gamma_{lv}\delta A_{LV} + \left(\gamma_{sl} - \gamma_{sv}\right)\delta A_{SL} = \left[\gamma_{lv}\cos\theta + \left(\gamma_{sl} - \gamma_{sv}\right)\right]\delta A_{SL} \tag{8.2}$$

$$\delta A_{LV} = \cos\theta \delta A_{SL} \tag{8.3}$$

In 1805, the British scientist Thomas Young [24] first proposed the equilibrium relationship between the static CA of droplets and the γ on an absolutely smooth solid surface, namely the classical Young's equation:

$$\cos\theta = \left(\gamma_{sv} - \gamma_{sl}\right)/\gamma_{lv} \tag{8.4}$$

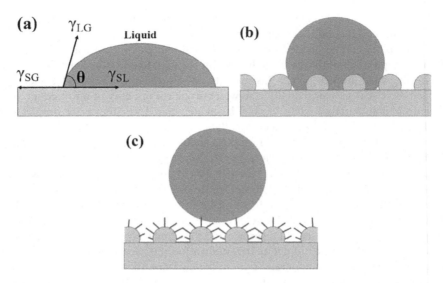

FIGURE 8.1 Wetting phenomenon of droplets on different solid surfaces. (a) Young model, (b) Wenzel model, and (c) Cassie-Baxter model.

where, A_{LV} and A_{SL} represent the contact area between liquid–gas and solid–liquid, respectively; θ represents the intrinsic static CA or equilibrium static CA of solid surface; γ_{sv}, γ_{sl}, and γ_{lv} represent the γ at solid–gas, solid–liquid and liquid–gas interface, respectively. Due to the attraction between molecules, the total energy of the system tends to be minimized to reach equilibrium; the equilibrium relationship is shown in Figure 8.1 (a).

According to the Young's equation, the surface free energy is reduced by changing the chemical composition of the solid surface (i.e. surface chemical modification), and the corresponding surface intrinsic CA is increased, which is beneficial to improve the surface hydrophobicity. It has been reported that fluorine-containing compounds can be obtained by replacing hydrogen atoms in polymer hydrocarbons with fluorine atoms, which can effectively reduce the surface free energy. Generally, the more hydrogen atoms are replaced, the lower the surface free energy is. Therefore, fluorine-containing compounds become the ideal choice to modify the surface and reduce the surface free energy to achieve superhydrophobicity. However, the superhydrophobic surface cannot be obtained on the smooth surface only by reducing the surface free energy, even the solid surface modified by –CF_3 group (6.7 mJ/m^2) with the lowest surface energy, the static CA is only about 120° [25].

8.2.2 Wenzel Model

Young's equation is only suitable for the ideal smooth surface. However, the substrate surface is not completely smooth in practical application, its surface chemical composition is inconsistent or has a certain rough structure, and often shows the phenomenon of CA hysteresis, which leads to the inapplicability of Young's equation. In 1936, the American scientist Wenzel extended the Young's equation for the rough surface, and first proposed the wetting model of a liquid on the rough solid surface, namely Wenzel model [26, 27].

Wenzel model assumes that liquid on rough surfaces can be completely submerged and filled with grooves on exposed surfaces, forming a uniform solid–liquid interface below the liquid (Figure 8.1 (b)). The main reason is to increase the solid surface area by increasing the surface roughness so as to realize the enhancement of hydrophobicity in the form of geometric series. CA of droplets on uniform interface is expressed by Wenzel's equation:

$$\cos\theta^* = r\cos\theta = r(\gamma_{sv} - \gamma_{sl})/\gamma_{lv} \tag{8.5}$$

Multifunctional Superhydrophobic 203

where θ^* is apparent CA when water droplets completely wetted rough surfaces. r represents rough factor, which is the ratio of actual surface area and projection area between liquid and substrate ($r > 1$), and θ is the intrinsic CA when the droplet is assumed to be on an ideal smooth surface. According to Wenzel theory, if $\theta < 90°$ and $\cos\theta > 0$, the change of $\cos\theta^*$ leads to the decrease of θ^* correspondingly when the roughness factor becomes larger, the hydrophilic surface becomes more hydrophilic with the increase of surface roughness, so as to achieve superhydrophilicity; if $\theta > 90°$ and $\cos\theta < 0$, $\cos\theta^*$ decreases when the roughness factor becomes larger, which makes θ^* rise correspondingly, the hydrophobicity of the hydrophobic surface enhances with the increase of surface roughness, which has a great influence on the surface wettability.

8.2.3 CASSIE-BAXTER MODEL

The Wenzel's equation can explain the relationship between apparent CA, intrinsic CA, and roughness on rough solid surface. However, the Wenzel's equation cannot explain the existence and mode of droplets on surfaces with minimal SA. This is because the droplets embedded on the solid surface must overcome a series of energy barriers to move or roll, which is inconsistent with the phenomenon of droplet rolling on the superhydrophobic surface of lotus effect.

In 1944, Cassie and Baxter further improved the Wenzel model and proposed the existence of composite CA on the solid surface [28]. It is assumed that the droplets do not penetrate into the rough structures of the solid surface, but suspend above its structure, so that a part of air is trapped at grooves on rough surfaces, and a composite solid–liquid–gas interface (Cassie interface) is formed under the droplet, which is composed of solid–liquid and liquid–gas interface.

The wettability of rough solid surface in Cassie wetting state is described by classical Cassie-Baxter equation for arbitrary two-component surface, as follows:

$$\cos\theta_c = f_1 \cos\theta_1 + f_2 \cos\theta_2 \tag{8.6}$$

where, θ_c is angle of liquid droplet completely suspended on rough surface (apparent CA); θ_1 and θ_2 are the intrinsic CAs of the two media on the solid surface, respectively; and f_1 and f_2 are the area fractions of water droplets on the surfaces of the two media ($f_1 + f_2 = 1$), respectively. Since the air is completely hydrophobic, θ_2 is equal to $180°$ and can be expressed as:

$$\cos\theta_c = f_1(\cos\theta_1 + 1) - 1 \tag{8.7}$$

Compared with Wenzel model, the equation of Cassie-Baxter (Figure 8.1 (c)) can more accurately represent the real surface system and well explain the wetting state of the "air cushion" surface with porous structure. From this equation, it can be seen that changing the solid surface roughness is beneficial to increase the gas–liquid contact area fraction on the interface during solid–liquid contact to obtain the superhydrophobic surface. The Cassie-Baxter model is suitable for both superhydrophobic and superhydrophilic conditions, which makes up for the shortcomings of Wenzel model.

In practice, it is very difficult to accurately measure f_1. The intrinsic CA and apparent CA are often measured by CA measuring instrument and then calculated. Since the solid surface represented by f_1 is generally not smooth, the r of wetted part in the solid–liquid contact area should be considered. For the gas–solid composite interface, the Cassie-Baxter equation of droplet in metastable state is as follows:

$$\cos\theta_c = f_1(r\cos\theta_1 + 1) - 1 \tag{8.8}$$

The equation shows that surface wettability is determined by surface microstructure and chemical composition. When $f_1 = 1$, the trapped air in groove is filled with liquid, the wetting state is transformed to Wenzel state.

In summary, the three formulas for predicting the CA are all exploring the relationship between the surface rough structure and the CA, which is expressed as the influence of the interfacial tension between the liquid/solid/gas three phases on the CA. In short, the interfacial tension between liquid/gas and liquid/solid should be as large as possible to meet the requirements of superhydrophobic surface formation, and the interfacial tension between liquid/solid should be greater than that of gas/solid so that the CA area of droplets on the substrate surface can be minimized, thus reducing the adsorption force between the droplets and the substrate, and suspending the droplet on the substrate surface.

Generally, the conical or needle-like rough structure is the most ideal structure for the construction of superhydrophobic surface, and the surface contacted by water droplets is required to reach more than 100 needle points to meet the requirements of surface superhydrophobicity. Unfortunately, this structure is not only difficult to achieve in preparation but also easy to be damaged under action of external mechanical forces (such as slightly touch or abrasion), resulting in a great reduction in superhydrophobicity, mainly attribute to its fragile and unstable structure.

8.2.4 CASSIE-WENZEL WETTING REGIME TRANSITION

The transition mechanism between Wenzel and Cassie-Baxter states is essential to achieve a stable composite interface. Although the energy barrier prevents the transition from Cassie to Wenzel states [29, 30], the Cassie-Wenzel wetting transition of droplets on rough surface can be induced by external stimuli, such as pressure [31], vibration [32], voltage [33] and light [34], which also occurs spontaneously during droplet evaporation [35]. However, this reverse wetting state transition (i.e. the Wenzel-Cassie transition) has never been discovered. Lafuma and QuéRé [36] suggested that the Wenzel-Cassie transition can occur only when the net surface energy of Cassie-Bacter and Wenzel state is equal, or when the predicted CA values of the two states are consistent. Later that year, they also found that this transition would not occur under the favorable conditions of free energy and considered that the wetting state of Cassie-Baxter was in a metastable state. Extrand [37] thought that the effect of droplets weight caused Cassie-Wenzel transition and proposed a contact line density model. Based on this model, this transition occurs when the droplet weight is greater than the surface tension corresponding to the three-phase boundary.

But other scholars have put forward different perspectives. Patankar [38] believed that the wetting state depends on the formation process of droplets. Quéré [39] pointed out that this transition can be changed by adjusting the curvature of the droplet (determined by the difference between the internal and external pressure of the droplet). Nosonovsky and Bhushan [40] considered the wetting transition to be a dynamic unstable process and proposed the influencing parameters of the unstable process. Some researchers also believed that the multi-scale roughness curvature determined the stability of the Cassie-Baxter wetting state [41, 42]. So far, many experimental results support these experimental methods, but it is not clear which conversion mechanism occupies the main position.

Subsequently, Nosonovsky and Bhushan [43] proposed a more detailed theory from the perspective of free energy. Due to the influence of adhesion hysteresis, the energy required for wetting process is less than that for non-wetting, so wetting and dewetting is not a symmetrical relationship. Adhesion hysteresis not only causes hysteresis in CA, but also hysteresis in the wetting state of Wenzel and Cassie-Baxter. Whether the droplet is in the Cassie state or the Wenzel state in contact with the rough surface can be described in Figure 8.2.

When the net surface energies of the two states are equal, there exists a critical CA (θ_0), which is expressed as $\cos\theta_0 = (f_1-1)/(r-f_1)$. Since $r > 1 > f_1$, $\cos\theta_0$ ranges from -1 to 0, that is, $\theta_0 > 90°$. When $90° < \theta_0 < \theta^*$, the air layer wrapped under the droplet can exist stably so that the droplet is in the most stable Cassie-Baxter state and there is no transition to the Wenzel state, and the compound contact is more conducive to the reduction of local energy. When $90° < \theta^* < \theta_0$, the air layer under the droplet cannot exist stably so that the droplet is in the most stable Wenzel state, resulting in extremely unstable Cassie-Baxter state, and the wetting contact is more conducive to the reduction

Multifunctional Superhydrophobic

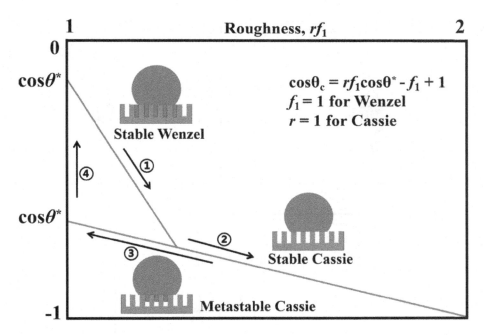

FIGURE 8.2 Transition between Cassie state and Wenzel state.

of local energy. Therefore, θ_0 must be reduced as much as possible in order to realize a stable self-cleaning superhydrophobic surface.

The transition process between Wenzel state and Cassie state can be better explained by analyzing the distribution of surface energy in the whole system. As shown in Figure 8.3(a), the CA is plotted by surface energy when a droplet is on a smooth solid surface. It is found that whether the droplet is in Wenzel state or Cassie state, only one minimum peak can be obtained on the curve, corresponding to the most stable intrinsic static CA. For Figure 8.3(b), the shape of the droplet cannot be treated as an ideal spherical or hemispherical when the microscale rough structure is introduced. At this time, the energy curve corresponds to many minimum energy peaks at the micro convex position, namely the droplet is in a metastable state on the rough surface, and the advancing CA and receding CA correspond to the maximum and minimum CA on the energy distribution, respectively.

Based on the above results, the curve relationship between surface energy and CA cannot explain the reason for the transition between the Cassie state and the Wenzel state, due to the fact that energy distribution of the two states is completely independent. In order to further study the transition between Cassie-Wenzel states, it is necessary to consider the relationship between surface energy and rough surface geometry. Assuming that the vertical average height of the gas–liquid interface under the droplet is h, the surface energy, h, CA and the shape of the droplet in different states are plotted, as shown in Figure 8.3(c). The lower and higher equilibrium lines in the graph correspond to Wenzel state and Cassie state, respectively. It can be seen from the figure that the state of droplet is mainly determined by the relationship between surface energy and h, and the energy of transition from Wenzel-Cassie state is much higher than that from Cassie-Wenzel state, which also explains the reason why Wenzel-Cassie state transition cannot be realized in the experiment.

8.2.5 Superhydrophobic Stability of Hierarchical Structure

Hierarchical structures play a vital role in inducing and maintaining stable superhydrophobic surfaces and self-cleaning surfaces. Both rough structure and inhomogeneity of solid surface can cause CA hysteresis phenomenon. If the surface roughness is accurately controlled at the molecular size

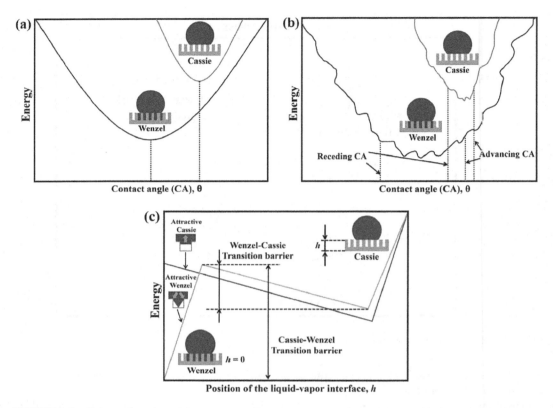

FIGURE 8.3 Schematics of net free-energy profiles. (a) Macroscale description, (b) microscale description with multiple energy minimums due to surface texture, and (c) origin of the two branches is found when a dependence of energy upon ***h*** is considered for the microscale description (blue line) and nanoscale imperfections (red line).

level, the CA hysteresis can be greatly reduced or even completely eliminated. The SA of water droplets corresponding to the solid surface with smaller CA hysteresis will also be smaller [44].

The following relationship between CA hysteresis and surface roughness of composite surface can be expressed as follows [45, 46]:

$$\theta_a - \theta_r = f_1 r \frac{\cos\theta_{a0} - \cos\theta_{r0}}{-\sin\theta} = \sqrt{f_1} r \frac{\cos\theta_{r0} - \cos\theta_{a0}}{\sqrt{2(r\cos\theta_0 + 1)}} \tag{8.9}$$

where, θ_{a0} and θ_{r0} are the advancing and receding angles of the smooth surface, respectively. For uniform interfaces, $f_1=1$; whereas for composite interfaces, f_1 is a smaller value. It can be inferred from the formula that the CA hysteresis will increase with the increase of roughness value for uniform interface, while the lower CA hysteresis and larger CA can be obtained by precisely controlling f_1 value for composite interface. Therefore, the composite interface is more conducive to the realization of superhydrophobicity and self-cleaning performance.

Composite interface is a multi-dimensional phenomenon whose formation depends on relative size of droplets and rough structures. Usually, the composite interface is unstable, and it will irreversibly change into a relatively stable uniform interface and the superhydrophobicity of interface is destroyed under the interference of external factors. Therefore, in order to obtain a stable composite interface, some unstable factors should be weakened or eliminated as much as possible, such as nano-droplet condensation, capillary fluctuation, and surface unevenness. Because the formation of

Multifunctional Superhydrophobic   207

the liquid–gas interface depends on the ratio of the distance between the adjacent protruding structures and the droplet diameter, while the nano-rough structure surface can obtain a smaller value of "f_1", so it is the best structure to increase the area fraction of the liquid–gas interface. In addition, the micro- or nano-hierachical structure can produce a large surface roughness and a higher r value can be obtained, and the nano-sized bulge structure can adhere to the droplets, thus preventing the droplets from entering the gap between the bulges [47].

In addition, Nosononvsky and Bhushan [48, 49] calculated and analyzed the energy changes of the system and found that the micro/nano-hierarchical structure of the solid surface can effectively adhere to the droplets, improving the stability of the composite interface. They also pointed out that the curvature and weight of the droplet are also factors affecting the interface transition, and demonstrated that the micro/nano rough structure surface can not only prevent the instability of the composite interface but also prevent the formation of gaps between the convex structures. Furthermore, the effect of roughness on the surface wettability and the instability mechanism of composite interface have size effect. This size effect can be effectively limited or reduced by constructing multi-scale rough interface on solid surface, which is more conducive to the maintenance of hydrophobicity/superhydrophobicity.

8.3 DESIGN OF FUNCTIONAL SUPERHYDROPHOBIC NANOCOMPOSITE SURFACE

According to the wetting theoretical model, there are two main ways for constructing functional superhydrophobic surfaces: one is to construct a rough structure on the intrinsic hydrophobic surface, and the other is to modify the hydrophilic rough surface with low surface energy substances. In fact, since the CA on smooth hydrophobic surface are lower than 120°, increasing the surface roughness is a key factor for superhydrophobic surface [25]. Up to now, the commonly used methods to produce rough surfaces include sol-gel [50], layer-by-layer self-assembly [51], etching [52], electrochemical deposition [53], phase separation [54], dip-coating [55], templating [56], chemical vapor deposition [57], electrospinning [58], and so on. In actual application, functional superhydrophobic surfaces due to different work environments and modes are divided into two-dimensional (films or coatings) and three-dimensional (bulk) superhydrophobic surfaces, which are mainly composed of structural units from zero-dimensional micro/nanoparticles or one-dimensional micro/nano fibers. Herein, the design strategies of superhydrophobic surface are classified into the following five categories based on the dimensional difference of materials constructed by structural units.

8.3.1 ZERO-DIMENSIONAL STRUCTURAL UNIT TO CONSTRUCT TWO-DIMENSIONAL SUPERHYDROPHOBIC NANOCOMPOSITE SURFACE

Nano- or/and microscale particles with zero dimensional structural units can be stacked to construct two-dimensional superhydrophobic surface (Figure 8.4), which has both rough structure and low surface energy after modification. Currently, there are many strategies that are being practiced to construct two-dimensional superhydrophobic surface. For example, transparent superhydrophobic coatings were prepared on glass, polycarbonate, and polymethylmethacrylate by modified SiO_2, ZnO, and ITO nanoparticles with low surface energy chemicals [59]. Similarly, superhydrophobic nano-patterned surfaces were constructed by micro/nano diatomite or nano-TiO_2 particles modified by polydimethylsiloxane (PDMS) [60]. A breakthrough was achieved by Lu et al. [61], who fabricated a robust self-cleaning surface by combining superhydrophobic nano-medium and adhesives. Such surface with sandwich-like structure is composed of superhydrophobic coating-adhesive-substrate, in which the adhesive forms as the bonding layer between the coating and substrate, effectively improving its mechanical abrasion.

Rough structure surface can maintain superhydrophobicity, mainly due to the effect of nanoscale structures. Verho et al. [62] proposed that the robustness of superhydrophobic surfaces with only

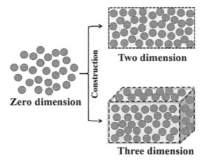

FIGURE 8.4 Zero-dimensional structural unit to construct two- or three-dimensional superhydrophobic nanocomposite surface.

single nano- or microstructures mainly depend on its own mechanical properties. Once the its surface nanostructures are damaged by abrasion, the droplets on the surface changed from Cassie state to Wenzel state and permanently loses its superhydrophobicity. In practical application, the robust superhydrophobic surfaces are prepared by mechanical compounding of polymer and micro/nanoparticles. The existence of polymer medium can effectively bond the micro/nanoparticles together, which helps to improve the adhesion between micro/nanoparticles and substrate surface. For example, Peng et al. [63] prepared an all-organic superhydrophobic nanocomposite coating by using the synergistic effect between polytetrafluoroethylene, fluorinated epoxy resin, and perfluoropolyether. The coating exhibits strong mechanochemical robustness under cyclic tape peels and Taber abrasion, and sustains exposure to highly corrosive media (namely, aqua regia and sodium hydroxide solution). In addition, the excellent mechanical flexibility of the coating enables to resist impalement during the impact of water jet at 35 m/s and can be applied to various substrates by spraying or brushing. Wang et al. [64] constructed superhydrophobic coatings on glass substrate by using fluorinated ethylene resin as matrix and perfluorosilane-modified SiO_2 nanoparticles as fillers. Then, the superhydrophobic coating on the substrate was demolded to form superhydrophobic film material which could resist at least 70 abrasion cycles with sandpaper.

Nano-polymers with high mechanical strength and good adhesion are selected as coating components, which can enhance both intrinsic strength and adhesion between coating and substrate. For example, Milionis et al. [65] prepared superhydrophobic nano coating by compounding acrylonitrile butadiene styrene rubber resin and hydrophobic SiO_2 nanoparticles, and its surface with micro/nano rough structure was almost intact and still showed good superhydrophobicity after 1700 abrasion cycles under the pressure of 20.5 kPa. Wang et al. [66] prepared polytetrafluoroethylene/polyvinylidene fluoride composite coating by powder mixing and hot pressing method and found that the abrasion resistance of the coating was significantly improved after adding polytetrafluoroethylene particles. Similarly, Zhu et al. [67] prepared polyphenylene sulfide/polytetrafluoroethylene superhydrophobic composite coating by spraying method, resulting that the addition of polytetrafluoroethylene could significantly reduce the adhesion of the composite coating and greatly improve its mechanical abrasion resistance.

8.3.2 Zero-dimensional Structural Unit to Construct Three-dimensional Superhydrophobic Nanocomposite Surface

Compared with the two-dimensional superhydrophobic surface, the three-dimensional surface is characterized by a certain thickness, which allows it to withstand greater abrasion (Figure 8.4). Most of the three-dimensional superhydrophobic surfaces are block material with uniform structure and possess good self-healing ability. After harsh mechanical abrasion, the newly exposed surfaces are highly similar to those before damage.

Multifunctional Superhydrophobic 209

Generally, the construction of bulk materials is the most ideal strategy to improve the key bottleneck of surface stability. Zero-dimensional structural units with low surface energy and micro/nanostructures are often selected to construct three-dimensional superhydrophobic surfaces. For example, Manna et al. [68] used branched polyethyleneimine, dipentaerythritol penta/hexaacrylate, and amino-functionalized graphene oxide as raw materials, based on the 1,4-conjugated addition reaction between amino and acrylate to prepare the bulk gel materials with excellent superhydrophobicity both inside and outside, showing excellent mechanical and chemical stability.

Davis et al. [69] added water droplets dropwise to uncross-linked PDMS and curing agent while mechanical stirring to form aqueous polymer emulsion. After the emulsion was pre-cured and cured, the moisture was removed from the emulsion to form a siloxane monomer with pore structure, obtaining the superhydrophobic bulk surface. Such surface exhibits fascinating mechanical stability after knife-scratch, cyclic tape peels, and finger-wiper tests. Moreover, the surface of pores in micro/nanostructure remained intact even after its top layer was exposed to repeated abrasion cycles with 240 mesh sandpaper. Zhang et al. [70] prepared robust superhydrophobic surface through a mixture of hydrophobic polypropylene, TiO_2 nanorods, SiO_2 nanoparticles, and PDMS. The results showed that abrasion surface and cross-section of the material were superhydrophobic, and the micro/nanostructure with low surface energy ran through the entire superhydrophobic material, which can reserve good superhydrophobicity before being completely worn off. In addition, Zhu et al. [71] constructed rough structure surface by pressing copper powder particles into the surface of ultra-high molecular weight polyethylene sheet, and then modified it with perfluorosilane to reduce its surface energy to obtain robust superhydrophobic surface.

8.3.3 ONE-DIMENSIONAL STRUCTURAL UNIT TO CONSTRUCT TWO-DIMENSIONAL SUPERHYDROPHOBIC NANOCOMPOSITE SURFACE

As you know, micro/nanoparticles with low surface energy are easy to construct superhydrophobic surfaces; however, the microstructure and hydrophobic substance of these surfaces are easily worn off, causing their superhydrophobicity to be partially or completely lost immediately and cannot be restored, mainly because of the lack of intensity in their structures.

Based on this, researchers tried to construct superhydrophobic surfaces using one-dimensional instead of zero-dimensional structure units, whose advantage lies in that one-dimensional structure unit has larger length diameter ratio. For composites, it is more difficult to separate one-dimensional fillers from the matrix materials under the action of Van der Waals force; For fiber fabrics, one-dimensional fibers can be entangled and occluded with each other, which makes superhydrophobic surfaces have better scratch resistance and durability (Figure 8.5a). For example, Xue et al. [72] used alkali etching method to etch abundant concave structure on the fiber surface of polyethylene terephthalate monolayer fabric, showing excellent superhydrophobicity after being modified by PDMS, because its fabric is woven by one-dimensional structural unit fiber, which can effectively resist mechanical abrasion, washing, scraping, and so on. Jung et al. [73] used epoxy resin to replicate microstructures on the silicon surface, and then the carbon nanotube composites were uniformly deposited on the microstructure of the silicon surface by spray method to form a superhydrophobic composite surface, because the carbon nanotubes have strong bonding force to the microstructure, showing high mechanical strength and abrasion resistance. Baidya et al. [74] formed a superhydrophobic coating by fully combining fluorinated nanofibers (5~20 nm in width and 500 nm in length), 3-(2-aminoethyl) propyl trimethoxysilane, and 1H, 1H,2H,2H-perfluorooctanyltriethoxysilane in aqueous medium. Such surface remains its superhydrophobic properties after withstand a variety of harsh mechanical damage tests (such as knife-scratch, sandpaper abrasion, tape peeling, etc.), showing excellent mechanical robustness. Li et al. [75] prepared large-scale superhydrophobic composite fiber films by four-jet electrospinning, in which the mechanical properties of the composite surface can be enhanced by blending polystyrene and polyamide, and the surface roughness can be effectively changed by adjusting the mass ratio of polystyrene/polyamide.

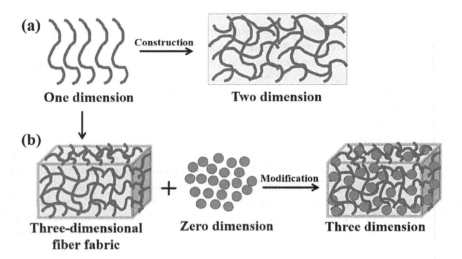

FIGURE 8.5 One-dimensional structural unit to construct two- or three-dimensional superhydrophobic nanocomposite surface.

8.3.4 ONE-DIMENSIONAL STRUCTURAL UNIT TO CONSTRUCT THREE-DIMENSIONAL SUPERHYDROPHOBIC NANOCOMPOSITE SURFACE

Three-dimensional fiber fabric formed by woven arrangement of one-dimensional fibers can be modified to obtain a three-dimensional superhydrophobic surface (Figure 8.5b). The modified substances are mainly divided into oligomers, polymer nanowires, nanoparticles, and composites.

For example, Zhou et al. [76] used fluorinated SiO_2 and fluorosilane-filled polydimethylsiloxane to prepare a super-robust superhydrophobic composite coating on the fabric substrate, which can still remain superhydrophobic after being damaged by strong acid, strong alkali, boiling water, and abrasion. The research group further found that fluorinated surfactants, polytetrafluoroethylene nanoparticles, and perfluorosilane were uniformly dispersed in water to form a solution, and then immersed cotton fabric in the solution to obtain superhydrophobic surface [77]. Similarly, Zimmermann et al. [78] modified the polyethylene terephthalate fabric with silicone nanowires to obtain superhydrophobic surface. Zhang et al. [79] introduced polyaniline and fluorosilane into the fabric through the vapor deposition process so that the fabric surface has both superhydrophobicity and superlipophilicity. The superhydrophobic surface can resist sandpaper abrasion and mechanical stretching because of the strong biding force between polymer and fiber. In addition, the benzene ring contained in aniline improves the rigidity of the molecule and enhances the mechanical strength of the surface. Therefore, the introduction of some functional reagents can achieve the coexistence of superhydrophobicity and high mechanical strength.

8.3.5 POROUS BULK STRUCTURAL UNIT TO CONSTRUCT THREE-DIMENSIONAL SUPERHYDROPHOBIC NANOCOMPOSITE SURFACE

In addition to the three-dimensional superhydrophobic surface constructed by zero-dimensional and one-dimensional structural units, porous blocks can also be used as templates to form three-dimensional superhydrophobic surfaces after being modified or filled with superhydrophobic medium. The abrasion-resistant microscale porous structures are selected as the barrier layer or sacrificial layer (usually hydrophilic medium), which can effectively resist the damage from external mechanical action. For example, Sun et al. [80] selected the porous fiber fabric with high mechanical strength as sacrificial layer, and adhered branched poly(ethyleneimine)/ammonium polyphosphate composite

Multifunctional Superhydrophobic

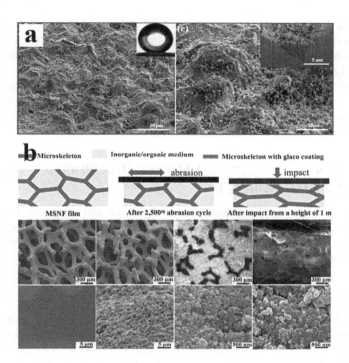

FIGURE 8.6 (a) Scanning electron micrographs (SEM) image of nanoparticles attached to microstructure surfaces. (b) Schematics showing that the MSNF film is unaffected by abrasion and impact, and SEM of porous iron, roughened, MSNF film surface. (From Zheng, Q.S. et al. *Adv. Funct. Mater.*, 30, 1910665, 2020. With permission).

to the fabric surface by dip coating, which greatly improved the mechanical stability of superhydrophobic surfaces. Highly inspired by this, Liu et al. [81] fabricated an ultrarobust self-cleaning surface with submicroscale rough structures on copper substrate using a nanoparticle-filled harder rough structure by electrodeposition. The surface still maintained its superhydrophobicity after several mechanical tests, including finger-wipe, knife-scratch, sand abrasion, and sandpaper abrasion. However, these surfaces are easily damaged under subjecting to repeated severe abrasion, mainly because their rough structures are prone to structural buckling or Euler instability (Figure 8.6a).

Recently, researchers have revolutionarily discovered that the mechanical superstable superhydrophobic surfaces can be obtained by completely filling the large-scale micromatrix with small-scale superhydrophobic nano-medium. Deng et al. [82] split the superhydrophobic and mechanical robustness into two different structural scales through the decoupling mechanism and proposed a new concept of microstructure "armor" to protect superhydrophobic nanomaterials from friction and abrasion. A series of microscale inverted pyramid grooves were etched on silicon wafers by photolithography, and the nanostructure "armor" distributed in the grooves greatly improves the mechanical robustness of the superhydrophobic surface. Moreover, this study not only integrates the comprehensive properties of high-strength mechanical robustness, anti-high speed jet impact, anti-condensation failure, and chemical and thermal degradation resistance but also realizes the high transmittance of the glass armored surface, which creates the necessary conditions for the surface to be applied to self-cleaning vehicle glass, solar cell cover plate, and building glass curtain wall.

Meanwhile, Zheng et al. [83] proposed to create a microskeleton-nanofiller (MSNF) film with exceptionally mechanical superstable superhydrophobicity (Figure 8.6b), which can be extended to a variety of porous materials (such as form ceramics or plastics). The resulting MSNF film can reserve a superhydrophobicity under not only continuous abrasion before the complete wearing off

Functional Thin Films Technology

the film but also heavy impact, and robustness increased greatly compared with reported literatures. Moreover, the MSNF film is also superamphiphobic and can reserve the superhydrophobicity under large bending or torsion. Furthermore, it is found that the requirement for fabricating ideal superhydrophobic surfaces needs to optimize the composite of the two-component materials, i.e. the ratio of the nanoparticles and the metal skeleton which play roles in enhancing the superhydrophobicity and the mechanical properties (wearability, strength, and flexibility). A higher intrinsic CA of the nanoparticle filler provides a better superhydrophobicity, and a smaller area fraction of the metal skeleton provides larger CA after mechanical abrasion.

8.4 APPLICATION OF FUNCTIONAL SUPERHYDROPHOBIC NANOCOMPOSITE SURFACE

With the continuous progress of micro/nanotechnology, more and more new functional superhydrophobic surfaces have emerged one after another. It not only has the single function of traditional superhydrophobicity but also has other novel functions, which can be widely used in specific fields of production and life.

8.4.1 ANTI-ICING

Icing causes great harm to people's daily life, industry, and transportation. Especially, icing can damage power transmission and industrial equipment, resulting in the interruption of transmission lines and satellite antenna signals that are poor or unable to work [84, 85]. Because the surface icing is formed by the freezing of water droplets, if the superhydrophobic coating is applied on the surface of these materials, the adhesion and diffusion of water droplets can be reduced and the icing problem can be effectively alleviated. Studies have confirmed that the freezing process is greatly delayed when water droplets are placed on an ice-cold superhydrophobic surface. The air layer intercepted by the surface micro/nanostructure can provide thermal insulation effect, which can greatly delay the freezing time of water droplets, and the water droplets will roll away immediately on the surface, and there is no time to freeze on the surface. On the contrary, it will spread and move on its surface when water droplets area placed on the copper substrate, leaving a water film layer on the surface and freezing immediately at the same time [86].

In addition, the icing experiment on superhydrophobic aluminum surface shows that once ice crystals frozen by water droplets appear in a local area on the surface, the water droplets develop around the ice crystals and eventually form a stable icing layer, and the icing condition is also closely related to the speed of water droplets freezing on the surface [87]. Based on the experimental study on the quantitative relationship between icing adhesion strength and hydrophilic/hydrophobic surface, resulting in that the adhesion strength of ice on hydrophobic surface was generally weaker than that of hydrophilic surface. In order to continuously reduce the adhesion strength of ice on the surface, it is necessary to prepare a higher hydrophobic surface [88].

Reducing the adhesion strength of ice on the surface or slowing down the process of surface icing cannot completely achieve the anti-icing; to explore a full anti-icing surface is still the ultimate goal. Gao et al. [89] added a kind of nano-polymer composite material on aluminum surface, which almost completely prevents supercooled water from freezing on its surface under laboratory and external atmospheric conditions. At the same time, they changed the size of nanoparticles in the coating and found that the less hydrophobic the surface (the higher the CA and receding CA), the better the anti-icing performance. This is because the size of nanoparticles affects the nucleation factors in the process of water crystallization. Aizenberg et al. [90] dropped normal temperature droplets form a certain height to low-temperature surfaces with different hydrophobicity and micro/nanostructures, resulting in specific micro/nanostructures to completely prevent icing, mainly attributing to the fact that superhydrophobic surface can repel falling droplets, causing them to leave the surface before nucleation and freezing.

Multifunctional Superhydrophobic

8.4.2 CONDENSATION

Condensation is a common phenomenon in industrial production, which is quite different from general wetting. The process of steam condensation starts from the nucleation of more than 10 nano-sized droplets, and the droplets gradually grow into general wetting form. The biggest thermal resistance in the condensation process comes from the liquor condensate itself, and adjusting the behavior of the condensate by controlling the surface structure and wettability is the main way to enhance the condensation heat transfer rate of steam [91]. Generally, the heat transfer coefficient of classical dropwise condensation on hydrophobic surface is several times or even tens of times higher than that of film condensation on hydrophilic surface. Because the superhydrophobic surface has extremely low adhesion to the droplets, the water droplets have small SA, strong movement, and easy to fall off on its surface, which gives birth to a new type of droplet merging and bouncing condensation, providing a new idea for enhancing the condensation heat transfer of steam.

However, the vapor nucleation scale (~10 nm) is lower than the structural characteristic scale of most superhydrophobic surface during the condensation process, and the condensed droplets will grow "bottom-up" between the micro and nanostructures, which will cause damage to the formed solid–gas–liquid composite interface, resulting in the failure of superhydrophobicity. This will greatly increase the heat transfer resistance of steam condensation and lead to the sharp deterioration of heat transfer performance [92]. Although the surface of micro/nanostructure has more stable hydrophobicity than single-level nanostructure, the influence of microstructure on condensation heat transfer has been controversial. On the one hand, the Lapace force generated by large-scale droplets with help of micro structures to produce the self-transporation behavior, which helps to maintain droplet condensation [93]; on the other hand, the introduction of micro structure increases the viscosity of droplets on the surface, resulting in a larger CA hysteresis than that of the pure nanostructures, which is unfavorable for the drop shedding [94].

It was found that the mechanism of superhydrophobicity failure lies in the nucleation, growth, and merging behavior of small-size droplets under condensation condition, and smaller-scale nanostructure can effectively regulate small-size droplets [95]. Therefore, designing reasonable nanostructures is the key to improve the wetting stability of superhydrophobic surface under condensation condition, and also is the basic way to enhance the condensation heat transfer ratio. Gao et al. [96] realized the high-density nucleation of condensed droplets by in situ constructing the superhydrophobic surface with cluster-shaped prismatic nano-needle structure, which significantly improved the condensation heat transfer ratio of steam. Yang et al. [97] firstly achieved stable and efficient droplet bouncing condensation heat transfer on the superhydrophobic surface of three-dimensional copper nanowire structure by controlling the initial nucleation of condensed droplets. More importantly, the hydrophilic/hydrophobic interphase surface [98], the superhydrophilic bottom/hydrophobic top layer composite structure surface [99], and the lubricating fluid-immersed composite structure surface [100] can be prepared by adjusting the surface structure and wettability, which can effectively improve the dynamic behavior of condensation droplets, and laid a foundation for further design and development of new enhanced condensation heat transfer technology.

8.4.3 OIL/WATER SEPARATION

With the vigorous development of global industry, many problems have emerged one after another, such as oil leakage, industrial production of oil from wastewater, and other serious environmental pollution. Addressing these problems, it is of great significance to study oil/water separation. Traditional buoyancy and gravity separation methods have good effects on oil/water separation. However, these traditional separation methods are inefficient and costly and are difficult to separate for oil/water mixed emulsion.

If the interfacial tension of a solid is between the surface tension of water and oil, such surfaces have both hydrophobic and lipophilic properties. Besides, the material with both superhydrophobic and superlipophilic properties can be achieved if the appropriate rough structure is constructed or chemically modified on the surface [101, 102]. Currently, according to different oil/water separation methods, it can be divided into superhydrophobic–superlipophilic surface (mixture separation of heavy oil and water), superhydrophilic–superhydrophobic surface (mixed separation between light oil and water), and intelligent surface of controllable wettability [103]. Chen et al. [104] firstly modified $CaCO_3$ nanoparticles with fluorosilane, and then coated them on a glued stainless steel grid to prepare superhydrophobic–superhydrophilic coatings. This stainless steel mesh coating has better separation efficiency for the mixture of water and different oils (such as chloroform, toluene, dichloromethane, and gasoline). Liu et al. [105] combined low surface energy molecules and layered double hydroxides to modify commercial fabrics to obtain superhydrophobic/superlipophilic functional fabric surface, which can not only separate oil and water (the separation efficiency is greater than 97%) but also realize selective oil adsorption. Xue et al. [106] prepared a new type of superhydrophilic/superhydrophobic mesh surface wrapped by polyacrylamide hydrogel. This special surface is composed of micron-sized porous structure and nano-rough structured hydrogel, which can selectively separate oil from oil/water mixture with higher separation efficiency and easy recycling. Gondal et al. [107] deposited on stainless steel grid by nano-TiO_2 spraying and prepared the porous structure surface of superhydrophilic and underwater superhydrophobic, and the oil/water separation efficiency can be as high as 99% when TiO_2 is coated on stainless steel mesh with pore size of 50~100 μm. Cheng et al. [108] self-assembled the mixed mercaptans ($HS(CH_2)_9CH_3$ and $HS(CH_2)_{11}OH$) onto the nano copper substrate to prepare the surface with controllable oil wettability surface in water. The controllability of such surface for underwater superlipophilic/superoleophobic can realize the separation of different oil/water mixtures.

8.4.4 SUPERAMPHIPHOBICITY

Most superhydrophobic surfaces cannot achieve oil resistance, and even lose superhydrophobicity after being contaminated by oil. As the upgrading and expansion of superhydrophobic surface, the superamphiphobic surface can repel water droplets and oil droplets, and it has a wider application prospect and is a research hotspot in recent years. Because the interfacial tension of oil is smaller than that of water, or even much smaller. Therefore, the preparation of superamphiphobic surface requires more subtle microstructure and lower surface energy.

As early as 1997, Tsujii et al. [109] used the electrochemical corrosion method to prepare superamphiphobic surfaces. They corroded the anode aluminum sheet in sulfuric acid electrolyte, formed a rough structure on the aluminum sheets, and then modified it with low surface energy materials to obtain the superamphiphobic surface, which made the CA of water greater than 170° and the CA of vegetable oil greater than 150°. Similarly, Tian et al. [110] prepared pyramid shaped thin films on gold substrates by electrochemical deposition, and then modified them with single molecular layer with low surface energy materials to obtain superamphiphobic surfaces.

In addition, Kanamori et al. [111] prepared macroporous silicon blocks with superamphiphobic properties by combining thiol-ene click reaction and sol-coagulation method. The CAs of organic liquids (such as ethylene glycol, formamide, 1-bromonaphthalene, and diiodomethane) on the material surface are all greater than 150°, showing good superhydrophobicity. It was found that this material could float above organic liquid or water and could be applied to three-dimensional superamphiphobic surface, selective separation media of organic liquid, and gas permeable membrane. Liu et al. [112] prepared superomniphobic surface with a specific doubly reentrant structure that enabled very low liquid–solid contact fraction, rendering bounce off all available liquids, even including perfluorohexane ($\gamma = 10$ mN/m).

Furthermore, Lee et al. [113] prepared semi-transparent superamphiphobic surface by Stober method. Firstly, SiO_2 fluoropolymer hybrid nanoparticles were prepared and sprayed on the substrate surface without any pretreatment to organic/inorganic coatings with micro/nanostructure. The CAs of the coatings to water and n-hexane were 163° and 151°, respectively, and exhibited a small SA.

Multifunctional Superhydrophobic 215

This superamphiphobic coating can be used on a variety of substrates to make it self-cleaning, such as LED lamps, solar cells and microfluidic systems. Zhou et al. [114] blended poly(vinylidene fluoride-hexafluoropropylene), fluorosilane and hydrophobic SiO_2 nanoparticles and prepared superamphiphobic fabric with self-repairing ability through two-step dip-coating method. Next, the research group also studied the characteristics of spontaneous and directional liquid transportation on the superamphiphobic fabric surface [115], which can be used to realize directional transportation of oil and water by ultraviolet irradiation and heating fabric. The intelligent fabric is expected to be applied in the field of biological cell separation or membrane separation.

8.4.5 CORROSION RESISTANCE

Metals are easily oxidized and corroded in humid and corrosive environment, which not only causes huge economic losses but also brings waste of resources, environmental pollution and potential hazards of safety. Therefore, it is urgent to take effective measures for metal anti-corrosion. The direct contact between corrosive liquids and metal surfaces was reduced by constructing superhydrophobic surface on metal materials to isolate water and mold and play an excellent anticorrosion [116–118].

In recent years, some self-assembled monolayers, such as silyl, alkyl alcohols, can effectively protect the substrate from being damaged in harsh corrosive environments by forming a thin and tightly arranged barrier layer on the substrate [119, 120]. In addition, the chemical bond between the modified component and the substrate can also improve the adhesion of the modified coating on the substrate. For example, Duan et al. [121] used ion exchange method to prepare lauric acid film with layered double hydroxide on anodic alumina/aluminum substrate. The film has good superhydrophobicity, and its micro/nano-hierarchical structure serves as a barrier layer, which provides long-term corrosion protection for aluminum substrate. Yuan et al. [122] first etched the copper substrate with the same volume of hydrogen peroxide and concentrated nitric acid to form a rough surface with protruding island structure, and then introduced active functional groups and 2,2,3,4,4,4-hexafluorobutyl acrylate to form a fluorine-containing polymerization superhydrophobic film after graft polymerization. Electrochemical tests show that after soaking in simulated seawater solution for 21 days, the fluoropolymerized superhydrophobic film still shows a good shielding effect on aggressive Cl- and greatly improves the corrosion resistance of copper substrate. Yeh et al. [123] used PDMS elastomer as soft template to replicate the structure similar to the lotus leaf surface. Under the condition of template imprinting, polyaniline solution was poured on PDMS soft template and solidified to form a superhydrophobic surface with micron papillary structure, and the CA is 156°. The results showed that the superhydrophobic surface of polyaniline coated with papillae structure has strong water resistance, which can effectively prevent the direct contact between erosive ions and metal substrate and inhibit the crevice corrosion of metal materials. Li et al. [124] used fluorine-containing polyurethane, nano-SiO_2, and silane coupling agent to form a superhydrophobic coating on the aluminum alloy substrate, resulting in that such coating prevents corrosive ions from contacting the metal surface through water, showing excellent corrosion resistance.

8.4.6 WETTABILITY CONVERSION

Normally, a single wettable surface is difficult to meet the requirements of practical application. Recently many scholars have begun to study controllable wettability superhydrophobic surfaces, which mainly change surface physical chemistry properties by external stimuli (such as pH, light, heat, electricity, and so on) to adjust the surface wetting behavior.

The structure and properties of pH response surface are regulated by changing pH, which basically presents superhydrophobicity at low pH and superhydrophilicity at high pH. Jiang et al. [125] modified the rough surface of the gold micro/nanostructure by pH-responsive molecules to prepare a pH-responsive surface with wettability. The CA was about 145° at pH=1 and close to 0°

at pH=13. In order to prepare a novel pH stimulus-responsive surface, Zhang et al. [126] designed and synthesized an alkyl mercaptan derivative with the end of malachite green group, which was superhydrophobic at high pH and superhydrophilic at low pH. The main reason is that malachite green is a stimulus-responsive material. Malachite green is hydrophobic without external stimulation and becomes a hydrophilic group with positive charge under the external pH stimulation.

The wetting state of photosensitive surface can be changed by chemical reaction under the action of light. Cho et al. [127] used a method that combines facile surface roughness control with an electrostatic self-assembly process and photoresponsive molecular switching of fluorinated azobenzene molecules, realizing the photoswitchable surface with wettability that can be reversibly switched between superhydrophobicity and superhydrophilicity under UV/visible irradiation, which is mainly due to the change in the configuration of azobenzene molecules under UV irradiation. Specially, some inorganic photosensitive materials can also achieve the above functions. Wang et al. [128] coated fluorinated TiO_2 on the fabric substrate to prepare a superhydrophobic surface, showing the reversible switching between superhydrophobicity and superhydrophilicity can be realized by alternately UV irradiation and stored in the dark.

The wettability conversion of functionalized surfaces prepared above can be realized by combining stimuli-responsive materials and certain surface roughness. Most of these surfaces can only respond to one stimulus, which limits their application in complex environments, so it is necessary to develop multifunctional superhydrophobic surfaces that respond to multiple stimuli. Jiang et al. [129] prepared a polymer surface responsive to pH and temperature, which achieves reversible superhydrophobic to superhydrophilic conversion in a wide range of pH, and even a very narrow temperature range. The results show that this double responsiveness was caused by the change of surface rough structure and chemical composition. Furthermore, the patterned wetting transition was realized on the hydrophobic surface of structured nanocomposite rod array by photoelectric method [130]. The droplet remains spherical (in Cassie state) on the superhydrophobic surface when the applied voltage is lower than the critical value. However, the critical value of electrowetting transformation was reduced by photosensitive materials when light irradiated the "H" pattern on the surface, resulting in the electrowetting phenomenon occurred after illumination, and the droplets were still intact where they were not illuminated.

8.4.7 OTHER APPLICATION

In addition to anti-icing, condensation, superamphiphobicity, oil–water separation, anti-corrosion and wettability conversion, functional superhydrophobic surfaces also have many other unique properties, such as adhesion, fog collection, transparency, antibacterial properties, and so on.

The anisotropic viscous superhydrophobic surface is a typical feature of some exotic organisms. For example, the viscous behavior of water droplets on spider silk has a specific directivity. Imitating the unique structure of spider silk, Zheng et al. [131] prepared a novel artificial hydrophilic fiber which can condense water vapor from mist to its surface like spider silk. Studies found that there are many spindle like protuberances composed of nanofibers on the "trunk" of the artificial fiber. The tiny protuberances gather to form large drops of dew when the moisture in the air condenses to the fiber surface, providing the fiber surface with continuous source of life water. For the water shortage areas, such as islands, deserts and drought, the fiber material can collect fresh water from fog to solve the shortage of water resources, and has important application prospects.

Optical transparency or anti-reflection is an important requirement of many materials or devices, and the realization of transparent self-cleaning surface has very important application value. Zhang et al. [132] first used nano-filament silane to construct different microstructures on the substrate, then treated and modified them with plasma and perfluorodecyltrichlorosilane, respectively, and prepared superhydrophobic anti-reflection coatings with good stability and transparency on glass

Multifunctional Superhydrophobic

slides. In general, the prepared superhydrophobic surfaces are easily infiltrated by organic solvents (such as surfactants, alkanes, or ethanol, etc.). Deng et al. [133] broke the conventional method and prepared superhydrophobic surface by burning the slide with a candle flame and depositing a layer of soot. The porous deposit of candle soot is wrapped in a silicon shell with a thickness of 25 nm, which was calcined at 600°C and silanized into a black transparent coating. The as-prepared surface can not only bounce oil droplets but also has a high transparency, high temperature resistance and mechanical robustness, which can be used for touch screens, goggles, and metal surfaces.

The antibacterial superhydrophobic coating can reduce the contact between the bacteria in the liquid and the coating, and the adhesion of bacteria to the coating. Li et al. [134] prepared superhydrophobic surface on AZ91D magnesium alloy by hydrothermal method, with a CA of 155° and a SA as low as 2°, which can effectively inhibit the corrosion of the alloy and reduce the probability of bacterial infection after being implanted in organisms. On the contrary, the untreated magnesium alloy was easy to be corroded and propagated by bacteria after being implanted into the organism. Moreover, Wong et al. [135] constructed nanostructured "pillars" on silicon wafers, and then poured a layer of liquid lubricant into them to obtain a new type of nanosuperhydrophobic material free from water vapor. This material can not only improve the condensation heat transfer performance of heat exchangers in power plants but also prevent safety accidents caused by wing icing and frosting.

8.5   SUMMARY

Functionalized superhydrophobic surface has a great application prospect and has become an important development direction in the world. In recent years, the theoretical model of superhydrophobic surface has been basically perfected, and various preparation methods have been developed. Unfortunately, the superhydrophobic surface will lose its stable superhydrophobic state due to steam condensation or mechanical action, which restricts its real large-scale industrial application. In fact, under the condition of supercooled steam condensation, the condensed droplets on the superhydrophobic surface change from the Cassie state suspended on the structure to the Wenzel state trapped in the structure, triggering the pinning of the droplets lead to disappearance of superhydrophobicity. Moreover, the external mechanical action can also cause the surface microstructure to bend or collapse, leading to the weakening or even failure of superhydrophobicity. Therefore, how to develop long-term stability superhydrophobic surface has become the research emphasis and hotspot at this stage.

In a word, the industrialization of superhydrophobic surface still has a long way to go, and the future work can focus on the following aspects in this field: (i) strengthen theoretical research while developing new preparation technology, and evolve from theoretical verification experiment to theoretical prediction experiment to provide technical support for the development of highly stable superhydrophobic surface; (ii) develop environment-friendly hydrophobic modifiers to replace fluorine-containing low surface energy substances which are widely used at present; (iii) develop a smooth superhydrophobic surface (intrinsic CA is greater than 150° and SA is less than 10°), which can be understood as a non-structured superhydrophobic surface. If this idea can be realized, it is expected that the problems of poor structure stability (such as abrasion or impact) and wetting state stability (such as condensation or freezing) will be solved thoroughly, and the functional superhydrophobic surface will be applied on a large scale.

ACKNOWLEDGMENTS

This work was supported by the Fundamental Research Funds for the Central Universities (2020GFYD006) and National Natural Science Foundation of China (52001062).

REFERENCES

[1] Barthlott W and Neinhuis C, Purity of the sacred lotus, or escape from contamination in biological surfaces. *Planta* 202 (1997) 1–8.

[2] Feng XJ and Jiang L, Design and creation of superwetting/antiwetting surfaces. *Advanced Materials* 18 (2006) 3063–3078.

[3] Bixler GD and Bhushan B, Bioinspired rice leaf and butterfly wing surface structures combining shark skin and lotus effects. *Soft Matter* 8 (2012) 11271–11284.

[4] Bai H, Ju J, Zheng Y, et al., Functional fibers with unique wettability inspired by spider silks. *Advanced Materials* 24 (2012) 2786–2791.

[5] Parker AR, and Lawrence CR, Water capture by a desert beetle. *Nature* 414 (2001) 33–34.

[6] Zhang J, Wu L, Li B, et al., Evaporation-induced transition from nepenthes pitcher-inspired slippery surfaces to lotus leaf-inspired superoleophobic surfaces. *Langmuir* 30 (2014) 14292–14299.

[7] Feng L, Zhang Y, Xi J, et al., Petal effect: a superhydrophobic state with high adhesive force. *Langmuir* 24 (2008) 4114–4119.

[8] Xue CH, Guo XJ, Ma JZ, et al., Fabrication of robust and antifouling superhydrophobic surfaces via surface-initiated atom transfer radical polymerization. *ACS Applied Materials & Interfaces* 7 (2015) 8251–8259.

[9] Qian H, Xu D, Du C, et al. Dual-action smart coatings with a self-healing superhydrophobic surface and anti-corrosion properties. *Journal of Materials Chemistry A* 5 (2017) 2355–2364.

[10] Sojoudi H, Wang M, Boscher ND, et al., Durable and scalable icephobic surfaces: similarities and distinctions from superhydrophobic surfaces. *Soft Matter* 12 (2016) 1938–1963.

[11] Wang Y, Xue J, Wang Q, et al., Verification of icephobic/anti-icing properties of a superhydrophobic surface. *ACS Applied Materials & Interfaces* 5 (2013) 3370–3381.

[12] Jin M, Feng X, Feng L, et al., Superhydrophobic aligned polystyrene nanotube films with high adhesive force. *Advanced Materials* 17 (2005) 1977–1981.

[13] Wu D, Wu SZ, Chen QD, et al., Curvature-driven reversible in situ switching between pinned and rolldown superhydrophobic states for water droplet transportation. *Advanced Materials* 23 (2011) 545–549.

[14] Truesdell R, Mammoli A, Vorobieff P, et al., Drag reduction on a patterned superhydrophobic surface. *Physical Review Letters* 97 (2006) 044504.

[15] Cheng M, Song M, Dong H, et al., Surface adhesive forces: a metric describing the drag-reducing effects of superhydrophobic coatings. *Small* 11 (2015) 1665–1671.

[16] Dong Z, Wu L, Wang J, et al., Superwettability controlled overflow. *Advanced Materials* 27 (2015) 1745–1750.

[17] Teisala H, Tuominen M, Kuusipalo J, et al., Superhydrophobic coatings on cellulose-based materials: fabrication, properties, and applications. *Advanced Materials Interfaces* 1 (2014) 1300026.

[18] Good R J, Contact angle, wetting, and adhesion: a critical review. *Journal of Adhesion Science and Technology* 6 (1992) 1269–1302.

[19] Jiang L, Wang R, Yang B, et al., Binary cooperative complementary nanoscale interfacial materials. *Pure and Applied Chemistry* 72 (2000) 73–81.

[20] Herminghaus S, Brinkmann M, Seemann R, et al., Wetting and dewetting of complex surface geometries. *Annual Review Materials Research* 38 (2008) 101–121.

[21] McHale G and Newton M I, Frenkel's method and the dynamic wetting of heterogeneous planar surfaces. *Colloids and Surfaces A: Physicochemical and Engineering Aspects* 206 (2002) 193–201.

[22] Chen W, Fadeev A Y, Hsieh M C, et al., Ultrahydrophobic and ultralyophobic surfaces: some comments and examples. *Langmuir* 15 (1999) 3395–3399.

[23] Furmidge CGL. Studies at phase interfaces. I. The sliding of liquid drops on solid surfaces and a theory for spray retention. *Journal of Colloid Science* 17 (1962) 309–324.

[24] Young T, An essay on the cohesion of fluids. *Philosophical Transactions of The Royal Society of London* 95 (1805) 65–87.

[25] Nishino T, Meguro M, Nakamae K, et al., The lowest surface free energy based on -CF_3 alignment. *Langmuir* 15 (1999) 4321–4323.

[26] Wenzel RN, Resistance of solid surfaces to wetting by water. *Industrial & Engineering Chemistry* 28 (1936) 988–994.

[27] Wenzel RN, Surface roughness and contact angle. *The Journal of Physical Chemistry* 53 (1949) 1466–1467.

[28] Cassie ABD and Baxter S, Wettability of porous surfaces. *Transactions of the Faraday Society* 40 (1944) 546–551.

[29] Nosonovsky M and Bhushan B, Biomimetic superhydrophobic surfaces: multiscale approach. *Nano Letters* 7 (2007) 2633–2637.

[30] Roy R, Weibel JA, Garimella SV. Re-entrant cavities enhance resilience to the Cassie-to-Wenzel state transition on superhydrophobic surfaces during electrowetting. *Langmuir* 34 (2018), 12787–12793.

[31] Jung YC and Bhushan B, Dynamic effects of bouncing water droplets on superhydrophobic surfaces. *Langmuir* 24 (2008) 6262–6269.

[32] Bormashenko E, Pogreb R, Whyman G, et al., Cassie-Wenzel wetting transition in vibrating drops deposited on rough surfaces: is the dynamic Cassie-Wenzel wetting transition a 2D or 1D affair? *Langmuir* 23 (2007) 6501–6503.

[33] Manukyan G, Oh JM, Van Den Ende D, et al., Electrical switching of wetting states on superhydrophobic surfaces: a route towards reversible Cassie-to-Wenzel transitions. *Physical Review Letters* 106 (2011) 014501.

[34] Groten J, Bunte C, RRhe J, et al., Light-induced switching of surfaces at wetting transitions through photoisomerization of polymer monolayers. *Langmuir* 28 (2012) 15038–15046.

[35] Susarrey-Arce A, Marin AG, Nair H, et al., Absence of an evaporation-driven wetting transition on omniphobic surfaces. *Soft Matter* 8 (2012) 9765–9770.

[36] Lafuma A and Quéré D, Superhydrophobic states. *Nature Materials* 2 (2003) 457–460.

[37] Extrand CW, Contact angles and hysteresis on surfaces with chemically heterogeneous islands. *Langmuir* 19 (2003) 3793–3796.

[38] Patankar NA, Transition between superhydrophobic states on rough surfaces. *Langmuir* 20 (2004) 7097–7102.

[39] Quéré D, Non-sticking drops. *Reports on Progress in Physics* 68 (2005) 2495.

[40] Nosonovsky M and Bhushan B, Stochastic model for metastable wetting of roughness-induced superhydrophobic surfaces. *Microsystem Technologies* 12 (2006) 231–237.

[41] Bormashenko E, Bormashenko Y, Stein T, et al., Environmental scanning electron microscopy study of the fine structure of the triple line and Cassie-Wenzel wetting transition for sessile drops deposited on rough polymer substrates. *Langmuir* 23 (2007) 4378–4382.

[42] Nosonovsky M, Multiscale roughness and stability of superhydrophobic biomimetic interfaces. *Langmuir* 23 (2007) 3157–3161.

[43] Nosonovsky M and Bhushan B, Patterned nonadhesive surfaces: superhydrophobicity and wetting regime transitions. *Langmuir* 24 (2008) 1525–1533.

[44] Gupta P, Ulman A, Fanfan S, et al., Mixed self-assembled monolayers of alkanethiolates on ultrasmooth gold do not exhibit contact-angle hysteresis. *Journal of the American Chemical Society* 127 (2005) 4–5.

[45] Bhushan B, Jung YC, Koch K, et al., Micro-, nano- and hierarchical structures for superhydrophobicity, self-cleaning and low adhesion. *Philosophical Transactions of the Royal Society of London A: Mathematical, Physical and Engineering Sciences* 367 (2009) 1631–1672.

[46] Bhushan B, Nosonovsky M, Jung YC, et al., Towards optimization of patterned superhydrophobic surfaces. *Journal of The Royal Society Interface* 4 (2007) 643–648.

[47] Nosonovsky M and Bhushan B, Roughness-induced superhydrophobicity: a way to design non-adhesive surfaces. *Journal of Physics: Condensed Matter* 20 (2008) 225009.

[48] Michael N and Bhushan B, Hierarchical roughness makes superhydrophobic states stable. *Microelectronic Engineering* 84 (2007) 382–386.

[49] Nosonovsky M and Bhushan B, Biologically inspired surfaces: broadening the scope of roughness. *Advanced Functional Materials* 18 (2008) 843–855.

[50] Qing YQ, Yang CN, Yu N, et al., Superhydrophobic TiO_2/polyvinylidene fluoride composite surface with reversible wettability switching and corrosion resistance. *Chemical Engineering Journal* 290 (2016) 37–44.

[51] Song Y, Nair RP, Zou M, et al., Superhydrophobic surfaces produced by applying a self-assembled monolayer to silicon micro/nano-textured surfaces. *Nano Research* 2 (2009) 143–150.

[52] Jiang D, Fan P, Gong D, et al., High-temperature imprinting and superhydrophobicity of micro/nano surface structures on metals using molds fabricated by ultrafast laser ablation. *Journal of Materials Processing Technology* 236 (2016) 56–63.

[53] Vengatesh P and Kulandainathan MA, Hierarchically ordered self-lubricating superhydrophobic anodized aluminum surfaces with enhanced corrosion resistance. *ACS Applied Materials & Interfaces* 7 (2015) 1516–1526.

[54] Erbil HY, Demirel AL, Avci Y, et al., Transformation of a simple plastic into a superhydrophobic surface. *Science* 299 (2003) 1377–1380.

[55] Deng Z, Liu L, Li Y, et al., In situ and ex situ pH-responsive coatings with switchable wettability for controllable oil/water separation. *ACS Applied Materials & Interfaces* 8 (2016) 31281–31288.

[56] Xu QF, Mondal B, Lyons AM, et al., Fabricating superhydrophobic polymer surfaces with excellent abrasion resistance by a simple lamination templating method. *ACS Applied Materials & Interfaces* 3 (2011) 3508–3514.

[57] Yao L, Zheng M, He S, et al., Preparation and properties of ZnS superhydrophobic surface with hierarchical structure. *Applied Surface Science* 257 (2011) 2955–2959.

[58] Arslan O, Aytac Z, Uyar T, et al., Superhydrophobic, hybrid, electrospun cellulose acetate nanofibrous mats for oil/water separation by tailored surface modification. *ACS Applied Materials & Interfaces* 8 (2016) 19747–19754.

[59] Ebert D and Bhushan B, Transparent, superhydrophobic, and wear-resistant coatings on glass and polymer substrates using SiO_2, ZnO, and ITO nanoparticles. *Langmuir* 28 (2012) 11391–11399.

[60] Nine MJ, Cole MA, Johnson L, et al., Robust superhydrophobic graphene-based composite coatings with self-cleaning and corrosion barrier properties. *ACS Applied Materials & Interfaces* 7 (2015) 28482–28493.

[61] Lu Y, Sathasivam S, Song J, et al., Robust self-cleaning surfaces that function when exposed to either air or oil. *Science* 347 (2015) 1132–1135.

[62] Verho T, Bower C, Andrew P, et al., Mechanically durable superhydrophobic surfaces. *Advanced Materials* 23 (2011) 673–678.

[63] Peng C, Chen Z, Tiwari MK, et al., All-organic superhydrophobic coatings with mechanochemical robustness and liquid impalement resistance. *Nature Materials* 17 (2018) 355.

[64] Wang P, Chen M, Han H, et al., Transparent and abrasion-resistant superhydrophobic coating with robust self-cleaning function in either air or oil. *Journal of Materials Chemistry A* 4 (2016) 7869–7874.

[65] Milionis A, Languasco J, Loth E, et al., Analysis of wear abrasion resistance of superhydrophobic acrylonitrile butadiene styrene rubber (ABS) nanocomposites. *Chemical Engineering Journal* 281 (2015) 730–738.

[66] Wang FJ, Lei S, Ou JF, et al., Superhydrophobic surfaces with excellent mechanical durability and easy repairability. *Applied Surface Science* 276 (2013) 397–400.

[67] Wang H, Zhao J, Zhu Y, et al., The fabrication, nano/micro-structure, heat-and wear-resistance of the superhydrophobic PPS/PTFE composite coatings. *Journal of Colloid and Interface Science* 402 (2013) 253–258.

[68] Das A, Deka J, Rather AM, et al., Strategic formulation of graphene oxide sheets for flexible monoliths and robust polymeric coatings embedded with durable bioinspired wettability. *ACS Applied Materials & Interfaces* 9 (2017) 42354–42365.

[69] Davis A, Surdo S, Caputo G, et al., Environmentally benign production of stretchable and robust superhydrophobic silicone monoliths. *ACS Applied Materials & Interfaces* 10 (2018) 2907–2917.

[70] Zhang X, Guo Y, Chen H, et al., A novel damage-tolerant superhydrophobic and superoleophilic material. *Journal of Materials Chemistry A* 2 (2014) 9002–9006.

[71] Zhu X, Zhang Z, Men X, et al., Robust superhydrophobic surfaces with mechanical durability and easy repairability. *Journal of Materials Chemistry* 21 (2011) 15793–15797.

[72] Xue CH, Li YR, Zhang P, et al., Washable and wear-resistant superhydrophobic surfaces with self-cleaning property by chemical etching of fibers and hydrophobization. *ACS Applied Materials & Interfaces* 6 (2014) 10153–10161.

[73] Jung YC and Bhushan B, Mechanically durable carbon nanotube-composite hierarchical structures with superhydrophobicity, self-cleaning, and low-drag. *ACS Nano* 3 (2009) 4155–4163.

[74] Baidya A, Ganayee MA, Jakka RS, et al., Organic solvent-free fabrication of durable and multifunctional superhydrophobic paper from waterborne fluorinated cellulose nanofiber building blocks. *ACS Nano* 11 (2017) 11091–11099.

[75] Li XH, Ding B, Lin JY, et al., Enhanced mechanical properties of superhydrophobic microfibrous polystyrene mats via polyamide 6 nanofibers. *The Journal of Physical Chemistry C* 113 (2009) 20452–20457.

Multifunctional Superhydrophobic

[76] Zhou H, Wang H, Niu H, et al., Fluoroalkyl silane modified silicone rubber/nanoparticle composite: a super durable, robust superhydrophobic fabric coating. *Advanced Materials* 24 (2012) 2409–2412.

[77] Zhou H, Wang H, Niu H, et al., A waterborne coating system for preparing robust, self-healing, superamphiphobic surfaces. *Advanced Functional Materials* 27 (2017) 1604261.

[78] Zimmermann J, Reifler FA, Fortunato G, et al., A simple, one-step approach to durable and robust superhydrophobic textiles. *Advanced Functional Materials* 18 (2010) 3662–3669.

[79] Zhou XY, Zhang ZZ, Xu XH, et al., Robust and durable superhydrophobic cotton fabrics for oil/water separation. *ACS Applied Materials & Interfaces* 5 (2013) 7208–7214.

[80] Chen S, Li X, Li Y, et al., Intumescent flame-retardant and self-healing superhydrophobic coatings on cotton fabric. *ACS Nano* 9 (2015) 4070–4076.

[81] Qing YQ, Hu CB, Yang C, et al., Rough structure of electrodeposition as a template for an ultrarobust self-cleaning surface. *ACS Applied Materials & Interfaces* 9 (2017) 16571–16580.

[82] Wang D, Sun Q, Hokkanen MJ, et al., Design of robust superhydrophobic surfaces. *Nature* 582 (2020) 55–59.

[83] Qing YQ, Shi SY, Lv CJ, et al., Microskeleton-nanofiller composite with mechanical super-robust superhydrophobicity against abrasion and impact. *Advanced Functional Materials* 30 (2020) 1910665.

[84] Frankenstein S and Tuthill AM, Ice adhesion to locks and dams: past work; future directions? *Journal of Cold Regions Engineering* 16 (2002) 83–96.

[85] Dow JP, Understanding the stall-recovery procedure for turboprop airplanes in icing conditions. *Flight Safety Digest* 24 (2005).

[86] Tourkine P, Le Merrer M, Quéré D, et al., Delayed freezing on water repellent materials. *Langmuir* 25 (2009) 7214–7216.

[87] Yin L, Xia Q, Xue J, et al., In situ investigation of ice formation on surfaces with representative wettability. *Applied Surface Science* 256 (2010) 6764–6769.

[88] Kulinich SA and Farzaneh M, Ice adhesion on super-hydrophobic surfaces. *Applied Surface Science* 255 (2009) 8153–8157.

[89] Cao L, Jones A K, Sikka V K, et al., Anti-icing superhydrophobic coatings. *Langmuir* 25 (2009) 12444–12448.

[90] Mishchenko L, Hatton B, Bahadur V, et al., Design of ice-free nanostructured surfaces based on repulsion of impacting water droplets. *ACS Nano* 4 (2010) 7699–7707.

[91] Wen R, Ma X, Lee YC, et al., Liquid-vapor phase-change heat transfer on functionalized nanowired surfaces and beyond. *Joule* 2 (2018) 2307–2347.

[92] Mouterde T, Lehoucq G, Xavier S, et al., Antifogging abilities of model nanotextures. *Nature Materials* 16 (2017) 658–663.

[93] Sharma CS, Combe J, Giger M, et al., Growth rates and spontaneous navigation of condensate droplets through randomly structured textures. *ACS Nano* 11 (2017) 1673–1682.

[94] Chen X, Patel RS, Weibel JA, et al., Coalescence-induced jumping of multiple condensate droplets on hierarchical superhydrophobic surfaces. *Scientific Reports* 6 (2016) 18649.

[95] Rykaczewski K, Osborn WA, Chinn J, et al., How nanorough is rough enough to make a surface superhydrophobic during water condensation? *Soft Matter* 8 (2012) 8786–8794.

[96] Zhu J, Luo Y, Tian J, et al., Clustered ribbed-nanoneedle structured copper surfaces with high-efficiency dropwise condensation heat transfer performance. *ACS Applied Materials & Interfaces* 7 (2015) 10660–10665.

[97] Wen R, Xu S, Ma X, et al., Three-dimensional superhydrophobic nanowire networks for enhancing condensation heat transfer. *Joule* 2 (2018) 269–279.

[98] Peng B, Ma X, Lan, Z, et al., Analysis of condensation heat transfer enhancement with dropwise-filmwise hybrid surface: droplet sizes effect. *International Journal of Heat and Mass Transfer* 77 (2014) 785–794.

[99] Oh J, Zhang R, Shetty PP, et al., Thin film condensation on nanostructured surfaces. *Advanced Functional Materials* 28 (2018) 1707000.

100. Dai X, Sun N, Nielsen SO, et al., Hydrophilic directional slippery rough surfaces for water harvesting. *Science Advances* 4 (2018) eaaq0919.

[101] Wang CF, Tzeng FS, Chen HG, et al., Ultraviolet-durable superhydrophobic zinc oxide-coated mesh films for surface and underwater oil capture and transportation. *Langmuir* 28 (2012) 10015–10019.

[102] Zhang JP and Seeger S, Polyester materials with superwetting silicone nanofilaments for oil/water separation and selective oil absorption. *Advanced Functional Materials* 21 (2011) 4699–4704.

[103] Jiale Y, Jinglan H, Feng C, et al., Oil/water separation based on natural materials with super-wettability: recent advances. *Physical Chemistry Chemical Physics* 20 (2018) 25140–25163.

[104] Chen B, Qiu J, Sakai E, et al., Robust and superhydrophobic surface modification by a "paint + adhesive" method: applications in self-cleaning after oil contamination and oil–water separation. *ACS Applied Materials & Interfaces* 8 (2016) 17659–17667.

[105] Liu X, Ge L, Li W, et al., Layered double hydroxide functionalized textile for effective oil/water separation and selective oil adsorption. *ACS Applied Materials & Interfaces* 7 (2015) 791–800.

[106] Xue Z, Wang S, Lin L, et al., A novel superhydrophilic and underwater superoleophobic hydrogel-coated mesh for oil/water separation. *Advanced Materials* 23 (2011) 4270–4273.

[107] Gondal MA, Sadullah MS, Dastageer MA, et al., Study of factors governing oil-water separation process using TiO$_2$ films prepared by spray deposition of nanoparticle dispersions. *ACS Applied Materials & Interfaces.* 6 (2014) 13422–13429.

[108] Cheng Z, Lai H, Du Y, et al., Underwater superoleophilic to superoleophobic wetting control on the nanostructured copper substrates. *ACS Applied Materials & Interfaces* 5 (2013) 11363–11370.

[109] Tsujii K, Yamamoto T, Onda T, et al., Super oil-repellent surfaces. *Angewandte Chemie International Edition* 36 (1997) 1011–1012.

[110] Tian Y, Liu H, Deng Z, et al., Electrochemical growth of gold pyramidal nanostructures: toward super-amphiphobic surfaces. *Chemistry of Materials* 18 (2006) 5820–5822.

[111] Hayase G, Kanamori K, Hasegawa G, et al., A superamphiphobic macroporous silicone monolith with marshmallow-like flexibility. *Angewandte Chemie International Edition* 52 (2013) 10788–10791.

[112] Liu T and Kim C, Turning a surface superrepellent even to completely wetting liquids, *Science* 346 (2014) 1096–1100.

[113] Lee SG, Ham DS, Lee DY, et al., Transparent superhydrophobic/translucent superamphiphobic coatings based on silica-fluoropolymer hybrid nanoparticles. *Langmuir* 29 (2013) 15051–15057.

[114] Zhou H, Wang H, Niu H, et al., Robust, self-healing superamphiphobic fabrics prepared by two-step coating of fluoro-containing polymer, fluoroalkyl silane, and modified silica nanoparticles. *Advanced Functional Materials* 23 (2013) 1664–1670.

[115] Zhou H, Wang H, Niu H, et al., Superphobicity/philicity janus fabrics with switchable, spontaneous, directional transport ability to water and oil fluids. *Scientific Reports* 3 (2013) 2964.

[116] DY Yu, JT Tian, JH Dai, et al., Corrosion resistance of three-layer superhydrophobic composite coating on carbon steel in seawater. *Electrochimica Acta* 97 (2013) 409–419.

[117] Yang TI, Peng CW, Lin YL, et al., Synergistic effect of electroactivity and hydrophobicity on the anticorrosion property of room-temperature-cured epoxy coatings with multi-scale structures mimicking the surface of Xanthosoma sagittifolium leaf. *Journal of Materials Chemistry* 22 (2012) 15845–15852.

[118] Weng CJ, Chang CH, Peng CW, et al., Advanced anticorrosive coatings prepared from the mimicked xanthosoma sagittifolium-leaf-like electroactive epoxy with synergistic effects of superhydrophobicity and redox catalytic capability. *Chemistry of Materials* 23 (2011) 2075–2083.

[119] Zamborini FP, Crooks RM., Corrosion passivation of gold by n-alkanethiol self-assembled monolayers: effect of chain length and end group. *Langmuir* 3279 (1998) 3279–3286.

[120] Hintze PE, Calle LM., Electrochemical properties and corrosion protection of organosilane self-assembled monolayers on aluminum 2024-T3. *Electrochimica Acta* 51 (2006) 1761–1766.

[121] Zhang F, Zhao L, Chen H, et al., Corrosion resistance of superhydrophobic layered double hydroxide films on aluminum. *Angewandte Chemie International Edition* 47 (2008) 2466–2469.

[122] Yuan S, Pehkonen SO, Liang B, et al., Superhydrophobic fluoropolymer-modified copper surface via surface graft polymerisation for corrosion protection. *Corrosion Science* 53 (2011) 2738–2747.

[123] Peng CW, Chang KC, Weng CJ, et al., Nano-casting technique to prepare polyaniline surface with biomimetic superhydrophobic structures for anticorrosion application. *Electrochimica Acta* 95 (2013) 192–199.

[124] Li SM, Wang YG, Liu JH, et al., Preparation of superhydrophobic coating on aluminum alloy with its anti-corrosion property. *Acta Physica Sinica* 23 (2007) 1631–1636.

[125] Jiang YG, Wang ZQ, Yu X, et al., Self-assembled monolayers of dendron thiols for electrodeposition of gold nanostructures: toward fabrication of superhydrophobic/superhydrophilic surfaces and pH-responsive surfaces. *Langmuir* 21 (2005) 1986.

Multifunctional Superhydrophobic

[126] Jiang YG, Wan PB, Wang ZQ, et al., Supramolecular assembly and soft material. *Proceeding of the 26th Annual Meeting of Chinese Chemical Society*, 54 (2008).

[127] Lim HS, Han JT, Kwak D, et al., Photoreversibly switchable superhydrophobic surface with erasable and rewritable pattern. *Journal of the American Chemical Society* 128 (2006) 14458.

[128] Yin YJ, Guo N, Wang CX, et al., Alterable superhydrophobic–superhydrophilic wettability of fabric substrates decorated with Ion–TiO_2 coating via ultraviolet radiation. *Industrial & Engineering Chemistry Research* 53 (2014) 14322.

[129] Xia F, Feng L, Wang S, et al., Dual-responsive surfaces that switch between superhydrophilicity and superhydrophobicity. *Advanced Materials* 18 (2006) 432–436.

[130] Tian D, Chen Q, Nie FQ, et al., Patterned wettability transition by photoelectric cooperative and anisotropic wetting for liquid reprography. *Advanced Materials* 21 (2009) 3744–3749.

[131] Zheng Y, Bai H, Huang Z, et al., Directional water collection on wetted spider silk. *Nature* 463 (2010) 640–643.

[132] Zhang J, Seeger S., Superoleophobic coatings with ultralow sliding angles based on silicone nanofilaments. *Angewandte Chemie International Edition* 50 (2011) 6652–6656.

[133] Deng X, Mammen L, Butt H J, et al., Candle soot as a template for a transparent robust superamphiphobic coating. *Science*, 335 (2012) 67–70.

[134] Wang Z, Su Y, Li Q, et al., Researching a highly anti-corrosion superhydrophobic film fabricated on AZ91D magnesium alloy and its anti-bacteria adhesion effect. *Materials Characterization* 99 (2015) 200–209.

[135] Dai X, Stogin BB, Yang S, et al., Slippery Wenzel state. *ACS Nano* 9 (2015) 9260–9267.

9 Solution-Processed Oxide-Semiconductor Films and Devices

Bui Nguyen Quoc Trinh

Vietnam National University, Hanoi, VNU Vietnam-Japan University, Vietnam

Vietnam National University, Hanoi, VNU University of Engineering and Technology, Vietnam

Endah Kinarya Palupi and Akihiko Fujiwara

Kwansei Gakuin University, Japan

CONTENTS

9.1 Introduction 226
    9.1.1 Current Status 226
        9.1.1.1 Key Semiconductor Devices with Thin-film Semiconductors 226
        9.1.1.2 Advantage of Thin-film Semiconductor Devices 227
        9.1.1.3 Solution Processing for Thin-film Semiconductor Devices 228
    9.1.2 Perspectives 229
9.2 Materials 230
9.3 Recent Development of N-type Semiconductor Thin Films and Devices 231
    9.3.1 Silicon-based Semiconductor Devices 231
    9.3.2 Indium-based Oxide Thin Films and Devices 232
    9.3.3 Indium-free Oxide Thin Films and Devices 234
9.4 Potential Development of P-type Thin Films 237
    9.4.1 Cu-doped ZnO Thin Films 237
    9.4.2 CuO Thin Films 238
    9.4.3 Prospective P-type Thin Films 242
9.5 Device Applications 242
    9.5.1 Thin-Film Transistors 242
    9.5.2 Solar Cells 245
        9.5.2.1 Absorption Figure of Merit 245
        9.5.2.2 Simple Solar Cell Structure ....`246
9.6 Conclusions 248
References 248

9.1 INTRODUCTION

9.1.1 CURRENT STATUS

9.1.1.1 Key Semiconductor Devices with Thin-film Semiconductors

One of the most popular applications of functional thin films is in semiconductor devices which are indispensable for smart society based on information and communication technology (ICT) in cooperation with sustainable development goals (SDGs). Among semiconductor devices, metal-oxide-semiconductor field-effect transistor (MOSFET) is widely used in ICT, such as sensing (data input), communication (data transmission), calculation (data handling), display (data output) etc. On the other hand, a simple p-n junction is used in solar cells, which is one of the most popular energy harvesting devices.

Transistors are semiconductor devices used to amplify or switch signals or power. It is currently embedded in integrated circuits being a key component of any electronic device, such as mobile phones, computers etc. In 1947, J. Bardeen and W. Brattain invented the first transistor known as the point-contact transistor (Figure 9.1(a)) which amplifies currents [1, 2]. Despite having unprecedented impact on the electronic industry, it was a relatively bulky device: bulky germanium was used as a semiconductor with two gold electrodes, which were an issue to be solved for mass production and integration. In 1960, M. M. Atalla and D. Kahng reported a MOSFET (Figure 9.1(b)), which provides a solution to the above-mentioned issue and has been a most important building block of modern electronics [3, 4]. In 1962, the thin-film transistor (TFT) (Figure 9.1(c)), a type of MOSFET, was developed by P. K. Weimer [5]. Since active semiconductor layers and other components are formed by deposition in TFTs, the quality of components including crystallinity is lower than that of the original MOSFET made of single crystalline wafers. On the other hand, owing to flexibility of the fabrication process, a wide variety of materials can be used as active semiconductor layers from

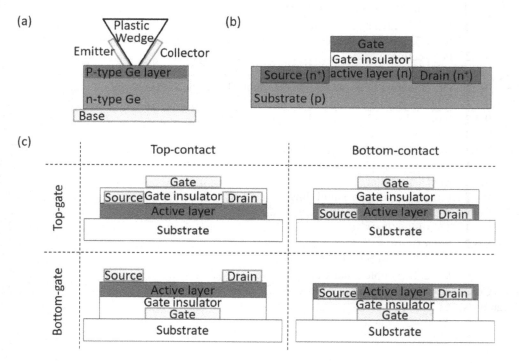

FIGURE 9.1 Schematics of (a) the first solar cell, (b) n-type MOSFET (MOSFET) structure and (c) several types of thin-film transistor structures.

Solution-Processed Oxide-Semiconductor 227

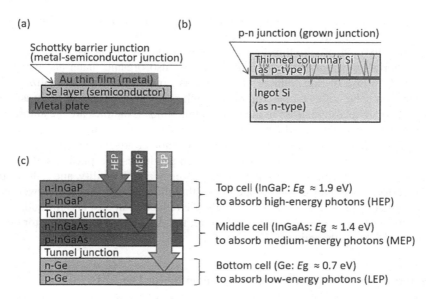

FIGURE 9.2 Schematics of conceptual structures of (a) the first solid-state photovoltaic cell made of a metal layer of gold and a semiconductor layer of selenium, (b) vertically stacked p-n junction invented by Russell Ohl and (c) a multi-junction solar cell.

elemental semiconductors to compound semiconductors including oxides. In addition, TFTs can be fabricated not only by deposition methods under high vacuum condition, such as molecular beam epitaxy (MBE), chemical vapor deposition (CVD) and laser ablation, but also by solution methods.

Solar cells are semiconductor devices which convert electromagnetic wave (light) energy directly into electricity by the photovoltaic effect. The first solid-state photovoltaic cell was discovered by C. E. Fritts in 1883 [6, 7]. The device was made of a layer of selenium overlaid with a thin film of gold (Figure 9.2(a)) [8], which is now regarded as Schottky barrier junction type photodiode. The photovoltaic effect at the p-n junction, consisting of p- and n-type semiconductors (Figure 9.2(b)), was first observed by R. S. Ohl [9]. The efficiency of these solar cells was as low as around 1%, which was not enough for the application. First practical solar cell with an efficiency of about 6% was demonstrated using a single-crystal sample by C. S. Fuller, D. Chapin and G. Pearson in 1954. Following wafer-based solar cells with crystalline materials (first-generation solar cells), thin-film solar cells (second-generation solar cells) are developed. For these types of devices, a wide variety of materials, such as amorphous silicon (a-Si), cadmium telluride (CdTe), copper indium gallium selenide (CIGS) etc., can be used. The efficiency of the first- and second-generation solar cells has been improved and is now more than 20%. For higher efficiency, multi-junction solar cells (third-generation solar cells), consisting of a number of p-n junctions (Figure 9.2(c)), have been investigated, and the cell efficiency is close to 50%.

In both cases, transistors and solar cells, thin-film technology is one of the key technologies for modern electronics.

9.1.1.2 Advantage of Thin-film Semiconductor Devices

Galena (the natural mineral form of lead (II) sulfide, PbS) was used in crystal detectors as the first semiconductor device, and germanium (Ge) was mainly used due to its higher carrier mobility in the early days of semiconductor devices. In the early stage of semiconductor research, the surface states in which the carrier trapped were the problem. After solving the problem with surface treatment, Si has become the most widely used in semiconductor devices. For high-speed devices, gallium arsenide (GaAs) is recently used. For these three major semiconductor materials, bulky single crystals are grown and the wafers sliced from ingot are used. With this method, the variation of semiconductor materials is limited only to materials capable of growing ingot.

For semiconductor devices with thin-film semiconductor active layers, on the other hand, a wide variation of materials, not only inorganic materials but also organic materials, can be adopted. Another merit of the use of thin films is multiple-layer construction, which realizes highly efficient multi-junction solar cells and heterojunction high-performance transistors. In addition, non-equilibrium states can be also realized in the form of thin films. Examples are the amorphous phase and special material compositions which cannot be obtained in bulk form. These advantages enrich the variety of semiconductor devices characteristics. As another advantage, thin-film semiconductors can be grown on a wide variety of substrates, such as glasses and plastic. This also enriches the functions of semiconductor devices, such as transparency and flexibility, and reduces their cost. Although the performance of electronic properties of thin films is, in most cases, lower than that of wafer-based single-crystal semiconductor materials, the reduction in cost and the addition of functions shown above are of great importance. Therefore, thin film based semiconductor devices play an important role in the forthcoming ICT society.

9.1.1.3 Solution Processing for Thin-film Semiconductor Devices

There are many methods for thin-film fabrication as shown in Table 9.1. In conventional wafer-based semiconductor devices, surface modification techniques for doping and insulator fabrication are used. For high-performance thin-film semiconductor devices, the gas phase deposition methods have been intensively studied and are already being used in commercial products. On the other hand, the liquid phase deposition methods are not fully exploited, although spin coating, a class of

TABLE 9.1
Methods for thin-film fabrication

Deposition	Gas Phase	Physical Vapor Deposition (PVD)	Evaporation	Thermal Evaporation
				Electron Beam Evaporation
				Arch Evaporation
				Laser Ablation
			Sputtering	RF Sputtering
				DC Sputtering
		Chemical Vapor Deposition (CVD)	Thermal CVD	
			Plasma-Enhanced CVD	
			Metal-Organic CVD	
	Liquid Phase	Spin Coating		
		Spray Pyrolysis		
		Drop Casting		
		Imprinting		
		Ink-jet Printing		
		Langmuir-Blodgett		
		Electroplating		
		Liquid Phase Epitaxy		
Surface Modification		Oxidization		
		Diffusion		
		Ion Implantation		
Others		Extrusion Molding		
		Rapid Quenching		

Solution-Processed Oxide-Semiconductor

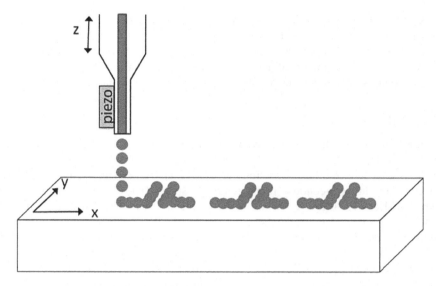

FIGURE 9.3 Schematic of ink-jet printing system.

liquid phase deposition, is partially used in lithography during the device fabrication process. This is because the electronic properties of thin films fabricated by the liquid phase deposition methods tend to be lower than those by the gas phase deposition methods. However, conventional gas phase deposition methods, such as thermal evaporation, electron beam evaporation, arch evaporation, laser ablation and sputtering, use large vacuum-based facilities that are expensive and waste most of the raw materials in the production process. On the other hand, printing technologies, such as ink-jet printing and imprinting, are expected to be a promising approach to fulfill energy/material saving for a sustainable society, because in these methods all raw materials are completely used without waste during the fabrication process (Figure 9.3) [10]. Solution processing including printing technology is a method wherein the active materials are solved into solvents for liquid phase deposition followed by removal of the solvent to form a solid thin film. Since solution processing is widely adopted as a method of liquid phase deposition, discussion is focused on the solution processing in this chapter.

In order to fully utilize the solution processing for thin-film semiconductor devices, the following issues must be solved.

1) Improvement of electronic properties:
   One of the most popular applications of thin-film semiconductor devices is high precision displays. For this, mobility, one of the main characteristics of semiconductor, is required to be more than about 10 cm^2/Vs [11].
2) Control of wettability of solution:
   Unlike the gas phase deposition methods, the wettability of the solution liquid to the substrates is an additional key factor on the quality of fabricated thin films. High wettability is preferred for uniform thin films. On the other hand, for patterned deposition, precise wettability control is required.

9.1.2 Perspectives

As shown in the previous section, thin-film semiconductor devices fabricated by solution processing have enormous potential as next-generation high-performance and low-cost semiconductor devices in sustainable ICT society. One of the main purposes is all printed devices. In order to achieve this, intensive investigations into the research for materials, fabrication of devices with n- and p-type semiconductor films without a precise patterning was carried out. In parallel with development of materials and devices, improvements of device performance are continuously being made. This will be further developed to printed devices for eliminating waste of materials.

9.2 MATERIALS

Figure 9.4 shows the band structure of conventional semiconductors (Ge, Si and GaAs) and oxide semiconductors (In_2O_3, ZnO, Cu_2O and NiO). The band gap energy value (Eg), which is defined as the energy difference between the conduction band minimum (CBM) and valence band maximum (VBM) for conventional semiconductors are relatively small (0.5–1.5 eV), and the energy levels of CBM and VBM are close to the Fermi energy of metal electrodes. This means that electrons and holes are easily injected into or extracted from conduction band and valence band of these semiconductors, respectively, by adjusting their Fermi levels of semiconductors to Fermi energy of electrodes via doping. Therefore, doping, substitution of constituent element(s) with other elements, is used to control the type and density of carriers. These characteristics of conventional semiconductors are quite important for semiconductor engineering. However, these conventional semiconductors have not been intensively investigated for solution processing because it is hard to prepare precursor solutions for these materials, and oxidation of the materials during the solution processing is almost unavoidable.

Eg values of metal oxides are larger than 2 eV, and the energy levels of CBM and VBM vary depending on the metal constituent elements. Oxides with CBM close to Fermi energy of electrodes tend to be n-type semiconductors, while oxides with VBM close to Fermi energy of electrodes tend to be p-type semiconductors. Typical examples of former are indium(III) oxide (In_2O_3) and zinc oxide (ZnO), the latter copper(I) oxide (Cu_2O) and Nickel(II) oxide (NiO). Although the control of electronic structure has not yet been well established, the carrier type, n- or p-type, can be determined by selection of materials. A similar approach is taken for organic semiconductors. In both cases of oxide and organic semiconductors, n-type semiconductors have been extensively studied

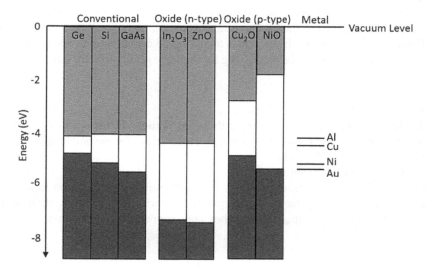

FIGURE 9.4 Band structure of conventional semiconductors, n- and p-type semiconductors. Light and dark gray boxes show conduction and valence band of semiconductors. For reference, Fermi level of metals used as electrodes are also shown. Since values depend on the way of estimation, materials' form etc., rough relative relation is to be focused in this figure.

Solution-Processed Oxide-Semiconductor

and many applications with these materials to TFTs were reported, while not many good candidates for p-type semiconductors have been found. Therefore, research on n-type semiconductors is mainly focused on device development, and research on p-type focused on materials search.

The advantages of oxide semiconductors to organic semiconductors are high performance and high stability in surrounding environment. In addition, oxide can be easily prepared by solution processing, and heat treatments for drying and synthesis after solution processing can be performed at the ambient environment in most cases. For the above reasons, oxide semiconductors are one of the most promising materials for solution-processed thin-film semiconductor devices.

9.3 RECENT DEVELOPMENT OF N-TYPE SEMICONDUCTOR THIN FILMS AND DEVICES

9.3.1 SILICON-BASED SEMICONDUCTOR DEVICES

In spite of difficulty in fabricating elemental semiconductor thin films by solution processing, challenging research on the fabrication of Si films by solution process was reported by T. Shimoda's group in 2006 [12]. They have succeeded in fabrication of TFTs with polycrystalline Si films by two kinds of solution processing, spin coating and ink-jet printing techniques (Figure 9.5). The mobility achieved by the spin coating and ink-jet printing was 108 cm^2/Vs and 6.5 cm^2/Vs, respectively. Meanwhile, most of the previous studies only achieved mobility of around 1 cm^2/Vs for amorphous Si films formed by the solution process. The key to achieving high mobility in their research is a film with a very high Si purity and the polycrystalline form. High purity Si was obtained by photo-induced ring-opening polymerization of cyclopentasilane (CPS), Si_5H_{10}, in a nitrogen-filled dry-box with a residual oxygen concentration of less than 0.5 ppm.

Shimoda's group have further developed their specific solution method and succeeded in boron and phosphorus doping for p- and n-type Si films although the necessary fabrication conditions, especially the environmental purity of nitrogen, were quite strict. Based on the doping technique, p-i-n junction solar cells were fabricated by spin cast method (Figure 9.6) [13]. The highest values of photovoltaic parameters, the open-circuit voltage (Voc), short-circuit current (Jsc), fill factor (FF) and energy conversion efficiency (η) were 0.59 V, 2.46 mA/cm^2, 0.35 and 0.51%, respectively. Although the performance is lower than conventional solar cells, establishment of doping to Si by solution processing had a great impact on Si technology.

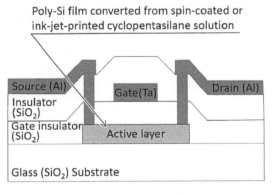

FIGURE 9.5 Schematic of a TFT with solution-processed active layer. Polycrystalline Si active layer was fabricated by spin coating or ink-jet printing, while SiO$_2$ insulating layer and electrodes (Al or Ta) were deposited by plasma chemical vapor deposition (CVD) and sputtering, respectively.

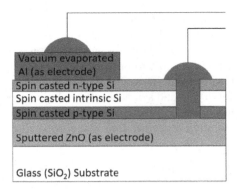

FIGURE 9.6 Schematic of a solution-processed a-Si:H solar cell structure.

9.3.2 INDIUM-BASED OXIDE THIN FILMS AND DEVICES

Since the discovery of gallium (Ga)- and zinc (Zn)-doped indium oxide, In-Ga-Zn-O, called IGZO, transparent amorphous oxide semiconductors (TAOSs) are now gaining worldwide attention as a replacement for Si-based semiconductors [14–19]. The advantage of TAOSs is high optical transparency in the visible-light region and high mobility more than 10 cm²/Vs in the amorphous form [20]. These characteristics are quite preferable for thin-film devices, and many other elements as dopants are explored. Typical examples of device performance of n-type TFTs with indium-based oxide thin films fabricated by conventional vapor deposition methods are extracted from [21, 22] and summarized in Table 9.2. The highest mobility exceeds 200 cm²/Vs.

One of the fatal drawbacks of most oxide semiconductors was oxygen-related defects such as oxygen vacancies that may induce stability issues under bias stress and illumination [23–29]. H.-D. Kim et al. reported that the O 1s peak in the X-ray photoelectron spectroscopy (XPS) spectrum of In_2O_3 was resolved into two sub-peaks originating from oxygen forming metal–oxygen (M-O) bonds and near oxygen vacant sites (V_O), and incorporation of nitrogen (N) into In_2O_3 suppressed V_O [26]. An example of XPS O 1s spectrum of indium-based oxide consisting of the two peaks is shown in Figure 9.7. K. Tsukagoshi et al. adopted Si, titanium (Ti) and tungsten (W) to suppress V_O

TABLE 9.2
Device performance of n-type TFTs with indium-based oxide thin film fabricated by conventional vapor deposition methods [21, 22]

Channel material	μ (cm²/Vs)	I_{ON}/I_{OFF}	Deposition method	Reference
In_2O_3	41.8	10^7	Atomic layer deposition	22
In-Si-O	17	10^9	Sputtering	21
In-Ti-O	32	10^9	Sputtering	21
In-Zn-O	157	10^{10}	Sputtering	22
In-Ga-O	9.45	10^8	Atomic layer deposition	22
In-W-O	39	10^{10}	Sputtering	22
In-Al-Zn-O	20.65	10^6	Sputtering	22
In-Ga-Zn-O	38.29	10^6	Sputtering	22
In-Sn-Zn-O	37.2	$\sim 10^7$	Sputtering	21
In-Hf-Zn-O	10	10^8	Sputtering	21
In-W-Ti-O	27.8	10^{11}	Atomic layer deposition	22
In-W-Zn-O	22.3	10^8	Sputtering	22
Al-In-Zn-Sn-O	30.2	10^9	Sputtering	21
In-Sn-O/Al-Sn-Zn-O	246	10^8	Sputtering	22

Solution-Processed Oxide-Semiconductor 233

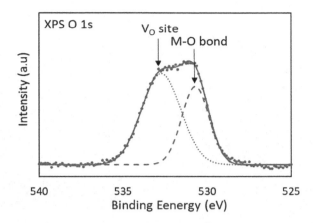

FIGURE 9.7 An example of XPS O 1s spectrum consisting of two peaks originating from metal–oxygen (M-O) bonds and near oxygen vacant (V_O) sites. Filled circle, solid curve, dashed curve and dotted curve denote experimental data, fitting result, M-O band contribution of fitting and V_O site contribution of fitting, respectively.

[30–34]. All elements used for suppression of V_O have higher values of bond dissociation energy with oxygen (N-O: 631 kJ/mol, Ti-O: 666 kJ/mol, W-O: 720 kJ/mol, Si-O: 799 kJ/mol) than the value of oxygen (O-O: 498 kJ/mol), while the elements used in IGZO have lower values (Zn-O: 250 kJ/mol, In-O: 346 kJ/mol, Ga-O: 374 kJ/mol) [35].

Reports of high device performance of TAOSs by conventional deposition methods have accelerated development of solution-processed TFTs. In general, low device performance of solution-processed TFTs is mainly caused by low quality of thin films, especially low crystallinity. However, the film quality of amorphous films is almost independent of the fabrication method. In addition, tetraethyl orthosilicate (TEOS), as a good candidate of raw material for constituent element of Si for reduction of V_O, is available for solution processing. Below, research on TFTs with Si-doped indium oxide, In-Si-O, is focused as a promising candidate of high-quality and high-stability (oxygen defect free) TFTs fabricated by solution processing [36–40]. As discussed in Section 9.1.1.3, for selective deposition of patterned circuits, controlling the wettability of precursor solution to the substrate, namely, interaction between the precursor solution and the substrate is quite important [41–43]. As a first approach, uniform In-Si-O thin-films both on hydrophilic and hydrophobic substrates were formed, followed by fabrication of TFTs.

For hydrophilic substrates, a precursor solution prepared by dissolving indium chloride and TEOS into acetonitrile and ethylene glycol was spin coated on NaOH-treated Si/SiO$_2$ substrates. Thin films were annealed at 300–600 °C in ambient or oxygen atmosphere. To fabricate a transistor, source and drain electrodes of 200-nm-thick aluminum (Al) were deposited via the thermal evaporation method through a stencil shadow mask with various sizes. The channel width, W, was fixed as 1000 μm while channel length, L, varied from 50 μm to 350 μm. A microscope image and schematic cross-sectional view of the In-Si-O TFT are shown in Figure 9.8. From the perspective of mobility, the optimal performance of solution-processed In-Si-O TFTs on the hydrophilic substrate corresponded to 8.1 cm^2/Vs for the In-Si-O TFT ($L = 100$ μm, $W = 1000$ μm) annealed at 400 °C in oxygen atmosphere, although the drain current I_D did not vanish even at gate-source voltage $V_{GS} = -40$ V: output characteristics (I_D versus drain-source voltage V_{DS}) and transfer characteristics (I_D versus V_{GS}) are shown in Figure 9.9. Reducing the annealing temperature resulted in reduction of the I_D at off states and increase of the current on/off ratio, I_{ON}/I_{OFF} (Figure 9.10). Maximum current on/off ratio was 6×10^6 for the In-Si-O TFT annealed at 300 °C in air.

For hydrophobic substrates, a precursor solution by dissolving indium(III) isopropoxide and TEOS into 2-(diethylamino)ethanol was spin on hexamethyldisilazane (HMDS)-treated Si/SiO$_2$ substrates. Annealing and fabrication process of In-Si-O TFTs were the same as that for TFTs on the hydrophilic

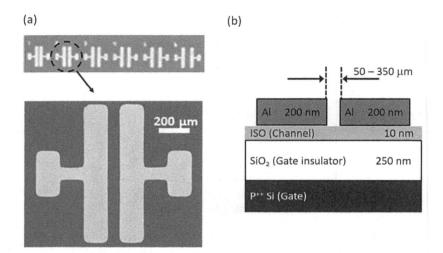

FIGURE 9.8 (a) Microscope image of In-Si-O TFTs. (b) Schematic cross-sectional view of the In-Si-O TFT. Reprinted from [40] with permission from Elsevier.

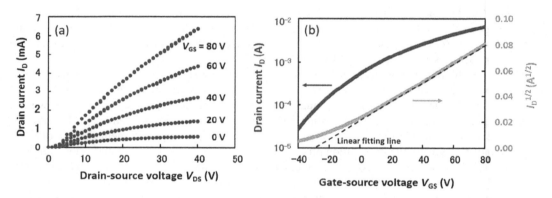

FIGURE 9.9 (a) Output and (b) transfer characteristics of In-Si-O TFT with $L = 100$ μm and $W = 1000$ μm fabricated by solution A on the hydrophilic substrate and annealed at 400 °C in oxygen atmosphere. Blue and red marks denote the experimental data, and dashed line denotes liner fitting results of the experimental data for $40\ \text{V} \leq V_{GS} \leq 80\ \text{V}$. Reprinted from [40] with permission from Elsevier.

substrates. The highest mobility corresponded to 3.4×10^{-1} cm^2/Vs (Figure 9.11). Although the number of trap sites for electrons of the hydrophobic substrates tends to be fewer than that of the hydrophilic substrates in general, the highest mobility of the In-Si-O TFT on hydrophobic substrate was lower than that of the TFT on hydrophilic substrate. The results suggest that mobility of the In-Si-O TFT on the hydrophobic substrate would be further increased by optimizing fabrication conditions.

The device performance of In$_2$O$_3$ and I-X-O (X: Si, Zn, Ga, Sc, Y, La) TFTs fabricated by the solution process are summarized in Table 9.3. Further improvements of device performance, such as mobility and I_{ON}/I_{OFF}, and direct puttering of TFT device have been investigated.

9.3.3 Indium-free Oxide Thin Films and Devices

Zinc oxide (ZnO) is a very familiar material, which is used as an additive in many products, such as cosmetics, ointments, foods, food supplements, paints, rubbers etc. It is also used as an electronic material. Recently, extremely high electron mobility and the resulting quantum phenomena

Solution-Processed Oxide-Semiconductor

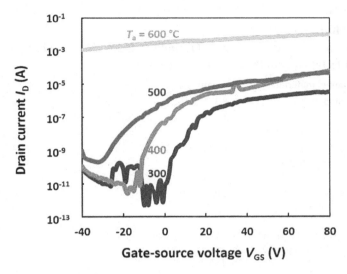

FIGURE 9.10 Transfer characteristics of In-Si-O TFTs with $L = 50$ μm and $W = 1000$ μm fabricated by solution A on the hydrophilic substrate and annealed at various temperatures in air. Reprinted from [40] with permission from Elsevier.

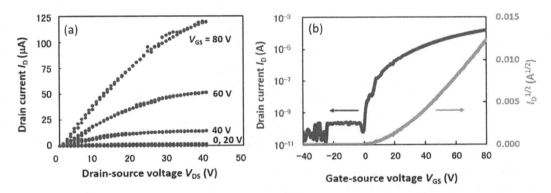

FIGURE 9.11 (a) Output and (b) transfer characteristics of ISO TFT with $L = 50$ μm and $W = 1000$ μm fabricated via solution B on the hydrophobic substrate and annealed at 500 °C in air. Reprinted from [40] with permission from Elsevier.

were observed [49–54]. High mobility up to 1×10^6 cm^2/Vs has been achieved in Mg$_x$Zn$_{1-x}$O/ZnO hetero-structures. In Mg$_x$Zn$_{1-x}$O/ZnO hetero-structures, charge carriers are induced in ZnO as two-dimensional electron gas (2DEG) by electric fields induced at interface of the two layers due to the difference in the spontaneous polarization between Mg$_x$Zn$_{1-x}$O and ZnO perpendicular to the hetero-layers (Figure 9.12). This encouraged application of ZnO to solution-processed TFTs.

A precursor solution was made by dissolving zinc acetate dihydrate into acetone with small amount of monoethanolamine. After spin coating of the precursor solution on a NaOH-treated Si/SiO$_2$ substrate, the thin films were annealed at 300–1000 °C in air. Different from In-Si-O films, ZnO films after annealing comprised polycrystalline form, and the crystallite size increased with increasing annealing temperature up to 600 °C and saturated to about 100 nm at higher temperatures. Fabrication of TFTs was the same as that of In-Si-O TFTs. A preliminary result of ZnO TFTs by solution processing showed a clear n-type operation in transfer characteristics with the mobility of 3.8×10^{-1} cm^2/Vs and the I_{ON}/I_{OFF} of 4.4×10^2 (Figure 9.13). Increase of mobility and reduction of off current $I_{D,\,min}$ being investigated. For improvement of mobility, increase of

TABLE 9.3
Device performance of In_2O_3 and I-X-O (X: Zn, Ga, Sc, Y, La) TFTs fabricated by the solution process

Channel material	μ (cm^2/Vs)	I_{ON}/I_{OFF}	Reference
In-Si-O	**8.3**	> 10^2	[40][a]
	1.9×10^{-3}	**6×106**	[40][b]
In_2O_3	0.7 (43.7[c])	10^6	[44]
$In_{0.7}Zn_{0.3}O_{1.35}$	9.78	10^4	[45]
$In_{1.00}Ga_{0.11}Zn_{0.29}O_x$	19.1	~10^3–10^5	[46]
In-Sc-O	~9	-	[47]
In-Y-O	~10	-	[47]
In-La-O	9.7	10^7	[46], [48]

a: TFT annealed at 400 °C in oxygen atmosphere as best μ.
b: TFT annealed at 300 °C in air as highest I_{ON}/I_{OFF}.
c: TFT with self-assembled nanodielectric as gate electric.

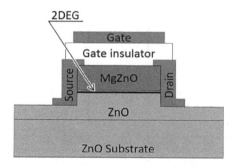

FIGURE 9.12 Schematic cross-sectional view of the field-effect transistor using $Mg_xZn_{1-x}O$/ZnO heterostructure with atomic-layer-deposited Al_2O_3 gate insulator.

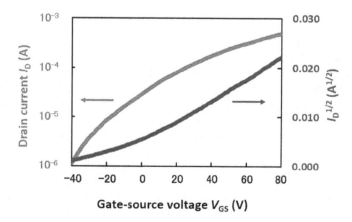

FIGURE 9.13 Transfer characteristics of polycrystalline ZnO TFT with L = 50 μm and W = 1000 μm fabricated on the hydrophilic substrate and annealed at 800 °C in air.

crystallite size by optimizing annealing conditions and reduction of the channel length so as to be close to each other is a promising approach.

9.4 POTENTIAL DEVELOPMENT OF P-TYPE THIN FILMS

9.4.1 Cu-doped ZnO Thin Films

In this part, we present on Cu-doped ZnO (CZO) thin films whose Cu doping concentrations were varied as 0%, 0.5%, 1%, 1.5% and 2%; and the annealing temperature and the annealing time were set to be 500°C and 30 min, respectively. Figure 9.14 shows XRD patterns for various doping concentrations of CZO thin films. Here, the concentration of Zn^{2+} salt ion in precursor solution is 0.5 M, and the molar ratio between Zn^{2+} salt and monoethanolamine is 1:1. It can be seen that the CZO thin films were oriented along with (100), (002) and (101) preferred planes at 31.79°, 34.43° and 36.27°, respectively. No traces of residue copper metal or zinc oxide were detected; this result confirms that Cu ions successfully replaced Zn in crystal lattice. Based on the determination of the preferential growth and the investigation on the variation of the crystallites size when the Cu doping concentration increases, we obtained that the best crystallization corresponds to the case of 0.5% dopant concentration.

Figure 9.15 (a), (b), (c), (d) and (e) show the surface morphology of CZO thin films, corresponding to the Cu-doping concentration of 0%, 0.5%, 1%, 1.5% and 2%. The white areas illustrate the crystal particles and the black spaces indicate the porosity of the surface morphology. SEM observation indicates that the morphology of CZO thin films significantly depends on Cu-doped concentration. It can be also seen from Figure 9.15 that the crystal density was relatively distributed on the substrate surface, and the grain size was quite uniform, that is, whole films have homogeneously smooth surfaces. According to this figure, once the Cu doping level was raised, the porous spaces in the surface morphology of the CZO thin films reduced, whereas the grain density increased. It means

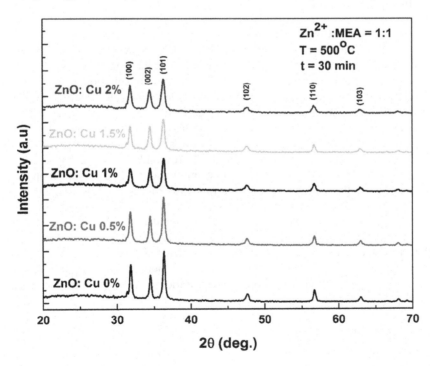

FIGURE 9.14 XRD patterns of Cu-doped ZnO thin films with various doping concentrations: 0%, 0.5%, 1%, 1.5% and 2%.

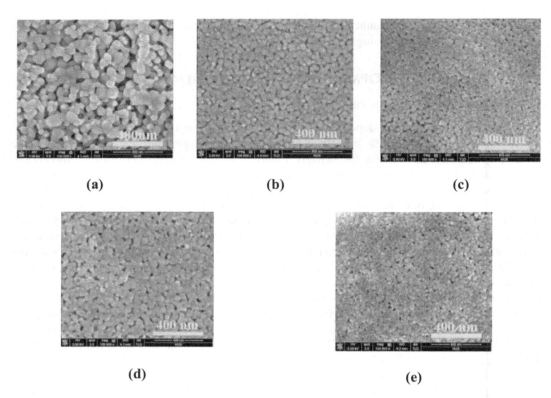

FIGURE 9.15 SEM micrographs of Cu-doped ZnO thin films with various doping concentrations: (a) 0%, (b) 0.5%, (c) 1%, (d) 1.5% and (e) 2%.

that the grain size decreased with rising Cu doping concentration. The degradation of the crystallinity caused by the change in concentration of Cu can be explained in terms of nucleation centers increasing with introduction Cu atoms, the latter acting as a nucleation center. Based on the XRD analysis, it can be concluded that the increase of Cu doping concentration leads to the reduction of the grain size, thus affecting the microstructure and the surface morphology of the CZO thin films.

Figure 9.16 illustrates the absorbance spectra of CZO thin films fabricated at different Cu-doped concentrations at 0%, 1%, 1.5% and 2%. We can see that CZO thin films had a high absorbance in the visible region. The absorbance of CZO thin films was optimal at wavelengths from 360 to 375 nm. The highest absorbance of the CZO thin films corresponds to the Cu-doped concentration of 0.5%. It can be further noticed that the doping concentration of Cu had an impact on the absorbance of CZO thin films. As the Cu doping concentration was raised, the absorbance spectrum peaks gradually shifted toward shorter wavelengths. As a result, the electronic energy (eV) increased, and the bandgap energy was augmented. Hence, it can be concluded that the increase of Cu doping concentration enhances the absorbance of the CZO thin films.

Table 9.4 shows the bandgap energy and transmittance of CZO thin films corresponding to different Cu-doped concentrations. It can be concluded that the optical properties of CZO thin films are not stable. The CZO thin film of 0.5% doped concentration is the most optical with both bandgap and transmittance.

9.4.2 CuO Thin Films

The crystalline quality of the CuO thin films was analyzed using an X-ray diffractometer (XRD, Bruker D5005) with Cu-Kα radiation at room temperature. XRD patterns of the CuO thin film samples prepared with varying Cu^{2+} concentrations of precursors are shown in Figure 9.17. Typical

Solution-Processed Oxide-Semiconductor

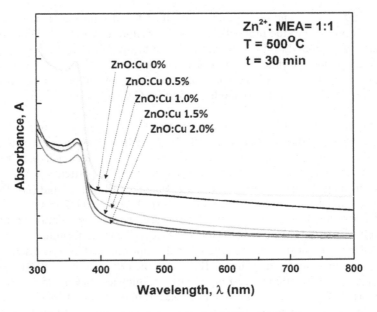

FIGURE 9.16 The absorbance spectra of Cu-doped ZnO thin films with various doping concentrations: 0%, 0.5%, 1%, 1.5% and 2%.

TABLE 9.4
The bandgap energy and transmission of CZO thin films with various Cu-doped concentrations at 500°C

Doping concentration	0%	0.5%	1%	1.5%	2%
E_g (eV)	3.20	3.16	3.21v	3.23	3.13
T%	74.49	87.68	67.58	44.52	75.33

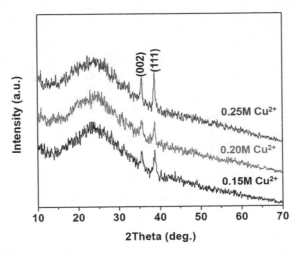

FIGURE 9.17 XRD patterns of CuO thin films on glass substrates, with different Cu^{2+} ion concentrations. Reprinted from [70] with permission from Elsevier.

peaks appear at $2\theta = 35.6°$ and $38.7°$, and are well matched with the standard (002) and (111) orientations corresponding to the monoclinic crystal structure of CuO (JCPDS41-0254). No other significant phases were found in the XRD patterns, confirming that all samples mainly exhibit the CuO phase. Furthermore, the ratio of the peak intensities at 35.6° and 38.7° concurs with that of the peak intensities at (002) and (111) orientations based on standard patterns. This implies that the samples are polycrystalline, similar to another report on CuO thin films prepared using the sol-gel method [55].

The morphology of the CuO thin films was observed via scanning electron microscopy (SEM, Nova NANOSEM-450) as shown in Figure. 9.18. To determine the average particle size of CuO, we utilized a free software, ImageJ, to obtain the dependence of the percentage distribution on the particle size. Thereafter, using Gaussian fitting functions, the average particle size (D_{mean}) and standard deviation (ΔD) of the particle sizes were extracted (Figure. 9.18 (d), (e) and (f)). D_{mean} was 28.4 ± 6.9 nm, 40.4 ±5.4 nm and 48.6 ± 5.9 nm for the 0.15 M, 0.20 M and 0.25 M samples, respectively. Observably, the particle sizes derived by the ImageJ software and the Gaussian fitting functions, D_{mean}, agree well with those calculated by using the Debye–Scherrer formula. From these results, it can be inferred that the grain size of the particles on the thin-film surface increases with the nominal Cu^{2+} ion concentration. Inset (i) in Figure. 9.18a shows the cross-sectional SEM image acquired to determine the thickness of the thin films. Empirically, the thickness of CuO thin films increases from 74 nm to 95 nm as the Cu^{2+} ion concentration increases from 0.15 M to 0.25 M. In addition, the CuO thin films become less porous as the film thickness increases, that is, the surface of the thin films becomes denser. Therefore, if the Cu^{2+} ion concentration is chosen as 0.25 M, a CuO thin film with a non-porous surface can be achieved. This will be discussed subsequently. During the first step of gel formation, the Cu^{2+} ions react with MEA and form a 4-species complex [56] linking copper ions together. For the same volume, a lower concentration of the Cu^{2+} ion produces a lower concentration of the complex. After post-heating at 90 °C and annealing at 550 °C, CuO particles with smaller sizes are formed. Consequently, these cannot fill the spaces in the thin films. On the other hand, at a higher nominal concentration of Cu^{2+} ions, the CuO particles are larger and denser. At 0.25 M, the

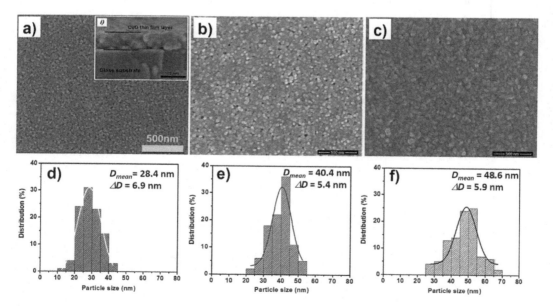

FIGURE 9.18 SEM micrographs of CuO thin films for various Cu^{2+} ion concentrations: (a) 0.15 M, (b) 0.20 M and (c) 0.25 M. Dependence of percentage distribution on particle size calculated by using ImageJ software and fitted by Gaussian functions: (d) 0.15M, (e) 0.20M and (f) 0.25M. Inset of (a) shows a cross-section for the case of 0.15M sample. Reprinted from [70] with permission from Elsevier.

Solution-Processed Oxide-Semiconductor 241

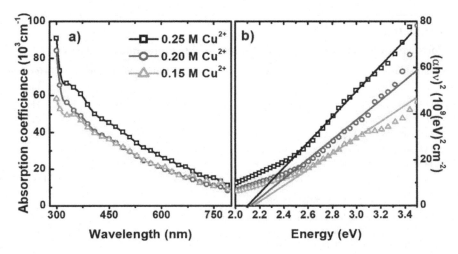

FIGURE 9.19 Variation of absorbance spectra (a) and Tauc graphs (b) plotted as a function of Cu^{2+} ion concentrations. Reprinted from [70] with permission from Elsevier.

concentration of the complexes completely fills up the voids within the thin films. Hence, dense thin films with low porosity are well formed.

The optical transmittance of the CuO thin films was determined using a UV/VIS spectrophotometer (UV 2450-PC, Shimadzu). Because the transmittance of thin films is sensitive to the film thickness, only the highest average transmittance in the visible wavelength region was considered for the sample with 0.25 M. The absorption coefficient of the thin films was calculated using the Beer-Lambert formula [57], that is,

$$\alpha = -\frac{\ln T}{d}, \quad (9.1)$$

where, T and d are the transmittance and thickness of the thin film, respectively. This computation is plotted in Figure 9.19(a). At the region of interest, it can be visualized that the absorption coefficient of the thin films prepared from 0.25 M Cu^{2+} ion concentration is higher than those of the thin films from a concentration of either 0.15 M or 0.20 M. The results can be qualitatively explained using the relation,

$$\alpha = C\varepsilon, \quad (9.2)$$

where ε is the characteristic extinction coefficient of the thin-film material, and C is the concentration of the precursor solution indicative of how dense the thin film would be. The higher absorption coefficient of the 0.25 M sample signifies that it has higher opacity than the samples prepared from lower Cu^{2+} ion concentration. The optical bandgap energy can be determined by using an extrapolation function, for which the Tauc relationship was used as follows [58]:

$$\alpha h\upsilon = A\left(h\upsilon - E_g\right)^2 \quad (9.3)$$

Here, α is the absorption coefficient, h is Planck's constant, υ is the photon frequency and E_g is the optical direct bandgap energy. An extrapolation of the linear region to the x axis, plotted in the graph of $(\alpha h\upsilon)^2$ versus photon energy, $h\upsilon$, gives the value of E_g, as shown in Figure 9.19(b).

From Figure 9.19, E_g can be extracted to be 2.15 eV for the 0.15 M sample. However, this is reduced to 2.10 eV for the 0.25 M sample. Although CuO is well known as an electrical p-type semiconductor with indirect bandgap [59–61], a previous study reveals that the optical direct bandgap

TABLE 9.5

Performance comparison of the p-type oxide-semiconductor TFTs

Channel layer	μsat (cm2V−1s−1)	On/off current ratio	Reference
Cu_2O	4.30	3.0×10^6	[40]
SnO	0.90	5.2×10^4	[13]
NiO	0.48	1.8×10^3	[12]
CuO (sputtering)	0.01	1.0×10^4	[39]
[a]CuO (solution process)	0.26 – 0.78	$10^5 – 10^6$	[38]
[b]CuO (solution process)	1.2×10^{-2}	2.0×10^4	[2]
[c]CuO (solution process)	6.24×10^{-4}	7.81×10^2	This work

Reprinted from [70] with permission from Elsevier.

a: CuO prepared at the low annealing temperature of 300°C.
b: CuO optimized with the annealing time.
c: CuO optimized with the precursor concentration.

can be calculated. This suggests a directly allowed inter-band transition of the material. The E_g of CuO has been reported to be approximately 2.1 eV [62, 63], which shows that our findings are comparable with others. In addition, in our case, the decrease of the bandgap indicates the improvement of the thin films in electrical properties. Furthermore, the shifts of E_g might be due to the decrease of the defects in the thin films.

9.4.3 PROSPECTIVE P-TYPE THIN FILMS

Table 9.5 presents a performance comparison of typical p-type oxide-semiconductor TFTs in the recent years. The performance of Cu_2O TFT appears to be more promising than that of SnO, NiO and CuO TFTs, because of higher field-effect mobility and on/off current ratio. Only few studies exist, in which the CuO TFTs were fabricated using a solution process. Although the performance of CuO TFT in this work is poorer than those reported by other studies [64], some possible routes can be considered to improve the transistor operation, for instance, optimization of the annealing time, annealing temperature, or the use of HfO_2 or ScO_x as a gate insulator, instead of SiO_2.

9.5 DEVICE APPLICATIONS

9.5.1 THIN-FILM TRANSISTORS

Figure 9.20 depicts the schematic drawing of the fabricated CuO TFT structure, in which a commercial Si substrate with heavy hole doping was used as the gate electrode, and a 250-nm-thick SiO_2 thin film formed by thermal oxidation technique was used as the gate insulator. The 40-nm-thick CuO thin film was deposited as a channel layer via a solution process with optimized conditions as mentioned in the ref. [71]. A 200 nm thick aluminum was patterned using the thermal evaporation technique for the source and drain electrodes of the CuO TFTs with stencil shadow masks of various sizes.

Figure 9.21 depicts the transfer characteristics of CuO TFTs for different L annealed at 550 °C in air, measured at a drain-source voltage of $V_{DS} = -80$ V. The inset of Figure 9.21 presents a typical microscopic image of the fabricated CuO TFTs. The CuO TFTs showed a clear p-type operation with an on/off current ratio of 10^2. In the saturation regime, I_D is determined as follows:

$$I_D = \frac{W.\mu.C_{SiO_2}}{2L}\left(V_{GS} - V_T\right)^2 \tag{9.4}$$

Solution-Processed Oxide-Semiconductor 243

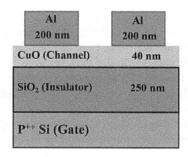

FIGURE 9.20 Schematic drawing of the CuO TFT structure. Reprinted from [71] with permission from Elsevier.

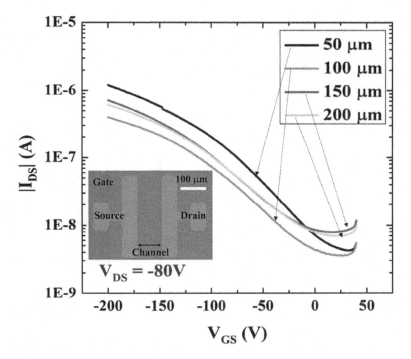

FIGURE 9.21 Transfer characteristics of the CuO TFTs with various channel lengths. The inset is a microscope image of the CuO TFT fabricated. Reprinted from [71] with permission from Elsevier.

where, μ is the field-effect mobility, and V_T is the threshold voltage. C_{SiO2} is the gate SiO$_2$ capacitance [65]. Additionally, the field-effect mobility of CuO TFTs can be estimated as follows:

$$\mu = \frac{2L}{WC_{SiO_2}} \left(\frac{\partial \sqrt{I_{DS}}}{\partial V_G} \right)^2 \qquad (9.5)$$

In consequence, we can estimate μ and V_T from the slope and intercept of the fitting line, according to the $\sqrt{I_{DS}}$ versus V_G characteristics. The μ, V_T and on/off current ratio obtained for the CuO TFTs with different channel lengths ranging from 50 to 200 μm with a step of 50 μm are listed in Table 9.4. The optimum values of μ and on/off current ratio for CuO TFTs were estimated to be 10^{-4} cm^{2}V^{-1}s^{-1} and 10^{-2}, respectively. In another study, the μ of 0.012 ~ 1.6 × 10^{-4} cm^{2}V^{-1}s^{-1} and the on/off current ratio of 10^3–10^4 were reported for the CuO TFTs fabricated via the spin-coating method [66], which is quite similar to our work. For typical solution-processed CuO TFTs, the field-effect mobility and the on/off current ratio were found to be 0.26 cm^{2}V^{-1}s^{-1} and 10^5, respectively. However, the best transistor performance with a μ of 0.78 cm^{2}V^{-1}s^{-1} and on/off current ratio of 10^5 was reported by A. Liu et al. for the CuO TFTs based on the ScO$_x$ dielectric [64]. Therefore, it is suggested that high-k materials should be potentially used to enhance the field-effect mobility and the on/off current ratio, in future studies.

In addition to the field-effect mobility and the on/off current ratio, the sub-threshold swing (*SS*) is also a key parameter of TFTs for the evaluation of switching ability, and it is determined as follows:

$$SS = \frac{\partial V_{GS}}{\partial \log I_{DS}} \tag{9.6}$$

The estimated values of *SS* for CuO TFTs with various channel lengths are 65.83, 77.04, 85.11 and 84.25 V/dec with different channel lengths ranging from 50 to 200 μm, with a step of 50 μm. CuO TFT with a channel length of 50 μm exhibits faster transition between the off state (low-leveled current) and on state (high-leveled current) as compared to other channel lengths.

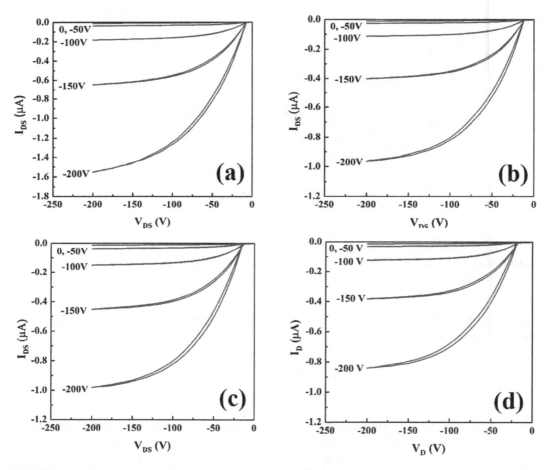

FIGURE 9.22 Output characteristics of the CuO TFTs with various channel lengths: a) 50 μm, b) 100 μm, c) 150 μm and d) 200 μm. Reprinted from [71] with permission from Elsevier.

Solution-Processed Oxide-Semiconductor 245

Figure 9.22 describes the output characteristics of CuO TFTs for different channel lengths annealed at 550°C in air. In this measurement, the V_{GS} was scanned from 0 to −200 V with a step size of −50V. The V_{DS} was also scanned from 0 to −200V. The output-characteristic curves showed clear linear and saturation regions associated with the field-effect transistor, essentially belonging to the p-type semiconducting operation, which should be interesting considering the known non-toxic character of the CuO. At a low V_{DS}, the magnitude of I_D increases linearly, thereby indicating that the existence of an injection barrier between the source electrodes and the channel of CuO thin film was in ohmic contact, which implies a well-switched on current originating from the hole carriers, as expected.

In this work, the thickness of the gate insulator layer was 250 nm. When the applied voltage was 200 V, the electric field was 800 kV/cm, which was relatively close to the breakdown electric field of 1000 kV/cm for SiO_2 material; however, this is not a serious problem. The surface of the SiO_2 should be well treated to avoid the presence of any interface layers between the SiO_2 and the CuO, although it was etched in 2% HF acid for 30 s. The presence of any interface layers may lead to the dropping of the applied voltage, and therefore, this leads to the application of a voltage smaller than the actual value to the SiO_2 layer. This causes the obtained subthreshold swing to be high. Therefore, the SiO_2 surface must be further processed by changing the dipping time or the acid concentration.

9.5.2 Solar Cells

9.5.2.1 Absorption Figure of Merit

Figure 9.23 shows the absorption rate (in percentage) of the thin films under UV–visible light. The average absorption rate of CuO thin films is relatively high at 55–80% at high photon energy, but is only 10–20% in the NIR region. In this work, we considered the global spectrum of sunlight at 37 °C. The data were obtained from the American Society for Testing and Materials (ASTM) for photovoltaic performance evaluation. The sun illuminates less flux in the UV region and maximum flux at

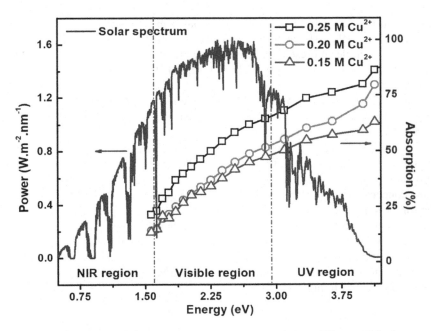

FIGURE 9.23 Absorption spectra of CuO thin films corresponded to different Cu^{2+} ion concentrations measured from 300 nm to 800 nm (point-line experimental data) and global solar spectrum (brown curve), referred from the G137 standard spectrum of American Society for Testing and Materials (ASTM) for the evaluation of photovoltaic performance. Reprinted from [70] with permission from Elsevier.

the visible region, but decreases as the photon energy decreases. Ideally, the optimal absorption for harvesting of natural light should have the same spectra with the solar flux. Figure 9.23 shows how the CuO thin films displayed a fine absorption ability in the visible and NIR region, but a highly unwanted absorption in the UV region. Therefore, in order to evaluate the light harvesting ability of the conductive oxide thin films, the absorption length L_α is considered using the following formula [67]:

$$\frac{1}{L_\alpha} = \frac{\int_{Eg}^{\infty} \alpha(E) u_{ph}(E) dE}{\int_{Eg}^{\infty} u_{ph}(E) dE} \tag{9.7}$$

L_α is a characteristic factor indicative of which material can harvest solar light effectively. $\alpha(E)$ is the absorption coefficient as a function of the photon energy, and $u_{ph}(E)$ is the photon flux.

Apart from L_α, the figure of merit was also determined by [68, 69]:

$$F = \left(-\rho \ln T\right)^{-1} = \left(\rho \alpha d\right)^{-1} \tag{9.8}$$

where, ρ is the electrical resistivity. However, the figure of merit computation excludes the solar flux spectrum, implying that it is not related to the nature of light harvesting ability. Therefore, we introduce an absorption figure of merit (so-called a-FOM) for non-solar cells, which includes the solar flux spectrum. By combining Equations (9.7) and (9.8), the following equation can be obtained:

$$a - FOM = < \rho \alpha d >^{-1} = \left(\frac{\int_{Eg}^{\infty} \rho d\alpha(E) u_{ph}(E) dE}{\int_{Eg}^{\infty} u_{ph}(E) dE}\right)^{-1} = \left(\rho d \frac{1}{L_\alpha}\right)^{-1} = \frac{L_\alpha}{\rho d} \tag{9.9}$$

The a-FOM must be favorable for the absorption layer in dark surfaces, like CuO thin films in photonic devices. Figure 9.24 shows the dependence of both L_α and a-FOM on the Cu^{2+} ion concentration. The black circles represent the L_α values, while the blue squares are for a-FOM. The L_α value was 122 nm for the 0.15 M sample, 121 nm for the 0.20 M sample and 99 nm for the 0.25 M sample. On the other hand, a-FOM showed a positive correlation with a value of 1.02 $\Omega^{-1}cm^{-1}$ for the 0.15 M sample, 11.39 $\Omega^{-1}cm^{-1}$ for the 0.20 M sample and 12.79 $\Omega^{-1}cm^{-1}$ for the 0.25 M sample. Additionally, using an optimum concentration of 0.25 M, the values of a-FOM and L_α were evaluated at various annealing temperatures of 350 °C, 450 °C and 550 °C, as shown in the inset of Figure 9.24. However, detailed data regarding the crystallization, surface morphology and optical and electrical properties will be presented elsewhere. The a-FOM of the sample (i.e., the 0.25 M sample) at 550 °C is higher than that obtained at 350 °C and 450 °C. According to these results, a temperature of 550 °C and concentration of 0.25 M is preferred for the deposition of high-quality CuO thin films. Accounting for the small difference in bandgap energy and a clear improvement in conductivity, the increment of a-FOM with Cu^{2+} ion concentration and annealing temperature is mainly attributed to the lower resistivity of denser thin films. Thus, the experimental results reveal that a Cu^{2+} ion concentration of 0.25 M and an annealing temperature of 550 °C are optimum for the fabrication of high-quality CuO thin films. The large value of a-FOM reflects a p-type semiconductor layer best used for harvesting natural light [70].

9.5.2.2 Simple Solar Cell Structure

Figure 9.25a shows the layered structure of a simple solar cell by using all solution processes. Here, the heterojunction was created between CuO and Al-doped ZnO (ZnO:Al) thin films, while the bottom electrode was ITO thin film and the top electrode was LNO thin film. One noted that Al thin-film layer was deposited on the LNO electrode to ensure a low resistance when evaluating the

Solution-Processed Oxide-Semiconductor 247

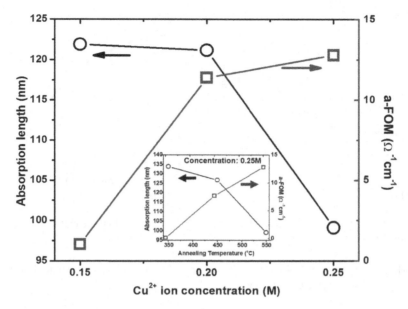

FIGURE 9.24 Absorption length (circle) and absorption figure of merit (square) of CuO thin films depend on Cu^{2+} ion concentration. The inset is dependence of absorption length and absorption figure of merit on annealing temperature for the concentration of 0.25 M. Reprinted from [70] with permission from Elsevier.

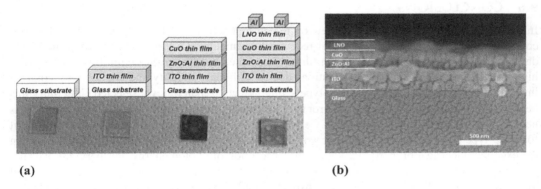

FIGURE 9.25 (a) Layers formation and (b) cross-sectional SEM of the fabricated oxide-semiconductor solar cell.

photonic-electric conversion. Figure 9.25b indicates the cross-sectional SEM of fabricated solar cell structure. The formation of each layer is relatively clear even under the solution process. Figure 9.26 exhibits a voltage-dark current (*I-V*) characteristic measured at room temperature. A non-linear behavior of *I-V* curve implies a typical characteristic of p-n junction, which formed between p-type semiconducting layer (CuO thin film) [71] and n-type semiconducting layer (ZnO:Al thin film). In this measurement the sun power was used to be 4 sun, corresponding to the power density of 400 W/cm^2. From the power-voltage (*P-V*) graph as shown in Figure 9.26, we can determine $V_{max} = 2.39$ V and $I_{max} = 9.55 \times 10^{-5}$ A at the maximum power. From *I-V* graph, we can obtain the open-circuit voltage $V_{oc} = 4.66$ V and the short-circuit current $I_{sc} = 1.61 \times 10^{-4}$ A. Therefore, we can calculate that the fill factor is as below.

$$FF = \frac{V_{max} \times I_{max}}{V_{op} \times I_{sc}} = 30.39\%$$

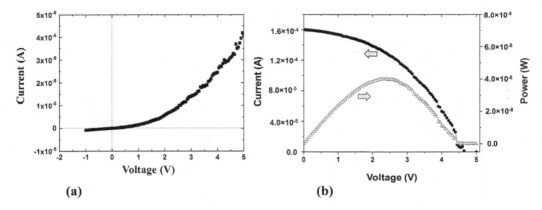

FIGURE 9.26 (a) A dark current-voltage (*I-V*) characteristic and (b) *P-V* characteristic of the fabricated oxide-semiconductor solar cell

In this work, the size of glass substrate is 20 × 20 cm². As a result, we estimate that the efficiency of all-solution-processed solar cell will be:

$$\eta = \frac{V_{oc} \times I_{oc} \times FF}{P_{in}} = 1.43\%$$

9.6 CONCLUSIONS

In this chapter, we presented an overview on the current status of solution-processed oxide-semiconductor materials and devices, and our recent achievements on n-type and p-type thin films and devices. For n-type semiconducting thin films, we pointed out that In-Si-O thin films possessed a high potential toward electronic applications due to its higher mobility and conductivity comparing with other materials. Besides, for p-type semiconducting material, we selected CuO thin films because of non-toxic and friendly environmental aspects, even though the electrical property is necessary to be improved further. Also, we proposed a new concept of absorption figure of merit for the p-type semiconducting layer that would be useful for selecting a trade-off material between the excellent electrical property and the strong absorption of photon energy. Furthermore, we successfully verified that thin-film transistor using In-Si-O operated as an n-channel transistor and that using CuO worked as a p-channel transistor. Finally, for the very first time, we demonstrated a simple solar cell with ZnO:Al/CuO junction structure, under all solution processes. Although the energy-conversion efficiency is not high, it should be promising for a new selectivity between two traditional categories of Si-based and organic solar cells.

REFERENCES

[1] "November 17 – December 23, 1947: Invention of the First Transistor". APS NEWS November 2000 (Volume 9, Number 10), This Month in Physics History, https://www.aps.org/publications/apsnews/200011/history.cfm

[2] J. Bardeen, W. H. Brattain, The transistor, a semi-conductor triode, *Phys. Rev.* 74 (1948) 230–231, https://journals.aps.org/pr/abstract/10.1103/PhysRev.74.230

[3] M. M. Atalla, Semiconductor Devices Having Dielectric Coatings, U.S. Patent 3,206,670. Filed 1960. Granted 1965.

[4] D. Kahng, M. M. Atalla, *Silicon-silicon dioxide field field induced surface devices*, IRE-AIEE Solid-State Device Research Conference, Pittsburgh, PA, 1960.

[5] P. K. Weimer, The TFT a new thin-film transistor, *Proc. IRE* 50 (1962) 1462–1469.

Solution-Processed Oxide-Semiconductor 249

[6] C E Fritts, On a new form of selenium cell, and some electrical discoveries made by its use, *Am. J. Sci.* 26 (1883) 465–472, doi:10.2475/ajs.s3-26.156.465.

[7] C. Fritts, On the Fritts selenium cell and batteries, *J. Frankl. Inst.* 119, (1885) 221–232, doi:10.1016/0016-0032(85)90426-0

[8] From PV Education, 2016. PVCDROM. Taken from M.A. Green, P. Benjamin, J.N. Shive, and M. Wolf. Available at: https://www.pveducation.org/pvcdrom/manufacturing-si-cells/first-photovoltaic-devices

[9] R. S. Ohl, Light-sensitive electric device, Patent 2,402,662. Filed 1941. Granted 1946.

[10] S. Chung, K. Cho, T. Lee, Recent progress in inkjet-printed thin-film transistors, *Adv. Sci.* 6 (2019) 1801445, doi:10.1002/advs.201801445.

[11] Y. Matsueda, *Required characteristics of TFTs for next generation flat panel display backplanes*, The *Proceedings of the 6th International Thin-Film Transistor Conference* (2010) 314.

[12] T. Shimoda, Y. Matsuki, M. Furusawa, T. Aoki, I. Yudasaka, H. Tanaka, H. Iwasawa, D. Wang, M. Miyasaka, Y. Takeuchi, Solution-processed silicon films and transistors. *Nature* 440 (2006) 783–786, doi:10.1038/nature04613.

[13] T. Masuda, N. Sotani, H. Hamada, Y. Matsuki, T. Shimoda, Fabrication of solution-processed hydrogenated amorphous silicon single-junction solar cells. *Appl. Phys. Lett.* 100 (2012) 253908, doi:10.1063/1.4730614.

[14] K. Nomura, H. Ohta, A. Takagi, T. Kamiya, M. Hirano, H. Hosono, Room-temperature fabrication of transparent flexible thin-film transistors using amorphous oxide semiconductors, *Nature* 432 (2004) 488–492, doi:10.1038/nature03090.

[15] K. Nomura, A. Takagi, T. Kamiya, H. Ohta, M. Hirano, H. Hosono, Amophous oxide semiconductor for high performance flexible thin film transistors, *Jpn. J. Appl. Phys.* 45 (2006) 4303–4308, doi:10.1143/JJAP.45.4303.

[16] H.Q. Chiang, B.R. McFarlane, D. Hong, R.E. Presley, J.F. Wager, Processing effects on the stability of amorphous indium gallium zinc oxide thin-film transistors, *J. Non-Cryst. Solids* 354 (2008) 2826–2830, doi:10.1016/j.jnoncrysol.2007.10.105.

[17] T. Kamiya, K. Nomura, H. Hosono, Present status of amorphous In-Ga-Zn-O thin-film transistors, *Sci. Technol. Adv. Mater.* 11 (2010) 044305, doi:10.1088/1468-6996/11/4/044305.

[18] T. Kamiya, H. Hosono, Material characteristics and applications of transparent amorphous oxide semiconductors, *NPG Asia Mater.* 2 (2010) 15–22, doi:10.1038/asiamat.2010.5.

[19] Y. Ueoka, Y. Ishikawa, N. Maejima, F. Matsui, H. Matsui, H. Yamazaki, S. Urakawa, M. Horita, H. Daimon, Y. Uraoka, Analysis of electronic structure of amorphous InGaZnO/SiO2 interface by angle-resolved X-ray photoelectron spectroscopy, *J. Appl. Phys.* 114 (2013) 163713, doi:10.1063/1.4828869.

[20] D. B. Buchholz, Q. Ma, D. Alducin, A. Ponce, M. Jose-Yacaman, R. Khanal, J. E. Medvedeva, R. P. H. Chang, The structure and properties of amorphous indium oxide, *Chem. Mater.* 26 (2014) 5401–5411, doi:10.1021/cm502689x.

[21] S. Parthiban, J.-Y. Kwona, Role of dopants as a carrier suppressor and strong oxygen binder in amorphous indium-oxide-based field effect transistor, *J. Mater. Res.* 29 (2014) 1585–1596, doi:10.1557/jmr.2014.187.

[22] N. Tiwari, A. Nirmal, M.R. Kulkani, R.A. John, N. Mathews. Enabling high performance n-type metal oxide semiconductors at low temperatures for thin film transistors. *Inorg. Chem. Front.* 7 (2020) 1822. doi:10.1039/D0QI00038H.

[23] P. K. Nayak, M. N. Hedhili, D. Cha, H. N. Alshareef, High performance In2O3 thin film transistors using chemically derived aluminum oxide dielectric, *Appl. Phys. Lett.* 103 (2013) 033518, doi:10.1063/1.4816060.

[24] J. E. Medvedeva, D.C. Buchholz, R.P.H. Chang, Recent advances in understanding the structure and properties of amorphous oxide semiconductors, *Adv. Electron. Mater.* 3 (2017) 1700082, doi:10.1002/aelm.201700082.

[25] D. G. Yang, H. D. Kim, J. H. Kim, S. W. Lee, J. Park, Y. J. Kim, H.-S. Kim, The effect of sputter growth conditions on the charge transport and stability of In-Ga-Zn-O semiconductors, *Thin Solid Films* 638 (2017) 361–366, doi:10.1016/j.tsf.2017.08.008.

[26] H.-D. Kim, J.H. Kim, K. Park, Y.C. Park, S. Kim, Y.J. Kim, J. Park, H-S Kim, Highly stable thin film transistor based on Indium Oxynitride Semiconductor, *ACS Appl. Mater. Interfaces* 10 (2018) 15873–15879, doi:10.1021/acsami.8b02678.

[27] K.-Y. Chen, C.-C. Yang, Y.-K. Su, Z.-H. Wang, H.-C. Yu, Impact of oxygen vacancy on the photo-electrical properties of In_2O_3-based thin-film transistor by doping Ga, *Materials* 12 (2019) 737, doi:10.3390/ma12050737.

[28] C. W. Shih, A. Chin, Remarkably high mobility thin-film transistor on flexible substrate by novel passivation material, *Sci. Rep.* 7 (2017) 1147, doi:10.1038/s41598-017-01231-3.

[29] J. Y. Choi, K. Heo, K.-S. Cho, S. W. Hwang, J.G. Chung, S. Kim, B. H. Lee, S. Y. Lee, Effect of Si on the energy band gap modulation and performance of silicon indium zinc oxide thin-film transistors, *Sci. Rep.* 7 (2017) 15392, doi:10.1038/s41598-017-15331-7.

[30] N. Mitoma, S. Aikawa, X. Gao, T. Kizu, M. Shimizu, M.-F. Lin, T. Nabatame, K. Tsukagoshi, Stable amorphous In_2O_3-based thin-film transistors by incorporating SiO_2 to suppress oxygen vacancies, *Appl. Phys. Lett.* 104 (2014) 102103, doi:10.1063/1.4868303.

[31] N. Mitoma, S. Aikawa, W. Ou-Yang, X. Gao, T. Kizu, M.-F. Lin, A. Fujiwara, T. Nabatame, K. Tsukagoshi, Dopant selection for control of charge carrier density and mobility in amorphous indium oxide thin-film transistors: Comparison between Si- and W-dopants, *Appl. Phys. Lett.* 106 (2015) 042106, doi:10.1063/1.4907285.

[32] S. Aikawa, T. Nabatame, K. Tsukagoshi, Effects of dopants in InO_x-based amorphous oxide semiconductors for thin-film transistor applications, *Appl. Phys. Lett.* 103 (2013) 172105, doi:10.1063/1.4822175.

[33] S. Aikawa, N. Mitoma, T. Kizu, T. Nabatame, K. Tsukagoshi, Suppression of excess oxygen for environmentally stable amorphous In-Si-O thin-film transistors, *Appl. Phys. Lett.* 106 (2015) 192103, doi:10.1063/1.4921054.

[34] T. Kizu, S. Aikawa, T. Nabatame, A. Fujiwara, K. Ito, M. Takahashi, K. Tsukagoshi, Homogeneous double-layer amorphous Si-doped indium oxide thin-film transistors for control of turn-on voltage, *J. Appl. Phys.* 120 (2016) 045702, doi:10.1063/1.4959822.

[35] Yu-Ran Luo, Bond Dissociation Energies, In *Comprehensive Handbook of Chemical Bond Energies* (CRC Press, New York, 2007).

[36] H. E. Jan, T. Nakamura, T. Koga, T. Ina, T. Uruga, T. Kizu, K. Tsukagoshi, T. Nabatame, A. Fujiwara, Amorphous In-Si-O film fabricated via solution process. *J. Electron. Mater.* 46 (2017) 3610–3614, doi:10.1007/s11664-017-5506-9.

[37] H. Hoang, T. Hori, T. Yasuda, T. Kizu, K. Tsukagoshi, T. Nabatame, B.N.Q. Trinh, A. Fujiwara, *Investigation on solution-processed In-Si-O thin film transistor via spin-coating method, IEEE Xplore Proceedings of 25th International Workshop on Active-Matrix Flatpanel Displays and Devices (AM-FPD)* (2018) P-12, doi:10.23919/AM-FPD.2018.8437420.

[38] H. Hoang, T. Hori, T. Yasuda, T. Kizu, K. Tsukagoshi, T. Nabatame, B.N.Q. Trinh, A. Fujiwara, Si-doping effect on solution-processed In-O thin-film transistors, *Mater. Res. Express* 6 (2019) 026410, doi:10.1088/2053-1591/aaecf9.

[39] H. Hoang, K. Sasaki, T. Hori, K. Tsukagoshi, T. Nabatame, B.N.Q. Trinh, A. Fujiwara, Silicon-doped indium oxide – a promising amorphous oxide semiconductor material for thin-film transistors fabricated by spin coating method, *IOP Conf. Ser.: Mater. Sci. Eng.* 625 (2019) 012002, doi:10.1088/1757-899X/625/1/012002.

[40] H. Hoang, Y. Ueta, K. Tsukagoshi, T. Nabatame, B.N.Q. Trinh, A. Fujiwara, Solution processed In-Si-O thin film transistors on hydrophilic and hydrophobic substrates, *Thin Solid Films* 698 (2020) 137860, doi:10.1016/j.tsf.2020.137860

[41] T. Minari, M. Kano, T. Miyadera, S.-D. Wang, Y. Aoyagi, K. Tsukagoshi, Surface selective deposition of molecular semiconductors for solution-based integration of organic field-effect transistors, *Appl. Phys. Lett.* 94 (2009) 093307, doi:10.1063/1.3095665.

[42] Y. Li, C. Liu, A. Kumatani, P. Darmawan, T. Minaril, K. Tsukagoshi, Patterning solution-processed organic single-crystal transistors with high device performance, *AIP Adv.* 1 (2011) 022149, doi:10.1063/1.3608793

[43] Y. Li, C. Liu, Y. Wang, Y. Yang, X. Wang, Y. Shi, K. Tsukagoshi, Flexible field-effect transistor arrays with patterned solution-processed organic crystals, *AIP Adv.* 3 (2013) 052123, doi:10.1063/1.4807669.

[44] H. S. Kim. P. D. Byrne, A. Facchetti, T. J. Marks, High performance solution-processed indium oxide thin-film transistors, *J. Am. Chem. Soc.* 130 (2008) 12580–12581, doi:10.1021/ja804262z.

[45] M.G. Kim, M.G. Kanatzidis, A. Facchetti, T.J. Marks, Low-temperature fabrication of high-performance metal oxide thin-film electronics via combustion processing, *Nat. Mater.* 10 (2011) 382–388, doi:10.1038/nmat3011.

[46] X. Yu, J. Smith, N. Zhou, L. Zeng, P. Guo, Y. Xia, A. Albarez, S. Aghion, H. Lin, J. Yu, R. P. H. Chang, M. J. Bedzyk, R. Ferragut, T. J. Marks, A. Facchetti, Spray-combustion synthesis: efficient solution route to high-performance oxide transistors, *Proc. Natl Acad. Sci. USA* 112 (2015) 3217–3222, doi:10.1073/pnas.1501548112.

Solution-Processed Oxide-Semiconductor

[47] J. Smith, L. Zeng, R. Khanal, K. Stallings, A. Facchetti, J.E. Medvedeva, M.J. Bedzyk, T.J. Marks, Cation size effects on the electronic and structural properties of solution-processed In-X-O thin films, *Adv. Electron. Mater.* 1 (2015) 1500146, doi:10.1002/aelm.201500146.

[48] X. Yu, T.J. Marks, A. Facchetti, Metal oxide for optoelectronic applications, *Nature Mater.* 15 (2016) 383–396, doi:10.1038/nmat4599.

[49] J. Falson, Y. Kozuka, J. H. Smet, T. Arima, A. Tsukazaki, M. Kawasaki, Electron scattering times in ZnO based polar heterostructures, *Appl. Phys. Lett.* 107 (2015) 082102, doi:10.1063/1.4929381.

[50] J. Falson, Y. Kozuka, M. Uchida, J. H. Smet, T. Arima, A. Tsukazaki, M. Kawasaki, MgZnO/ZnO heterostructures with electron mobility exceeding $1 \times 10^6 \, cm^2/Vs$, *Sci. Rep.* 6 (2016) 26598, doi:10.1038/srep26598.

[51] Y. Kozuka, A. Tsukazaki, M. Kawasaki, Challenges and opportunities of ZnO-related single crystalline heterostructures, *Appl. Phys. Rev.* 1 (2014) 011303, doi:10.1063/1.4853535.

[52] D. Takamizu, Y. Nishimoto, S. Akasaka, H. Yuji, K. Tamura, K. Nakahara, T. Tanabe, H. Takasu, T. Onuma, M. Kawasaki, and S. F. Chichibu, Direct correlation between the internal quantum efficiency and photoluminescence lifetime in undoped ZnO epilayers grown on Zn-polar ZnO substrates by plasma-assisted molecular beam epitaxy, *J. Appl. Phys.* 103 (2008) 063502, doi:10.1063/1.2841199.

[53] Y. Nishimoto, K. Nakahara, D. Takamizu, A. Sasaki, K. Tamura, S. Akasaka, H. Yuji, T. Fujii, T. Tanabe, H. Takasu, A. Tsukazaki, A. Ohtomo, T. Onuma, S. F. Chichibu, and M. Kawasaki, Plasma-assisted molecular beam epitaxy of high optical quality MgZnO Films on Zn-polar ZnO substrates, *Appl. Phys. Express* 1 (2008) 091202, doi:10.1143/APEX.1.091202.

[54] Y. Kozuka, J. Falson, Y. Segawa, T. Makino, A. Tsukazaki, and M. Kawasaki, Precise calibration of Mg concentration in $Mg_xZn_{1-x}O$ thin films grown on ZnO substrates, *J. Appl. Phys.* 112 (2012) 043515, doi:10.1063/1.4748306.

[55] J.J. Loferski, Thin films and solar energy applications, *Surf. Sci.* 86 (1979) 424-443, doi:10.1016/0039-6028(79)90422-9.

[56] E. Casassas, L. L. Gustems, R. Tauler, Spectrophotometric study of complex formation in copper(II) mono-, di-, and tri-ethanolamine systems, *J. Chem. Soc. Dalton Trans.* (1989) 569-573, doi:10.1039/DT9890000569.

[57] F. Alharbi, J.D. Bass, A. Salhi, A. Alyamani, H.C. Kim, R.D. Miller, Abundant non-toxic materials for thin film solar cells: Alternative to conventional materials, *Renew. Energy* 36 (2011) 2753-2758, doi:10.1016/j.renene.2011.03.010.

[58] S.S. Shariffudin, S.S. Khalid, N.M. Sahat, M.S.P. Sarah, H. Hashim, Preparation and characterization of nanostructured CuO thin films using sol-gel dip coating, *IOP Conf. Series: Mater. Sci. and Eng.* 99 (2015) 012007, doi:10.1088/1757-899X/99/1/012007.

[59] A. Filippetti and V. Fiorentini, Magnetic ordering in CuO from first principles: A cuprate antiferromagnet with fully three-dimensional exchange interactions, *Phys. Rev. Lett.* 95 (2005) 086405, doi:10.1103/PhysRevLett.95.086405.

[60] M. Nolan and S.D. Elliott, The p-type conduction mechanism in Cu2O: a first principles study, *Phys. Chem. Chem. Phys.* 8 (2006) 5350–5358, doi:10.1039/B611969G .

[61] F. P. Koffyberg and F. A. Benko, A photoelectrochemical determination of the position of the conduction and valence band edges of p-type CuO, *J. Appl. Phys.* 53 (1982) 1173, doi:10.1063/1.330567.

[62] T. Dimopoulos, A. Peic, P. Mullner, M. Neuschitzer, R. Resel, S. Abermann, M. Postl, E. J. W. List, S. Yakunin, W. Heiss, H. Bruckl, Photovoltaic properties of thin film heterojunctions with cupric oxide absorber, *J. Renew. Sustain. Energy* 5 (2013) 011205, doi:10.1063/1.4791779 .

[63] K. S. Wanjala, W. K. Njoroge, N. E. Makori, J. M. Ngaruiya, Optical and electrical characterization of CuO thin films as absorber material for solar cell applications, *Amer. J. Conden. Matter. Phys.* 6 (2016) 1-6, doi:10.5923/j.ajcmp.20160601.01.

[64] A. Liu, G. Liu, H. Zhu, H. Song, B. Shin, E. Fortunato, R. Martins, F. Shan, Water-induced scandium oxide dielectric for low-operating voltage n-and p-type metal-oxide thin-film transistors, *Adv. Funct. Mater.* 25 (2015) 7180-7188, doi:10.1002/adfm.201502612.

[65] Y. Matsuoka, K. Uno, N. Takahashi, A. Maeda, N. Inami, E. Shikoh, Y. Yamamoto, H. Hori, A. Fujiwara, Intrinsic transport and contact resistance effect in C_{60} field-effect transistors, *Appl. Phys. Lett.* 89 (2006) 173510, doi:10.1063/1.2372596.

[66] Y. Yang, J. Yang, W. Yin, F. Huang, A. Cui, D. Zhang, W. Li, Z. Hu, J. Chu, Annealing time modulated the film microstructures and electrical properties of P-type CuO field effect transistors, *Appl. Surf. Sci.* 481 (2019) 632-636, doi:10.1016/j.apsusc.2019.03.130.

[67] F. H. Alshammari, P. K. Nayak, Z. Wang, H. N. Alshareef, Enhanced ZnO thin-film transistor performance using bilayer gate dielectrics, *ACS Appl. Mater. Interfaces* 8 (2016) 22751-22755, doi:10.1021/acsami.6b06498.

[68] K. Matsuzaki, K. Nomura, H. Yanagi, T. Kamiya, M. Hirano, H. Hosono, Epitaxial growth of high mobility Cu_2O thin films and application to p-channel thin film transistor, *Appl. Phys. Lett.* 93 (2008) 202107, doi:10.1063/1.3026539.

[69] S. Y. Sung, S. Y. Kim, K. M. Jo, J. H. Lee, J. J. Kim, S. G. Kim, K. H. Chai, S. Pearton, D. Norton, Y. W. Heo, Fabrication of p-channel thin-film transistors using CuO active layers deposited at low temperature, *Appl. Phys. Lett.* 97 (2010) 222109, doi:10.1063/1.3521310.

[70] H.Q. Nguyen, D.V. Nguyen, A. Fujiwara, and B.N.Q. Trinh, Solution-processed CuO thin films with various Cu^{2+} ion concentrations, *Thin Solid Films*, 660 (2018) 819-823, doi:10.1016/j.tsf.2018.03.036.

[71] B.N.Q. Trinh, N.V. Dung, N.Q. Hoa, N.H. Duc, D.H. Minh, and A. Fujiwara, Solution-Processed Cupric Oxide P-type Channel Thin-Film Transistors, *Thin Solid Films*, 704 (2020) 137991, doi:10.1016/j.tsf.2020.137991.

10 Gold Nanocrystal-built Films for SERS-based Detection of Trace Organochlorine Pesticides

Xia Zhou, Hongwen Zhang, and Weiping Cai
Institute of Solid State Physics, HFIPS,
Chinese Academy of Sciences, P.R. China

CONTENTS

10.1 Introduction 254
10.2 Temperature Regulation Growth of Au Nanocrystals 254
    10.2.1 Concave Polyhedral Nanocrystals 255
    10.2.2 Reaction Temperature-dependent Morphology 256
    10.2.3 Temperature-controlled Preferential Growth 259
10.3 Seeds' Addition Rate-controlled Growth 260
    10.3.1 Morphology and Structure 260
    10.3.2 Influence Factors 262
        10.3.2.1 Addition Rate 262
        10.3.2.2 The Addition Amount 263
    10.3.3 Formation of Chestnut-like Au Nanocrystals 263
10.4 Metal-Organic Framework-wrapping of Au Nanocrystals 266
    10.4.1 ZIF-8 Wrapping of UAANs 266
    10.4.2 Morphology and Structure of the Wrapped UAANs 266
    10.4.3 The Formation of ZIF-8 Wrapping Shell 269
10.5 Au Nanocrystals-built Films and Their SERS Activities 269
    10.5.1 Au Nanocrystals-built Films 269
    10.5.2 SERS Activity 270
10.6 Applications in Trace Detection of Hexachlorocyclohexane 273
    10.6.1 Concave Au Trisoctahedral Nanocrystals-based SERS Chips 274
        10.6.1.1 Concentration-Dependent Raman Spectra 274
        10.6.1.2 Freundlich Adsorption-induced Raman Spectra 275
    10.6.2 The Chestnut-like Au Nanocrystal-built SERS Chips 276
    10.6.3 ZIF-8 Wrapped Au Nanocrystals-built SERS Chips 277
        10.6.3.1 ShellThickness-dependent SERS 277
        10.6.3.2 Concentration-dependent SERS 277
        10.6.3.3 Wrapping-induced HCH Enrichment 279
        10.6.3.4 The Size Selectivity of Molecules 280
10.7 Conclusions 281
Acknowledgments 282
References 282

10.1 INTRODUCTION

The nanocrystals with different shapes such as polyhedrons, spheres, bipyramids, cubes, plates, wires, and rods have extensively been reported [1–3]. These nanocrystals were all enclosed with the convex surfaces or low-index facets. In contrast, the dendritic and concave nanocrystals possess sharp corners/edges and high-index facets and show different optical and catalytic performances [4, 5]. For noble metals, these concave and dendritic nanocrystals have attracted much attention in recent years since there exist important potential applications in plasmonics [6], surface-enhanced Raman spectroscopy (SERS) substrates [2, 7], catalysis [8], etc. Particularly, for the dendritic and concave Ag or Au nanocrystals, they could exhibit high SERS effect because of their large number density of edges/corners and tips which can generate the large local electric-field enhancement [9, 10], and hence serve as the "hot spots" for the SERS-based detection applications.

It is well known that the SERS-based detection is an ultrasensitive technique and has a lot of potential applications in fields such as biological sensing [11–13], environmental monitoring [14, 15], food safety [16, 17], biomedical science [18–20], and molecular imaging [21], etc. The fabrication of the SERS substrates or chips, with both stable structure and strong SERS effect or activity, is of importance in the SERS-based detection technique. Mostly, the SERS chips are made of the noble metals with nanostructure. The edges/corners and sharp tips on the chips are the most important enhancement structures [22–24]. Such structures can amplify the Raman signals since they not only can serve as the "super electromagnetic intensifiers" (like nano-antennas) or the "hot spots" but also are the preferential adsorption sites of some target molecules [25]. Therefore, the fabrication of the noble metal substrates, with the large number density of edges/corners and sharp tips, is an important way to obtain the high activity of SERS chips.

It has been reported that the noble metal SERS chips with the well-defined tipped and edged structures are mainly prepared by the seed-assisted growth methods and chemical reduction routes. For instance, by using a seed-assisted growth method, Wang et al. [3] synthesized the Au nanostars, which exhibited the strong SERS activity due to their high number density of "hot spots". Fang et al. [26] prepared the sea urchin-like Au microparticle arrays by reduction of $HAuCl_4$ in Fe suspension solution and showed that such arrays were of the high SERS activity with 1 or 2 orders of magnitude higher than the individual microparticles. By using the L-dopa as the reductant, they also prepared the Au-Ag alloy nanourchins with hollow structure and about 70–100 tips in each one. Such alloying nanourchins showed the significantly enhanced local electromagnetic field [27]. Generally, the seed-assisted growth methods are easy to operate and can be utilized to synthesize the nanostars with few tips, but it is difficult to obtain the nanocrystals with high number density of tips like the flower- or urchin-like nanocrystals. For the Fe suspension reduction method, it is difficult to form a homogeneous solution because of the magnetic properties of the Fe, and hence the uniformity of the reduction is limited. As for the L-dopa reduction method, it needs ammonia and formic acid to clean the solution for several times to remove the residual reducing agent.

In recent years, many progresses have been made in shape-controlled synthesis of noble metal nanocrystals, with the high number density of the edges/corners and tips, and applications of the nanocrystals-built thin films. In this chapter, we mainly introduce the fabrication of Au nanocrystals with controllable shape and size via the kinetically controlled growth of seeds, the wrapping of metal-organic framework on the nanocrystals, SERS activity of mono-dispersed Au nanocrystals-built films, and their applications in SERS-based detection of trace organochlorine pesticides, such as hexachlorocyclohexane (HCH).

10.2 TEMPERATURE REGULATION GROWTH OF AU NANOCRYSTALS

It has been found that the shape and morphology of Au nanocrystals can be controlled and tuned by temperature regulation of the precursor solution during a modified seed-assisted growth [28]. The Au seeds of about 3 nm in size were firstly synthesized via adding proper amounts of $NaBH_4$ in the

Gold Nanocrystal-built Films

HAuCl$_4$ aqueous solution containing hexadecyltrimethylammonium chloride (CTAC) [3]. Typically, 0.30 mL of ice-cold, freshly prepared NaBH$_4$ aqueous solution (10 mM) was injected into 10 mL of aqueous solution with HAuCl$_4$ (0.25 mM) and CTAC (0.10 M) at 25 °C and left undisturbed for 2 h. The Au seeds were thus obtained in the solution. The seeds' solution was diluted 1000 times with CTAC aqueous solution for further use.

Au nanocrystals were then synthesized via the seed-assisted growth method at different temperatures. Typically, 10 mL of the growth solution with HAuCl$_4$ (0.50 mM), CTAC (0.10 M) and L-ascorbic acid (AA) (10 mM) was prepared in a flask. 0.05 mL of the diluted Au seeds' solution was added in the flask and undisturbed in a water bath with a certain temperature (from 5 °C to 100 °C). After reaction for 4h, the products in the solution were washed with deionized water twice through centrifugation before they were dispersed in 1.0 mL of water to obtain the colloidal solution of the Au nanocrystals.

10.2.1 Concave Polyhedral Nanocrystals

The inset of Figure 10.1a shows the colloidal solution obtained after the growth solution was dropped with 0.05 mL of the diluted Au seeds' solution at 25 °C for 4 h. Figure 10.1a gives the corresponding optical absorbance spectrum. There exists an obvious absorption peak around 600 nm, which is attributed to the local surface plasmon resonance (SPR) of Au nanocrystals [29]. This means that Au nanocrystals have been formed in the solution. The scanning electron microscopic (SEM) observation reveals that the corresponding products in the colloidal solution are comprised of the particles with about 100 nm in size and polyhedral shape, as typically shown in Figure 10.1b. These polyhedral particles are actually generated from an octahedron by "pulling out" the centers of the eight triangular {111} facets, and assume the concave trisoctahedrons consisting of eight trigonal pyramids [30]. The 24 facets on the trisoctahedron are all assigned to the crystal planes in the family of the <110> zone axis [10].

It was found that the optical absorption peak's position of the Au colloidal solution could be controlled and tuned by the addition amount of Au seeds. When the added Au seeds' solution amount was reduced from 0.1 mL to 0.01 mL, the SPR was shifted from 575 nm to 780 nm (Figure 10.2), while the nanocrystal's size was significantly increased the from 80 nm to 135 nm and the nanocrystal's shape nearly unchanged (still concave trisoctahedron), as demonstrated in Figure 10.3. This indicates that the red-shift of the Au concave trisoctahedral nanocrystals in local SPR is induced by the increasing particle size.

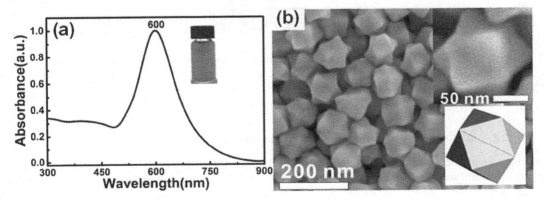

FIGURE 10.1 (a) The optical absorbance spectrum of the as-prepared colloidal solution obtained by adding 0.05 mL Au seed solution into the growth solution at 25 °C. The inset is the photo of the colloidal solution. (b) The SEM image corresponding to the products in (a). The inset is the SEM image and geometric model of an individual Au concave trisoctahedron. Reproduced with permission from [28].

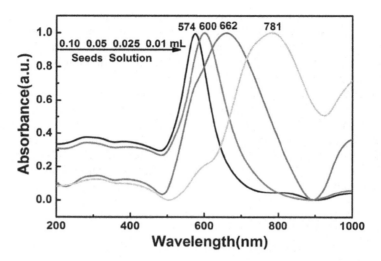

FIGURE 10.2 Optical absorbance spectra of Au colloidal solutions obtained by addition of different Au seeds' solution amounts at 25°C, respectively. Reproduced with permission from [28].

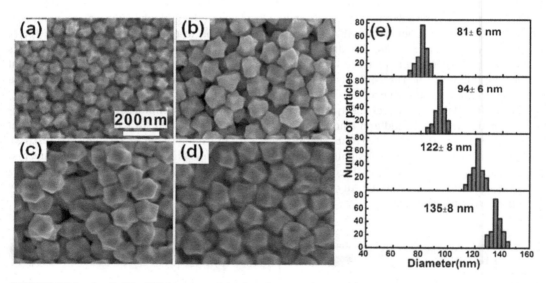

FIGURE 10.3 (a–d): The SEM images of the Au nanocrystals prepared by adding 0.10 mL, 0.05 mL, 0.025 mL and 0.01 mL of Au seeds' solution, respectively, at 25°C. (e): The size distribution histograms of the particles. The frames from top to bottom: correspond to the samples in panels a–d, respectively. Reproduced with permission from [28].

10.2.2 Reaction Temperature-dependent Morphology

If the reaction temperature was higher than 25 °C up to 100 °C, the products were still Au concave trisoctahedral nanocrystals, although the edges/corners tended to be truncated or smooth when the reaction temperature was increasing, as typically shown in Figure 10.4, while the nanocrystals were nearly unchanged in size.

When the reaction temperature was lower than 25°C, however, the Au nanocrystals were very sensitive to the temperature in morphology, and even a small change in temperature would greatly influence their shape. When the reaction temperature decreased from 25°C to 20°C, the morphology

Gold Nanocrystal-built Films 257

FIGURE 10.4 The SEM images of the Au nanocrystals obtained at different reaction temperatures by adding 0.05 mL of Au seeds' solution. (a): 40°C, (b): 60°C, (c): 80°C, and (d):100°C. Reproduced with permission from [28].

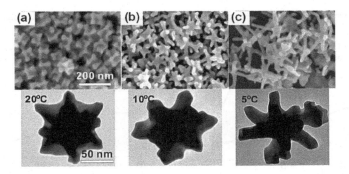

FIGURE 10.5 SEM (top) and TEM (down) images of the Au nanocrystals prepared by adding 0.05 mL of seeds' solution at different temperatures. (a) 20°C, (b) 10°C, and (c) 5°C. Reproduced with permission from [28].

of the Au nanocrystals was obviously changed from concave trisoctahedron to the calyptriform but the size was nearly unchanged (about 100 nm), as illustrated in Figure 10.5a. Further decrease of the temperature to 10°C and 5°C led to the formation of the coral and even dendritic nanocrystals, respectively, as shown in Figure 10.5b and 5c. These Au nanocrystals prepared at different reaction temperatures were uniform in morphology and size. Further, the transmission electron microscopic (TEM) examinations also showed clearly the shapes of the nanocrystals obtained at different temperatures, as typically illustrated in Figure 10.5. Here, it should be mentioned that the reaction temperature could not be below 5 °C, since the CTAC in solution would crystallize and precipitate at this low temperature.

Similarly, the increase of the Au seeds' addition amount would decrease the size of Au nanocrystals but nearly un-change the shape for each reaction temperature, as typically shown in Figure 10.6, which corresponds to the Au nanocrystals obtained at 20 °C.

Figure 10.7 clearly shows the red-shift of the SPR with the decreasing Au seeds' addition or the increasing particle's size, for the corresponding Au nanocrystals. Typically, for the Au nanocrystals prepared at 20 °C, the SPR peak was red-shifted from 581 nm to 776 nm, and the corresponding particle size increased from about 90 nm to 190 nm, when the addition amount was decreased from 0.1 mL to 0.01 mL. So, we can easily control and tune the nanocrystals' size without obvious morphological change just via changing the seeds' amount. Furthermore, the dipolar SPR became increasingly broadened with the increasing nanocrystals' size, while a narrower quadrupolar SPR appeared at a shorter wavelength than the dipolar one and was increasingly pronounced with the rising nanocrystals' size because of the phase retardation effects [31].

258　　　　　　　　　　　　　　　　　　　　　　　　　　　Functional Thin Films Technology

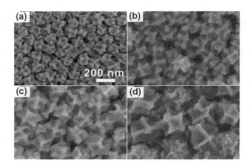

FIGURE 10.6 The SEM images of the Au nanocrystals obtained at 20°C by adding different amounts of Au seeds' solution. (a) 0.10 mL, (b) 0.05mL, (c) 0.025mL, and (d) 0.01 mL. Reproduced with permission from [28].

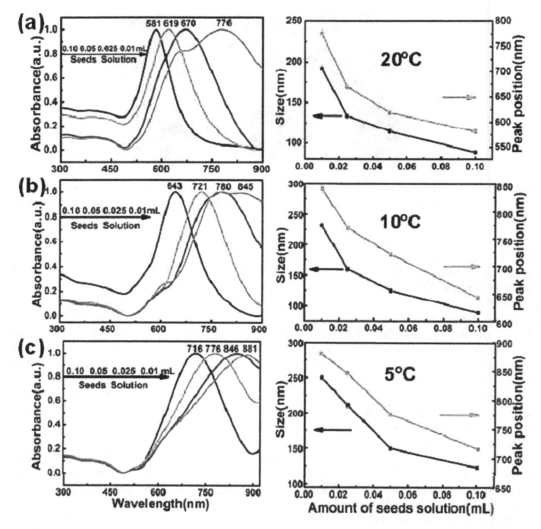

FIGURE 10.7 Influence of the added amount of Au seeds' solution on the optical absorbance spectra of the colloidal solution prepared at different temperatures (left panels), and nanocrystals' size and SPR position (right panels). (a): 20°C, (b): 10°C, and (c): 5°C. Reproduced with permission from [28].

Gold Nanocrystal-built Films

10.2.3 Temperature-Controlled Preferential Growth

The reaction temperature-dependent shape of the Au nanocrystals could be attributed to temperature-controlled preferential growth of the Au seeds. In the seeds' solution, $NaBH_4$ would reduce the Au^{3+} ions to Au^0, or the following reaction would take place.

$$AuCl_4^- + 3e^- \rightarrow Au^0 + 4Cl^- \tag{10.1}$$

However AA in the growth solution is weaker than that of $NaBH_4$ in reduction capacity [32–34]. So, the Au^{3+} ions were only reduced to Au^+ instead of Au^0, or the following reaction took place:

$$AuCl_4^- + 2e^- \rightarrow AuCl_2^- + 2Cl^- \tag{10.2}$$

When the growth solution was added with the Au seeds, the Au^+ ions in the solution would be reduced to Au^0 because of the plenty of electrons provided by excess reducing agent or $NaBH_4$ in the seeds' solution [32–34], according to the following reaction.

$$AuCl_2^- + e^- \rightarrow Au^0 + 2Cl^- \tag{10.3}$$

The partial Au^0 would form Au^0-CTAC complex colloids with the surfactants [32]. So the reduced free Au^0 atoms would co-exist with the Au^0-CTAC complexes in the growth solution. Due to the presence of the surfactant CTAC, such complexes could preferentially be adsorbed on the {110} planes of the seeds, while the free Au^0 atoms would be preferentially deposited on {111} plane [34]. So, the seeds grow mainly along <110> and <111> directions. Evidently, the growth rates should be temperature-dependent and different for both directions, leading to the Au nanocrystals with reaction temperature-dependent shapes.

When the reaction temperature was at 25 °C, the difference between both directions in growth rate was small, leading to the formation of the concave trisoctahedron. At > 25 °C, the values of the growth rates in both directions were getting closer. The Au nanocrystals were thus tending to be smooth or truncated with the increasing reaction temperature. If the reaction temperature was lower than 25 °C, however, due to the preferential adsorption of the surfactant CTAC, the growth rate in <110> direction is increasingly faster than that of <111> direction, leading to the formation of calyptriform, coral, and dendritic structures with the decreasing temperature. Further, the TEM examination of the dendritic structure obtained at 5 °C has also confirmed such preferential growth along <110>, as shown in Figure 10.8.

The added Au seeds correspond to the nuclei of Au nanocrystals in the growth solution. So, the decrease in the seeds' addition amount would reduce the number of the Au nanocrystals in the solution and increase the size of the final nanocrystals with insignificant change in morphology at a given temperature. Also the increase in size would lead to red-shift of the SPR [3].

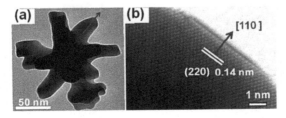

FIGURE 10.8 (a): TEM image of a single dendritic Au nanocrystal prepared at 5°C. (b): High-resolution TEM image for the area of the trunk's top marked in (a). Reproduced with permission from [28].

10.3 SEEDS' ADDITION RATE-CONTROLLED GROWTH

As mentioned above, shapes of the nanocrystals are determined by the growth rates of the different crystal directions. In the seed-assisted growth method, the growth kinetics of nanocrystals can also be tuned and controlled by the seeds' addition rate, in addition to the reaction temperature. Here, based on the kinetically controlled growth, a modified seed growth route is demonstrated for the Au nanocrystals with high number density of tips or edges via tuning the Au seeds' addition rate in the growth solution [35].

Firstly, Au seeds and diluted seeds' solution were prepared as mentioned in the above section. 0.1 mL of $AgNO_3$ (10 mM), 0.5 mL of $HAuCl_4$ (10 mM), 0.5 mL of AA (0.02 M), and 0.2 mL of HCl (1.0 M) were then added into 10 mL of CTAC (0.1M) aqueous solution to form the growth solution. Here, for decreasing the growth rate, the small amount of Ag^+ ions was used [3]. 50 μL of the diluted Au seeds' solution was then injected into the growth solution at 25 μL/min using a micro-injection pump. The mixed solution was undisturbed at 20 °C for 4h and changed from colorless to purple color. The Au nanocrystals were thus formed in the solution. Finally, products were collected and washed with deionized water twice through centrifugation, and dispersed in 1.0 mL of water. Such Au colloidal solution shows purple color, as illustrated in the inset of Figure 10.9. The corresponding optical absorbance spectrum shows that there is a strong peak at 738 nm, in addition to a small shoulder around 570 nm, which should correspond to dipolar localized SPR of Au nanoparticles [26]. This indicates the existence of Au nanocrystals in the solution.

10.3.1 MORPHOLOGY AND STRUCTURE

Figure 10.10 is the SEM image of the products in the solution. The products consist of the particles with about 100 nm in size and the radial nanoneedles about 30–50 nm in length (Figure 10.10a). These particles are chestnut-like in morphology (Figure 10.10b). In addition to the many sharp protrusions and nanogaps, there are about 120–150 tips or nanoneedles for each chestnut-like particle. The small optical absorbance peak around 570 nm originates from the hybridization of plasmons localized in the core parts of the Au nanocrystals due to the chestnut-like morphology, while the main peak around 738 nm results from the increased aspect ratios of the radial nanoneedles, as previously reported [26, 27].

Figure 10.11a shows the XRD results of the as-prepared chestnut-like particles. There exist four characteristic diffraction peaks at 38.2°, 44.4°, 64.6°, 77.5°, respectively, corresponding to the planes (111), (200), (220), and (311) planes of face-centered cubic Au (JCPDS, No. 96-901-1613). Further, Figure 10.11b presents the TEM image for a typical chestnut-like particle. The needles are <10 nm in

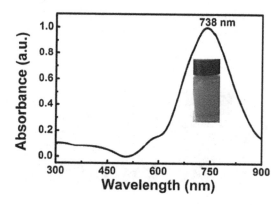

FIGURE 10.9 The optical absorbance spectrum of the as-prepared Au nanocrystal colloidal solution. The inset: the photo of the colloidal solution in a bottle. Reproduced with permission from [35].

Gold Nanocrystal-built Films 261

FIGURE 10.10 The SEM images of the as-prepared Au nanocrystals. (a): Low magnification. (b): The magnified image of a single particle in (a). The inset is the photo of one chestnut for reference. Reproduced with permission from [35].

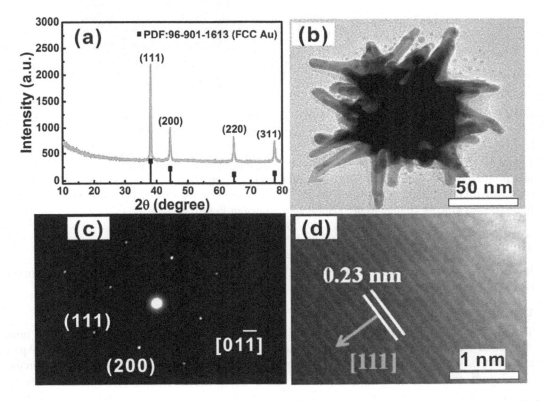

FIGURE 10.11 (a): The XRD results of the chestnut-like particles. The line spectrum is the standard diffraction of Au powders. (b): The TEM image of a single chestnut-like particle. (c): The SAED pattern of the particle in (b). (d): HRTEM image of the area in the nanoneedle marked with arrow in (b). The arrows in (b) and (d) are parallel. Reproduced with permission from [35].

width and about 50 nm in length. The selected area electron diffraction (SAED) has confirmed that the chestnut-like particles are single crystalline in structure (Figure 10.11c). Figure 10.11d shows the high-resolution TEM (HRTEM) observation has revealed that the nanoneedles are [111]-oriented.

The morphological evolution with the reaction time was examined for the formation process of the chestnut-like Au nanocrystals. Figure 10.12 shows the morphological evolution of the products with the reaction time. When the reaction time was 1 hour or less, the concave nanocubes were formed (Figure 10.12a). With increase of the reaction time to 2 h, these nanocubes were tending to be more concave with emergence of the ultrafine needles in the cubes, as typically illustrated in

FIGURE 10.12 SEM images of the Au nanocrystals obtained after reaction for different durations. (a) 1 h, (b) 2 h, (c) 3 h, and (d) 6 h. The scale bars are 200 nm. Reproduced with permission from [35].

Figure 10.12b. When the reaction time reached 3 h, many concave nanocubes became chestnut-like in morphology (Figure 10.12c). After reaction for 4 h, all particles grew into the chestnut-like nanocrystals with high number density of tips (or nanoneedles) (Figure 10.10). The further reaction only induced insignificant change in morphology, as shown in Figure 10.12d.

10.3.2 Influence Factors

It was found that the Au seeds' addition conditions, including adding rate and amounts, were important to the formation of the chestnut-like Au nanocrystals.

10.3.2.1 Addition Rate

The addition rate of the Au seeds' solution could significantly influence the morphology of the final Au nanocrystals. When 50 μL of the diluted Au seeds' solution was added at a low rate (say, 10 μL/min) into the growth solution, we could obtain the concave Au nanocubes with about 100 nm in the edge length after the reaction (Figure 10.13). Obviously, such concave nanocubes are enclosed by {111} facets and <100> edges [4, 10], and completely different from the chestnut-like nanocrystals obtained at a moderate addition rate (about 25 μL/min) or shown in Figure 10.10.

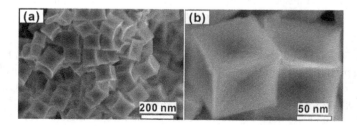

FIGURE 10.13 The SEM images of the Au nanocrystals prepared by adding 50 μL of the diluted Au seeds' solution in the growth solution at 10 μL/min. (a): Low magnification. (b): High magnification. Reproduced with permission from [35].

Gold Nanocrystal-built Films

FIGURE 10.14 The morphology and microstructure of the Au nanocrystals prepared by adding 50 µL of the diluted Au seeds' solution into the growth solution at the rate of 50 µL/min. (a): The SEM image. (b): The TEM image of a single Au nanospindle. (c): The HRTEM image of the area marked with arrow in (b). The arrows in (c) and (b) are parallel. Reproduced with permission from [35].

Contrarily, when 50 µL of the diluted Au seeds' solution was added into the growth solution at an enough fast rate (say, 50 µL/min), the formed Au nanocrystals were spindle-like in shape with about 150 nm in length, as typically shown in Figure 10.14a. The HRTEM examination has confirmed that the long axis of the nanospindles is parallel to <111> direction, as clearly shown in Figure 10.14b and 14c. Only when the seeds' solution was added into the growth solution at a moderate rate (about 25 µL/min), the chestnut-like Au nanocrystals with high number density of tips/edges could be fabricated, as shown in Figure 10.10.

10.3.2.2 The Addition Amount

In addition, the addition amount of the seeds' solution can change the size of the Au nanocrystals but not affect their morphology. Keeping the addition rate of 25 µL/min, when the addition amount of the diluted seeds' solution was decreased from 100 µL to 10 µL, the Au nanocrystals were mostly chestnut-like in morphology but the mean sizes increased from 80 nm to 140 nm, as shown in Figure 10.15. The SPR peak of the corresponding colloidal solutions was red-shifted from 642 nm to 805 nm due to the increasing particles' size [36], as shown in Figure 10.16. Further, if the addition rate was slow (say, 10 µL/min), the concave nanocubes were obtained (Figure 10.13), but the evolutions in the size with the addition amount were similar to those above, as illustrated in Figure 10.17.

10.3.3 FORMATION OF CHESTNUT-LIKE AU NANOCRYSTALS

The formation of the chestnut-like Au nanocrystals could be attributed to the much higher growth rate along <111> than those along the other directions. After the growth solution was added with the Au seeds' solution, the gold ions would be reduced to Au^0 atoms, as described in Section 10.2.3. The reduced free Au^0 atoms would be preferentially deposited or attached on the {111} facets of the added seeds, while the surfactant CTAC in the growth solution is preferentially adsorbed on the {110} facets [34]. So the order of the growth rates V, for the Au seeds along the directions <111>, <100> and <110>, should be $V_{<111>} > V_{<100>} > V_{<110>}$. It means that the seeds would grow preferentially along the direction <111>, followed by <100>. The final morphology of the Au nanocrystals depends on the magnitude and difference of the growth rates in different directions.

When the seeds' solution was added into the growth solution at a moderate rate (about 25 µL/min), the growth rate of the seeds should also be moderate, and the chestnut-like nanocrystals would thus be formed, as shown in Figure 10.18. After the seeds' solution was dropped in the growth solution, reaction (10.3) took place and the free Au^0 atoms were continuously produced in the solution (Step I), leading to the growth of the added seeds (Figure 10.18a and 18b). The seeds would preferentially grow along <111> (Step II). In the initial stage of the growth (<1 h), the Au seeds

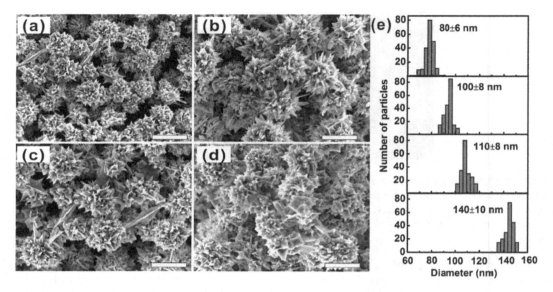

FIGURE 10.15 SEM images of the products obtained by adding (a) 100 µL, (b) 50 µL, (c) 25 µL and (d) 10 µL of the diluted Au seeds' solution into the growth solution, at a moderate addition rate (25 µL/min). (e): The size distribution histograms of the chestnut-like Au nanocrystals. The panels from top to bottom correspond to (a–d). The scale bar is 100 nm. Reproduced with permission from [35].

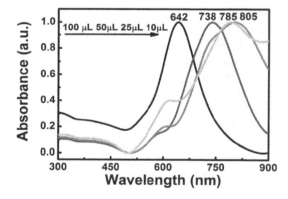

FIGURE 10.16 The optical absorbance spectra of the colloidal solutions prepared with the different addition amounts of the diluted Au seeds' solution in the growth solution at 25 µL/min. Reproduced with permission from [35].

would grow from the initial equi-axial shape into the concave nanocubes due to $V_{<111>} > V_{<100>}$ (Figure 10.18c). Such concave nanocubes are enclosed by the facets {111} and the edges <100>, as shown in Figure 10.13a. With the reaction going on and ever-growing preferentially along <111>, the concave nanocubes would evolve into the nanocrystals with the radial <111>-oriented nanoneedles (Figure 10.18d). Further, the more and more branches would also grow along <111> at the defect sites on the pre-formed nanoneedles (Figure 10.18e). Finally, Au nanocrystals grow into the chestnut-like shape with high number density of tips (Figure 10.18f).

Based on the discussion above, the reaction rate is crucial. Evidently, the low reaction rate should be beneficial to formation of the concave nanocubes due to the growth along both <111> and <100>, while the large reaction rate would only form the nanoneedles due to the large $V_{<111>}$. The seeds'

Gold Nanocrystal-built Films

FIGURE 10.17 SEM images of the Au nanocrystals prepared by adding (a) 100 μL, (b) 50 μL, (c) 25μL, and (d) 10 μL of the diluted Au seeds' solution into the growth solution at a rate of 10 μL/min. (e) The size distribution histograms of the Au concave nanocubes. The panels from top to bottom correspond to (a–d). The scale bar is 200 nm. Reproduced with permission from [35].

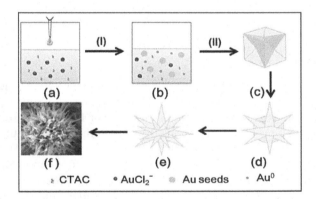

FIGURE 10.18 Schematic illustration for the formation of the chestnut-like Au nanocrystals. (a): The growth solution was added with seeds. (b): The continuous formation of free Au⁰ atoms and the growth of added Au seeds in the growth solution. (c) Formation of concave nanocubes during the initial growth period. (d) The radial growth of <111>–oriented nanoneedles. (e) The branched growth on the formed nanoneedles. (f) The chestnut-like Au nanocrystal. Step (I): The reduction of gold ions and preferential adsorption of CATA on seeds. Step (II): The preferential growth of the seeds along <111>. Reproduced with permission from [35].

addition rate could determine the reaction rate and hence affect the growth kinetics at the given temperature. When the growth solution was added with Au seeds' solution at a low rate (say, 10 μL/min), the Au seeds would slowly grow along <111> and <100> due to the low reaction rate, leading to formation of the concave nanocubes (Figure 10.13). On the contrary, when the seeds' solution was added at a large rate (say, 50 μL/min), the reaction rate would be greatly increased. In this case, the Au seeds would fast and preferentially grow along <111>, forming the Au nanospindles (Figure 10.14). As for the influence of the seeds' amount, it can be attributed to the correspondence between the seeds and nuclei of the Au nanocrystals, as mentioned in Section 10.2.3.

10.4 METAL-ORGANIC FRAMEWORK-WRAPPING OF AU NANOCRYSTALS

The metal-organic framework (MOF) compounds are of the large internal surface area, adjustable pore size, high structural flexibility, and unique chemical properties [37–41], and have the applications in many fields such as drug delivery [42], catalysis [43], gas storage/separation [44–46], and chemical sensing [47]. There are many MOF compounds. The zeolite imidazole framework (ZIF) is an important MOF's category among them. Typically, the ZIF-8 ($Zn(MeIM)_2$, MeIM-methylimidazole) has the large internal surface area and a pore size of 11.6 Å [48, 49]. Because of the confinements of porous space and geometry, such porous ZIF-8 could play a role in catching or enriching some molecules with suitable size (<11.6 Å) [48]. If this porous ZIF-8 is wrapped on the above Au nanocrystals to form a core–shell configuration (or Au@ZIF-8 nanocrystal), it is expected that such core–shell nanocrystals can show the strong enrichment and the high SERS performances to some molecules (such as HCH) with <11.6 Å in size.

There are some reports on the porous ZIF-8 wrapped metal nanoparticles. For instance, Zheng et al. [48] coated the porous ZIF-8 layer on the Au nanoparticles with arbitrary shapes by using the quaternary ammonium surfactant as a linker, employed such ZIF-8 wrapped nanoparticles to study the diffusion of the liquid molecules with the size smaller than the pore size in ZIF-8. Qiao et al. [43] wrapped the ZIF-8 on Au nanoparticles in the solution with dodecyltrimethylammonium bromide (DTAB) for the enhanced adsorption to volatile organic compound molecules and showed that the ZIF-8 wrapping could adsorb and enrich the molecules smaller than its pore size [46, 47]. In this section, we introduce the wrapping of Au nanocrystals with porous ZIF-8 [50]. In addition, the urchin-like Au-Ag alloying nanocrystals (UAANs) with many sharp tips and edges can produce large number density of "hot spots", which could generate the strong electromagnetic field enhancement up to about 10^9 in order of magnitude [27]. Here we first chose such UAANs as the core parts to demonstrate ZIF-8 wrapping of Au nanocrystals.

10.4.1 ZIF-8 WRAPPING OF UAANS

UAANs were prepared via the Ag seeds' growth in the L-Dopa-contained $HAuCl_4$ solution, as previously reported [27, 50]. Typically, 100 mL of the boiling $AgNO_3$ (1.1 mM) aqueous solution was added with 2.4 mL of sodium citrate solution (34 mM). After keeping the solution at room temperature for 15 min, the Ag seeds with 25 nm in size were formed in the solution. Such Ag seeds' solution was the grayish-green in color. Then, a flask filled with 13.4 mL of $HAuCl_4$ aqueous solution (3.58 mM) was placed in the water bath at 15°C and stirred for 10 min. Subsequently, 4.8 mL of L-Dopa aqueous solution (10 mM) and 1.8 mL of Ag seeds' solution were added into the flask for reaction. After 15 min, the UAANs or the products were formed in the solution, collected and cleaned. Finally, the cleaned UAANs were dispersed in 10 mL of water to obtain the UAAN colloidal solution.

The ZIF-8 wrapping of the UAANs were conducted via adding the pre-formed UAANs into the ZIF-8 precursor solution with hexadecyltrimethyl ammonium bromide (CTAB). Typically, 1 mL of UAAN colloidal solution was dropped into 3.5 mL of the precursor solution containing 7.0 mM $Zn(NO_3)_2$, 0.43 M 2-MeIM and 0.10 mM CTAB and put in the water bath at 40°C for 3 h. The ZIF-8 wrapped UAANs were thus formed in the solution. Finally, the wrapped UAANs were collected, washed, and dispersed in methanol.

10.4.2 MORPHOLOGY AND STRUCTURE OF THE WRAPPED UAANS

Figure 10.19a shows the UAANs after the seeds' growth in the L-Dopa-contained $HAuCl_4$ solution, which are about 100 nm in mean size, nearly spherical in shape, and of many spokewise tips on the surface with several tens of nanometers in length and few nanometers in thickness. After the precursor solution was added with the UAANs and kept at 40°C for 3 h, the wrapped particles with about

Gold Nanocrystal-built Films

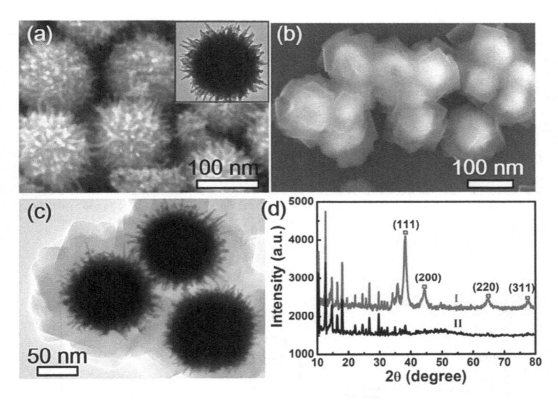

FIGURE 10.19 The morphology and structure of the as-prepared products. (a): SEM image of the UAANs. The inset: TEM image of a single UAAN. (b) and (c): SEM and TEM images of the wrapped particles, respectively. (d): XRD patterns of the wrapped particles (I) and the pure ZIF-8. Curves I and II correspond to the sample in (b) and the pure ZIF-8 (II). Reproduced with permission from [50].

140 nm in mean size were obtained, as illustrated in Figure 10.19b. TEM examination revealed that the wrapped particles are of core–shell structure with about 20 nm in shell-layer thickness and the core parts are the UAANs, as demonstrated in Figure 10.19c. The corresponding XRD pattern is shown in Figure 10.19d. The spectrum in the range from 10° to 35° is in agreement with the sodalite zeolite-type crystal structure of ZIF-8 [51], while diffraction peaks at 35°~80° are attributed to the UAANs [27].

Further, the energy-dispersive X-ray spectroscopy (EDS) measurement shows that such ZIF-8-wrapped UAANs contain the elements C, N, Zn, in addition to Ag and Au, as shown in Figure 10.20. Further, the element mappings of the wrapped particles show that the components Ag and Au are homogenously distributed in the core region, indicating the Au-Ag alloyed cores [52, 53], while the elements C, N, Zn are distributed in the shell layers (Figure 10.20b-g). It can thus be concluded that the wrapped particle is built of the ZIF-8 shell layer and the UAAN core.

The N_2 isothermal sorption was measured for such wrapped particles, as illustrated in Figure 10.21a. Their specific surface area is 1150 $m^2 \cdot g^{-1}$, which is slightly less than that of the pure ZIF-8 (1215 $m^2 \cdot g^{-1}$) due to the UAANs in the core parts. Such high specific surface area indicates the porous shell layer. From the micropore analysis, the pores are mostly around 11.6 Å and very few around 16.5 Å in size (Figure 10.21b).

Finally, the ZIF-8 shell layer could be tuned and controlled in thickness just via changing the CTAB concentration in the precursor solution [48]. Figure 10.22 shows the results that the shell thickness increases from 10 nm to 40 nm with decreasing CTAB concentration from 0.5 mM to 0.05 mM. Similarly, the ZIF-8 could be also wrapped on the other Au nanocrystals, such as Au

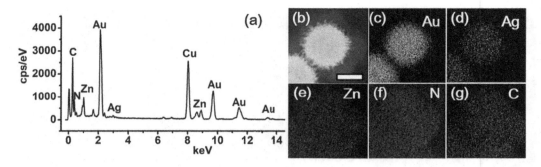

FIGURE 10.20 (a): EDS spectrum of all elements for the as-prepared ZIF-8 wrapped UAANs. The Cu signal is from the copper mesh used in TEM characterization. (b): Scanning TEM image of the wrapped particles and (c–g) the corresponding EDS elemental mappings. The scale bar in (b) is 50 nm. Reproduced with permission from [50].

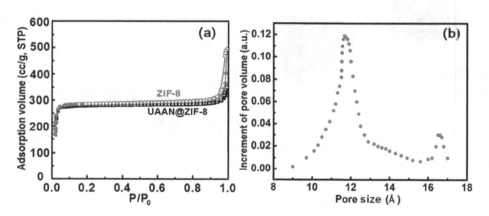

FIGURE 10.21 (a) N_2 adsorption–desorption isotherms for the pure ZIF-8 and ZIF-8 wrapped UAANs. (b) The distribution of the pore size for the ZIF-8 wrapped UAANs. Reproduced with permission from [50].

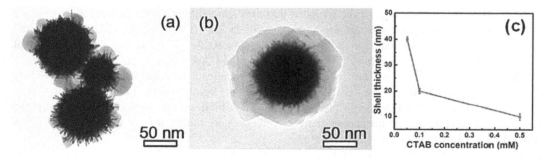

FIGURE 10.22 The typical TEM images of UAAN@ZIF-8 under different CTAB concentrations. (a) 0.5 mM, and (b) 0.05 mM. (c) The ZIF-8 shell thickness as a function of the CTAB concentration in the precursor solution.

Gold Nanocrystal-built Films

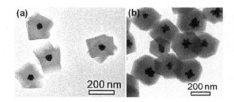

FIGURE 10.23 The TEM images of the ZIF-8 wrapped particles. (a): The ZIF-8 wrapped Au concave trisoctahedrons and (b): the ZIF-8 wrapped Au calyptriform nanocrystals. Reproduced with permission from [50].

calyptriform or Au concave trisoctahedral nanocrystals, with tunable shell thickness, by addition of the Au nanocrystals in the CTAB-contained precursor solutions at 40 °C (Figure 10.23).

10.4.3 THE FORMATION OF ZIF-8 WRAPPING SHELL

The ZIF-8 wrapping on the plasmonic nanoparticles could be attributed to the linker function of the surfactant CTAB [48]. Here, the precursor solution was prepared with CTAB, $Zn(NO_3)_2$ and 2-MeIM, and due to the reaction between 2-MeIM and $Zn(NO_3)_2$, the ZIF-8 molecules and even crystal nuclei would be formed in the precursor solution [48]. When the precursor solution was added with the pre-formed Au nanocrystals under the condition of 40 °C, the CTAB could be adsorbed on the Au nanocrystals as the packaging agent, due to its bilayer structure [54]. The ZIF-8 molecules and crystal nuclei in the solution would be preferentially attached on the CTAB-adsorbed Au nanocrystals, due to the strong interaction between the ZIF-8 and the adsorbed CTAB [55]. As the reaction was going on, the ZIF-8 crystals would grow on the Au nanocrystals, leading to the wrapping of the ZIF-8 layer on the Au nanocrystals [56].

10.5 AU NANOCRYSTALS-BUILT FILMS AND THEIR SERS ACTIVITIES

The above-mentioned Au nanocrystals are of high number density of tips and edges/corners, which can generate the strong local-field enhancements at these areas under the external field excitation and could exhibit strong SERS effect [3, 26, 27, 57]. So, such nanocrystals are the ideal building blocks of the high-performance SERS substrates or chips. In this section, the different nanocrystals-built films and their SERS activities are introduced.

10.5.1 AU NANOCRYSTALS-BUILT FILMS

The films built of the different Au nanocrystals were first prepared for the SERS substrates or chips via alternatively dropping the corresponding as-prepared colloidal solutions on silicon wafers and drying. Typically, Figure 10.24 shows the thin films built of the Au concave nanocrystals (shown in Figure 10.1b and Figure 10.5) prepared under different temperatures. Using such alternatively dropping and drying, we can prepare the nanocrystals' films with a certain thickness and uniform structure. The thickness of the films can be controlled by the alternative number of the dropping and drying. Figure 10.25 gives the cross-sectional views of the Au concave trisoctahedral nanocrystals-built films with different thicknesses. Similarly, the thin films built of the Au nanocrystals, such as chestnut-like crystals, concave nanocubes and nanospindles (mentioned in Section 3) were also fabricated on silicon wafers as typically illustrated in Figure 10.26.

These nanocrystals-built films are uniform in structure on the microscale. So, if they are used as SERS chips, the good reproducibility in the Raman spectral measurements should be achieved. Here, 4-Aminothiophenol (4-ATP) was chosen as the probe molecules for test of such reproducibility in spectral measurements. Typically, the chestnut-like Au nanocrystals-built film shown in Figure 10.26a was soaked in the 4-ATP solution with 10^{-5} M and then dried. The Raman spectra

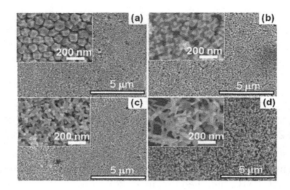

FIGURE 10.24 SEM images of the films built of the Au nanocrystals obtained by adding 0.05 mL of seeds' solution at different reaction temperatures. (a): 25°C (concave trisoctahedral nanocrystals), (b): 20°C (calyptriform crystals), (c): 10°C (coral-like crystals), and (d): 5°C (dendritic crystals). The insets: the corresponding local magnified images. Reproduced with permission from [28].

FIGURE 10.25 The SEM images of the cross-sections for the Au concave trisoctahedral nanocrystals-built films with different thicknesses. Reproduced with permission from [28].

FIGURE 10.26 The SEM images of the films built of (a): the chestnut-like Au nanocrystals, (b): the Au nanospindles and (c): the concave Au nanocubes. The insets are the corresponding local magnified images.

of the 4-ATP molecules from the 20 random spots on this film were measured, as shown in Figure 10.27a. The relative standard deviation (RSD) value of the main peak intensity at 1078cm^{-1} is only 8.7%, showing good reproducibility (Figure 10.27b). Similarly, for the other thin films, the RSD values were all below 10%, as typically shown in Figure 10.27c, d, corresponding to the Au nanospindles' film and the concave Au nanocubes-built film.

10.5.2 SERS Activity

The influence of the excitation wavelength was firstly examined for the Au nanocrystals-built films. There exists an optimal excitation wavelength at 785 nm for all the Au nanocrystals-built films, as typically shown in Figure 10.28a corresponding to the Raman spectra of 4-ATP molecules on the

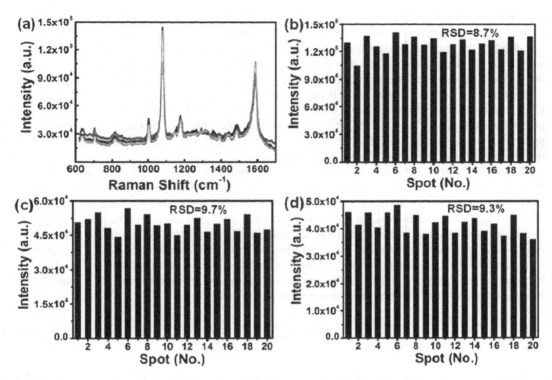

FIGURE 10.27 The reproducibility in the Raman spectral measurements after soaking the Au nanocrystals-built films in the 4-ATP solution with 10^{-5} M and drying (excited at 785 nm). (a): Raman spectra from 20 random spots on the chestnut-like Au nanocrystals-built film. (b), (c), and (d): The intensities of the main peak at 1078 cm^{-1} from each spot on the chestnut-like Au nanocrystals-built film (the data from (a)), the Au nanospindles' film and the concave Au nanocubes-built film, respectively. Reproduced with permission from [35].

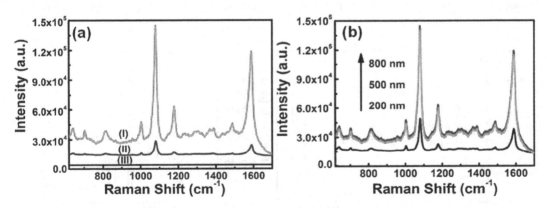

FIGURE 10.28 The Raman spectra of the chestnut-like Au nanocrystals-built film after soaking in the 4-ATP solution with 10^{-5}M and drying. (a): The soaked films with 800 nm in thickness and excitation at different wavelengths. Curve (I): 785 nm; Curve (II): 633 nm; Curve (III): 532 nm. (b): The soaked films with different thicknesses and excitation at 785 nm. Reproduced with permission from [35].

chestnut-like Au nanocrystals-built film exited at 532, 633, and 785 nm. Furthermore, the thickness of the thin film would also influence the SERS performances when the film is less than the penetration depth of the laser beam to the thin film. It has been confirmed that the Raman intensity increases with the film's thickness up to 500 nm. When the films are 500 nm or larger in thickness; however, the Raman signals are independent of the thickness, as typically shown in Figure 10.28b.

For quantification analysis of the SERS activity for these Au nanocrystals-built films as the SERS chips, the enhancement factors (EFs) were estimated. Here, we also chose 4-ATP as the Raman probe molecules, that's because it is a non-resonant molecule with minimal chemical enhancements under near-infrared excitation [58]. In addition, the Au nanocrystals-built films with >500 nm in thickness and the excitation at 785 nm were employed.

Figure 10.29a shows the Raman spectra of the chestnut-like nanocrystals-, Au nanocubes- and Au nanospindles-built films after dropping 20 μL of 10^{-5} M 4-ATP solution and drying. All these films show the same Raman spectral pattern, which is in good agreement with that of solid 4-ATP (Figure 10.29b). The main peaks at 1578 cm^{-1} and 1078 cm^{-1} are assigned to the phenol ring C–C stretching mode and the C–S stretching mode, respectively [59]. From Figure 10.29a, the intensities of main peaks are comparable, for these films, to that of the pure solid 4-ATP, indicating the significantly enhanced effect. Further, the Raman peaks of the 4-ATP molecules on the chestnut-like nanocrystals are the strongest, while the Au nanocubes and Au nanospindles show much lower Raman intensities. The EF value is determined by the formula [60]

$$EF = \frac{I_{SERE}/N_{SERS}}{I_{RS}/N_{RS}} \quad (10.4)$$

where I_{SERS} and I_{RS} are the intensities of the selected Raman peaks of 4-ATP molecules on the Au nanocrystals-built films and Si wafer, respectively. N_{SERS} and N_{RS} are the corresponding numbers of the 4-ATP molecules within the laser spot areas (ϕ1 μm) on the films. The Raman spectra were measured by dropping the 20 μL of 10^{-5} M and 50 μL of 0.1M 4-ATP solutions on the Au nanocrystals' film (with the circular coverage area about ϕ3 mm in diameter) and a Si wafer with 3 mm × 3 mm in coverage area, respectively, as shown in Figure 10.29a and Figure 10.30a. Here, it is assumed that all the 4-ATP molecules are uniformly adsorbed on the films and Si substrate after drying. The values of N_{SERS} and N_{RS} were thus estimated to be about 1.3×10^7 and 3.3×10^{11}, respectively. From Equation (10.4), the EF values for the different Au nanocrystals-built films could be obtained for the

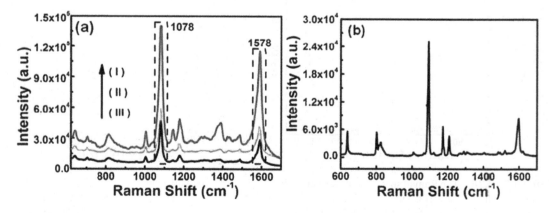

FIGURE 10.29 (a): The Raman spectra of the different Au nanocrystals-built films after dropping 20 μL of 10^{-5} M 4-ATP solution on them and drying. Curves (I), (II) and (III) correspond to the films built of the chestnut-like Au nanocrystals, the Au nanospindles and the concave Au nanocubes, respectively. (b): The Raman spectrum of the pure solid 4-ATP. Reproduced with permission from [35].

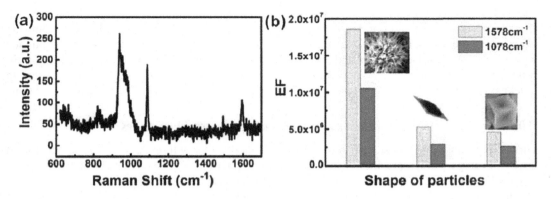

FIGURE 10.30 (a): The Raman spectrum measured after dropping 50 μL of 4-ATP solution (0.1M) on Si wafer and drying. The peak around 930–950 cm^{-1} is from Si. (b): The EF values at 1078 cm^{-1} and 1578 cm^{-1} for the different Au nanocrystals-built films. Reproduced with permission from [35].

TABLE 10.1
The EF values of the films built of Au nanocrystals with different shapes (for the peak at 1078 cm^{-1})

	Concave trisoctahedron	Calyptriform	Coral-like	dendrite
EF	1.12×10^7	1.07×10^7	1.18×10^6	1.03×10^6

vibrations at 1078 cm^{-1} and 1578 cm^{-1}. Figure 10.30b shows the corresponding results. The chestnut-like Au nanocrystals possess the highest SERS activity and the EF values higher than 10^7, while the Au nanospindles- and concave nanocubes-built films show much lower EF values. Similarly, the EF values for the concave trisoctahedral, calyptriform, coral-like and dendritic nanocrystals-built films shown in Figure 10.24 were obtained, as listed in Table 10.1. The EF values of the Au trisoctahedral and calyptriform nanocrystals are one order of magnitude higher than that of the dendritic and coral-like nanocrystals.

The high EF values for the Au nanocrystals with chestnut-like, concave trisoctahedral and calyptriform shapes can be attributed to their special structures. These nanocrystals-built films have the high number density of the tips, sharp edges/corners, which can generate the strong electromagnetic field enhancement due to the tip-effect and local SPR [61, 62]. In addition, due to the surface plasmonic coupling effects, the nano-gaps among the nanocrystals in the film are also the "hot spots" with significant local-field enhancement effect [60, 63–67]. It is the high number density of the "hot spots" that these Au nanocrystals-built films exhibit the strong SERS activity.

10.6 APPLICATIONS IN TRACE DETECTION OF HEXACHLOROCYCLOHEXANE

Hexachlorocyclohexane (HCH or $C_6H_6Cl_6$), as a broad-spectrum insecticide, has eight isomers. They are of the high fat-solubility, small water solubility, high structural stability and low volatility, and hence not easily degraded in the environment. Among the eight isomers, the γ isomer (γ-HCH) or lindane and the α isomer (α-HCH) have the highest and second highest toxicity, respectively [68, 69]. The HCH is usually accumulated in the stable media, such as bodies of water, sediments, soils, and organisms. Therefore, the efficient (sensitive, fast, portable) detection and fingerprint recognition of HCH residues is very important. Currently, they are mostly detected and analyzed using the conventional methods, such as gas chromatography–mass spectrometry (GC-MS) coupling technique and gas chromatography equipped with electron capture detection (GC/ECD) [70, 71]. Such methods are time-consuming and complex. The SERS-based method could be a promising route to the quick and ultrasensitive detection of the toxic HCH molecules [72–74]. As mentioned above, the concave trisoctahedral and chestnut-like Au nanocrystals-built films possess the high EF values. Here we introduce the application of these Au nanocrystals-built films with > 500 nm in thickness as SERS chips in detection of trace HCH by using the excitation wavelength of 785 nm.

10.6.1 CONCAVE AU TRISOCTAHEDRAL NANOCRYSTALS-BASED SERS CHIPS

Firstly, the concave Au trisoctahedral nanocrystals-built films with enough thickness were used as SERS chips to measure the concentration-dependent Raman spectra of γ-HCH (or lindane). Figure 10.31a presents the corresponding results after soaking the chips in the lindane solutions with different concentrations and drying. Except the peak intensity, all spectra are of the same spectral pattern and peak positions, which are in good agreement with those of solid lindane (Figure 10.31b) [74–76]. The main peak at 345 cm^{-1} is assigned to the C–Cl stretching vibrations, and the peaks in 400–1400 cm^{-1} are assigned to the C–C and C–H stretching, and CH$_2$ bending from the aliphatic cyclic structure of the analyzed molecules [74–76].

10.6.1.1 Concentration-Dependent Raman Spectra

From Figure 10.31a, we can see that all Raman peaks increase in the intensity with increasing lindane concentration. The detection limit was below 10^{-7} M (~30 ppb). So, such Au trisoctahedral nanocrystals-built SERS chip can be employed to detect the trace lindane molecules in solutions. It should be mentioned that the dependence of the Raman peak intensity I on the lindane concentration C can be described by a double logarithmic linear relation, or

$$Log I = A + B \cdot \log(C) \tag{10.5}$$

where A and B are the constants independent of C (in molar concentration). Representatively, Figure 10.32a gives the plot of the intensity of the main peaks at 345 cm^{-1} vs the lindane concentration in logarithmic scale, showing a good linear relation from 10^{-7} M to 10^{-3} M (or 30 ppb to 300 ppm). The constants A and B are estimated to be 3.51 and 0.28, respectively, by fitting. Equation (10.5) can thus be rewritten as a power function, or

$$I = 3236 \times C^{0.28} \tag{10.6}$$

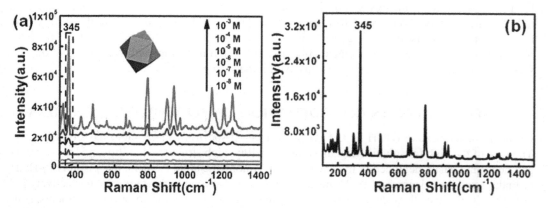

FIGURE 10.31 (a): The Raman spectra of the concave Au trisoctahedral nanocrystals-built SERS chips after soaking in the lindane ethanol solutions with different concentrations and drying. (b): The Raman spectrum of the solid lindane. Reproduced with permission from [38].

Gold Nanocrystal-built Films 275

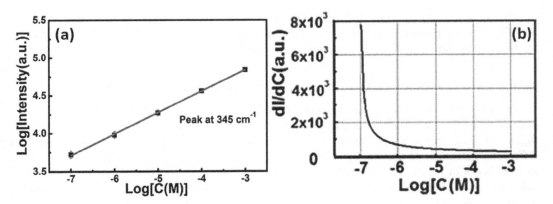

FIGURE 10.32 (a): The plot of the peak intensity (*I*) at 345 cm^{-1} vs the lindane concentration (*C*) in logarithmic scale [data are from Figure 10.31(a)]. The straight-line is the linear fitting results. (b): The sensitivity of Raman signals at 345 cm^{-1} versus the lindane concentration. Reproduced with permission from [28].

Also, the concentration dependence of the sensitivity dI/dC can be written as

$$\frac{dI}{dC} = 906 \times C^{-0.72} \tag{10.7}$$

Figure 10.32b shows the dI/dC versus the lindane concentration. Obviously, the lower the concentration is, the higher the sensitivity. The concave Au trisoctahedral nanocrystals-built SERS chips are thus suitable for detection of trace lindane molecules.

10.6.1.2 Freundlich Adsorption-induced Raman Spectra

The logarithmic linear relation of the peak intensity with the lindane concentration is associated with the concentration-dependent adsorption. When the SERS chip was soaked in the solution for an enough time, the adsorption of the lindane molecules on the Au nanocrystals would reach equilibrium. The Raman peak intensity should be proportional to the number of the molecules adsorbed on the chip within the laser spot's area, or the equilibrium adsorption amount q_e:

$$I = K_1 q_e \tag{10.8}$$

where K_1 is the constant. It is evident that the higher lindane concentration in the solution would lead to the more lindane molecules adsorbed on the chip and hence the stronger Raman signals, exhibiting the increasing Raman peak intensity with the lindane concentration (Figure 10.31a). It is well known that, according to Freundlich adsorption model [77, 78], if molecules are adsorbed on a heterogeneous surface, the equilibrium adsorption amount could be expressed as

$$q_e = K_F \cdot C^{\frac{1}{n}} \tag{10.9}$$

where K_F and n are the parameters reflecting the adsorption capacity and adsorption intensity, respectively. Combining Equations. (10.8) and (10.9), the relationship between the Raman intensity and the lindane concentration were obtained, or

$$I = K \cdot C^{\frac{1}{n}} \tag{10.10}$$

where $K = K_1 \cdot K_F$. Obviously, Equation (10.10) is in agreement with Equation (10.6). Such agreement has also indicated that the lindane molecules' adsorption on the Au nanocrystals follows Freundlich model. According to Equations. (10.6) and (10.10), the value of the parameter n was determined to be 3.57. This also provides a simple way to determine the adsorption parameters, which are usually obtained by the time-consuming adsorption isothermal measurements.

10.6.2 The Chestnut-like Au Nanocrystal-built SERS Chips

Alternatively, the above-mentioned chestnut-like Au nanocrystals-built SERS chips, which have the high EF value (Figure 10.30b), can also be used for efficient detection of organochlorine molecules. Figure 10.33a gives the Raman spectra of the chips after soaking in lindane ethanol solutions with different concentrations and drying. The concentration-dependent spectral evolution was similar to that shown in Figure 10.31a but the detection limit was lower and below 3.44×10^{-8} M (or 10 ppb). The peak intensity versus the concentration is also subject to Equation (10.5) or the double logarithmic linear relation. Figure 10.33b shows the corresponding results.

Here, it should be mentioned that the isomers of HCH can also be differentiated using such SERS chips. Typically, Figure 10.34 shows the Raman spectra of γ-HCH and α-HCH molecules on the chestnut-like Au nanocrystal-built SERS chip. In spite of their similar molecule's structure, the Raman spectral patterns are different from each other. This is attributed to the fact that there are

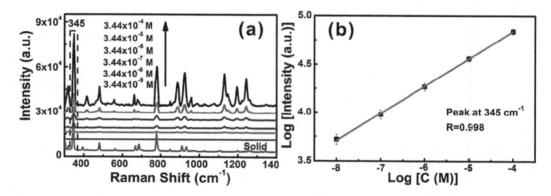

FIGURE 10.33 (a): The Raman spectra of the concave chestnut-like Au nanocrystals-built chips after soaking in the lindane ethanol solutions with different concentrations and drying. The bottom curve is the spectrum of solid γ-HCH. (b): The plot of peak intensity at 345 cm^{-1} vs concentration C in logarithmic scale. Reproduced with permission from [35].

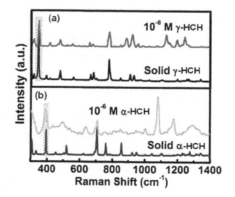

FIGURE 10.34 The Raman spectra of the chestnut-like Au nanocrystals-built SERS chips after soaking in (a) γ-HCH and (b) α-HCH solutions (10^{-6} M) and drying. The curves at the bottom in (a) and (b): for the solid HCHs. Reproduced with permission from [35].

Gold Nanocrystal-built Films 277

significant differences between and γ-HCH and α-HCH molecules in the vibrations of C-Cl and C-C bonds [70], since the Raman signals are very sensitive to the molecular vibrations.

So, these chestnut-like and concave trisoctahedral Au nanocrystals-built films can be used as the efficient SERS chips for recognition and detection of trace HCH in solutions. Using such SERS chips, we could also quantitatively detect HCH in a large concentration range according to Equation (10.5).

10.6.3 ZIF-8 Wrapped Au Nanocrystals-built SERS Chips

For the ZIF-8 wrapped Au nanocrystals-built film, it could be used as the SERS chips with enrichment effect and selectivity for detection of trace HCH molecules, since the ZIF-8 shell layer is porous with 11.6 Å in pores size, which is slightly larger but comparable to the size of HCH molecules (~7 Å) [46]. Such chips could effectively enrich the HCH molecules. Here, the ZIF-8 wrapped UAANs-built films are utilized as the SERS chips to demonstrate their superiority in the trace detection of HCHs.

10.6.3.1 ShellThickness-dependent SERS

The influence of the ZIF-8 shell's thickness on the SERS performances of the UAANs to HCH molecules was studied. Figure 10.35 shows the Raman spectra of the UAANs wrapped with different thicknesses of ZIF-8 shell after immersion in the 10^{-7} M γ-HCH (and α-HCH) solutions. The typical Raman peaks are at 393 cm^{-1} for α-HCH and at 345 cm^{-1} for γ-HCH, which are in agreement with those of the pure solid α-HCH and γ-HCH (curves I) [28, 74]. The ZIF-8 wrapped UAANs with 20 nm of shell thickness exhibit the strongest Raman signals to α-HCH and γ-HCH (curves IV). The thicker or thinner ZIF-8 shells would lead to the relatively lower SERS signals, as shown in curves III or V of Figure 10.35. All wrapped UAANs-built chips (Curves III-V) show much stronger Raman signals than the bare ones (Curves II). It should be mentioned that compared with the ZIF-8 wrapped concave Au calyptriform and trisoctahedral and nanocrystals, the ZIF-8 wrapped UAANs are the strongest in the SERS activity to the α-HCH and γ-HCH due to their high-density tips.

10.6.3.2 Concentration-dependent SERS

Further, the HCH concentration-dependent SERS spectral measurements were carried out for the 20 nm ZIF-8 shell-wrapped UAANs to show the importance of the ZIF-8 shell layer in the SERS-based detection of HCH. Typically, Figure 10.36a shows the corresponding results for the detection

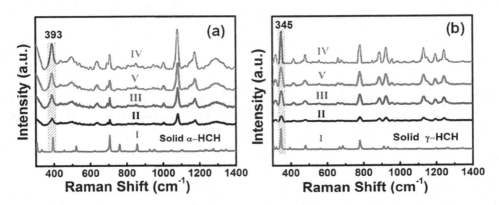

FIGURE 10.35 The Raman spectra of the UAANs wrapped with different thicknesses of ZIF-8 shells after soaking in the (a): α-HCH and (b): γ-HCH solutions with 10^{-7} M. Curves I: The Raman spectra of the pure solid α-HCH and γ-HCH. Curves II, III, IV, V: The Raman spectra of the samples with shell thickness: 0, 10 nm, 20 nm, 40 nm, respectively. Reproduced with permission from [50].

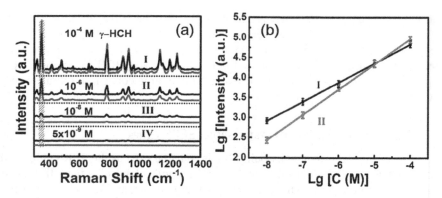

FIGURE 10.36 (a): The Raman spectra of the ZIF-8 wrapped (black) and bare (gray) UAANs-built chips after soaking in the solutions with different γ-HCH concentrations. Curves I, II, III, and IV: 1×10^{-4}, 1×10^{-6}, 1×10^{-8}, and 5×10^{-9} M, respectively. The shadowed bands are the main peak (at 345 cm^{-1}) of γ-HCH. (b): The double-logarithmic plots of the peak intensity at 345 cm^{-1} vs γ-HCH concentrations for the wrapped (I) and the bare (II) UAANs. Reproduced with permission from [50].

of γ-HCH. The Raman peaks increase in intensity with increasing HCH concentration for both the wrapped and the bare UAANs. When the γ-HCH concentration was below 1×10^{-5} M, the wrapped UAANs showed the stronger Raman signals than the bare UAANs (curves II, III, IV in Figure 10.36a) and had the detection limit below 5×10^{-9} M (1.5 ppb). For the γ-HCH concentration higher than 1×10^{-5} M, however, the opposite was true and the ZIF-8 wrapping induced the decrease of the Raman peaks in intensity, compared with the bare UAANs, as illustrated in curves I of Figure 10.36a.

Also, for both the wrapped and bare UAANs-built chips, the Raman peak intensity I shows the linear relation with the γ-HCH concentration C in logarithmic scale, or is subject to Equation (10.5). Figure 10.36b presents the double-logarithmic plots of I vs C for both chips at 345 cm^{-1}, assuming the good linear relations. By linear fitting, the values of the constants A and B in Equation (10.5) were estimated to be 6.75, 0.48 for the wrapped UAANs-built chip, and 7.50, 0.63 for the bare UAANs-built chip, respectively. The slope of the plot for the wrapped UAANs is smaller than that for the bare UAANs. There exists an intersection point between these two plots at C ≈ 1×10^{-5} M. This means that the difference between both chips in the peak intensity is γ-HCH concentration-dependent. For measurement of the ZIF-8 wrapping-enhanced SERS effect, the intensity ratio R of the Raman peak for the wrapped UAANs-built chip to that for the bare UAANs-built chip was used. According to Equation (10.5), this intensity ratio R can be written as

$$R = P \cdot C^{-m} \tag{10.11}$$

where m and P are the constants and depend on the parameters A and B. When R is much larger than 1 or >>1, it indicates that the ZIF-8 wrapping-enhanced SERS effect is significant. Otherwise, if $R<1$, the ZIF-8 wrapping would decrease the SERS performance, compared with the bare UAANs-built chip. For the Raman peak at 345 cm^{-1},

$$R \approx 0.18 \cdot C^{-0.15} \tag{10.12}$$

Figure 10.37 shows the corresponding R values at 345 cm^{-1} vs γ-HCH concentration (curve I), and the ever-increasing R value with decreasing concentration. For the low γ-HCH concentration (say, below 1.4×10^{-5} M), the ratio is larger than 1, meaning the ZIF-8 wrapping–enhanced effect which increases with the decreasing concentration. As for α-HCH, the results are very similar to γ-HCH (curve II in Figure 10.37).

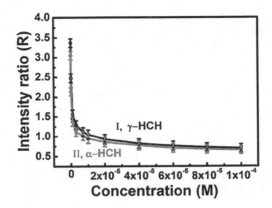

FIGURE 10.37 The intensity ratio *R* of the Raman peaks for the wrapped UAANs-built chip to that for the bare UAANs-built chip vs HCH concentration. Curves (I) and (II): at 345cm^{-1} for γ-HCH and at 393 cm^{-1} for α-HCH, respectively. (The solid lines were obtained according to Eq. 11). Reproduced with permission from [50].

10.6.3.3 Wrapping-induced HCH Enrichment

As mentioned above, ZIF-8 wrapping could enhance SERS performance of the Au nanocrystals to HCH in low concentration. This could mainly originate from the HCH enrichment in the ZIF-8 shell layer. The molecular diffusion and the pores' confinement of ZIF-8 could induce the HCH enrichment. When the bare UAANs-built chip was soaked into the HCH solution, due to the weak interaction between Au and HCH, only relatively less HCH molecules were adsorbed on the chip, leading to the low Raman signal. When the porous ZIF-8 wrapped UAANs-built chip was immersed in the solution, however, the HCH molecules would be ceaselessly diffused into the porous shell layer till the equilibrium concentration was reached, and confined within the pores' space. After subsequently drying, these confined molecules were staying inside the shell or enriched around the UAANs. Such enrichment would result in the higher Raman intensity compared with that of the immersed bare UAANs-built chip, as illustrated in Figure 10.36a.

Obviously, the thicker the shell layer is, the more molecules are enriched around the UAANs. So, the Raman peak intensity increases with the rising shell's thickness. If the shell layer was too thick (say, >20 nm), however, the laser absorption by the ZIF-8 layer during the Raman spectral measurements became non-negligible, while the number of HCH molecules, within the strong electromagnetic field enhancement space (10–20 nm in thickness) above the UAANs, would no more increase with the shell thickness. In this case, the Raman intensity would be decreased with the rising shell's thickness, as shown in curves V of Figure 10.35.

As for the concentration dependence of wrapping effect, it is associated with the porous structure of the ZIF-8 and diffusion-induced blockage. When the HCH concentration is low enough (<1×10^{-5} M), the lower the concentration in solution is, the easier the diffusion of the HCH molecules into the pores of the shell layer, and the higher the ZIF-8 wrapping-enhanced SERS effect, showing the increase of *R* values with decreasing concentration (Figure 10.37). Contrarily, when the HCH concentration in solution was high, the HCH molecules could easily diffuse into the pores of ZIF-8 in the initial stage, but as the soaking was going on, the diffusion-induced blockage of pores would take place and the diffusion become harder and harder because the HCH molecules are comparable to the pores in size. Therefore, when the HCH concentration was high enough (say, >1×10^{-5} M), the obtained Raman signals for the ZIF-8 wrapped UAANs-built chip were even weaker than those of the bare UAANs-built chip. In a word, the porous ZIF-8 wrapping layer is beneficial to the detection of trace HCH.

As mentioned above, the HCH molecules could be enriched around Au nanocrystals by diffusion into the pores of the ZIF-8 shell. In this case, the Raman intensity is soaking time-dependent

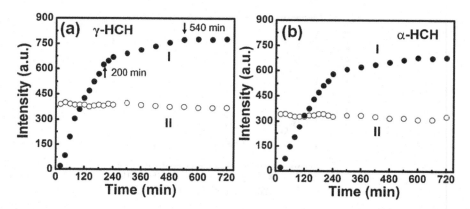

FIGURE 10.38 The intensities of Raman main peak for (I) the wrapped and (II) the bare UAANs-built films versus the soaking time in the 10^{-8} M γ-HCH (a) and α-HCH (b) solutions. Reproduced with permission from [50].

in an enough long period. Figure 10.38a shows the evolution of the Raman peak intensity at 345 cm^{-1} (for γ-HCH) with the soaking time for the wrapped UAANs-built chip (curve I). The intensity of the peak at 345 cm^{-1} was rapidly increased with the soaking time up to 200 min, and then rised slowly. After soaking for 540 min, the peak intensity was nearly unchanged. Obviously, in the initial stage, due to the high concentration gradient and smaller size than the pores of ZIF-8 (11.6 Å), the HCH molecules could fast diffuse from the solution into the pores in the ZIF-8 shells [48], leading to the rapidly increased Raman intensity. However, the diffusion would slow down due to the blockage effect induced by the rising concentration within the shell layer. After soaking for enough time (>540 min), the equilibrium was reached, leading to the unchanged Raman intensity. For the bare UAANs-built chip, however, the maximum intensity was reached in a short time (<20 min) (curve II of Figure 10.38a). For the detection of α-HCH using both chips, the results are similar, as demonstrated in Figure 10.38b.

10.6.3.4 The Size Selectivity of Molecules

In addition, it should be mentioned that the ZIF-8-wrapping shell was of good molecular size selectivity. Here, Rhodamine 6G (R6G) and 4-ATP, which are larger and smaller in size than the pores in the ZIF-8, respectively, were chosen as the test molecules to show the size selectivity. Figure 10.39a presents the Raman spectra of the ZIF-8 wrapped and the bare UAANs-built chips

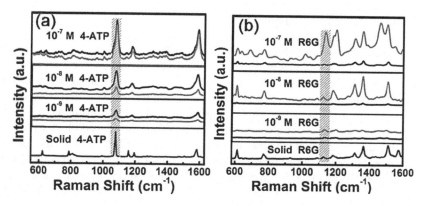

FIGURE 10.39 The Raman spectra of the ZIF-8 wrapped (black) and bare (gray) UAANs-built chips after soaking in (a): 4-ATP and (b): R6G solutions with different concentrations for 8h. The bottom lines are the Raman spectra of the pure solid 4-ATP and R6G for references. Reproduced with permission from [50].

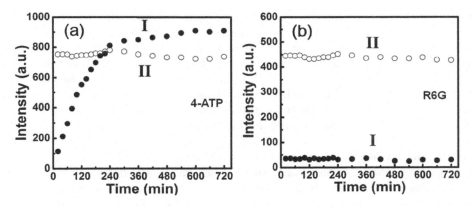

FIGURE 10.40 The intensity of the Raman peak (a): at 1078 cm^{-1} for 4-ATP and (b): at 1134 cm^{-1} for R6G molecules on the (I) wrapped and (II) bare UAANs-built chips vs the soaking time in the 10^{-9} M solutions. Reproduced with permission from [50].

after soaking in the 4-ATP solutions with different concentrations. Obviously, the ZIF-8 wrapped UAANs-built SERS chip can detect 4-ATP, exhibiting the much higher Raman signals than that of the bare UAANs-built chip. For test molecule R6G, however, the wrapped UAANs-built chip shows very weak Raman signals compared with the bare one, as illustrated in Figure 10.39b. These indicate that R6G molecules cannot diffuse into the ZIF-8 shells while 4-ATP can enter. Correspondingly, the evolution of the Raman peak intensity for 4-ATP with the time is similar to that of HCH, showing the ever-increasing peak intensity with the soaking time in an enough long period (curve I of Figure 10.40a). But for R6G, the wrapped UAANs-built chip shows the spectral evolution with the soaking time similar to the bare UAANs-built chip, as illustrated in Figure 10.40b.

10.7 CONCLUSIONS

In summary, the Au nanocrystals with various shapes could be synthesized based on the kinetically controlled seed growth strategy. The kinetically controlled growth can be achieved by adjusting the reaction temperature or the addition rate of the Au seeds' solution. (i) The relative growth rates along <110> and <111> directions of Au seeds were controlled via adjusting the temperature of the growth solution, and hence shapes of Au nanocrystals were tuned. When the temperature was decreased from 25 °C down to 5 °C, the shapes of the nanocrystals evolved from concave trisoctahedral to calyptriform, coral, dendritic nanocrystals, which is attributed to the difference between <111> and <110> directions in temperature-dependent growth rates. (ii) Alternatively, the growth kinetics could also be controlled by the Au seeds' addition rate into the growth solution, which can determine growth kinetics of crystal nuclei along <111>. Under a proper addition rate, the chestnut-like Au nanocrystals would be obtained, while the much lower or faster addition rate induced the formation of the concave Au nanocubes or the Au nanospindles. As for the seeds' addition amount, it could determine the nanocrystals' size but only insignificantly change the morphology since the growth kinetics was nearly unchanged under the given addition rate. These Au nanocrystals could be controllably wrapped with the porous ZIF-8 shell layer via addition of the pre-formed Au nanocrystals in the ZIF-8 precursor solution with CTAB.

Due to the high number density of sharp corners/edges and tips, the thin films built of the Au concave trisoctahedron, chestnut-like and calyptriform Au nanocrystals have exhibited the high SERS activity with EF>10^7. It has been demonstrated that such Au nanocrystals-built films can be used as SERS chips for efficient detection and recognition of trace HCH, and the detection limit could be below 10 ppb. Typically for γ-HCH, the Raman intensity at main peak was subject to the logarithmic linear relation with the concentration up to 100 ppm. Such double-logarithmic linear relation originates from Freundlich-typed adsorption of HCH molecules on the Au nanocrystals.

Further, because of the porous structure with an appropriate and defined pore's size, ZIF-8 wrapped Au nanocrystals-built films can enrich the HCH molecules around Au nanocrystals, and hence be utilized as the high-performance SERS chips for the detection of trace HCH molecules. An appropriate thickness of the ZIF-8 shell layer can significantly increase the SERS performance of the Au nanocrystals to the trace α-HCH and γ-HCH, showing wrapping-enhanced SERS effect. The detection limit could be lower than 1.5 ppb. Finally, the ZIF-8 shell layer has a good size selectivity for the target molecules due to its well-defined pore size.

In a word, in this chapter, we have not only introduced the facile fabrication routes of the Au nanocrystals with various specific shapes via adjusting the growth kinetics of the seeds but also demonstrated the possibility that the thin films built of the Au nanocrystals with high density of tips and/or corners/edges can be used as the high-performance SERS chips for the trace recognition and quantitative detection of the toxic organochlorine pesticides.

ACKNOWLEDGMENTS

This work is financially supported by the National Key Research and Development Program of China (Grant No 2017YFA0207101), Natural Science Foundation of China (Grant No.11974352).

REFERENCES

[1] Zhang, Y. G., Lu, F., Yager, K. G., van der Lelie, D., Gang, O. 2013. A general strategy for the DNA-mediated self-assembly of functional nanoparticles into heterogeneous systems. *Nat. Nanotechnol.* 8: 865–872.

[2] Zhang, Q. F., Zhou, Y. D., Villarreal, E., Lin, Y., Zou, S. L., Wang, H. 2015. Faceted gold nanorods: nanocuboids, convex nanocuboids, and concave nanocuboids. *Nano Lett.* 15: 4161–4169.

[3] Zhang, Q., Large, N., Wang, H. 2014. Gold nanoparticles with tipped surface structures as substrates for single-particle surface-enhanced Raman spectroscopy: concave nanocubes, nanotrisoctahedra, and nanostars. *ACS Appl. Mater. Interfaces* 6: 17255–17267.

[4] Zhang, J. A., Langille, M. R., Personick, M. L., Zhang, K., Li, S. Y., Mirkin, C. A. 2010. Concave cubic gold nanocrystals with high-index facets. *J. Am. Chem. Soc.* 132: 14012–14014.

[5] Zhang, J., Winget, S. A., Wu, Y.; Su, D., Sun, X., Xie, Z. X., Qin, D. 2016. Ag@Au Concave cuboctahedra: a unique probe for monitoring Au-catalyzed reduction and oxidation reactions by surface-enhanced Raman spectroscopy. *ACS Nano* 10: 2607–2616.

[6] Rycenga, M., Cobley, C. M., Zeng, J., Li, W. Y., Moran, C. H., Zhang, Q., Qin, D., Xia, Y. N. 2011. Controlling the synthesis and assembly of silver nanostructures for plasmonic applications. *Chem. Rev.* 111: 3669–3712.

[7] Hong, J. W., Lee, S. U., Lee, Y. W., Han, S. W. 2012. Hexoctahedral Au nanocrystals with high-index facets and their optical and surface-enhanced Raman scattering properties. *J. Am. Chem. Soc.* 134: 4565–4568.

[8] Zhang, H., Jin, M. S., Xia, Y. N. 2012. Noble-metal nanocrystals with concave surfaces: synthesis and applications. *Angew. Chem. Int. Ed.* 51: 7656–7673.

[9] Qi, W. K., de Graaf, J., Qiao, F., Marras, S., Manna, L., Dijkstra, M. 2012. Ordered two-dimensional superstructures of colloidal octapod-shaped nanocrystals on flat substrates. *Nano Letters* 12: 5299–5303.

[10] Xia, X. H., Zeng, J., McDearmon, B., Zheng, Y. Q., Li, Q. G., Xia, Y. N. 2011. Silver nanocrystals with concave surfaces and their optical and surface-enhanced Raman scattering properties. *Angew. Chem. Int. Ed.* 50: 12542–12546.

[11] Xu, L. G., Zhao, S., Ma, W., Wu, X. L., Li, S., Kuang, H., Wang, L. B., Xu, C. L. 2016. Multigaps embedded nanoassemblies enhance in situ Raman spectroscopy for intracellular telomerase activity sensing. *Adv. Funct. Mater.* 26: 1602–1608.

Gold Nanocrystal-built Films   283

[12] Henry, A. I., Sharma, B., Cardinal, M. F., Kurouski, D., Van Duyne, R. P. 2016. Surface-enhanced Raman spectroscopy biosensing: in vivo diagnostics and multimodal imaging. *Anal. Chem.* 88: 6638–6647.

[13] Zhou, B., Li, X., Tang, X., Li, P., Yang, L., Liu, J. 2017. Highly selective and repeatable surface-enhanced resonance Raman scattering detection for epinephrine in serum based on interface self-assembled 2D nanoparticles arrays. *ACS Appl. Mater. Interfaces* 9: 7772–7779.

[14] Ben-Jaber, S., Peveler, W. J., Quesada-Cabrera, R., Cortes, E., Sotelo-Vazquez, C., Abdul-Karim, N., Maier, S. A., Parkin, I. P. 2016. Photo-induced enhanced Raman spectroscopy for universal ultra-trace detection of explosives, pollutants and biomolecules. *Nat. Commun.* 7: 12189–12195.

[15] Cui, L., Zhang, Y. J., Huang, W. E., Zhang, B. F., Martin, F. L., Li, J. Y., Zhang, K. S., Zhu, Y. G. 2016. Surface-enhanced Raman spectroscopy for identification of heavy metal arsenic(V)-mediated enhancing effect on antibiotic resistance. *Anal. Chem.* 88: 3164–3170.

[16] Wang, P., Wu, L., Lu, Z. C., Li, Q., Yin, W. M., Ding, F., Han, H. Y. 2017. Gecko-inspired nanotentacle surface-enhanced Raman spectroscopy substrate for sampling and reliable detection of pesticide residues in fruits and vegetables. *Anal. Chem.* 89: 2424–2431.

[17] Craig, A. P., Franca, A. S., Irudayaraj, J. 2013. Surface-enhanced Raman spectroscopy applied to food safety. *Annu. Rev. Food. Sci. Techonl.* 4: 369–380.

[18] Zhou, Q. F., Zheng, J., Qing, Z. H., Zheng, M. J., Yang, J. F., Yang, S., Ying, L., Yang, R. H. 2016. Detection of circulating tumor DNA in human blood via DNA-mediated surface-enhanced Raman spectroscopy of single-walled carbon nanotubes. *Anal. Chem.* 88: 4759–4765.

[19] Zaleski, S., Clark, K. A., Smith, M. M., Eilert, J. Y., Doty, M., Van Duyne, R. P. 2017. Identification and quantification of intravenous therapy drugs using normal Raman spectroscopy and electrochemical surface enhanced Raman spectroscopy. *Anal. Chem.* 89: 2497–2504.

[20] Zhou, B. B., Li, S. F., Tang, X. H., Li, P., Cao, X. M., Yu, B. R., Yang, L. B., Liu, J. H. 2017. Real-time monitoring of plasmon-induced proton transfer of hypoxanthine in serum. *Nanoscale* 9: 12307–12310.

[21] Jin, Q. R.; Li, M., Polat, B., Paidi, S. K., Dai, A., Zhang, A., Pagaduan, J. V., Barman, I., Gracias, D. H. 2017. Mechanical trap surface-enhanced Raman spectroscopy for three-dimensional surface molecular imaging of single live cells. *Angew. Chem. Int. Ed.* 56: 3822–3826.

[22] Pazos-Perez, N., Barbosa, S., Rodriguez-Lorenzo, L., Aldeanueva-Potel, P., Perez-Juste, J., Pastoriza-Santos, I., Alvarez-Puebla, R. A., Liz-Marzan, L. M. 2010. Growth of sharp tips on gold nanowires leads to increased surface-enhanced Raman scattering activity. *J. Phys. Chem. Lett.* 1: 24–27.

[23] Pradhan, M., Chowdhury, J., Sarkar, S., Sinha, A. K., Pal, T. 2012. Hierarchical gold flower with sharp tips from controlled galvanic replacement reaction for high surface enhanced Raman scattering activity. *J. Phys. Chem. C* 116: 24301–24313.

[24] Lee, J., Hua, B., Park, S., Ha, M., Lee, Y., Fan, Z., Ko, H. 2014. Tailoring surface plasmons of high-density gold nanostar assemblies on metal films for surface-enhanced Raman spectroscopy. *Nanoscale* 6: 616–623.

[25] Liu, D. Q., Wang, X., He, D. Y., Dao, T. D., Nagao, T., Weng, Q. H., Tang, D. M., Wang, X. B., Tian, W., Golberg, D., Bando, Y. 2014. Magnetically assembled Ni@Ag urchin-like ensembles with ultra-sharp tips and numerous gaps for SERS applications. *Small* 10: 2564–2569.

[26] Fang, J., Du, S., Lebedkin, S., Li, Z., Kruk, R., Kappes, M., Hahn, H. 2010. Gold mesostructures with tailored surface topography and their self-assembly arrays for surface-enhanced Raman spectroscopy. *Nano Lett.* 10: 5006–5013.

[27] Liu, Z., Yang, Z., Peng, B., Cao, C., Zhang, C., You, H., Xiong, Q., Li, Z., Fang, J. 2014. Highly sensitive, uniform, and reproducible surface-enhanced Raman spectroscopy from hollow Au-Ag alloy nanourchins. *Adv. Mater.* 26: 2431–2439.

[28] Zhou, X., Zhao, Q., Liu, G. Q., Zhang, H. W., Li, Y., Cai, W. P. 2017. Temperature regulation growth of Au nanocrystals: from concave trisoctahedron to dendritic structures and their ultrasensitive SERS-based detection of lindane. *J. Mater. Chem. C* 5: 10399–10405.

[29] Ma, Y. Y., Kuang, Q., Jiang, Z. Y., Xie, Z. X., Huang, R. B., Zheng, L. S. 2008. Synthesis of trisoctahedral gold nanocrystals with exposed high-index facets by a facile chemical method. *Angew. Chem. Int. Ed.* 47: 8901–8904.

[30] Proussevitch, A. A., Sahagian, D. L. 2001. Recognition and separation of discrete objects within complex 3D voxelized structures. *Comput. Geosci.* 27: 441–454.

[31] Wang, H., Halas, N. J. 2008. Mesoscopic Au "meatball" particles. *Adv. Mater.* 20: 820–821.

[32] Wang, Z. L., Gao, R. P., Nikoobakht, B., El-Sayed, M. A. 2000. Surface reconstruction of the unstable {110} surface in gold nanorods. *J. Phys. Chem. B* 104: 5417–5420.

[33] Perez-Juste, J., Liz-Marzan, L. M., Carnie, S., Chan, D. Y. C., Mulvaney, P. 2004. Electric-field-directed growth of gold nanorods in aqueous surfactant solutions. *Adv. Funct. Mater.* 14: 571–579.

[34] Murphy, C. J., San, T. K., Gole, A. M., Orendorff, C. J., Gao, J. X., Gou, L., Hunyadi, S. E., Li, T. 2005. Anisotropic metal nanoparticles: Synthesis, assembly, and optical applications. *J. Phys. Chem. B* 109: 13857–13870.

[35] Zhou, X., Zhao, Q., Liu, G. Q., Zhang, H. W., Li, Y., Cai, W. P. 2018. Kinetically-controlled growth of chestnut-like Au nanocrystals with high-density tips and their high SERS performances on organochlorine pesticides. *Nanomaterials* 8: 560–574.

[36] Rycenga, M., Langille, M. R., Personick, M. L., Ozel, T., Mirkin, C. A. 2012. Chemically isolating hot spots on concave nanocubes. *Nano Lett.* 12: 6218–6222.

[37] Li, P. Z., Wang, X. J., Liu, J., Phang, H. S., Li, Y. X., Zhao, Y. L. 2017. Highly effective carbon fixation via catalytic conversion of CO_2 by an acylamide-containing metal-organic framework. *Chem. Mater.* 29: 9256–9261.

[38] Luz, I., Soukri, M., Lail, M. 2017. Confining metal-organic framework nanocrystals within mesoporous materials: a general approach via "solid-state" synthesis. *Chem. Mater.* 29: 9628–9638.

[39] Noh, H., Kung, C. W., Islamoglu, T., Peters, A. W., Liao, Y. J., Li, P., Garibay, S. J., Zhang, X., DeStefano, M. R., Hupp, J. T., Farha, O. K. 2018. Room temperature synthesis of an 8-connected Zr-based metal-organic framework for top-down nanoparticle encapsulation. *Chem. Mater.* 30: 2193–2197.

[40] Lee, H. C., Hwang, J., Schilde, U., Antonietti, M., Matyjaszewski, K., Schmidt, B. 2018. Toward ultimate control of radical polymerization: functionalized metal-organic frameworks as a robust environment for metal-catalyzed polymerizations. *Chem. Mater.* 39: 2983–2994.

[41] Sugikawa, K., Nagata, S., Furukawa, Y., Kokado, K., Sada, K. 2013. Stable and functional gold nanorod composites with a metal-organic framework crystalline shell. *Chem. Mater.* 25: 2565–2570.

[42] Wang, Y. X., Zhao, M. T., Ping, J. F., Chen, B., Cao, X. H., Huang, Y., Tan, C. L., Ma, Q. L., Wu, S. X., Yu, Y. F., Lu, Q. P., Chen, J. Z., Zhao, W., Ying, Y. B., Zhang, H. 2016. Bioinspired design of ultrathin 2D bimetallic metal-organic-framework nanosheets used as biomimetic enzymes. *Adv. Mater.* 28: 4149–4155.

[43] Huang, L., Zhang, X. P., Han, Y. J., Wang, Q. Q., Fang, Y. X., Dong, S. J. 2017. In situ synthesis of ultrathin metal-organic framework nanosheets: a new method for 2D metal-based nanoporous carbon electrocatalysts. *J. Mater. Chem. A* 5: 18610–18617.

[44] Furukawa, H., Cordova, K. E., O'Keeffe, M., Yaghi, O. M. 2013. The chemistry and applications of metal-organic frameworks. *Science* 341: 974–975.

[45] He, L., Liu, Y., Liu, J., Xiong, Y., Zheng, J., Liu, Y., Tang, Z. 2013. Core-shell noble-metal@metal-organic-framework nanoparticles with highly selective sensing property. *Angew. Chem. Int. Ed. Engl.* 52: 3741–3745.

[46] Lee, H. K., Lee, Y. H., Morabito, J. V., Liu, Y., Koh, C. S. L., Phang, I. Y., Pedireddy, S., Han, X., Chou, L. Y., Tsung, C. K., Ling, X. Y. 2017. Driving CO_2 to a quasi-condensed phase at the interface between a nanoparticle surface and a metal-organic framework at 1 bar and 298 K. *J. Am. Chem. Soc.* 139: 11513–11518.

[47] Kreno, L. E., Leong, K., Farha, O. K., Allendorf, M., Van Duyne, R. P., Hupp, J. T. 2012. Metal-organic framework materials as chemical sensors. *Chem. Rev.* 112: 1105–1125.

[48] Zheng, G. C., de Marchi, S., Lopez-Puente, V., Sentosun, K., Polavarapu, L., Perez-Juste, I., Hill, E. H., Bals, S., Liz-Marzan, L. M., Pastoriza-Santos, I., Perez-Juste, J. 2016. Encapsulation of single plasmonic nanoparticles within ZIF-8 and SERS analysis of the MOF flexibility. *Small* 12: 3935–3943.

[49] Qiao, X., Su, B., Liu, C., Song, Q., Luo, D., Mo, G., Wang, T. 2018. Selective surface enhanced Raman scattering for quantitative detection of lung cancer biomarkers in superparticle@MOF structure. *Adv. Mater.* 30: 1702275–1702282.

[50] Zhou, X., Liu, G. Q., Zhang, H. W., Li, Y., Cai. W. P. 2019. Porous zeolite imidazole framework-wrapped urchin-like Au-Ag nanocrystals for SERS detection of trace hexachlorocyclohexane pesticides via efficient enrichment. *J. Hazard. Mater.* 368: 429–435.

[51] Venna, S. R., Jasinski, J. B., Carreon, M. A. 2010. Structural evolution of zeolitic imidazolate framework-8. *J. Am. Chem. Soc.* 132: 18030–18033.

Gold Nanocrystal-built Films

[52] Li, Z., Zeng, H. C. 2013. Surface and bulk integrations of single-layered Au or Ag nanoparticles onto designated crystal planes {110} or {100} of ZIF-8. *Chem. Mater.* 25: 1761–1768.

[53] Chen, L. Y., Peng, Y., Wang, H., Gua, Z. Z., Duana, C. Y. 2014. Synthesis of Au@ZIF-8 single- or multi-core-shell structures for photocatalysis. *Chem. Commun.* 50: 8651–8654.

[54] Gomez-Grana, S., Hubert, F., Testard, F., Guerrero-Martinez, A., Grillo, I., Liz-Marzan, L. M., Spalla, O. 2012. Surfactant (Bi) layers on gold nanorods. *Langmuir* 28: 1453–1459.

[55] Pan, Y. C., Heryadi, D., Zhou, F., Zhao, L., Lestari, G., Su, H. B., Lai, Z. P. 2011. Tuning the crystal morphology and size of zeolitic imidazolate framework-8 in aqueous solution by surfactants. *Crystengcomm* 13: 6937–6940.

[56] Hu, P., Zhuang, J., Chou, L. Y., Lee, H. K., Ling, X. Y., Chuang, Y. C., Tsung, C. K. 2014. Surfactant-directed atomic to mesoscale alignment: metal nanocrystals encased individually in single-crystalline porous nanostructures. *J. Am. Chem. Soc.* 136: 10561–10564.

[57] Li, C. C., Shuford, K. L., Park, Q. H., Cai, W. P., Li, Y., Lee, E. J., Cho, S. O. 2007. High-yield synthesis of single-crystalline gold nano-octahedra. *Angew. Chem. Int. Ed.* 46: 3264–3268.

[58] Mohri, N., Matsushita, S., Inoue, M., Yoshikawa, K. 1998. Desorption of 4-aminobenzenethiol bound to a gold surface. *Langmuir* 14: 2343–2347.

[59] Barbosa, S., Agrawal, A., Rodriguez-Lorenzo, L., Pastoriza-Santos, I., Alvarez-Puebla, R. A., Kornowski, A., Weller, H., Liz-Marzan, L. M. 2010. Tuning size and sensing properties in colloidal gold nanostars. *Langmuir* 26: 14943–14950.

[60] McLean, T. M., Cleland, D., Gordon, K. C., Telfer, S. G., Waterland, M. R. 2011. Raman spectroscopy of dipyrrins: nonresonant, resonant and surface-enhanced cross-sections and enhancement factors. *J. Raman. Spectrosc.* 42: 2154–2164.

[61] Camden, J. P., Dieringer, J. A., Zhao, J., Van Duyne, R. P. 2008. Controlled plasmonic nanostructures for surface-enhanced spectroscopy and sensing. *Accounts Chem. Res.* 41: 1653–1661.

[62] He, X., Yue, C., Zang, Y. S., Yin, J., Sun, S. B., Li, J., Kang, J. Y. 2013. Multi-hot spot configuration on urchin-like Ag nanoparticle/ZnO hollow nanosphere arrays for highly sensitive SERS. *J. Mater. Chem. A* 1: 15010–15015.

[63] Fang, Y. R., Li, Y. Z., Xu, H. X., Sun, M. T. 2010. Ascertaining p, p '-dimercaptoazobenzene produced from p-aminothiophenol by selective catalytic coupling reaction on Silver nanoparticles. *Langmuir* 26: 7737–7746.

[64] Kleinman, S. L., Sharma, B., Blaber, M. G., Henry, A. I., Valley, N., Freeman, R. G., Natan, M. J., Schatz, G. C., Van Duyne, R. P. 2013. Structure enhancement factor relationships in single gold nanoantennas by surface-enhanced Raman excitation spectroscopy. *J. Am. Chem. Soc.* 135: 301–308.

[65] Wang, H., Levin, C. S., Halas, N. J. 2005. Nanosphere arrays with controlled sub-10-nm gaps as surface-enhanced Raman spectroscopy substrates. *J. Am. Chem. Soc.* 127: 14992–14993.

[66] Lee, S. J., Morrill, A. R., Moskovits, M. 2006. Hot spots in silver nanowire bundles for surface-enhanced Raman spectroscopy. *J. Am. Chem. Soc.* 128: 2200–2201.

[67] Wang, H. H., Liu, C. Y., Wu, S. B., Liu, N. W., Peng, C. Y., Chan, T. H., Hsu, C. F., Wang, J. K., Wang, Y. L. 2006. Highly Raman-enhancing substrates based on silver nanoparticle arrays with tunable sub-10 nm gaps. *Adv. Mater.* 18: 491–492.

[68] Candeias, M., Pita, T., Alves-Pereira, I., Ferreira, R. 2015. Comparative study of toxicological effects of lindane and isoproturon pesticides in the Saccharomyces cerevisiae. *Toxicol. Letter.* 238: 328–328.

[69] Willett, K. L., Ulrich, E. M., Hites, R. A. 1998. Differential toxicity and environmental fates of hexachlorocyclohexane isomers. *Environ. Sci. Technology.* 32: 2197–2207.

[70] Xu, X. Q., Yang, H. G., Li, Q. L., Yang, B. J., Wang, X. R., Lee, F. S. C. 2007. Residues of organochlorine pesticides in near shore waters of LaiZhou Bay and JiaoZhou Bay, Shandong Peninsula, China. *Chemosphere* 68: 126–139.

[71] Ali, M., Kazmi, A. A., Ahmed, N. 2014. Study on effects of temperature, moisture and pH in degradation and degradation kinetics of aldrin, endosulfan, lindane pesticides during full-scale continuous rotary drum composting. *Chemosphere* 102: 68–75.

[72] Izquierdo-Lorenzo, I., Kubackova, J., Manchon, D., Mosset, A., Cottancin, E., Sanchez-Cortes, S. 2013. Linking Ag nanoparticles by aliphatic alpha, omega-dithiols: a study of the aggregation and formation of interparticle hot spots. *J. Phys. Chem. C* 117: 16203–16212.

[73] Kubackova, J., Izquierdo-Lorenzo, I., Jancura, D., Miskovsky, P., Sanchez-Cortes, S. 2014. Adsorption of linear aliphatic alpha, omega-dithiols on plasmonic metal nanoparticles: a structural study based on surface-enhanced Raman spectra. *Phys. Chem. Chem. Phys.* 16: 11461–11470.

[74] Kubackova, J., Fabriciova, G., Miskovsky, P., Jancura, D., Sanchez-Cortes, S. 2015. Sensitive surface-enhanced Raman spectroscopy (SERS) detection of organochlorine pesticides by alkyl dithiol-functionalized metal nanoparticles-induced plasmonic hot spots. *Anal. Chem.* 87: 663–669.

[75] Guerrini, L., Aliaga, A. E., Carcamo, J., Gomez-Jeria, J. S., Sanchez-Cortes, S., Campos-Vallette, M. M., Garcia-Ramos, J. V. 2008. Functionalization of Ag nanoparticles with the bis-acridinium lucigenin as a chemical assembler in the detection of persistent organic pollutants by surface-enhanced Raman scattering. *Anal. Chim. Acta* 624: 286–293.

[76] Mishra, S., Vallet, V., Poluyanov, L. V., Domcke, W. 2006. Calculation of the vibronic structure of the photodetachment spectra of CCCl- and CCBr. *J. Chem. Phys.* 125: 164327–164334.

[77] Liu, G., Cai, W., Kong, L., Duan, G., Li, Y., Wang, J., Cheng, Z. 2013. Trace detection of cyanide based on SERS effect of Ag nanoplate-built hollow microsphere arrays. *J. Hazard. Mater.* 248–249: 435–441.

[78] Zhao, Q., Liu, G., Zhang, H., Zhou, F., Li, Y., Cai, W. 2017. SERS-based ultrasensitive detection of organophosphorus nerve agents via substrate's surface modification. *J. Hazard. Mater.* 324: 194–202.

11 The Effect of Deposition Parameters on the Mechanical and Transport Properties in Nanostructured Cu/W Multilayer Coatings

Alexander M. Korsunsky and León Romano Brandt
University of Oxford, UK

CONTENTS

11.1 Introduction...288
11.2 Materials and Methods 288
    11.2.1 Materials: Cu and W...288
    11.2.2 Magnetron Sputtering Deposition 289
    11.2.3 Ion Beam Deposition...289
    11.2.4 FIB-DIC Ring-core Depth Profiling..289
    11.2.5 Synchrotron XRD 291
    11.2.6 Micro-Cantilever Deflection..291
    11.2.7 Energy-Dispersive X-Ray Spectroscopy 292
    11.2.8 Transient Grating Spectroscopy 292
    11.2.9 In situ X-Ray Reflectivity Measurements 293
    11.2.10 In situ Grazing-Incidence Small-Angle X-Ray Scattering Measurements 294
    11.2.11 Transmission Electron Microscopy 294
    11.2.12 Atomic Force Microscopy..295
    11.2.13 In situ SEM Heating...295
11.3 The Effect of Magnetron Sputtering Deposition Conditions on Residual Stress
and Thermal Diffusivity in Cu/W Nano-Multilayer Coatings 295
    11.3.1 Experimental Approach and Interpretation 296
        11.3.1.1 FIB-DIC Micro-ring-core Milling 296
        11.3.1.2 $Sin^2(\Psi)$ Analysis of XRD Data.......................................298
        11.3.1.3 Micro-cantilever Deflection 301
    11.3.2 Results and Discussion...302
    11.3.3 Summary 303
11.4 Stress-assisted Thermal Diffusion Barrier Breakdown in Ion Beam Deposited Cu/W
Nano-multilayers on Si Substrate at Elevated Temperatures....................................304
    11.4.1 Experimental Approach and Interpretation 304
    11.4.2 Results and Discussion...304
        11.4.2.1 Reference State Characterisation 304

		11.4.2.2	Structural Evolution Upon Thermal Treatment	305
		11.4.2.3	Diffusion Modelling	308
	11.4.3	Summary		313
11.5	Conclusion			313
References				314

11.1 INTRODUCTION

Nano-multilayered coatings (NML) allow the design of materials with unique mechanical, electrical, optical, and functional properties by combining the advantages of different materials in a layered structure [1]. Copper and tungsten (Cu/W) multilayers, for instance, are promising candidates for the thermal management at the micro- and nanoscale, combining the excellent thermal and electrical conductivity of Cu with the low thermal expansion and high mechanical strength of tungsten [2–4]. This combination makes the mutually immiscible Cu/W system particularly interesting for plasma-facing components in fusion reactors [5] and protective coatings in nuclear fission reactors, as the interfacial structure of nano-granular W can additionally serve as a sink for radiation-induced defects [6].

Magnetron Sputter Deposition (MSD) and Ion Beam Deposition (IBD) as two of the most popular Physical Vapour Deposition (PVD) methods include a wide range of process parameters, such as the argon working pressure, sputtering power, and the substrate bias and temperature. These parameters can be used to fine-tune the stresses, microstructure and thermal diffusivity at the nanoscale during deposition [7]. Influencing factors have been systematically studied for decades [8]. However, the coating thickness used in technology applications reduced significantly in recent years from several micrometre thick coatings towards the (sub-) nanometre scale, while the demands for thermal and mechanical reliability have increased. To bridge this gap, thorough knowledge of the relationship between deposition parameters and spatially resolved thin film properties are required. At the same time, analysis methods have improved significantly, allowing deeper insights into material property distributions as ever before.

In this chapter, the effect of the deposition parameters on the microstructural properties such as the preferred crystal orientation (texture), grain size and shape, residual stress, density, defect structure, and also the individual layer thickness determining the volume fraction of the material at or near interfaces will be discussed. Materials and methods are described in the next Section 11.2. The effect of the deposition chamber pressure on these properties and their link to the thermal diffusivity and stability of Cu/W NML coatings will be demonstrated and analysed by means of a series of advanced complementary electron, ion, and X-ray analysis techniques. Starting from the reconstruction of residual stress depth profiles in Section 11.3, the mechanisms underlying the stress build-up are discussed, followed by an analysis of the microstructural drivers for the variation of thermal diffusivity with varying MSD chamber pressure in the Cu/W nano-multilayers. The following Section 11.4 presents the characterisation of the thermal stability of Cu/W NMLs deposited onto Si substrate by IBD via the observation of diffusion phenomena within and between buried coating layers using synchrotron X-Ray Diffraction (XRD), Transmission Electron Microscopy (TEM), and transmission Energy-Dispersive X-Ray Spectroscopy (t-EDX). The experimental observations are used to build an integrated model of the system.

11.2 MATERIALS AND METHODS

11.2.1 MATERIALS: CU AND W

For the present study, Cu/W nano-multilayer systems were chosen. The source of materials were CP (commercially pure) targets of copper and tungsten used within PVD systems.

Two different Cu/W NML systems were deposited using Magnetron Sputter Deposition (MSD) and Ion Beam Deposiition (IBD), respectively. MSD samples were used for the determination of

Properties in Nanostructured Cu/W Multilayer Coatings

289

residual stress depth profiles and thermal diffusivity, while IBD samples were employed for the study of dynamic diffusion behaviour at elevated temperatures. An overview of the deposition parameters and all experimental methods is provided in this section.

11.2.2 MAGNETRON SPUTTERING DEPOSITION

Five multilayer systems were deposited onto naturally oxidised monocrystalline silicon (001) wafers which were cleaned prior to deposition by acetone and ethanol rinsing, and dried under nitrogen flow. The working pressure (P) was varied from sample to sample using the following chamber pressures: $P_1 = 0.50$ Pa, $P_2 = 0.66$ Pa, $P_3 = 0.74$ Pa, $P_4 = 0.81$ Pa, and $P_5 = 0.89$ Pa. Before deposition, the target was clean sputtered while the substrates were protected by a shutter. Each multilayer consisting of 10 bilayers of Cu/W 50/50 nm thin films was deposited at room temperature (RT) by planar magnetron sputtering in DC mode using Ar as working gas in a high vacuum system (base pressure $\sim 1 \cdot 10^{-5}$ Pa), resulting in a total film thickness of 1 μm. The sputtering rate was calibrated using X-ray reflectometry (XRR). Each of the W and Cu targets were 7.6 cm in diameter, with a purity 4N (i.e., 99.99%). The thin films were deposited using unbalanced magnetron on grounded rotating substrates at a constant target power of 130 W. The target-to-substrate distance was 18 cm.

11.2.3 ION BEAM DEPOSITION

Ten bilayers of Cu/W were deposited onto 600 μm single crystalline Si wafers obtained from an industrial source with [001] orientation, starting with W. The deposition was carried out at RT by ion beam sputtering with a focused Ar + ion gun at 1.2 keV multicusp radio-frequency source in a NORDIKO-3000 system [1]. During film growth the ion gun was supplied with constant radio-frequency power of 165 W and constant Ar flux of 10 standard cubic centimetre per minute (sccm). The 150 mm diameter targets were sputtered during 10 min allowing both ion gun stabilisation and target pollution cleaning caused by vacuum break. The ion gun axis was inclined at 45° to the normal of the surface target while the sample surface was parallel to the target plane. The sputtering chamber was pumped down to a base pressure of $2 \cdot 10^{-6}$ Pa while the working pressure during film growth was $\sim 10^{-2}$ Pa. Substrates were preliminarily cleaned with acetone and ethanol and finally dried with an argon gas jet prior to their introduction in the deposition chamber.

11.2.4 FIB-DIC RING-CORE DEPTH PROFILING

Recent advances in the resolution of combined Scanning Electron Microscopy (SEM) and Focused Ion Beam (FIB) devices have allowed for the development of new methods for the spatially resolved study of residual stresses at the nanoscale [9]. Two main approaches have been presented recently: cantilever-based [10, 11] and ring-core based [12, 13]. The first type relies on deflection measurement of a coated cantilever, while removing individual material layers with the FIB and observing the change of deflection, thus allowing to calculate the stress state of the coating at a resolution of around 50 nm. The advantage of this method lies in the fact that no surface preparation is required, as the deflection is easily measurable. A clear disadvantage, however, is the fact that the cantilever is usually being milled on an edge of the sample, which requires sample cutting at the risk of stress field modifications. Sectioning the sample in preparation for cantilever milling thus carries the risk of disturbing the pre-existing stress field. The FIB-DIC ring-core method, on the other side, is a well-established and thoroughly verified method for the semi-destructive probing of residual stresses at the micron scale. It is based on the gradual removal of material in a ring-core shape shown in Figure 11.1 with the intermediate acquisition of SEM frames at each milling steps. Mechanically disconnecting the central material 'island' from the surrounding strain field leads to a full elastic strain relief. This can be observed by tracking surface points of the central island using DIC in post-processing at an accuracy of around $\varepsilon \approx 10^{-5}$, which can then be used to back-calculate the

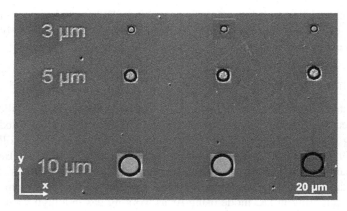

FIGURE 11.1 Arrangement of FIB markers for DIC depth-profiling analysis. Ring-core diameter indicated on the left.

average residual stresses present in the probed volume prior to milling, if the elastic properties of the material are known. Recent advances in this technique allow to extract the depth-dependent stress and strain profiles in milling direction, based on the FIB-DIC strain relief profiles and by employing the eigenstrain theory [14].

As the DIC requires a trackable surface pattern to correlate images from different milling steps without altering the residual stress state of the samples measurably, a 4.1 nm thin Au/Pd layer was deposited on top of the multilayers using a commercial sputter coater (SC7620, Quorum Technologies) at a sputtering rate of 0.092 nm·s^{-1}. This step was required to modify the smooth sample surface, which decreases the tracking accuracy of the DIC significantly. Furthermore, damaging the multilayer surface without protective coating for contrast generation might affect the residual stress state due to surface layer amorphisation. For this reason, a fixed target-to-sample distance of 45 mm was chosen, and the process current was adjusted to 18 mA. Subsequently, the surface was exposed to the FIB at a low beam current of 0.20 nA at 30 kV, creating a nanoscale surface speckle pattern [15] without affecting the multilayer beneath. This was confirmed by cross-sectional high-resolution SEM imaging. To confirm that the Au/Pd contrast layer did not alter the residual stress state in the multilayer systems, a 30 nm thin layer of Au/Pd was deposited on a 75 μm Kapton substrate, while the curvature was recorded in situ. The substrate did not show any change in curvature, confirming that the residual stress within the DIC contrast layer can be neglected for further analysis.

Subsequently, ring-cores of three different diameters were milled step-wise into the sample using FIB. The diameter of the ring-core was chosen according to the desired depth-sensitivity, as the sensitivity for strain reconstruction is limited to a depth-to-diameter ratio of h/d = 0.4 [16, 17]. In the final analysis, a maximum of h/d = 0.25 was used. As a consequence, the chosen ring-core diameters for this analysis were D_1 =10 μm, D_2 = 5 μm, and D_3 = 3 μm, and for each diameter three ring-cores were produced to minimise the statistical scatter of results, as shown in Figure 11.1. While D_3 returned the best depth resolution for the coating, D_1 probed further into the substrate and was used to analyse the interaction between substrate and coating. D_2 as intermediate diameter improved the statics of both the coating and the upper part of the substrate. The FIB was operated at a beam energy of 30 keV and a beam current of 0.20 nA for the D_1 and D_2 ring-cores. For the D_3 ring-cores, however, a lower beam current of 0.04 nA was chosen to minimise the ion beam damage [18–21] of the smaller central island surface area. After each milling step, five SEM images were acquired for DIC analysis to improve the tracking statistics. As each ring-core affects the residual stress field due to local strain relaxation, the correct spacing needs to be chosen in between measurement points. It was shown that a distance in between ring-cores of five times the ring-core diameter is sufficient to avoid any mechanical interaction between points [9]. Therefore, the distance in between points corresponds to 5·D of the larger ring-core, respectively, as shown in Figure 11.1. Please note that the darker grey value of the third 10 μm diameter point was caused by FIB imaging after FIB-DIC acquisition, which did therefore not affect the obtained stress reconstruction.

11.2.5 Synchrotron XRD

Major advances regarding the availability, brilliance, and focusing capability of synchrotron light sources in the recent years have allowed to perform spatially resolved stress measurements of thin films and coatings by nano-diffraction. Albeit very complex sample preparation, this method permitted reliable insights into the residual stress state of coatings [22]. To confirm and expand the insights gained from the FIB-SEM methods, we employed classic Wide-Angle X-Ray Scattering (WAXS) for $\sin^2(\psi)$ analysis [23–25] to extract averaged residual stress magnitude present in the MSD-produced multilayer. The experiment was performed at the B16 beamline at Diamond Light Source (UK) in grazing angle scattering geometry. The chosen photon energy was 18 keV with a beam size of 250×250 μm^2, collimated by crossed slits, and a grazing angle of $\alpha = 1°$, resulting in a large interaction volume of the beam with the coating and therefore improved pattern statistics. The WAXS patterns were acquired for 360 s using an Image Star 9000 detector at a sample-detector distance of 66×10^{-3} m, which was calibrated using a Lanthanum hexaboride (LaB$_6$) sample. The setup is shown in Figure 11.2.

The obtained WAXS diffraction patterns are shown in Figure 11.3, along with the identified phases and crystal orientations. Only the upper half of the Debye-Scherrer rings was used for further analysis, as the lower half was mostly absorbed by the substrate and did not contain any additional information for strain reconstruction due to the two-fold symmetry of powder diffraction rings.

Further to the above XRD characterisation of MSD samples, a Wide-Angle X-Ray scattering (WAXS) pattern of the IBD sample was acquired under identical experimental conditions for the study of grain texture properties. The obtained pattern is shown in Figure 11.14b in Section 11.4.2.

11.2.6 Micro-Cantilever Deflection

To cross-validate the measured residual stresses within the coating by a further independent method, a micro-cantilever was milled at the centre of the sample deposited at $P_1 = 0.5$ Pa, using the FIB at 30 keV energy and 0.2 nA beam current. In this way, it was ensured that the analysed stress field was undisturbed, as opposed to machining a micro-cantilever at a cut cross-section. One arising

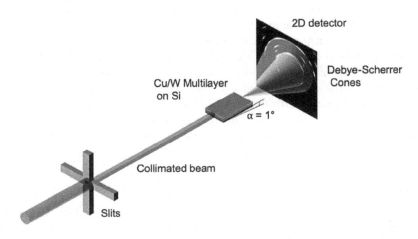

FIGURE 11.2 Synchrotron WAXS setup using collimated X-Ray beam and 2D area detector.

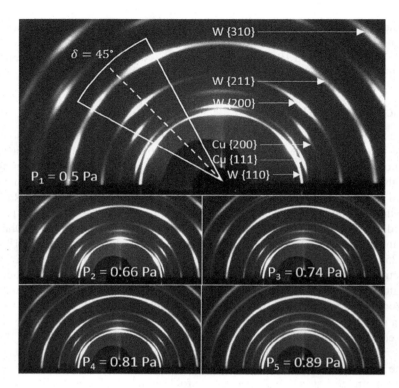

FIGURE 11.3 Synchrotron WAXS patterns of Cu/W 50/50 nano-multilayers deposited at different magnetron Ar working pressures. Crystal orientations are identified in the top figure.

consequence, however, is the non-square shape of the cantilever cross-section caused by the inclined angle during milling, which needs to be accounted for an accurate residual stress reconstruction. Subsequently, SEM frames were acquired at an electron beam energy of 5 keV. Shape, dimensions, and deflection of the cantilever are shown in Figure 11.4, alongside the measurements extracted from the SEM frames shown in Figure 11.4b.

11.2.7 Energy-Dispersive X-Ray Spectroscopy

High-resolution cross-sectional SEM imaging was combined with Energy-Dispersive X-ray spectroscopy of all three samples at an electron energy of 10 keV, after FIB milling of trenches using a beam current of 0.2 nA and an accelerating voltage of 30 keV. The SEM images shown in Figure 11.5a–c indicate a decrease in interface sharpness and increase in layer waviness with increasing deposition pressure. This was further confirmed by EDX line profiles that show a uniform intensity versus depth for the sample deposited at P = 0.5 Pa as can be seen in Figure 11.5a, while a decreasing intensity amplitude towards the substrate-coating interface is observed at higher deposition pressures in Figure 11.5b,c. The copper distribution in the near-interface layers of the sample deposited at P = 0.89 Pa is nearly uniform, indicating a poor interface quality, which can be linked to an increased roughness between immiscible Cu and W layers.

11.2.8 Transient Grating Spectroscopy

Non-destructive mapping of the thermal diffusivity was carried out using laser-induced Transient Grating Spectroscopy (TGS) [26–30]. For this purpose, two laser pulses are crossed at a fixed angle at the sample surface creating a spatially periodic interference pattern with a wavelength of 2.758 ±

Properties in Nanostructured Cu/W Multilayer Coatings

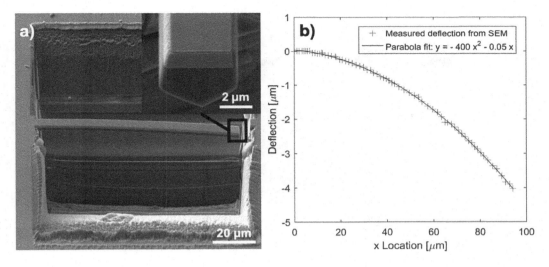

FIGURE 11.4 (a) Deflected micro-cantilever prepared by FIB milling. Top right corner: shape and dimensions of the cross-section, showing a clear contrast between Cu/W coating and Si substrate. (b) Deflection measured from SEM data, plotted alongside parabolic fit. Box: cross-section of cantilever from SEM frame.

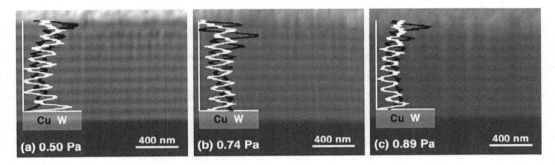

FIGURE 11.5 (a)–(c). High-resolution cross-sectional SEM frames of W/Cu NML deposited at different magnetron chamber pressures, alongside nominal EDX signals for Cu and W distribution along the coating depth.

0.002 μm [31]. This results in a spatially periodic heating of the sample, as shown in Figure 11.6. Thermal expansion subsequently leads to sample surface displacement, thus creating a transient displacement grating. A third continuous probe beam, diffracted by this transient temperature and displacement grating, is used to determine the decay period of the temperature and displacement grating, from which the thermal diffusivity can be deduced. For the selected TGS wavelength, the thermal diffusivity signal is dominated by material up to a depth of ~0.9 μm, thus ensuring averaging coating measurements without a significant contribution of the Si substrate. The thermal diffusivity was averaged over a grid of 6 by 6 points with a spatial resolution of 100 μm for each sample.

11.2.9 IN SITU X-RAY REFLECTIVITY MEASUREMENTS

To quantify the thickness, roughness, and density evolution of the IBD Cu/W NMLs at different temperatures, X-Ray Reflectivity (XRR) measurements were carried out at the BM28 (XMaS) beamline at the European Synchrotron Radiation Facility (ESRF, Grenoble, France). A photon energy of 12.4 keV and beam size of 0.5 × 0.2 mm² were chosen. The sample was attached to an 8 mm diameter

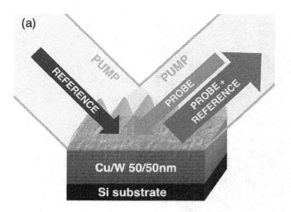

FIGURE 11.6 Schematic of the transient grating spectroscopy setup used to determine the thermal diffusivity within the multilayer coating.

heater with PEEK lid using ceramic adhesive, which was then mounted onto a diffractometer. The heater remained under high vacuum throughout the experiment to avoid sample oxidation. Data collection in a 2θ-range between 0° and 6° was carried out at RT using an Avalanche PhotoDiode (APD) mounted onto a KF16 flange, followed by one scan at 250 °C. GenX software [32] was used to fit a sample model matching the experimental XRR data and extract the sample parameters at both temperatures.

11.2.10 IN SITU GRAZING-INCIDENCE SMALL-ANGLE X-RAY SCATTERING MEASUREMENTS

Grazing-Incidence Small-Angle X-Ray Scattering (GISAXS) is a surface and interface scattering technique that is ideally suited to detect nanoscale periodic features buried within or present on the surface of samples [33]. Due to the very shallow angle of incidence between beam and sample surface this technique provides large-scale averaged geometric distribution statistics. Temperature-dependent GISAXS patterns during a single heat ramp from 30 °C to 400 °C were acquired in steps of 10 °C at the BM28 beamline at ESRF. An incidence angle slightly above the conventional GISAXS range (0.05° and 0.5°) of 0.66° was chosen, as it revealed additional lateral and vertical roughness correlation information compared to shallower angles at which the GISAXS pattern was averaged to one large feature without distinct layer separation. The sample environment was identical to the one used for the XRR analysis. An X-ray photon energy of 8 keV with a beam size of 100 × 80 µm² was chosen to achieve the best possible resolution of the GISAXS patterns on the detector. A PILATUS3 300K area detector with a pixel size of 172 × 172 µm² and a resolution of 487 × 619 pixels split into three detector modules was used. A sample-detector distance of 1.56 m was determined using a AgBh standard calibration sample, followed by analysis in the DAWN software package [34]. GISAXS patterns were simulated and fitted using the BornAgain software package [35, 36].

11.2.11 TRANSMISSION ELECTRON MICROSCOPY

TEM lamellas of a pristine sample and the sample used for the in situ GISAXS experiment were extracted and thinned using a Tescan Lyra3 Focused Ion Beam (FIB) – Scanning Electron Microscope (SEM). The lamellas were sectioned using Ga FIB at a voltage of 30 kV and beam current of 11 nA, followed by fine polishing and thinning at 0.2 nA beam current. TEM images were acquired at an electron beam energy of 200 keV using a JEOL JEM-2100 TEM at the Research Complex at Harwell (RCaH, Harwell, UK). In addition, Scanning Transmission Electron Microscopy (STEM) on the sample obtained from the GISAXS experiment was combined with transmission Energy-Dispersive

Properties in Nanostructured Cu/W Multilayer Coatings 295

X-ray Spectroscopy (t-EDX) inside the Tescan Lyra3 FIB-SEM at an electron beam energy of 20 keV and a working distance of 5 mm, which enabled the local identification of materials based on their characteristic X-ray emission energy at a resolution of 40×40 nm^2.

11.2.12 ATOMIC FORCE MICROSCOPY

Atomic Force Microscopy (AFM) measurements of the Cu/W NML surface in as-deposited state and after in situ GISAXS heating were acquired at the Laboratory for In Situ Microscopy (LIMA, Department of Engineering Science, University of Oxford) using a Veeco Dimension 3100 in tapping mode. A surface area of 0.6×0.6 µm^2 was scanned for the sample in reference state, while a map of the surface area of 5×5 µm^2 was acquired for the heated sample, in order to capture the newly formed large surface features. ProfilmOnline [37] was used for AFM data processing and rendering.

11.2.13 IN SITU SEM HEATING

Finally, to enable sample surface observations during heating, in situ SEM analysis of a Cu/W NML was carried out using a Zeiss Evo LS15 SEM in combination with the heating module of a 5kN tensile stage (Kammrath & Weiss, Germany). The multilayer sample was placed on the surface of the heating module, which was then heated in a stepwise fashion from RT up to 400°C, while SEM frames were acquired. The working distance and beam energy were 20 mm and 10 keV, respectively.

11.3 THE EFFECT OF MAGNETRON SPUTTERING DEPOSITION CONDITIONS ON RESIDUAL STRESS AND THERMAL DIFFUSIVITY IN CU/W NANO-MULTILAYER COATINGS

Accurate estimations of residual stress and thermal diffusivity are key factors for the reliable design of nanostructured thin films and coatings [38]. Many applications in microelectronics, such as Micro-Electromechanical Systems (MEMS) and sensors rely on the structural integrity of coatings, which are often only a few nanometres thin. The complex internal stress fields and thermal diffusivity properties that arise as a consequence of PVD are difficult to predict accurately due to the number of interlinked atomic level mechanisms involved [23]. Experimental measurements are therefore of crucial importance for the understanding and design of the residual stress and thermal properties in thin films.

The residual stress evolution during nucleation, grain boundary formation, single film coating and multi-layer deposition is complex and combines several microstructural mechanisms and effects. During nucleation and island growth on the substrate, the global average stresses are close to zero due to the lack of formation of a continuous film [23]. After the onset of island coalescence the residual stress becomes tensile, as the free spaces between islands become filled in with disordered material characterised by large free volume and ultimately convert into grain boundaries. During this process, it is of energetic advantage for the islands to accommodate internal tensile strains in order to form a uniform film and share grain boundaries with neighbouring islands [39]. As soon as the film has formed a continuous layer, the effects of energetic particle bombardment begin to dominate the stress state. The phenomenon known as 'atomic peening' affects the residual stress state and defect population significantly [40]. Highly energetic particles impacting the surface lead to recoil implantation of coating surface atoms and integration of argon atoms into the films. More specifically, the impact of particles on the surface also incorporates defects into the structure through displacement sequences within the grains [23]. The momentum propagation during magnetron sputter deposition can lead not only to both the motion of atoms to energetically favourable sites, such as vacancies, but also to the creation of point defects such as interstitials, along with Frenkel pairs. Overall, these effects lead to the formation of a highly densified layer with stored compressive residual stress that improves the thermal transport within the material.

It has been well known for many years that changing the chamber pressure modifies the stress state, defect structure, and thermal diffusivity significantly [8]. Several phenomena contribute to this effect. Firstly, the larger quantity of Ar atoms in between the sputtering target and the sample surface results in a significantly lower mean free path length for the sputtered Cu and W atoms. The atomic peening effect is therefore reduced by a significant amount, as there is a larger momentum transfer between the working gas and the sputter material, resulting in a lower impact energy at the sample surface. This also leads to a reduced level of atomic implantation in between the grain boundaries. Finally, in addition to the lack of atomic peening and grain boundary implantation, a high working pressure inside the magnetron chamber leads to the tensile stress due to the formation of an underdense columnar growth morphology, which is typical for refractory metals such as tungsten [23], resulting in a decrease in thermal diffusivity. The experimental details for the analysis of stress state and thermal diffusivity in MSD samples will be presented in the following sections and discussed in the context of the existing literature data.

11.3.1 Experimental Approach and Interpretation

11.3.1.1 FIB-DIC Micro-ring-core Milling

To obtain full-field displacement and thus strain relief profiles, a MATLAB-based DIC software was used to track surface points [41]. Subsequently, the strain relief profiles for the three measurement points of each ring-core diameter were averaged to a single profile. From this averaged strain relaxation profile, the residual elastic strain depth profile was reconstructed by following the approach described in [16]. The analysis was performed by analysing information provided by the strain relief curves lying within the normalised range $0.015 < h/D < 0.25$, where h is the milling depth and D is the ring-core diameter [16]. An example of strain relief profiles for the sample deposited at $P_5 = 0.89$ Pa is shown in Figure 11.7 for all three ring-core diameters against the FIB material removal depth, h.

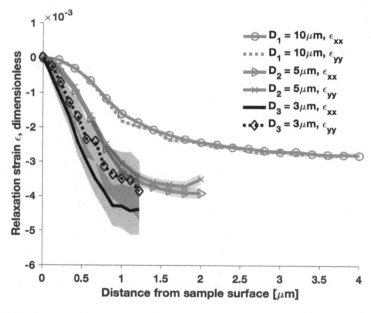

FIGURE 11.7 Relief strain for ring-cores of different diameters on the sample deposited at a magnetron working pressure of $P_1 = 0.89$ Pa. The shaded areas around the plot lines represent the 95% confidence interval.

The 95% confidence interval shown as shaded area around the relaxation profile is determined by two error sources: 1) the uncertainty of the DIC process δ_{DIC}, which is determined by the tracking error of pixels which are smoothed by a Gaussian filter; 2) in order to extract the average strain relief taking place in the central ring-core island monitored by DIC, a first-order polynomial is fitted to the displacement across the pillar diameter along the sought direction. The slope of this polynomial corresponds to the relaxation strain, leading to a slope fitting error δ_{fit}. These two errors are then combined by following the appropriate error summation rules.

Converting the obtained elastic strains to stresses [16] requires the Young's modulus E and Poisson's ratio ν, which were assumed to be 102 GPa and 0.34 for nanocrystalline Cu and 338 GPa and 0.36 for nanocrystalline W, respectively [1]. These properties were averaged by following the rule of mixture and considering a material ratio of Cu:W of 1:1, resulting in average elastic properties of $\bar{E}_{Cu/W}$ = 220 GPa and $\bar{\nu}_{Cu/W}$ = 0.35. To calculate the stress within the substrate, a Young's modulus value of E_{Si} = 160 GPa and a Poisson's ratio of 0.27 were used [42], which explains the discontinuity of the residual stress profiles at a depth of 1 µm, caused by a change of elastic material properties used for calculation, as shown in Figure 11.8.

The resulting stress depth profiles are shown in Figure 11.8, which clearly reveal a transition in residual stress profiles from compression to tension with increasing magnetron chamber pressure during thin film deposition. The depth resolution of the shown stress graphs, which is directly linked to the milling step size, corresponds to the thickness of individual layers (50 nm). The 95% confidence boundaries resulting from averaging the probed stresses for several ring-core diameters and several measurement points per diameter are shown as shaded areas around the stress profiles. The larger error close to the multilayer surface can easily be explained by looking at the strain relaxation profiles in Figure 11.7. Smaller ring-core diameters increase the uncertainty significantly, due to their smaller DIC tracking surface area. As the smaller ring-core diameters probe the stresses close to the surface, the error increases in this region.

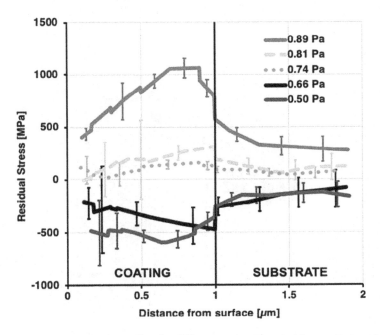

FIGURE 11.8 Residual stress depth profiles for different magnetron working pressures obtained from the FIB-DIC ring-core method.

11.3.1.2 Sin²(Ψ) Analysis of XRD Data

In order to extract the residual stress from the obtained WAXS patterns, the sin²(Ψ) technique was used. Having been developed in the early stages of experimental residual stress analysis, the sin²(Ψ) technique is often being referred to as a method of reference to determine residual stresses [43]. As an experimentally easy and non-destructive near-surface technique, it is mainly used to obtain averaged residual stress values, due to the comparably large beam size of laboratory X-Ray sources.

In contrast to the synchrotron XRD pattern acquisition chosen for this work, the traditional laboratory setup relies on a sample being illuminated with photons at an energy of around 8 keV. The obtained Bragg peaks are then collected by a point detector, resulting in a one-dimensional 2θ vs. intensity line profile. This traditional laboratory setup therefore requires tilting the sample around a line parallel to its surface (Ψ angle), to capture the stress induced lattice distortions can be observed by a shift in the measured 2θ peak position. The resulting sin²(ψ) vs. lattice parameter profiles can then be used to extract the average residual stress within the beam interaction volume.

A much faster way of capturing the sin²(Ψ) behaviour involves the use of high-energy WAXS with high photon flux, as presented in Section 11.2.4. The use of area detectors with small pixel size and at short sample-detector distance allows to collect the full diffraction ring, as the 2θ range is narrowed down by the high photon energy. As the Debye-Scherrer cone angle approaches 90°, the scattering vector k_ψ is moving on a plane perpendicular to the laboratory coordinate system, which is approximately perpendicular to the sample surface at sufficiently small angle of incidence α. This scattering geometry results in a residual stress probing direction normal to the sample surface, as shown in Figure 11.9.

To extract the sin²(Ψ) behaviour of the coating, the obtained patterns were integrated by azimuthal binning in 22.5° steps, starting from the central δ = 0° position up to $δ \lesssim ±90°$, resulting in a total number of nine measurement points. The comparably large step size for integration was chosen due to non-uniform patterns caused by texture. Fit2D was used for integration, after an initial calibration using the acquired LaB6 diffraction pattern. High-accuracy distortion correction was performed based on the calibration pattern prior to the experimental data analysis, ensuring the exclusion of spurious results. Four fully available Debye-Scherrer rings shown in Figure 11.3 were selected for azimuthal integration: W{110}, Cu{111}, Cu{200}, and W{200}, of which all showed a linear $\varepsilon_{\Psi\phi}^{hkl}\left(\sin^2\left(\Psi\right)\right)$ profile.

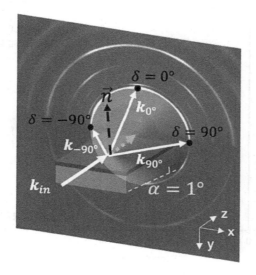

FIGURE 11.9 Diffraction geometry during synchrotron WAXS acquisition. Black: incoming and diffracted wave vectors $\mathbf{k}_{in}$ and $\mathbf{k}_\delta$. Beamline coordinate system in white.

Properties in Nanostructured Cu/W Multilayer Coatings

To calculate the angle Ψ based on the azimuthal angle δ, the following relationship was employed [24]:

$$\cos(\psi) = \sin(\alpha)\sin(\theta) + \cos(\alpha)\cos(\theta)\cos(\delta). \tag{11.1}$$

Depending on the residual stress state within the diffracting material volume, the $\varepsilon_{\Psi\phi}^{hkl}\left(\sin^2(\Psi)\right)$ behaviour can adopt different shapes [44]. To determine the residual stress state with the highest accuracy, the following relation based on an elliptical stress–strain relationship was adopted [45]:

$$\varepsilon_{\phi\Psi} = \frac{d_{\phi\Psi} - d_0}{d_0} = \frac{1+v}{E}\left\{\sigma_{11}\cos^2\phi + \sigma_{12}\sin(2\phi) + \sigma_{22}\sin^2\phi - \sigma_{33}\right\}\sin^2\Psi$$
$$+ \frac{1}{E}\left\{\sigma_{33} - v\sigma_{11} - v\sigma_{22}\right\} + \frac{1+v}{E}\left\{\sigma_{13}\cos(\phi) + \sigma_{23}\sin(\phi)\right\}\sin(2\Psi). \tag{11.2}$$

The angle ϕ refers to the rotation of the sample around the global y-axis shown in Figure 11.9. The same elastic properties E and v were used as shown above for the FIB-DIC depth profiling analysis. Further justification for this assumption despite stronger texture at lower deposition pressure will be provided later, based on the ratio of σ_{Cu} to σ_W. The reference lattice parameter d_0 was set to be the d-spacing at $\delta = 0$, which will be justified at a later point.

Based on the origin of residual stresses within nanometric multilayers through island growth and grain boundary atom implantation, we can assume an equi-biaxial residual stress field in the surface plane direction of the sample. This simplifies the problem by assuming that $\phi = 0$, $\sigma_{11} = \sigma_{22}$, which yields

$$\varepsilon_\Psi = \frac{d_\Psi - d_{\Psi=0}}{d_{\Psi=0}} = \frac{1+v}{E}\left\{\sigma_{11} - \sigma_{33}\right\}\sin^2\Psi + \frac{1}{E}\left\{\sigma_{33} - 2v\sigma_{11}\right\} + \frac{1+v}{E}\sigma_{13}\sin(2\Psi). \tag{11.3}$$

This can be rewritten as:

$$\varepsilon_\Psi = \frac{d_\Psi - d_{\Psi=0}}{d_{\Psi=0}} = A + B\sin^2\Psi + C\sin(2\Psi),$$

$$A = \frac{1}{E}\left\{\sigma_{33} - 2v\sigma_{11}\right\}, B = \frac{1+v}{E}\left\{\sigma_{11} - \sigma_{33}\right\}, C = \frac{1+v}{E}\sigma_{13}. \tag{11.4}$$

It can easily be seen that the only component showing a measurable sensitivity towards a change in $d_{\Psi=0}$ is the ellipse intercept with the y-axis **A**. For the further analysis, only the parameters **B** and **C** will be used, as they show no significant sensitivity towards a change in d_0. It is therefore possible to determine the composite residual stress $\{\sigma_{11} - \sigma_{33}\}$ and the shear stress σ_{13} without accurate estimation of d_0.

The out-of-plane stress (σ_{33}) and shear stress (σ_{13}) components were found to be negligible compared to the in-plane component. The calculated linear fits representing $\varepsilon_\Psi = B\sin^2\Psi$ are thus shown in Figure 11.10 alongside the experimental data. The presented data shows the evolution of the $\sin^2\Psi$ behaviour depending on different magnetron chamber deposition pressures. The data error bars indicate the 95% confidence intervals for the Gaussian fitting of the diffraction peaks, based on the following error calculation for strain:

$$\delta_{\varepsilon\psi} = \varepsilon_\Psi \cdot \sqrt{\left(\frac{\sqrt{\left(\delta_{d_\psi}\right)^2 + \left(\delta_{d_{\psi=0}}\right)^2}}{d_\psi - d_{\psi=0}}\right)^2 + \left(\frac{\delta_{d_{\psi=0}}}{d_{\psi=0}}\right)^2} \tag{11.5}$$

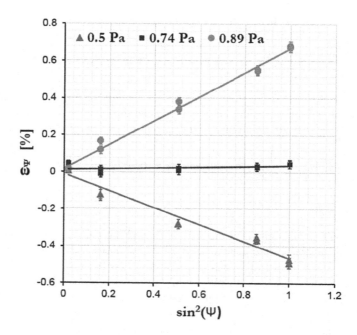

FIGURE 11.10 Sin²(Ψ) profiles of the W{110} reflection after deposition at three different magnetron chamber pressures. Points: experimental WAXS data. Dashed line: fitted line profiles.

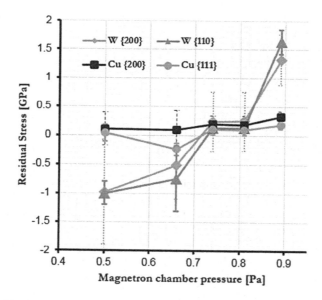

FIGURE 11.11 Residual stresses σ_{11}, based on ellipse fitting parameter B for individual W and Cu grain orientations.

The so obtained in-plane stress values ($\sigma_{11} - \sigma_{33}$) for different XRD peaks are shown in Figure 11.11. The data clearly indicates significantly higher magnitudes of residual stresses within W as compared to Cu. As this behaviour is clearly indicating a strong dominance of isotropic W on the averaged residual stress, considerations regarding the orientation-dependence of the elastic properties of Cu would not affect the final average measurably. Further confirmation for this

Properties in Nanostructured Cu/W Multilayer Coatings 301

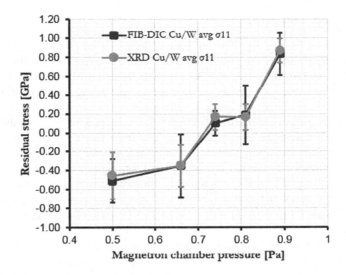

FIGURE 11.12 Comparison of residual stresses obtained from FIB-DIC and XRD measurements.

simplification can be drawn from the excellent agreement of experimental data between FIB-DIC and XRD analyses, as shown in Figure 11.12. Figure 11.11 shows a residual stress variation in W from $\sigma_{11} - \sigma_{33} = -1 \pm 0.46$ GPa at 0.5 Pa deposition chamber pressure to 1.45 ± 0.36 GPa at 0.89 Pa chamber pressure.

11.3.1.3 Micro-cantilever Deflection

In order to compare the measured FIB-DIC with the deflection of a micro-cantilever, a forward calculation approach was chosen, following the eigenstrain deflection model of Korsunsky et al. [46]. A micro-cantilever was modelled to contain the residual stress profile obtained from FIB-DIC analysis, while assuming a linear variation of residual stress within the substrate down to a depth of 3.38 μm, corresponding to the total thickness of the micro-cantilever. Based on this residual stress profile and the cross-sectional geometry obtained from the real micro-cantilever, a curvature was predicted. To describe the residual stress within the cantilever, the following formulation was chosen:

$$\sigma(y) = E^*(y) \cdot (a + b \cdot y) - E^*(y) \cdot \varepsilon^*(y), \tag{11.6}$$

in which y is the coordinate pointing from the coating surface downwards, $E^*(y)$ is Young's modulus depending on y, b is the curvature of the cantilever deflection, $\varepsilon*(y)$ is the eigenstrain profile as a function of depth, and a refers to the component of strain that is uniform across the section and corresponds to the compressive deformation of the cantilever [46]. Equilibrium conditions lead to the following relationships for resultant force and moment, using a beam width function $w(y)$:

$$F = \int \sigma(y) \cdot w(y) dy = 0 \tag{11.7}$$

$$M = \int \sigma(y) \cdot w(y) \cdot y dy = 0 \tag{11.8}$$

In the final step, the predicted curvature was compared to the one measured from the real micro-cantilever. Using the eigenstrain profile $\varepsilon*(y)$ obtained from FIB-DIC returned a predicted curvature of -754 ± 56 m^{-1}. This result is in excellent agreement with the experimental data presented in 11.2.5, which shows a curvature of -800 ± 18 m^{-1}, resulting in a relative difference of only ~6%.

11.3.2 Results and Discussion

The high spatial resolution of the FIB-DIC and eigenstrain nanoscale depth profiling method allowed to reveal the details of the stress variation within Cu/W nano-multilayers, as well as the stress-affected zone within the substrate. The residual stress profiles clearly indicate the location of the stress peak at around $0.75 \bullet t_c$ distance from the sample surface, with t_c being the thickness of the multilayer coating. While approaching zero stress towards the sample surface, the stress peak magnitude can lie significantly above the average stress measured by XRD, as shown by graph P_1 in Figure 11.8 – an information which is particularly useful for the mechanical design of coatings. A clear correlation between the argon working pressure and the multilayer stress was made, showing a transition from compressive residual stresses at low working pressures to tensile residual stresses at higher pressures. It should be noted that not only the average stress value magnitude increases for low/ high working pressures, but also the difference between the peak stress and the average stress, making it highly important to consider the detailed stress profiles for reliability design under these deposition conditions.

The observed stress transitions from tensile to compressive can be explained by considering the interaction between sputtered metal atoms and working gas [23]: a low Ar atmosphere pressure results in a high kinetic energy of sputtering atoms when reaching the sample surface. The mean velocities stated in the previous section correspond to a directed atomic kinetic energy of $E_{Cu} = 4.2$ eV for Cu and $E_W = 11.7$ eV for W at 0.50 Pa, assuming equal velocity for both elements, and to $E_{Cu} = 1.3$ eV for Cu and $E_W = 3.8$ eV per W atom at 0.89 Pa. In an attempt to compare these values with literature data stating an energy range for self-interstitial formation in Cu between 2.5 eV and 5.7 eV [47] and 8.9–9.6 eV in W [48], it becomes apparent that large-scale kinetic energy driven interstitial formation is possible at 0.50 Pa, but not at 0.89 Pa. In addition, collision cascades caused by a low proportion of highly energetic particles with kinetic energies several times higher than the mean value can result in vacancy elimination and layer densification [49]. While the formation of interstitials in a lattice with low vacancy density causes intragranular compressive residual stress at lattice scale [23], further insertion of highly energetic particles into the grain boundaries adds additional in-plane compressive residual stress. As a consequence of this 'atomic peening' process, pronounced compressive residual stress across the multilayer system can be observed at low deposition pressures. With increasing chamber pressure the kinetic energy of the sputtered atoms drops significantly, resulting in reduced adatom mobility, higher vacancy density, and the formation of a reduced density columnar morphology [50, 51]. The attracting forces between column boundaries without the 'atomic peening'-induced compressive residual stress result in a build-up of tensile residual stress [52] comparable to the effect observed during island coalescence in the early stages of thin film formation [53], also known as 'island zipping'.

Further insights include the stress transfer from coating to substrate. Starting from the substrate-coating interface, the residual stress asymptotically approaches the stress-free state at a depth of around 3 μm, which can be determined from extrapolation. Even though the performed XRD analysis can only provide averaged stress levels, it adds a significant amount of insights to this analysis, as the residual stress obtained by XRD and $\sin^2 \Psi$ can be calculated for Cu and W separately.

Micro-cantilever deflection measurements confirmed the residual stress level determined by FIB-DIC analysis and hence supported the calculations based on a combination of FIB-DIC and XRD analysis. The agreement between FIB-DIC and micro-cantilever measurements was excellent, with a difference of around 6%.

While strong variations in WAXS intensity along the azimuthal direction are visible for all Cu and W reflections, this effect decreases – particularly for the W{211} reflection – in the samples deposited at 0.74 Pa and 0.89 Pa, as can be seen in Figure 11.3. This trend is observable at reduced level for the Cu {111} and Cu {200} reflections, indicating the existence of mildly textured copper and randomly oriented tungsten at higher deposition pressures. These observations indicate a change in crystal growth dynamics during MSD depending on the mean free path length of the incoming Cu and W atoms, which affects their kinetic energy. Based on the findings presented in [54], the mean free path length of Cu and W atoms during sputtering at 0.5 Pa and 0.89 Pa magnetron chamber pressure can be estimated as 1.5 cm and 0.7 cm, respectively. As a consequence, the average number of interactions between sputtered metal and atmospheric Ar atoms doubles at 0.89 Pa magnetron chamber pressure as compared to 0.50 Pa. Further investigations into the thermalisation of sputtered atoms during MSD [55] indicate a sharp decrease in mean particle velocity from 3.5 km/s at 0.50 Pa to around 2 km/s at 0.89 Pa. Other studies have previously reported noticeable changes in texture for magnetron bias voltage variations [56], which again modify the kinetic energy of sputtered atoms. The additional kinetic energy of incoming particles at lower Ar pressure therefore leads to an increase in adatom surface mobility, ultimately resulting in the formation and growth of highly textured grains and sharper interfaces.

Finally, the TGS results are shown alongside their 95% confidence interval in Figure 11.13 [31]. Averaged coating stresses obtained from FIB-DIC depth profiling were added to the graph for comparison. A linear magnetron chamber pressure dependence can be observed for both the thermal diffusivity and average residual stress. While the residual stress transitions from compressive to tensile, the thermal diffusivity decreases from $3.3 \cdot 10^{-5}$ m^2/s at 0.50 Pa to $2.6 \cdot 10^{-5}$ m^2/s at 0.89 Pa. This significant reduction in thermal diffusivity of over 20% can be explained by microstructural changes induced by the deposition conditions. Lattice point defects incorporated during MSD at higher Ar pressure through Ar atom entrapment and the mechanisms mentioned in the previous section result in increased electron scattering rates [57, 58]. Increased roughness of Cu/W interfaces at higher magnetron pressures might further decrease the thermal diffusivity as compared to the sharper interfaces obtained at lower deposition pressures.

11.3.3 Summary

A hierarchical analysis of stresses has been proposed using three different independent and complementary techniques: firstly, XRD allowed assessing both Cu and W average through-thickness residual stress magnitudes. By using exclusively sin^2 Ψ ellipse fitting parameters which were independent of the stress-free lattice parameter d_0, a robust analysis method was developed and employed. Consequently, detailed insights into the strain accommodation mechanisms within different materials and grain orientations were obtained, as shown in Figure 11.11. Furthermore, the use of FIB-DIC ring-core milling permitted to study the stress accommodation within the coating multilayer-substrate systems spatially resolved at a high resolution of around 50 nm. Significant

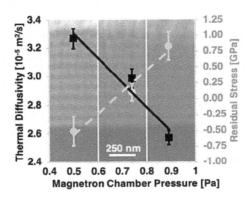

FIGURE 11.13 Thermal diffusivity and averaged in-plane residual stress measurements versus magnetron chamber pressure during deposition shown before their respective SEM micrograph. While a clear trend of linearly decreasing thermal diffusivity is visible for increasing Ar pressure, the average residual stress transitions from highly compressive to highly tensile.

stress peaks became apparent in Figure 11.8, especially for the samples deposited at P = 0.89 Pa, suggesting that under-dense columnar growth in combination with a reduction of atomic peening significantly increases the difference between average and peak stress within the coating. In addition, the residual stress profiles showed that peak stresses occur in proximity to the substrate-coating interface, which facilitates mechanical failures at the substrate-coating interface. A discontinuity in the mechanical properties across this interface might promote further mechanical failure. Finally, the deflection of a micro-cantilever was predicted, based on the residual strain relief profiles obtained from FIB-DIC, which matched the experimentally determined cantilever curvature with high accuracy. The proposed way of calculation takes into account the fact that the actual cantilever cross-section was not rectangular in the centre of the sample due to geometric constrains during ion beam milling. This technique turned out to be an ideal tool for the confirmation of experimental results obtained by novel experimental techniques.

11.4 STRESS-ASSISTED THERMAL DIFFUSION BARRIER BREAKDOWN IN ION BEAM DEPOSITED CU/W NANO-MULTILAYERS ON SI SUBSTRATE AT ELEVATED TEMPERATURES

11.4.1 Experimental Approach and Interpretation

Based on the insights into residual stress and thermal diffusivity gained in the previous section, we present an extensive multi-technique analysis of the diffusion and degradation in ion beam deposited (IBD) Cu/W NMLs on Si substrate, combining large-scale averaging techniques such as synchrotron in situ X-Ray Reflectometry (XRR), in situ Grazing-Incidence Small-Angle X-Ray Scattering (GISAXS) and Wide angle X-Ray Scattering (WAXS) with locally probing techniques such as TEM imaging, EDX analysis, and AFM [59]. This combination of methods permits the observation of diffusion mechanisms within and between buried layers of nanostructured coatings before any changes become apparent at the sample surface.

11.4.2 Results and Discussion

11.4.2.1 Reference State Characterisation

Prior to studying the thermal stability of the Cu/W NMLs, TEM- and XRD-based characterisation of as-deposited samples was performed to determine the layer thickness, density, and texture [59]. The TEM image in Figure 11.14a shows 10 bilayers of Cu and W on Si substrate with increasing layer waviness towards the sample surface. This phenomenon has been observed and described in the past and can largely be attributed to columnar growth during energetic film deposition. A formula for calculating the wavelength of the oscillation D = 3H+7 nm presented in [60] based on the thickness of the thicker layer H and returned an estimated wavelength of 46 nm for H_{Cu} = 13 nm, as estimated from TEM images. This estimate showed excellent agreement with TEM observations and indicates the presence of cylindrical columnar morphology with a diameter of around 46 nm.

Further to the direct cross-sectional TEM observations, WAXS data shown in Figure 11.14b was used for phase identification and texture analysis. Face-centred-cubic (FCC) Cu with {111} and {200} crystallographic orientations was confirmed from Debye-Scherrer rings, while the body-centred-cubic (BCC) α–W phase with {110} and {200} crystallographic orientations was present. Pronounced fibre texture was present in all phases, as can be seen from the strong azimuthal intensity variations of the Debye-Scherrer rings shown in Figure 11.14b. It was shown in the previous section that the presence of strong texture in magnetron sputtered Cu/W NMLs at low Ar pressure co-exists with in-plane compressive residual stresses and increased thermal diffusivity. Since the Ar pressure during IBD was two orders of magnitude below the magnetron sputtering chamber pressure, a reduced level of interaction between highly energetic sputtered metal atoms and Ar atoms inside the

Properties in Nanostructured Cu/W Multilayer Coatings

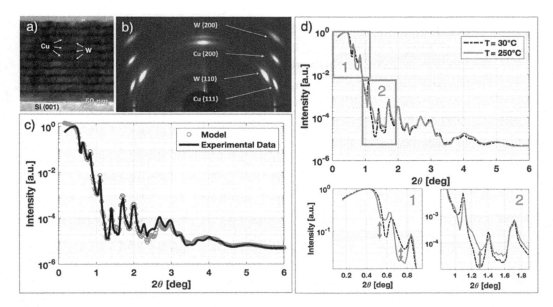

FIGURE 11.14 (a) Cross-sectional TEM image of the Cu/W nano-multilayer showing the Si substrate and individual Cu (bright) and W (dark) layers. (b) Debye-Scherrer rings of NML with identified reflections indicating highly textured material. (c) Experimental XRR profile (black line) and simulated model (red circles). (d) XRR profile before (black dashed line) and after heating (blue continuous line).

deposition chamber can be assumed. A previous comparative study confirmed this assumption [61], by showing that while magnetron sputter deposition emits particles in an energy range between 1 and 100 eV when reaching the sample surface, ion beam deposited particles can reach up to 1000 eV, thus enabling the implantation of self-interstitials in Cu [62] when considering the kinetic particle energy of around 20 eV [63] required for their formation. As a result, self-interstitial formation and 'atomic peening' [64], as well as atom insertion into grain boundaries [65], result in highly densified Cu and W layers under compressive residual stress. Indeed, the residual stress state of identical Cu/W NML systems deposited using the identical deposition system was evaluated in the past [66] and found to be −6.6 GPa for W and −0.4 GPa for Cu, as determined by $\sin^2(\Psi)$ analysis. The average film stress was determined by substrate curvature analysis and was found to be −1.0 GPa [66]. While a W stress of −6.6 GPa seems high at first, a recent study has shown that nanocrystalline tungsten with a crystal size of 10 nm is able to sustain considerably larger compressive stresses [67]. The depth-profile of the in-plane stress distribution within the Cu/W NML on Si substrate can be assumed to be near 0 GPa at the sample surface, while a pronounced stress peak near the NML-substrate interface is present [68].

The XRR profile of the Cu/W NML in reference state at RT is shown in Figure 11.14c alongside the fitted profile using GenX software. The obtained average layer parameters listed in Table 11.1 show the thickness and roughness of Cu and W layers. Furthermore, a high relative layer density found in both nanocrystalline Cu and W reached 94% of their bulk material density [69], which further provides supporting experimental evidence for the presence of self-interstitials arising from the above-described layer densification mechanisms, as well as vacancy defects, which lower the overall density below single crystal level.

11.4.2.2 Structural Evolution Upon Thermal Treatment

XRR intensity profiles were further acquired at 250°C, as shown in comparison with the reference scans in Figure 11.14d. While the overall intensity versus 2θ distribution is similar to the reference state, distinct differences are identifiable in the regions labelled '1' and '2'. Amplitude changes

TABLE 11.1

Cu/W NML average layer thickness, roughness, and density at 30°C versus 250°C as obtained from XRR profile fitting

Sample temperature	30°C	250°C
W thickness [nm]	5.86	5.25
W roughness [nm]	1.07	1.65
Cu thickness [nm]	12.97	13.59
Cu roughness [nm]	1.35	3.44
W density [g/cm^3]	18.13 (94% bulk)	16.94 (88% bulk)
Cu density [g/cm^3]	8.43 (94% bulk)	7.92 (88% bulk)

revealing differences in layer density and layer roughness [70] are highlighted by red arrows in the magnified regions shown underneath the graph. The layer thickness, density, and roughness at 250°C for both Cu and W are compared with the reference state in Table 11.1. A significant decrease in layer density by 6% upon heating was observed for both materials, while an increase in layer roughness by a factor of 1.5 and 2.5 was observed for W and Cu, respectively. One possible explanation for the decrease in density is the thermal activation of Cu self-interstitial motion, which was shown to require less thermal energy as compared to vacancy motion [62, 71]. Vacancy-interstitial annihilation can further contribute to the volumetric expansion and decrease in density [72]. These mechanisms, however, can only be speculated on and require further high-resolution TEM analysis. Thermal expansion of the layers, however, was found to have a minimal contribution to the observed changes in layer thickness, with a Cu expansion of 0.39 % for a temperature increase of 230°C, assuming a thermal expansion coefficient of $\alpha_{Cu}= 17.7 \cdot 10^{-6}$ 1/K [73]. Considering the lower thermal expansion coefficient of nanocrystalline W with $\alpha_W= 4.5 \cdot 10^{-6}$ 1/K [74], the in-plane thermal expansion of Cu is hindered by the W layers, resulting in an increase in in-plane compressive residual stress within Cu layers by around 300 MPa and a decrease in W in-plane residual stress. In summary, thermal activation of diffusion processes at 250°C resulting in measurable microstructural changes in both Cu and W were confirmed experimentally – an activation temperature significantly below other values reported in the literature based on surface observations during heating. Highly compressive in-plane residual stresses and a complex defect structure, both of which were introduced by ion beam deposition, are suggested as underlying reasons.

To gain further insights into the temperature-dependent evolution of the microstructure at higher temperatures, GISAXS patterns were collected between 30°C and 400°C. In its as-deposited state the NML exhibits a periodic vertical GISAXS pattern with a period corresponding to the Cu/W bilayer thickness, as shown in Figure 11.15a. Additional intensity contributions related to reflection and refraction phenomena at interfaces within the multilayer, however, make a detailed analysis of individual peaks challenging. The intensity loss of pattern features along the Q_Z direction is representative of the interface roughness [75], with a steeper decay towards higher Q_Z representing a higher degree of interface roughness. The specular peak is shown as red dot in the centre of the frame at $Q_Y = 0$ nm^{-1} and $Q_Z = 0.825$ nm^{-1}. Distinct lateral peaks are visible at $Q_Y = \pm 0.13$ nm^{-1}, corresponding to a lateral in-plane geometric ordering of 48 nm. This is in strong agreement with the estimated columnar grain diameter of 46 nm based on TEM images. Further in-plane ordering at low Q_Y between ± 0.0265 nm^{-1} is visible in the pattern centre, indicating ordering at larger length scales of ≥ 237 nm. Based on this information and combined with the data obtained from XRR fitting, a hierarchical microstructural multilayer model was implemented in the BornAgain software package, as illustrated in Figure 11.15c showing the experiment geometry and Figure 11.15d showing the model side view. For this purpose, columnar multilayer morphology was simulated by means of cylindrical layered mesocrystals consisting of alternating 40 nm diameter Cu and W particles with heights of 12.5 nm and 5.9 nm, respectively, as shown in Figure 11.15d. The Cu/W particles with

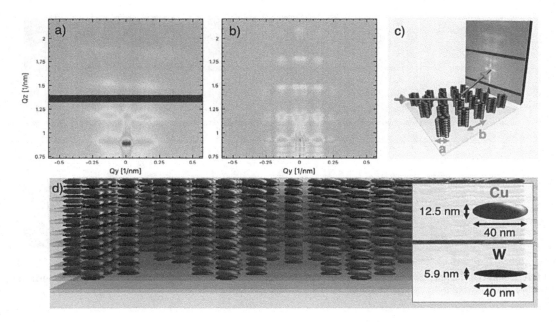

FIGURE 11.15 (a) GISAXS pattern of Cu/W NML at 30°C. (b) Simulated Cu/W GISAXS pattern based on BornAgain material model. (c) Overview of BornAgain model and scattering geometry. Cu/W columnar grains are shown on Si substrate, alongside incoming and reflected beam. While the Cu and W particles are embedded in Cu and W layers in the simulation, they have been removed from the illustration for the purpose of visibility. (d) Side view of rendering showing the simulated multilayer microstructure with columnar Cu/W growth. Dimensions are indicated on the right.

an intercolumnar spacing of 48 nm (characteristic length 'a' in Figure 11.15c) were embedded in a Cu/W multilayer structure which is indicated by the transparent layers in Figure 11.15d. To account for the lateral ordering at low Q_Y, the mesocrystals were arranged within the model in a hexagonal grid with a spacing of 237 nm (length 'b' in Figure 11.15c). While further model accuracy could be achieved by adding additional mesocrystal clusters with a variation of spacings instead of embedding a single cluster in continuous Cu and W layers, the present simulation delivered excellent agreement with the experimentally obtained GISAXS pattern, as shown in Figure 11.15b. A smooth and uniform sample surface in the sample reference state was confirmed by in situ imaging of the sample surface during heating, as shown in Figure 11.16a. It was therefore experimentally confirmed that ion beam deposition of Cu/W NMLs results in the formation of Columnar grain structures during film growth.

Upon heating, first changes in the scattering intensity distribution were observed at a temperature of 230 °C, as shown in the upper image of Figure 11.16b. Peak blurring along the Q_Z direction was observed, resulting in a decreased vertical spacing between individual peaks thus indicating an increase in layer thickness and decrease in density, as previously confirmed by XRR at 250°C. While these changes were observed within buried layers and interfaces, no changes on the sample surface were identified from in situ, in-SEM sample heating at the same temperature, as indicated in the lower row in Figure 11.16b. In contrast, the formation of randomly spaced surface crystals was observed at T = 280°C, as shown in Figure 11.16c. The crystal composition was confirmed as pure Cu from EDX spectroscopy. At the same temperature, the GISAXS pattern showed an increase in intensity at higher-order Q_Z peaks indicating a decrease in layer roughness. Particularly the third-order peak at Q_Z = 1.35 nm^{-1} notably increased in intensity in addition to feature sharpening. Combined with the observed Cu surface crystal formation, it can be assumed that the Cu atoms released from highly densified nanocrystals during heating that led to a layer thickness increase at 230°C started migrating in vertical direction towards the sample surface at 280°C. While the origin of the surface

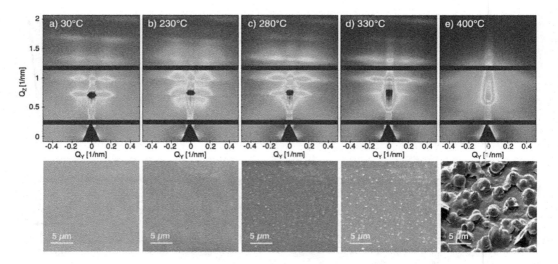

FIGURE 11.16 Upper row: temperature-dependent GISAXS patterns. Lower row: in situ in-SEM surface observations during temperature increase.

crystal Cu couldn't be attributed to specific layers, nanoscale transport mechanisms described in the literature [76–78] indicate that they are to a large extent composed of material from buried Cu layers due to the prevailing residual stress state in the sample. Furthermore, a shift of intensity from the lower lateral peaks towards lower Q_Y at the pattern centre can be observed, which can be interpreted as a large-scale transformation from a regularly spaced columnar structure towards larger and more randomly spaced and sized objects, for instance by columnar grain coalescence.

11.4.2.3 Diffusion Modelling

A feasible diffusion mechanism of Cu along W grain boundaries towards the free sample surface has been described recently [76]; however, the observed onset temperature of 280 °C for surface crystal formation in this study lies significantly below the 400 °C reported. While the materials and layer thicknesses involved are identical to the literature data, differences in residual stress state and defect structure are the possible drivers underlying the lower thermal stability. While the samples presented in [76, 78] were deposited using magnetron sputter deposition with a resulting residual stress state of −3 GPa and −0.5 GPa compressive residual stress for W and Cu, respectively, IBD employed for the samples presented in this section induces a more pronounced defect structure during deposition, as well as a higher level of compressive residual stress. Based on previous study it is known that the in-plane stress gradient along the vertical direction ranges from a stress-free surface layer to a compressive stress peak near the NML-substrate interface [68]. As a consequence, diffusional transport from highly compressive buried Cu layers towards the stress-free surface appears to be the underlying transport mechanism. The diffusion coefficient of Cu atoms along W grain boundaries with account taken of the residual stress in the W layers is expressed in [76] as follows:

$$D_b(\sigma) = D_0 \cdot e^{\frac{\sigma_{ii,W}\Omega}{3RT} - \frac{E_A^*}{kT}} \tag{11.9}$$

with a pre-exponential factor of the self-diffusivity of nanocrystalline Cu D_0, an in-plane tungsten interlayer first stress invariant of $\sigma_{ii,W} = \sigma_{11,W} + \sigma_{22,W}$ ($\sigma_{33} = 0$), a diffusion activation volume Ω, the universal gas constant R, temperature T, diffusion activation energy E_A^*, and the Boltzmann constant k.

Furthermore, micropillar compression of Cu/TiN NMLs at elevated temperatures [79] revealed anomalous plastic flow of Cu alongside a significant reduction in Cu 0.5% yield stress at a temperature of 200°C to around 0.81 GPa. This value lies close to the above estimated Cu residual stress

Properties in Nanostructured Cu/W Multilayer Coatings

at 250°C, indicating that the Cu layer has been compressed beyond yield stress at this point. The observed outflow of Cu from the micropillar boundary was explained using a Coble creep model depending on the stress state in Cu layers:

$$\frac{d\varepsilon}{dt} = \dot{\varepsilon} = A_c \frac{\delta'}{d^3} \frac{\sigma_{Cu} V_{Cu}}{kT} D_{GB}. \tag{11.10}$$

with the strain rate $\dot{\varepsilon}$, a geometric pre-factor A_c, the grain boundary width δ', the layer thickness d, the Cu atomic volume V_{Cu}, the residual stress in Cu layers $\sigma_{Cu} = \sigma_{11} = \sigma_{22}$, and the diffusion coefficient in the grain boundary D_{GB}.

The stress-driven Cu diffusion observed in the present study appears to be similar to the micropillar compression induced plastic flow [79], however, with in-plane compression generated by a combination of residual stress and thermal expansion coefficient mismatch, resulting in vertical diffusion as opposed to vertical nano-indenter-driven compression resulting in in-plane plastic flow. Equations (11.9) and (11.10) can therefore be combined to obtain a model for the stress-assisted migration of Cu towards the sample surface. While the model in [76] focuses on the effect of residual stress in the W interlayer on the diffusion of Cu across this interlayer, the model presented in [79] only considers the role of the applied stress on the Cu layer for diffusion analysis. Combining these two models therefore yields a model accounting for the implications of residual stresses in both Cu and W layers and the effect on the material transport towards the sample surface. As a result, an expression describing the residual stress-driven diffusion of in-plane compressed copper in vertical direction across W barrier layers was obtained:

$$\frac{d\varepsilon}{dt} = \dot{\varepsilon} = A_c \frac{\delta'}{d^3} \frac{\sigma_{Cu} V_{Cu}}{kT} D_0 \cdot e^{\frac{\sigma_{ii,W} \Omega}{3RT} \frac{E_A^*}{kT}} \tag{11.11}$$

For further calculation, the parameters shown in Table 11.2 were employed.

The resulting temperature-dependent Coble creep strain rate within a Cu layer was calculated and is shown in Figure 11.17. It becomes apparent that the residual stress state in the Cu layer has a strong effect on the strain rate and thus the nanoscale transport. For a residual stress of −0.4 GPa in the Cu layer and −6.6 GPa in the W layer, the Coble creep strain rate at 280°C was calculated as −4.35 • 10^{-3} 1/s, which provides a convincing explanation for the observed surface crystal formation temperature.

TABLE 11.2

Parameters used to calculate temperature-dependent Coble-creep strain rate

Parameter	Value
D_0 [80]	$3 \cdot 10^{-9}$ m²/s
$\sigma_{11,w} = \sigma_{22,w}$	$-6.6 \cdot 10^9$ Pa
Ω [76]	$7.69 \cdot 10^{-6}$ m³/mol
E_A^* [80]	$1.025 \cdot 10^{-19}$ J
A_c [79]	148
δ' [79]	$0.5 \cdot 10^{-10}$ m
d	$13 \cdot 10^{-9}$ m
σ_{Cu}	$-0.4 \cdot 10^9$ Pa
V_{Cu} [79]	$8.78 \cdot 10^{-30}$ m³

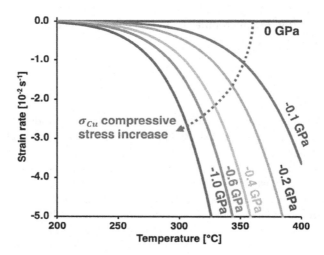

FIGURE 11.17 Evolution of Cu layer strain rate vs. temperature and layer stress.

Subsequently raising the sample temperature to 330°C resulted in a further significant intensity shift from $Q_Y = \pm 0.13$ nm^{-1} towards the pattern centre, indicating the gradual disappearance of lateral ordering and formation of random large-scale structures buried within the multilayer structure, while surface crystals grow further, as shown in Figure 11.16d. The further reduction in lateral peak intensity at $Q_Z = 0.68$ nm^{-1} may partly be caused by the obstruction of the exiting X-ray photons by the surface Cu crystals. A significant increase in low-Q_Y intensity (red peak) further confirms the transformation from evenly spaced columnar grains towards large-scale and more random ordering. Finally, increasing the temperature beyond 330°C resulted in a gradual collapse of the sample nanostructure both in lateral and vertical direction, as shown in the final GISAXS pattern in Figure 11.16e that shows single specular peak in the pattern centre at 400°C. Looking at the SEM frame acquired after heating the sample to 400°C clearly shows the significant microstructural transformation that has occurred during heating, with spheroidal features being buried underneath a thin surface layer.

To further investigate the composition of these micro-particles and understand the mechanisms leading to their formation, cross-sectional TEM and t-EDX analysis of the sample heated during the GISAXS experiment was carried out. For this purpose, a thin, electron transparent TEM lamella was created using Ga+ FIB, as described in Section 11.2 (Materials and Methods) above. T-EDX on TEM lamellae was used to limit the interaction volume between sample and electron beam, resulting in a significantly improved local resolution. The lamella cross-section of the sample after heating is shown in Figure 11.18. The material map in Figure 11.18a clearly shows the significant diffusion-driven transformation. The nanostructured Cu layers have coalesced into flat spheroidal microparticles buried underneath a thick tungsten layer. The identified surface Pt layer was deposited prior to lamella cutting for surface protection. The material distribution within the Cu micro-particles was identified based on quantitative stoichiometric analysis, as shown in the graph underneath the material map. The particle was composed of 75 atomic % Cu and 25 atomic % Si, indicated the presence of the orthorhombic or tetragonal [81] η''-Cu$_3$Si phase, while the presence of the η and η' phases are unlikely due to the higher formation temperature required [82]. TEM images shown in Figure 11.18b–d show the different stages of Cu$_3$Si microparticle formation: Cu penetration across the W diffusion barrier into the Si substrate results in the precipitation of Cu$_3$Si, leading to a significant volumetric expansion [83] as can be seen in Figure 11.18b. A reason for the delayed diffusion of Cu into Si as compared to previously described diffusion towards the free sample surface can be found by looking at the thermal expansion coefficient mismatch between W and Si. Since the W diffusion

Properties in Nanostructured Cu/W Multilayer Coatings 311

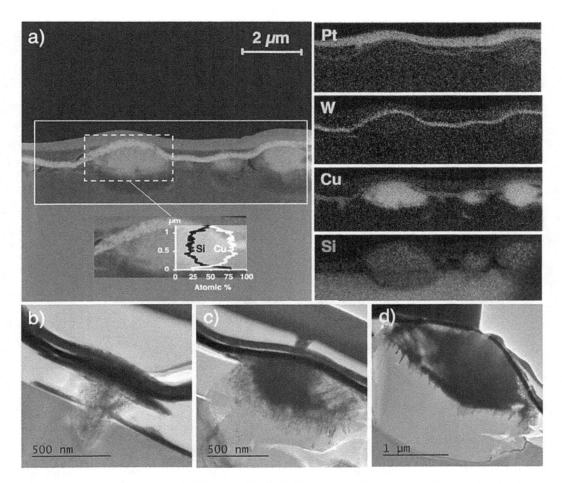

FIGURE 11.18 Cross-sectional TEM Lamella of Cu/W nano-multilayer on Si substrate after heating to 400°C, showing a full degradation of the layered structure. (a) Left: Transmission EDX map showing the material distribution after heating. Local atomic percentage is shown in the detail frame underneath. Right: Detailed maps for each element showing the exact local material distribution. (b–d) TEM images of spheroidal particles in different stages of growth from early stage (b) to final particle (d).

barrier layer showed excellent adhesion with the Si substrate, W expansion was hindered by interfacial adhesion. While near-surface W layers experienced a decrease in residual stress caused by Cu expansion, the first W layer on Si experienced an increase in compressive residual stress. According to Equation (11.1) this reduces the thermal diffusion significantly, resulting in a delayed nanoscale mass transport at higher temperature range, once the Cu compressive residual stress reaches the threshold value required for diffusion along the highly compressive W grain boundaries. Once Cu diffuses into the substrate, the expanding Cu-Si alloy breaks across the multilayer structure, permitting the absorption of further interlayer Cu, while additional Si from the substrate contributes to particle growth by sustaining the formation of Cu_3Si, as shown in Figure 11.18c. In the final stage, the multilayer structure has been entirely deprived of Cu, leaving a layered pure W structure on top of the Cu_3Si particles, as can be seen in Figure 11.18d.

Finally, the described large-scale transformation was confirmed by comparative AFM surface scans carried out before and after heating. The scan in reference state shown in Figure 11.19a shows regular surface peaks with an average spacing of around 50 nm, thus corresponding to the previously determined columnar grain spacing. A large contrast in both shape and size of the objects can be

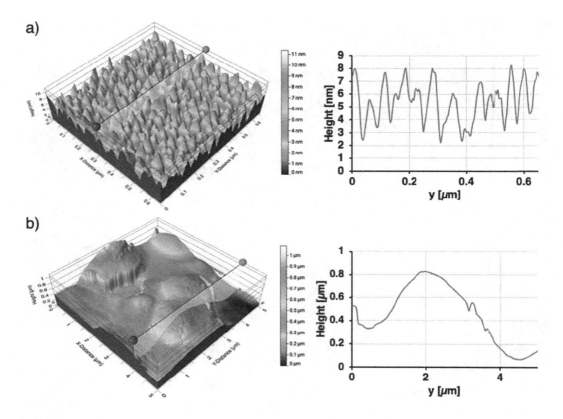

FIGURE 11.19 AFM scans before of the sample surface (a) in reference state and (b) after heating. Line profiles shown on the right are indicated by the black line on the left.

seen in the post heating scan shown in Figure 11.19b, which confirms the presence of microparticles with a height of around 1 μm and a diameter of around 3 μm.

In summary, based on the experimental insights described above, the following degradation mechanism of Cu/W NMLs on Si substrate was identified:

1) Ion beam deposition induced vacancy diffusion is activated at T > 230 °C, resulting in Cu layer thickness increase, roughness increase, and density decrease, as shown in Figure 11.20b.
2) Additional compressive stress in Cu layers at 280 °C exceeds the nanocrystalline yield stress. As a consequence, stress-induced Coble creep results in nanoscale Cu diffusion along W grain boundaries towards the sample surface. The formation of Cu nanocrystals can be observed on the sample surface, as shown in Figure 11.20c.

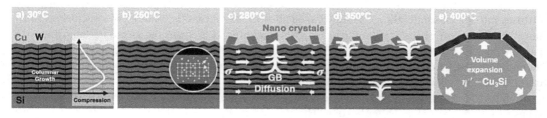

FIGURE 11.20 Illustration showing the mechanism leading to large-scale microstructural breakdown in Cu/W nano-multilayers on Si substrate.

Properties in Nanostructured Cu/W Multilayer Coatings

3) Beyond 330 °C a large-scale microstructural transformation can be observed: thermally induced stress in the lower Cu layers exceeds the diffusion threshold in the W diffusion barrier. As a consequence, Cu diffuses into Si substrate, resulting in η''- Cu_3Si formation associated with rapid volumetric expansion, thus breaking down the layered structure, resulting in Cu absorption and complete transformation into η''- Cu_3Si, as shown in Figure 11.20d,e.

11.4.3 SUMMARY

Three key results were presented in Section 11.4: firstly, a novel and powerful combination of advanced experimental methods was employed to obtain temperature-dependent insights into stress and diffusion mechanisms within buried layers of nanostructured materials. Secondly, the effect of highly energetic particle deposition methods on the thermal stability and residual stress state was discussed before the background of defect formation and diffusion. A clear link between deposition conditions, nanoscale defect structure and thermal instability was established. Finally, this new understanding was used to develop a detailed description and model of the Coble creep induced strain within Cu layers, resulting in vertical diffusion of Cu across W layers into the Si substrate. It was found that the pronounced compressive residual stresses and defects arising from ion beam deposition contribute to large extent to the degradation of the nanostructured coating through a mechanism of Cu diffusion into the Si substrate, ultimately resulting in a transformation of nanostructured Cu layers into in η''- Cu_3Si protrusions.

11.5 CONCLUSION

In this chapter a broad range of experimental techniques for the interrogation of the mechanical and thermal properties of Cu/W NMLs was presented and discussed before the background of kinetic atomic interaction during deposition and the arising defect structure and residual stress state. It was shown that the particle energy during MSD and IBD strongly affects the residual stress state, thermal diffusivity and stability.

In Section 11.3, a clear correlation between magnetron chamber pressure and the texture, residual stress, and thermal diffusivity of W/Cu nano-laminates was established. The interaction between the working gas and sputtered atoms was identified as key driver underlying the observed property changes: lower kinetic energy of particles reaching the substrate surface leads to the formation of low-density columnar growth in the nanoscale strata, resulting in defect accumulation. At lower pressures the 'atomic peening' effect in combination with the formation of defects enables the formation of highly densified and textured material by interstitial formation and vacancy filling through collision cascades during deposition. In addition, increased interface roughness between Cu and W layers might contribute to a reduction in thermal diffusivity.

In Section11.4, a mechanism highlighting the importance of the interaction between residual stress and nanoscale Cu transport in Cu/W NMLs at elevated temperatures was presented, modelled, and thoroughly discussed. It was found that the low Ar pressure during IBD gives rise to high kinetic particle energies, resulting in a highly compressive in-plane residual stress state with complex defect structure, with the same mechanisms as described for MSD. It was also found that the pronounced compressive residual stresses and defects arising from ion beam deposition contribute to large extent to the degradation of the nanostructured coating through a mechanism of Cu diffusion into the Si substrate, ultimately resulting in a transformation of nanostructured Cu layers into in η''- Cu_3Si protrusions.

If chosen carefully, residual stress, thermal diffusivity, and thermal stability become additional degrees of freedom available during engineering design process, not only by changing structural integrity properties but also functional ones. For instance, compressive residual stress in coatings can improve the tensile properties due to a pre-straining of the material. Not only the sign and magnitude of residual stress are important, but also gradients of residual stress can play an important role in modifying the final product's properties. A uniform stress distribution is usually preferred, in order to avoid stress localisations within the coating, which may result in failure below the expected level of loading. In this respect, using the lowest possible working argon pressure to create compressive residual stress would not be the best solution from the structural point of view, as a compressive peak is deposited in near-substrate region which might give rise to failure. Instead, choosing a slightly higher pressure might help the reliability of the coating by creating a uniform residual stress distribution. However, if maximum thermal diffusivity is required, a low MSD pressure would be of benefit, while thermal expansion in combination with compressive residual stress could lead to buckling. Ion beam deposited samples, on the other hand, are subjected to an elevated degree of residual stress and a more pronounced defect structure, resulting in a lower thermal stability. It was shown experimentally and based on an analytical model that compressive residual stress in Cu layers results in a promotion of diffusion at elevated temperature, while compressive residual stress in W diffusion barriers results in an increased protection from diffusional flow along grain boundaries. To maximise thermal stability at elevated temperature, tensile residual stress in Cu layers and compressive residual stress in W layers is desirable. However, these properties would result in a decrease of the mechanical performance under external of thermal expansion induced strain. These selected examples highlight the importance of application-specific material property tailoring via MSD and IBD parameter control. The combination of presented insights into practical aspects of nano-multilayer property tailoring and theoretical mechanisms such as defect formation and migration therefore enable the manufacture of optimised coating materials in their mechanical, thermal, and microstructural properties, tailored to the required field of application.

Future technological applications will require further reduction of layer thicknesses, which make nano-multilayers more susceptible to degradation by thermal diffusion and give rise to increased levels of residual stress. To maintain material reliability, a combination of further experimental technique development and in-depth theoretical understanding of the links between manufacture and material properties is of utmost importance. While the techniques and models presented in this chapter form a firm basis for the understanding of the mechanical and thermal properties of nano-structured coatings, further validation and modelling will be required – for instance by leveraging the potential of computational molecular dynamics simulations, which permit further understanding of energy exchange processes during PVD.

REFERENCES

[1] L. R. Brandt et al., "Probing the deformation and fracture properties of Cu/W nano-multilayers by in situ SEM and synchrotron XRD strain microscopy," *Surf. Coat. Technol.*, vol. 320, pp. 158–167, 2017.

[2] C. Wang, P. Brault, C. Zaepffel, J. Thiault, A. Pineau, and T. Sauvage, "Deposition and structure of W-Cu multilayer coatings by magnetron sputtering," *J. Phys. D. Appl. Phys.*, vol. 36, no. 21, pp. 2709–2713, 2003.

[3] S. Djaziri et al., "Controlled biaxial deformation of nanostructured W/Cu thin films studied by X-ray diffraction," *Surf. Coat. Technol.*, vol. 205, no. 5, pp. 1420–1425, 2010.

[4] B. Girault, P. Villain, E. Le Bourhis, P. Goudeau, and P. O. Renault, "X-ray diffraction analysis of the structure and residual stresses of W/Cu multilayers," *Surf. Coat. Technol.*, vol. 201, no. 7 SPEC. ISS., pp. 4372–4376, 2006.

[5] C. Linsmeier et al., "Development of advanced high heat flux and plasma-facing materials," *Nucl. Fusion*, vol. 57, no. 9, p. 092007, 2017.

[6] Y. Gao et al., "Radiation tolerance of Cu/W multilayered nanocomposites," *J. Nucl. Mater.*, vol. 413, no. 1, pp. 11–15, 2011.

[7] B. Tlili, C. Nouveau, G. Guillemot, A. Besnard, and A. Barkaoui, "Investigation of the effect of residual stress gradient on the wear behavior of PVD thin films," *J. Mater. Eng. Perform.*, vol. 27, no. 2, pp. 457–470, 2018.

[8] J. A. Thornton, J. Tabock, and D. W. Hoffman, "Internal stresses in metallic films deposited by cylindrical magnetron sputtering," *Thin Solid Films*, vol. 64, no. 1, pp. 111–119, 1979.

Properties in Nanostructured Cu/W Multilayer Coatings 315

[9] A. J. Lunt et al., "A state-of-the-art review of micron-scale spatially resolved residual stress analysis by FIB-DIC ring-core milling and other techniques," *J. Strain Anal. Eng. Des.*, vol. 50, no. 7, pp. 426–444, 2015.

[10] R. Treml et al., "High resolution determination of local residual stress gradients in single- and multilayer thin film systems," *Acta Mater.*, vol. 103, pp. 616–623, 2016.

[11] J. McCarthy, Z. Pei, M. Becker, and D. Atteridge, "FIB micromachined submicron thickness cantilevers for the study of thin film properties," *Thin Solid Films*, vol. 358, no. 1–2, pp. 146–151, 2000.

[12] E. Salvati and A. M. Korsunsky, "An analysis of macro- and micro-scale residual stresses of Type I, II and III using FIB-DIC micro-ring-core milling and crystal plasticity FE modelling," *Int. J. Plast.*, vol. 98, pp. 123–138, 2017.

[13] M. Renzelli, M. Z. Mughal, M. Sebastiani, and E. Bemporad, "Design , fabrication and characterization of multilayer Cr-CrN thin coatings with tailored residual stress profiles," *JMADE*, vol. 112, pp. 162–171, 2016.

[14] A. M. Korsunsky, *A Teaching Essay on Residual Stresses and Eigenstrains*. Butterworth-Heinemann, Oxford, 2017.

[15] Q. Wang, S. Kishimoto, Y. Tanaka, and K. Naito, "Fabrication of nanoscale speckle using broad ion beam milling on polymers for deformation analysis," *Theor. Appl. Mech. Lett.*, vol. 6, no. 4, pp. 157–161, 2016.

[16] A. M. Korsunsky et al., "Nanoscale residual stress depth profiling by Focused Ion Beam milling and eigenstrain analysis," *Mater. Des.*, vol. 145, pp. 55–64, 2018.

[17] E. Salvati, L. Romano-Brandt, M. Z. Mughal, M. Sebastiani, and A. M. Korsunsky, "Generalised residual stress depth profiling at the nanoscale using focused ion beam milling," *J. Mech. Phys. Solids*, vol. 125, pp. 488-501, 2019.

[18] E. Salvati, T. Sui, A. J. G. Lunt, and A. M. Korsunsky, "The effect of eigenstrain induced by ion beam damage on the apparent strain relief in FIB-DIC residual stress evaluation," *Mater. Des.*, vol. 92, pp. 649–658, 2016.

[19] E. Salvati et al., "Nanoscale structural damage due to focused ion beam milling of silicon with Ga ions," *Mater. Lett.*, vol. 213, pp. 346–349, 2018.

[20] A. M. Korsunsky et al., "Quantifying eigenstrain distributions induced by focused ion beam damage in silicon," *Mater. Lett.*, vol. 185, pp. 47–49, 2016.

[21] J. Guénolé, A. Prakash, and E. Bitzek, "Influence of intrinsic strain on irradiation induced damage: the role of threshold displacement and surface binding energies," *Mater. Des.*, vol. 111, pp. 405–413, 2016.

[22] A. Zeilinger et al., "In-situ observation of cross-sectional microstructural changes and stress distributions in fracturing TiN thin film during nanoindentation," *Sci. Rep.*, vol. 6, no. February, pp. 1–14, 2016.

[23] G. Abadias et al., "Review Article: stress in thin films and coatings: current status, challenges, and prospects," *J. Vac. Sci. Technol. A*, vol. 36, no. 2, p. 020801, 2018.

[24] M. Stefenelli et al., "X-ray analysis of residual stress gradients in TiN coatings by a Laplace space approach and cross-sectional nanodiffraction: a critical comparison," *J. Appl. Crystallogr.*, vol. 46, no. 5, pp. 1378–1385, 2013.

[25] U. Selvadurai, W. Tillmann, G. Fischer, and T. Sprute, "The influence of multilayer design on residual stress gradients in Ti/TiAlN systems," *Mater. Sci. Forum*, vol. 768–769, pp. 264–271, 2014.

[26] A. Reza et al., "Non-contact, non-destructive mapping of thermal diffusivity and surface acoustic wave speed using transient grating spectroscopy," *Rev. Sci. Instrum.*, vol. 91, no. 5, p. 054902, 2020.

[27] O. W. Käding, H. Skurk, A. A. Maznev, and E. Matthias, "Transient thermal gratings at surfaces for thermal characterization of bulk materials and thin films," *Appl. Phys. A Mater. Sci. Process.*, vol. 61, no. 3, pp. 253–261, 1995.

[28] J. A. Johnson et al., "Phase-controlled, heterodyne laser-induced transient grating measurements of thermal transport properties in opaque material," *J. Appl. Phys.*, vol. 111, no. 2, p. 023503, 2012.

[29] C. A. Dennett and M. P. Short, "Time-resolved, dual heterodyne phase collection transient grating spectroscopy," *Appl. Phys. Lett.*, vol. 110, no. 21, p. 211106, 2017.

[30] F. Hofmann, M. P. Short, and C. A. Dennett, "Transient grating spectroscopy: an ultrarapid, nondestructive materials evaluation technique," *MRS Bull.*, vol. 44, no. 5, pp. 392–402, 2019.

[31] L. Romano Brandt, A. Reza, E. Salvati, E. Le Bourhis, F. Hofmann, and A. M. Korsunsky, "Controlling thermal diffusivity, residual stress and texture in W/Cu Nano-multilayers by magnetron chamber pressure variation," *SSRN Electron. J.*, 2020.

[32] M. Björck and G. Andersson, "*GenX* : an extensible X-ray reflectivity refinement program utilizing differential evolution," *J. Appl. Crystallogr.*, vol. 40, no. 6, pp. 1174–1178, 2007.

[33] P. Müller-Buschbaum, "Grazing incidence small-angle X-ray scattering: an advanced scattering technique for the investigation of nanostructured polymer films," *Anal. Bioanal. Chem.*, vol. 376, no. 1, pp. 3–10, 2003.

[34] J. Filik et al., "Processing two-dimensional X-ray diffraction and small-angle scattering data in DAWN 2," *J. Appl. Crystallogr.*, vol. 50, no. 3, pp. 959–966, 2017.

[35] G. Pospelov et al., "BornAgain: software for simulating and fitting grazing-incidence small-angle scattering," *J. Appl. Crystallogr.*, vol. 53, no. 1, pp. 262–276, 2020.

[36] G. Pospelov, W. Van Herck, J. Burle, "BornAgain — Software for simulating and fitting X-ray and neutron small-angle scattering at grazing incidence." [Online]. Available: https://www.bornagainproject.org. [Accessed: 2020].

[37] KLA Corporation, "ProfilmOnline." [Online]. Available: https://www.profilmonline.com/. [Accessed: 06-Oct-2020].

[38] A. M. Engwall, Z. Rao, and E. Chason, "Origins of residual stress in thin films: interaction between microstructure and growth kinetics," *Mater. Des.*, vol. 110, pp. 616–623, 2016.

[39] E. Chason, J. W. Shin, S. J. Hearne, and L. B. Freund, "Kinetic model for dependence of thin film stress on growth rate, temperature, and microstructure," *J. Appl. Phys.*, vol. 111, no. 8, 2012.

[40] F. M. D'Heurle and J. M. Harper, "Note on the origin of intrinsic stresses in films deposited via evaporation and sputtering," *Thin Solid Films*, vol. 171, pp. 81–92, 1989.

[41] M. Senn, *Digital Image Correlation and Tracking.* Available: https://uk.mathworks.com/matlabcentral/fileexchange/50994-digital-image-correlation-and-tracking, 2015.

[42] "Material: Silicon (Si), bulk." [Online]. Available: https://www.memsnet.org/material/siliconsibulk/. [Accessed: 19-Oct-2018].

[43] A. M. Korsunsky, "A Critical discussion of the sin2ψ stress measurement technique," *Mater. Sci. Forum*, vol. 571–572, pp. 219–224, 2008.

[44] A. Benediktovich, I. Feranchuk, and A. Ulyanenkov, *Theoretical Concepts of X-Ray Nanoscale Analysis.* Springer, New York, vol. 183. 2014.

[45] G. S. Schajer, Ed., *Practical Residual Stress Measurement Methods.* Chichester, UK: John Wiley & Sons, Ltd, 2013.

[46] A. M. Korsunsky, S. Cherian, R. Raiteri, and R. Berger, "On the micromechanics of micro-cantilever sensors: property analysis and eigenstrain modeling," *Sensors Actuators A Phys.*, vol. 139, no. 1–2 SPEC. ISS., pp. 70–77, 2007.

[47] A. K. Bandyopadhyay and S. K. Sen, "Calculation of self-interstitial formation energy (Both split and non-split) in noble metals," *Phys. Status Solidi*, vol. 157, no. 2, pp. 519–530, 1990.

[48] I. M. Neklyudov, E. V. Sadanov, G. D. Tolstolutskaja, V. A. Ksenofontov, T. I. Mazilova, and I. M. Mikhailovskij, "Interstitial atoms in tungsten: interaction with free surface and in situ determination of formation energy," *Phys. Rev. B* 78, 115418, 2008.

[49] J. Dalla Torre et al., "Microstructure of thin tantalum films sputtered onto inclined substrates: experiments and atomistic simulations," *J. Appl. Phys.*, vol. 94, no. 1, pp. 263–271, 2003.

[50] R. Koch, "Stress in evaporated and sputtered thin films - a comparison," *Surf. Coat. Technol.*, vol. 204, no. 12–13, pp. 1973–1982, 2010.

[51] R. Koch, "The intrinsic stress of polycrystalline and epitaxial thin metal films," *J. Phys. Condens. Matter*, vol. 6, no. 45, pp. 9519–9550, 1994.

[52] G. Abadias, W. P. Leroy, S. Mahieu, and D. Depla, "Influence of particle and energy flux on stress and texture development in magnetron sputtered TiN films," *J. Phys. D. Appl. Phys.*, vol. 46, no. 5, pp. 55301–55310, 2013.

[53] S. C. Seel and C. V. Thompson, "Tensile stress generation during island coalescence for variable island-substrate contact angle," *J. Appl. Phys.*, vol. 93, no. 11, pp. 9038–9042, 2003.

[54] Z. J. Radzimski, "Directional copper deposition using dc magnetron self-sputtering," *J. Vac. Sci. Technol. B Microelectron. Nanom. Struct.*, vol. 16, no. 3, p. 1102, 1998.

[55] W. Z. Park et al., "Investigation of the thermalization of sputtered atoms in a magnetron discharge using laser-induced fluorescence," *Appl. Phys. Lett.*, vol. 58, no. 22, pp. 2564–2566, 1991.

[56] J. Keckes et al., "X-ray nanodiffraction reveals strain and microstructure evolution in nanocrystalline thin films," *Scr. Mater.*, vol. 67, no. 9, pp. 748–751, 2012.

[57] P. Nath and K. L. Chopra, "Thermal conductivity of copper films," *Thin Solid Films*, vol. 20, no. 1, pp. 53–62, 1974.

[58] Z. Zhou, C. Uher, A. Jewell, and T. Caillat, "Influence of point-defect scattering on the lattice thermal conductivity of solid solution Co(Sb1-xAsx)3," *Phys. Rev. B - Condens. Matter Mater. Phys.*, vol. 71, no. 23, p. 235209, 2005.

[59] L. Romano Brandt, E. Salvati, D. Wermeille, C. Papadaki, E. Le Bourhis, and A. M. Korsunsky, "Stress-assisted thermal diffusion barrier breakdown in Cu/W nano-multilayers on Si substrate observed by in situ GISAXS and transmission EDX," *ACS Appl. Mater. Interfaces*, vol. 13, no. 5, pp. 6795–6804, 2021.

[60] Z. Czigány and G. Radnóczi, "Columnar growth structure and evolution of wavy interface morphology in amorphous and polycrystalline multilayered thin films," *Thin Solid Films*, vol. 347, no. 1–2, pp. 133–145, 1999.

[61] C. Bundesmann and H. Neumann, "Tutorial: the systematics of ion beam sputtering for deposition of thin films with tailored properties," *J. Appl. Phys.*, vol. 124, no. 23, p. 231102, 2018.

[62] Y. N. Osetsky, D. J. Bacon, A. Serra, B. N. Singh, and S. I. Golubov, "One-dimensional atomic transport by clusters of self-interstitial atoms in iron and copper," *Philos. Mag.*, vol. 83, no. 1, pp. 61–91, 2003.

[63] C. M. Gilmore and J. A. Sprague, "Molecular dynamics simulation of defect formation during energetic Cu deposition," *Thin Solid Films*, vol. 419, no. 1–2, pp. 18–26, 2002.

[64] H. Windischmann, "Intrinsic stress in sputter-deposited thin films," *Critical Rev. Solid State und Mater. Sci.*, vol. 17, no. 6, pp. 547–596, 1992.

[65] D. Magnfält, G. Abadias, and K. Sarakinos, "Atom insertion into grain boundaries and stress generation in physically vapor deposited films," *Appl. Phys. Lett.*, vol. 103, no. 5, p. 051910, 2013.

[66] B. Girault, "Étude De L'Effet De Taille Et De Structure Sur L'Élasticité," *Thesis*, 2008.

[67] J. Yang et al., "Strength enhancement of nanocrystalline tungsten under high pressure," *Matter Radiat. Extrem.*, vol. 5, no. 5, p. 058401, 2020.

[68] L. Romano-Brandt, E. Salvati, E. Le Bourhis, T. Moxham, I. P. Dolbnya, and A. M. Korsunsky, "Nano-scale residual stress depth profiling in Cu/W nano-multilayers as a function of magnetron sputtering pressure," *Surf. Coat. Technol.*, vol. 381, no. November 2019, p. 125142, 2020.

[69] F. Foote and E. R. Jette, "The fundamental relation between lattice constants and density," *Phys. Rev.*, vol. 58, no. 1, pp. 81–86, 1940.

[70] E. Chason and T. M. Mayer, "Thin film and surface characterization by specular X-ray reflectivity," *Crit. Rev. Solid State Mater. Sci.*, vol. 22, no. 1, pp. 1–67, 1997.

[71] M. Bockstedte, A. Mattausch, and O. Pankratov, "Ab initio study of the annealing of vacancies and interstitials in cubic SiC: vacancy-interstitial recombination and aggregation of carbon interstitials," *Phys. Rev. B - Condens. Matter Mater. Phys.*, vol. 69, no. 23, p. 235202, 2004.

[72] G. J. Dienes and A. C. Damask, "Kinetics of vacancy-interstitial annihilation. III. Interstitial migration to sinks," *Phys. Rev.*, vol. 128, no. 6, pp. 2542–2546, 1962.

[73] J. H. Zhao, Y. Du, M. Morgen, and P. S. Ho, "Simultaneous measurement of Young's modulus, poisson ratio, and coefficient of thermal expansion of thin films on substrates," *J. Appl. Phys.*, vol. 87, no. 3, pp. 1575–1577, 2000.

[74] A. Lahav, K. A. Grim, and I. A. Blech, "Measurement of thermal expansion coefficients of W, WSi, WN, and WSiN thin film metallizations," *J. Appl. Phys.*, vol. 67, no. 2, pp. 734–738, 1990.

[75] H. Jiang et al., "In situ GISAXS study on the temperature-dependent performance of multilayer monochromators from the liquid nitrogen cooling temperature to 600 °C," *Appl. Surf. Sci.*, vol. 508, p. 144838, 2020.

[76] A. V. Druzhinin et al., "Effect of internal stress on short-circuit diffusion in thin films and nanolaminates: application to Cu/W nano-multilayers," *Appl. Surf. Sci.*, vol. 508, p. 145254, 2020.

[77] C. Cancellieri et al., "The effect of thermal treatment on the stress state and evolving microstructure of Cu/W nano-multilayers," *J. Appl. Phys.*, vol. 120, no. 101, 2016.

[78] F. Moszner et al., "Thermal stability of Cu/W nano-multilayers," *Acta Mater.*, vol. 107, pp. 345–353, 2016.

[79] R. Raghavan et al., "Mechanical behavior of Cu/TiN multilayers at ambient and elevated temperatures: stress-assisted diffusion of Cu," *Mater. Sci. Eng. A*, vol. 620, pp. 375–382, 2015.

[80] J. Horváth, R. Birringer, and H. Gleiter, "Diffusion in nanocrystalline material," *Solid State Commun.*, vol. 62, no. 5, pp. 319–322, 1987.

[81] J. K. Solberg, "The crystal structure of η-Cu3Si precipitates in silicon," *Acta Crystallogr. Sect. A*, vol. 34, no. 5, pp. 684–698, 1978.

[82] N. Ponweiser and K. W. Richter, "New investigation of phase equilibria in the system Al-Cu-Si," *J. Alloys Compd.*, vol. 512, no. 1, pp. 252–263, 2012.

[83] M. Seibt, M. Griess, A. A. Istratov, H. Hedemann, A. Sattler, and W. Schröter, "Formation and properties of copper silicide precipitates in silicon," *Phys. Status Solidi Appl. Res.*, vol. 166, no. 1, pp. 171–182, 1998.

Index

Page numbers in **bold** indicate tables, page numbers in *italic* indicate figures.

A

absorption figure of merit, 245
activities, 269
addition amount, 255, 257, 259, 263–264, 281
addition rate, *6*, 8–9, 22, 32, 260, 262–265, 281
adhesion strength, 212
AFM, 8–10, 15, 180, 295
AlN-TiN Nanocolumn Composite Composition
    Spreads, 132
ALD (atomic layer deposition), 86–87, 114
all-positive dispersion, 58, 77
amorphous, 178, 179, 186, 228
anatase, 186
anisotropic viscous superhydrophobic surface, 216
anodic aluminum oxide, 165, 166, 168, 169
applied bias photon-to-current efficiency (ABPE), 137
anti-reflection, 25
anti-reflective layer, 25
Au nanocrystals, 254–267, 269–272, 274–277, 279,
    281–282
Au nanocubes, 262, 270–272, 281
Au nanospindles, 265, 270–273, 281
autocorrelation curve, *69*, 71, *72*, *73*, 76

B

bandgap/bandgaps, 25, 52, 53, 146, 238, **239**, 241, 242, 246
band gap energy, 230, 238
barrier layer, 173, 175, 184
$BaZnO_2$, 132
$BiFeO_3$, 133
black phosphorus, 48, **78**

C

calyptriform, 257. 259, 269–270, 273, 277, 281
carbon dioxide reduction, 151, 158, 161, 162
carbon nanotube/carbon nanotubes, 10, 47, 48–51, 54–56,
    58, 77–81, 149, 163, 186, 193, 196, 209,
    220, 283
CCVD (combustion CVD), 111
cermet, 21, 24, 27–41, 44, 45
chestnut-like, 260–265, 269–274, 276–277, 281
chips, 254, 269, 272, 274–278, 280–282
circularity, 174–178
CNT/CNTs, *see* carbon nanotube/carbon nanotubes
coating, 21, 24–27, 30–32, *33*, 35, 36, 38, 41, 42, 45
Coble Creep, 309, 312, 313
combinatorial characterization, 8
combinatorial chemistry, 1, 4, 5
combinatorial pulsed laser deposition, 5
combinatorial reactive sputtering, 130
combinatorial thin film synthesis, 3–5, 13
complex oxide, 13
composite, 25, 26, 29, 35, 38, 45

compositional variation, 8, 9, 11, 15
concave, 254–257, 259, 261–265, 269–277, 281
concentrating solar power, 21
concentration-dependent, 274–277
conduction band minimum (CBM), 230
coordination design, 155, 156, *156*
coral, 257, 259, 270, 273, 281
Cu-doped ZnO (CZO), 237
Cu_2O, 230
CVD (chemical vapor deposition), 53, 54, 61, 85–87, 111,
    115–117, 227, 228, 231

D

DCA, 169–176, 190, 191
DC magnetron sputtering, 88–91
dendritic, 254, 257, 259, 270, 273, 281
deposited, 167, 168, 174, 178–183, 186–189, 191
deposition, 165, 168, 178, 180, 186–189
dielectric, 25–28, 35, 36, 39–41, 44
    breakdown strength, 139
    relaxation, 140
diffusion, 266, 279–280, 309–314
dip-coating, 97–101
dipolar polarization, 140
dissolution effect, 169, 170, 173, 190
distribution uniformity, 165, 171, 174, 175, 181, 182
double-layer anodization method, 184
dropwise condensation, 212
dry-pressing, 108–109

E

EDX, 288, 292–293, 295, 304, 307, 310–311
electric-field, 254
emittance, 21–36, 38, 41, 42, 45
encapsulates, 54
enhancement factors (EFs), 272
enrichment, 266, 277, 279
EPD (electrophoretic deposition), 102–105
equilibrium, 275, 279–280
ESD (electrostatic spray deposition), 111–112
etching, 168, 184, 185, 187
exclusion principle, 50, 52
external stimuli, 215

F

fabrication, 254, 282
FAVD (flame-assisted vapor deposition), 111
ferroelectric crystal, 141
ferroelectricity, 141
FIB-DIC, 289–291, 296–297, 299, 301–304
fiber fabric, 209
fiber/fibers, 50, 51, 53–61, 63–82
films, 254, 269–274, 277, 280–282

320 Index

fluorine mica, 82
formation, 257, 259, 261–265, 269, 281
functional superhydrophobic nanocomposite surface, 199, 207, 212
Freundlich, 275, 282

G

GISAXS, 294–295, 304–308, 310
graphene, 47, 48, 50–52, 54, 58, 59, 68, 75, 79–82
graphite, 52, 53
growth kinetics, 260, 265, 281–282
growth solution, 255, 259–260, 262–265, 281

H

heat generation, 166, 170–173
hexachlorocyclohexane (HCH), 254, 266, 273–274, 276–282
high aspect ratio, 167, 168
high-temperature, 21, 22, 36, 42–45
hot spots, 254, 266, 273
HPA, 165, 169–178, 184, 190, 191
hydrogen evolution reaction, 151, 158, 160, 161
hydrothermal (solvothermal) reactions, 132

I

Incident Photon-to-Current Efficiency (IPCE), 134
indium-based oxide, 232
indium(III) oxide (In$_2$O$_3$), 230
influence, 256, 258, 262, 265, 270, 272, 277
interfacial polarization, 140
interfacial tension, 200–202
interpore distance, 166, 168
ion beam deposition, 31, 89, 288, 306–307, 312–313

J

joules heat, 166, 169–174

K

Kelvin probe force microscopy (KFM), 10
kinetically, 254, 260, 281

L

lattice defects, 305, 311–312
linear relation, 274–276, 278, 282
liquid-phase exfoliation, 53
lithography, 112–115
logarithmic, 274–276, 278, 281–282
long-term stability, 217
low-purity, 165–167, 169, 171, 185

M

magnetron sputter deposition, 31, 288, 295, 305, 308
magnetron sputtering, 30, 45
materials informatics, 2, 5, 16

Maxwell–Wagner–Sillars (MWS) interfacial polarization, 143
mechanical abrasion, 207–209, 212
membrane detachment, 165, 184, 185, 191
metal-organic framework (MOF), 266
metal-oxide-semiconductor field-effect transistor (MOSFET), 226
metastable state, 201
micro-cantilever, 291, 293, 301, 302, 304
micro/nanoparticles, 207–209
micro/nano structure surface, 203
microskeleton-nanofiller, 210
μ-SOFC (micro-solid oxide fuel cell), 102, 112–114
mobility, 231, 243
mode-locking, 47, 48, 50, 72, 73, 76, 78, 80, 81
morphology, 254, 256–257, 259–269, 281
morphotropic phase boundary (MPB), 142
moving shutter, 131
multilayered, 21, 24–28, 41, 45

N

nanocomposite surface, 200, 207–210
nanocrystals-built, 254, 269–277, 281
nano-needle structure, 213
nano-polymers, 208
nanoporous, 165–169, 181, 182, 184, 186–191
nitride, 26, 32, 34, 40, 41
nitride thin films, 128
N-methylpyrrolidone, 75
noble metal, 254
nonlinear/nonlinearly, 48–50, 54, 56, 58–60, 64, 65, 70, 74, 75, 79–81
n-type semiconductor, 227, 230–231

O

one-dimensional structural unit, 209–210
organochlorine pesticides, 254, 282
output characteristics, 233
oxide thin film, 129
oxygen reduction reaction, 151, 158–160

P

particle size, 181, 182, 191
percolation threshold, 144
perfluorohexane, 214
performances, 254, 266, 272, 277
permittivity contrast, 144
photocatalysis, 166, 169, 186, 187
photovoltaic cell, 227
photovoltaics, 22
piezoelectricity, 128
piezoelectric stress constant, 139
piezophotocatalysis, 134
piezophotodegradation, 136
piezophotoelectrochemical (PPEC) water splitting, 137
piezopotential, 134
piezotronic-related property, 133
plasma spray deposition, 94–96
plasmonic, 254, 269, 273
PLD (pulse layer deposition), 91–94

Index

P-N junction, 226, 227, 247
point-contact transistor, 226
polarization-dependent, 54
polyhedral, 255
polymorphism, 141
poly(vinylidenefluoride-co-trifluoroethylene) (PVDF-TrFE), 142
pore, 266–267, *268*, 277, 279–280, 282
  diameter, 170, 174, 175, *177*, 186
  distribution, 174, 182
  structure, 166–168, 170, 174, 178, 191
  widening, 166, 178, 187, 188
porous, 266–267, 277, 279, 281–282
porous bulk structural unit, 210
precursor solution, 254, 266–269, 281
preferential growth, 259, 265
process and composition optimization, 13
P25 TiO_2, 186–191
p-type semiconductor, 227, 237
pulsed laser deposition, 59, 64, 81
pulse sequence, 59, *69*, 71–74, 76
pulse spectrum, 49, 50, 52, 59, 71–73, 76, 77
pulse voltage detachment, 184
pulse width/pulse widths, 48, 49, 52, 53, 58, 59, 61, 63, 66, 68, 70, 71, 73–78
pump lasers, 47

Q

Q-switched lasers, 48, 75
quantification, 272

R

radio frequency magnetron sputtering, 88–91
Raman spectra, 269–272, 274–280
rare-earth, 48
reflectance, 23–32, 34, 38, 39, 41–44
reflective index, 27
regulation, 254
relative, 270, 277, 279, 281
relaxor phase, 142
residual stress, 140, 168, 179–180, 288–292, 295–308, 311–314
reverse bias method, 184
RF spectrum, *69*, 72–74, 76
rutile, 186

S

sandwich-structured, 54
saturable absorber, 5, 47, 51, 79–82
Schottky barrier height (SBH) variation, 134
Schottky barrier junction, 227
Schottky behavior, 133
screen-printing, 101–102
seed-assisted, 254–255, 260
seeds, 254–260, 262–266, 270, 281
selectivity, 277, 280, 282
self-assembled monolayers, 215
self-mode-locked, 52
self-organization, 166, 168

SERS, 166, 169
shape, 254–257, 259–260, 263–264, 266, 273, 281–282
sharp, 254, 260, 266, 273, 281
shell, 266–269, 277, 279–282
signal-to-noise ratio, 58
sine-squared psi analysis, 298, 300–304
single-atom catalysts (SACs), 151–162
single-walled carbon nanotube, 51, 78, 79, 81
size, 254–260, 263–267, *268*, 277, 279–282
slurry coating, 96–97
solar absorptance, 21–23, 25–27, 29–32, 35, 36
solar cells, 227, 245
solar radiation, 21–23, 35
solar selective absorber, 21, 22, 25, 44, 45
sol-gel, 100–101
solution-based process, 132, 228, 231, 242
spectral selectivity, 25, 34, 44, 45
spin coating, 97–98, 228, 242
spray pyrolysis, 109–111
steam condensation, 212
stimuli-responsive materials, 216
subtle microstructure, 214
superhydrophobic medium, 210
superhydrophobic-superhydrophilic coatings, 214
surface energy, 205–207
surface-enhanced Raman spectroscopy (SERS), 254, 266, 269–270, 272–279, 281–282
surface free energy, 152, *152*
surface-to-volume ratio, 158
surface wettability, 200–201
structure, 254, 259–261, 263, 266–267, 269, 273–274, 276, 279, 282

T

tandem absorber, 21, 24, 25
tape-calendering, 31–32
tape-casting, 105–107
TEM, 288, 294, 304–306
temperature, 254–260, 265–266, 269–270, 281
theoretical model, 200, 207, 217
thermal diffusivity, 313–314
thermal stability, 25–27, 30, 31, 36–39, 41, 43–45
thickness, 266–272, 274, 277, 279, 282
thin, 254, 269–270, 272, 275, 277, 279–282
thin-film transistor (TFT), 226, 232, 242
3D AAO, 165, 187, 188, 191
through-hole structure, 184, 185
$Ti_xAl_{1-x}N$, 129
tips, 254, 260, 262–264, 266, 269, 273, 277, 281–282
topological insulators, 47, 48, 51, 52, 58, 59
toxic, 273–274, 282
trace, 254, 273–275, 277, 279, 281–282
transfer characteristics, 233
transient grating spectroscopy, 292, 294
transition mechanism, 204
transition metal dichalcogenides, 47, 51, 52
transparent amorphous oxide semiconductors (TAOSs), 232
transverse piezoelectric coefficient, 139
trisoctahedral, 255–256, 269–270, 273–274, 277, 281
turnover frequency, 161, 162
2D materials, 151, 156–158
two dimensional X-ray diffraction (2D-XRD), 8

Index

U

ultrasonication, 53
urchin-like Au–Ag alloying nanocrystals (UAANs), 266–267, *268*, 277–281

V

vacuum slip-casting, 100
valence band maximum (VBM), 230
various, 281–282
vinylidene fluoride (VDF), 142
Volmer–Heyrovsky step, 160
Volmer–Tafel step, 160

W

water crystallization, 212
wet-chemistry method, 151, 154
wide-angle X-ray scattering, 291, 304–305
wide band gap semiconductor, 13

widening effect, 174
wrapped, 266–269, 277–282

X

X-ray reflectivity, 293, 295, 310–311
XRD, 8, 9, 11, 29, 30, 37–38, 179

Y

Yb-doped, 47, 54, 57–59, 61, 63, 66, 68–70, 79, 82

Z

zero dimensional structural unit, 207
ZIF-8 ($Zn(MeIM)_2$, MeIM-methylimidazole), 266–269, 277–282
zinc oxide (ZnO), 230
$ZnSnN_2$, 128
$Zn–Sn_3N_4$ Composition Spreads, 131
$ZnSnO_3$, 129